# Climate Change and Sustainable Agro-ecology in Global Drylands

## CABI CLIMATE CHANGE SERIES

Climate change is a major environmental challenge to the world today, with significant threats to ecosystems, food security, water resources and economic stability overall. In order to understand and research ways to alleviate the effects of climate change, scientists need access to information that not only provides an overview of and background to the field, but also keeps them up to date with the latest research findings.

This series addresses many topics relating to climate change, including strategies to develop sustainable systems that minimize impact on climate and/or mitigate the effects of human activity on climate change. Coverage will encompass all areas of environmental and agricultural sciences. Aimed at researchers, upper level students and policy makers, titles in the series provide international coverage of topics related to climate change, including both a synthesis of facts and discussions of future research perspectives and possible solutions.

### Titles Available

1. Climate Change and Crop Production
   Edited by Matthew P. Reynolds

2. Crop Stress Management and Climate Change
   Edited by José L. Araus and Gustavo A. Slafer

3. Temperature Adaptation in a Changing Climate: Nature at Risk
   Edited by Kenneth Storey and Karen Tanino

4. Plant Genetic Resources and Climate Change
   Edited by Michael Jackson, Brian Ford-Lloyd and Martin Parry

5. Climate Change Impact and Adaptation in Agricultural Systems
   Edited by Jürg Fuhrer and Peter J. Gregory

6. Livestock Production and Climate Change
   Edited by Pradeep K. Malik, Raghavendra Bhatta, Junichi Takahashi, Richard A. Kohn and Cadaba S. Prasad

7. Climate Change and Insect Pests
   Edited by Christer Björkman and Pekka Niemelä

8. Climate Change and Agricultural Water Management in Developing Countries
   Edited by Chu Thai Hoanh, Vladimir Smakhtin and Robyn Johnston

9. Climate Change Challenges and Adaptations at Farm-level: Case Studies from Asia and Africa
   Edited by Naveen Prakash Singh, Cynthia Bantilan, Kattarkandi Byjesh and Swamikannu Nedumaran

10. Climate Change Impacts on Urban Pests
    Edited by Partho Dhang

11. Climate Change Impacts and Sustainability: Ecosystems of Tanzania
    Edited by Pius Z. Yanda, Claude G. Mung'ong'o and Edmund B. Mabhuye

12. Climate, Ticks and Disease
    Edited by Pat Nuttall

13. Climate Change, Adaptation and Gender: Policy, Practice and Methodological Underpinnings
    Mamta Mehar and Narayan Prasad

14. Climate Change and Sustainable Agro-ecology in Global Drylands
    Edited by Adel El-Beltagy, Rattan Lal and Kauser Malik

# Climate Change and Sustainable Agro-ecology in Global Drylands

**Prof. Dr. Adel S. El-Beltagy**

*Professor, Arid Land Agricultural Graduate Studies & Research Institute (ALARI), Ain Shams University, Egypt; Chair of the International Dryland Development Commission (IDDC)*

**Prof. Dr. Rattan Lal**

*Distinguished University Professor of Soil Sciences; Director of Rattan Lal Center for Carbon Sequestration and Management, School of Environment and Natural Resources, Ohio State University, Ohio, USA*

**Prof. Dr. Kauser Malik**

*Distinguished National Professor, School of Life Sciences, and Dean of Postgraduate Studies, Forman Christian College, Pakistan; Formerly Federal Minister (National Food Security and Research)*

CABI

**CABI is a trading name of CAB International**

CABI
Nosworthy Way
Wallingford
Oxfordshire OX10 8DE
UK

Tel: +44 (0)1491 832111
E-mail: info@cabi.org
Website: www.cabi.org

CABI
200 Portland Street
Boston
MA 02114
USA

Tel: +1 (617)682-9015
E-mail: cabi-nao@cabi.org



A catalogue record for this book is available from the British Library, London, UK.

ISBN-13: 9781800624856 (hardback)
9781800624863 (ePDF)
9781800624870 (ePub)

DOI: 10.1079/9781800624870.0000

Commissioning Editor: David Hemming
Editorial Assistant: Emma McCann
Production Editor: Shankari Wilford

Typeset by Straive, Pondicherry, India
**Printed and bound by CPI Group (UK) Ltd, Croydon, CR0 4YY**

# Contents

# Author Biographies

**Prof. Dr. Adel S. El-Beltagy** (Egypt) is Chair of the International Dryland Development Commission (IDDC); Professor, Arid Land Agriculture Graduate Studies & Research Institute (ALARI), Ain Shams University, Egypt; Chair, Food & Agriculture Research Council, Egyptian Academy of Science &Technology; Board member, Regional Action for Climate Change (RACC), Science & Technology in Society Forum (STS), Kyoto, Japan. He was President, Governing Board, International Centre for Advanced Mediterranean Agronomic Studies (CIHEAM/Paris) (2010–2015); Minister of Agriculture & Land Reclamation, Egypt (2014–2015); Board member, Bibliotheca Alexandrina (BA) (2006–2012); Chair, BA Executive Board (2015–2018); Vice-Chair, Board of the Global Crop Diversity Trust (GCDT) (2007–2011); Council member, Consultative Group on International Agricultural Research (CGIAR) (2006–2011); Chair, Global Forum on Agricultural Research (GFAR) (2006–2010); Member, FAO High-level Committee addressing the implementation of the UN Millennium Development Goals (MDGs) (2006–2011); Board member, UN Millennium Ecosystem Assessment Board (2004–2005); Chair, Global Consortium to Rebuild Agriculture in Afghanistan (2002–2006); Chairman, CGIAR Consortium for Sustainable Agricultural Development in Central Asia & the Caucasus (CAC) (1996–2006); Director-General, International Center for Agricultural Research in Dry Areas (ICARDA) (1995–2006); Member, CGIAR Genetic Resources Policy Committee (1994–1999); Chairman, Science and Technology Council of the International Sahara & Sahel Observatory (SSO) (1993–2002). **Awards**: Gold Medal of the International Centre for Advanced Mediterranean Agronomic Studies (CIHEAM); Commander of the Order of Agricultural Merit, in recognition of his efforts in international agriculture development, particularly in the Mediterranean countries, Government of France; The Sultan Qaboos 'First Class' Order for Culture, Sciences and Arts, Oman; Al-Istiklal Medal from His Majesty King Abdullah II bin Al Hussein of Jordan; Fellow of the University of Wales. **Publications**: authored/co-authored more than 200 publications. http://www.adelelbeltagy.org

**Prof. Dr. Rattan Lal** (USA) is Director of CFAES, Rattan Lal Center for Carbon Management and Sequestration, Ohio State University (1987– ); Chair in Soil Science and Goodwill Ambassador for Sustainability Issues for IICA, Costa Rica; Member, UN Food System Coordination Hub. He was a member of the 2021 UNFSS Science Committee and Action Track 3 with focus on 'Coalition 4 Action on Soil Health; soil physicist (1969–1987) at IITA, Ibadan, Nigeria; Senior Research Fellow (1968–1969), University of Sydney; President of the World Association of Soil and Water Conservation (1987–1990); International Soil & Tillage Research Organization (1987–1989); Soil Science Society of America (2006–2008); International Union of Soil Sciences (2017–2018). His research interests are soil carbon sequestration, soil physical properties and processes, soil health, conservation agriculture, soil restoration and sustainable management of soils, and food and nutritional security. He has mentored *c.*400 scholars and is a member of seven global professional academies. **Awards**: 9 honorary degrees from universities in the USA, Europe, India and South America; Laureate of the 2018 World Soil Prize, 2019 Japan Prize, 2020 World Food Prize, 2020 Arrell Innovation Prize, and 2021 Padma Shri Award from the President of India. **Publications**: *c.*1100 journal articles, 575 book chapters, 110 books (as author or editor).

**Prof. Dr. Kauser Abdulla Malik** (Pakistan) is former Federal Minister of National Food Security and Research (since 2023); Distinguished National Professor of Biotechnology, currently leading agricultural biotechnology programmes at KAM School of Life Sciences at Forman Christian College, Lahore, since 2008. He was Chairman of Pakistan Agricultural Research Council (1998–2001); Member, Biosciences, Pakistan Atomic Energy Commission (2001–2006); Member, (Food Agri) Planning Commission, Islamabad (2006–2008); Head of Soil Biology Division of Nuclear Institute for Agriculture and Biology (NIAB) (1971–1992); founding Director of the National Institute for Biotechnology and Genetic Engineering (NIBGE) at Faisalabad (1992–1998); Member, Standing Advisory Group for Nuclear Applications (SAGNA) of IAEA for 9 years; Member, Board of Trustees of International Foundation of Science (IFS), Stockholm; BoG Member, International Institute for Applied Systems Analysis (IIASA), Vienna. His earlier research was on biology of saline soils and biodynamics of organic matter formation. These led to the development of biosaline agriculture technology which was later picked up by FAO/IAEA Joint Division and implemented in several countries of the Middle East. He was a pioneer in reporting associative nitrogen fixation associated with roots of non-legumes, which led to the development of commercial biofertilizers for cereal crops. Subsequently, he developed an interest in plant biotechnology and in plant biotechnology related to biofortification and transgenic wheat for salt tolerance, drought tolerance, and increased fertilizer-use efficiency. **Awards**: 3 civil awards by the President of Pakistan for his services to science; Fellow of Pakistan Academy of Sciences (PAS) and The World Academy of Sciences (TWAS). **Publications**: over 250 research articles published in peer-reviewed journals; edited 5 books.

---

**Aly Abousabaa** (Egypt) is Regional Director, Central and West Asia and North Africa, CGIAR, as well as Director General of the International Center for Agricultural Research in the Dry Areas (ICARDA). He holds an MSc in civil engineering from Montana State University and a BSc in civil engineering from Alexandria University. He is a strategic leader in sustainable development, climate change advocacy, and agriculture research in global dry regions. His scope covers North Africa and the Middle East, Sub-Saharan Africa, and Central and West Asia. In addition to being an active reformer both within ICARDA and One CGIAR, he promotes the global scaling of dryland agriculture innovations to improve the livelihoods resilience of smallholder family farmers under a climate crisis. Prior to joining CGIAR to head ICARDA, he was Vice President of the African Development Bank (AfDB). He led the preparation and execution of the Bank's key strategies in the areas of agriculture, water, natural resources management, climate change, infrastructure, women, youth, and governance. Through mobilizing and managing a portfolio of US$11 billion, his work made a transformational impact on the lives of millions of people across Africa. At the AfDB, he also helped coordinate climate change negotiations in Africa and acted as Gender Champion. He continues to sit on several international boards and has built up a personal global network of international partners in development.

**Dr. Anuradha Agrawal** (India) is Project Director of ICAR-Directorate of Knowledge Management in Agriculture (DKMA), New Delhi, and the National Coordinator in the National Agricultural Higher Education Project (NAHEP), a joint mega-project of the Indian Council of Agricultural Research (ICAR) and World Bank. She has more than 30 years' experience in the area of plant genetic resources (PGR) management, with special reference to use of biotechnology in the *ex situ* conservation of 'difficult-to-conserve crops'. While working as Officer-in-Charge (2017–2022) at the Tissue Culture and Cryopreservation Unit (TCCU) of ICAR-National Bureau of Plant Genetic Resources (NBPGR) in New Delhi, she contributed significantly to advancing the research for protocol development for conservation of germplasm in the In Vitro Genebank and Cryo Genebank at ICAR-NBPGR, particularly for vegetatively propagated species in fruit, medicinal, spices and other crops. She has led or been part of 19 national and international research projects, including multi-institutional endeavors, besides imparting extensive teaching and training in PGR. **Awards**: Panjabrao Deshmukh Outstanding Woman Scientist Award 2011 from ICAR. **Publications**: More than 300 in PGR management, plant tissue culture, agriculture policy, agrobiodiversity, and biotechnology.

**Dr. Mario Allegri** (Uruguay) is an agronomist from the Faculty of Agronomy, University of the Republic (UdelaR), Uruguay. He is Gerald Thomas Chair, New Mexico State University and Member of the National Academy of Engineering of Uruguay. He holds an MSc from New Mexico State University, USA, and a PhD from Michigan State University, USA. He served on scientific international and national organizations, focused on sustainable development, agricultural research, and innovation; former President of the Forum of the Americas for Agricultural Technology Research and Development (FORAGRO); Member, Executive Committees of GFAR, CGIAR, GCARD, IAASTD; researcher, Director of Agricultural Experiment Stations; and National Director of the National Agricultural Research Institute (INIA), Uruguay. He is involved in scientific and technological activities in different countries.

**Prof. Dr. Wei Bai** (China) is Director, Institute of Tillage and Cultivation Research, Liaoning Academy of Agricultural Sciences (LAAS). He is a specialist of dryland farming and has received many national and provincial awards. His research focuses on dryland farming systems, water use and climate change. He has published more than 50 papers on dryland farming.

**Dr. Michael Baum** (Germany) is Acting Deputy Director General, Research Team Leader – Breeding and Scaling Improved Varieties of Dryland Cereals and Legumes and Morocco Platform Manager at ICARDA. He holds a PhD in plant breeding and cytogenetics from the University of Göttingen in Germany. He was guest professor at Guangzhou University, China (2009–2012). Since joining ICARDA in 1992 as a biotechnologist, he has overseen ICARDA's biodiversity and crop management work in a number of countries across the Middle East and North Africa. Since ICARDA relocated from Aleppo, Syria, in 1992, he has contributed to the development of the Moroccan dryland research platform. The establishment of the Rainfed Research Platform in Morocco in 2013 has allowed ICARDA to strengthen further the collaboration with INRA and other national education, research and development institutions (ONSSA, ONCA, LADA, IAV HII, ENA Meknes, and 6 universities) and with many partners around the world and establish a fully equipped experimental station at Marchouch (100 ha), and a drought and heat research platform at Sidi El-Aydi INRA station. Other INRA research stations were rehabilitated to conduct joint research activities. Rabat and Marchouch are also the locations for the breeding of durum and bread wheat, barley, and food legumes (chickpea, lentil, faba bean, grass pea) in Morocco. As part of the Accelerated Breeding Initiative, major product profiles for these crops have been developed that fit the needs of the farmers in WANA countries. A major training hub has been established in Morocco since the ICARDA decentralization plan occurred in 2013, where more than 130 regional and global training courses were organized benefitting 1372 participants from more than 30 countries (1083 participants from Morocco). In addition, over 144 Moroccan students (85 PhD and 59 MSc) have been co-advised by ICARDA researchers. **Awards**: 'Frosty Hill Fellowship' from Cornell University (2000). **Publications**: *c.*150 peer-reviewed journal articles, 30 book chapters, and 40 conference papers. h-index 52; citations 9162.

**Dr. Prakriti Bista** (New Mexico) is Research Scientist at New Mexico State University. Her research focuses on monitoring and modelling soil carbon, nutrient, and water dynamics of traditional, intensified and diversified cropping systems to enhance agricultural sustainability and mitigate climate change impacts. She holds a PhD in Agronomy from the University of Wyoming and has undertaken post-doctoral training from Oregon State University. Presented at many local, regional, and international conferences. **Publications**: 17 peer-reviewed papers, 4 book chapters, 12 extension articles and 3 YouTube videos.

**Dr. Zhe Chen** (China) is a Master's candidate at the College of Agronomy and Biotechnology, China Agricultural University. His research focuses on the effects of carbon sequestration and emission reduction under conservation tillage.

**Dr. Simeon Ehui** (USA, Cote d'Ivoire) is since August 2023, the Director General of the International Institute of Tropical Agriculture (IITA) based in Nigeria. Prior to this, he was the Regional Director for Sustainable Development for West and Central Africa at the World Bank. He formerly held the position of Director for the World Bank's Food and Agriculture Global Practice. He also served as the Bank's Manager of agriculture and water in South Asia and Africa, and was the Lead Economist and Sector Leader for the Sustainable Development Network in Nigeria, overseeing sectors like agriculture, environment, social, water, urban, transport, and energy. Dr Ehui also holds the position of CGIAR director for Continental Africa, effectively collaborating with major regional institutions. This role significantly enhances the scope, relevance, and impact of CGIAR's efforts across Africa. Dr Ehui is a member of CGIAR's Global Leadership Team, which helps guide the global partnership and alignment of efforts to boost food and nutrition security on the African Continent. Before his tenure at the World Bank, Dr Ehui spent 15 years at CGIAR overseeing various agricultural research and development projects in Africa and Asia for IITA and International Livestock Research Institute (ILRI).

**Prof. Dr. Abdel-Ghany M. El-Gindy** (Egypt) is Professor of Agriculture Engineering and Dean, Faculty of Desert Agriculture, Rector of the University Branch – Ras Sudr, South Sinai, Egypt; independent and international on-farm irrigation engineering and management expert and consultant; Professor of Agriculture Engineering, Faculty of Agriculture, Ain Shams University, Cairo; Dean of the Faculty of Agriculture of Ain Shams University, Deputy Director and Director of Agricultural Engineering Research Institute, ARC, Ministry of Agriculture, Egypt. He was the Secretary of the Committee of the Sector of Agricultural Studies, Supreme Council of Universities of Egypt (2003–2019); Irrigation and Drainage Expert and Consultant for Arab Organization for Agricultural Development (AOAD) European Union, FAO, ICARDA and the World Bank.

**Dr. Jim Falk** (Australia) is Honorary Professorial Fellow, Melbourne Sustainable Society Institute; Professor Emeritus, University of Wollongong. His research has focused on science and technology in their social contexts with particular emphasis on issues associated with globalization, climate change, technological change, nuclear technology and militarization, and information technology in society. His most recent research focuses on issues associated with climate change, geoengineering, and the history and social and policy implications of concepts of time. **Publications**: *Worlds in Transition: Evolving Governance across a Stressed Planet*, co-authored with Professor Joseph Camilleri.

**Dr. Hafez Ghanem** (France, Egypt) is Senior Fellow in the Global Economy and Development Program at the Brookings Institution. He holds a Master's degree in Economics from the American University in Cairo and a PhD in Economics from the University of California. He was Vice President of the World Bank for Africa. He leads an active regional portfolio consisting of over 600 projects totalling more than $71 billion. Under his leadership, the World Bank supports inclusive growth and poverty reduction by financing projects that boost human capital, raise agricultural productivity, and provide economic opportunities for youth. He was Vice President of the World Bank for the Middle East and North Africa (2015–2018); Assistant Director General, Food and Agriculture Organization of the United Nations (2007–2012). https://www.worldbank.org/en/about/people/h/hafez-ghanem#2

**Dr. Rajan Ghimire** (Nepal) is an Associate Professor of Cropping Systems and Soil Health at New Mexico State University. He leads a multidisciplinary research programme on soil organic carbon and nutrient cycling, greenhouse gas emissions, and soil health assessment and management. He has delivered over 25 invited talks and symposia, and served on the scientific and organizing committees of several national and international meetings. He served in various regional, national, and international professional organizations and advisory councils. He holds a BSc and MSc in Agriculture from Tribhuvan University, Nepal; PhD in Soil Science from the University of Wyoming; post-doctoral training from Oregon State University. He has authored/co-authored 87 refereed and *c.*130 non-refereed publications.

**Prof. Dr. Ayman Abou Hadid** (Egypt) is Emeritus Professor at the Arid Lands Agricultural Studies and Research Institute, Ain Shams University, Cairo; Minister of Agriculture in four cabinets; President of the Agricultural Research Center; Chairman of the Egyptian Union of Producers and Exporters of Horticultural Crops; and Executive President of the Egyptian Environmental Affairs Agency. He has established three new institutes in Egypt: Arid Land Institute in the Faculty of Agriculture of Ain Shams University; Central Laboratory for Agricultural Climate in the Agricultural Research Center; Climate Change Information Center and Renewable Energy in the Ministry of Agriculture. The International Society for Horticulture Sciences (ISHS) elected him as the Chairman of the Commission on Protected Cultivation. He is a member of the Executive Board of IDDC. **Awards**: Medal of Science and Arts First Grade of President of Egypt; National Recognition Award for Agricultural Sciences, Egyptian Academy.

**Dr. Cong He** (China) is a PhD candidate at the College of Agronomy and Biotechnology, China Agriculture University. His main research interest is soil organic carbon mineralization under conservation agriculture practices, and using kinetic fitting to study carbon mineralization under agro-ecosystems.

**Prof. M.L. Jat** (India) is a Principal Scientist/Systems Agronomist and Sustainable Agrifood System (SAS) Science lead at International Maize and Wheat Improvement Center (CIMMYT), Kathmandu. He has a PhD degree in Agronomy from ICAR-Indian Agricultural Research Institute (IARI), New Delhi, and served the Indian Council of Agricultural Research (ICAR) as a Systems Agronomist for 11 years before joining CIMMYT in 2009. He has devoted more than two decades to intensive work on basic and applied science in agronomy, soils and the environment to promote conservation agriculture (CA)-based sustainable intensification and climate-smart agriculture in smallholder systems of Asia. His research on CA has provided a scientifically sound basis and direction for promoting sustainable intensification through policy changes. **Awards**: Fellow of the National Academy of Agricultural Sciences (NAAS); several awards including ICAR's prestigious Rafi Ahmed Kidwai Award.

**Dr. Chakib Jenane** (Morocco) is the Regional Director for Sustainable Development for the Western and Central Africa Region (AFW) at the World Bank and is responsible for driving sustainable development initiatives and transformative projects in the region. He previously served in the same organization as the Practice Manager for the Agriculture and Food Practice for AFW. Prior to joining the World Bank, he held key positions at the United Nations Industrial Development Organization among which as Chief of the Agro-industry Technology Unit and deputy-Director of the Agribusiness Development Branch. He has also served as a university Professor at the Department of Agricultural Engineering at the Institute of Agronomy and Veterinary Medicine in Rabat, Morocco. He holds a Doctorate-es-Sciences degree from the Institute of Agronomy and Veterinary Medicine in Rabat and an MS in agricultural engineering from the University of Minnesota. He also completed a post-Doctorate within the Bio-systems Engineering Department of the University of Nebraska-Lincoln. He has published extensively on power and machinery and agro-industry for development.

**Dr. Charles Kennel** (USA) belongs to the National Academy of Sciences, the American Academy of Arts and Sciences, the American Philosophical Society, and the International Academy of Astronautics. He was Professor and Chair of the UCLA Physics Department and UCLA Executive Vice Chancellor; NASA Associate Administrator for Mission to Planet Earth; Director of the Scripps Institution of Oceanography; Chair US National Academy's Board on Physics and Astronomy and Space Studies and its committees on Global Change Research and Fusion Sciences; Member, NASA Advisory Council for 12 years; Distinguished Visiting Fellow of Christ's College, Cambridge, UK, and distinguished Professor Emeritus at Scripps.

**Prof. Dr. Iqrar Ahmad Khan** (Pakistan) is a graduate of the University of Agriculture, Faisalabad, Pakistan, and the University of California, Riverside, USA. He established several national and international research centres and institutes, including a Chinese Confucius Institute and a Bio-Energy Institute. He was COP/Director US-Pak Center of Advanced Studies (2017–2019) and Ela Bhatt Professor at the University of Kassel, Germany (2020). His research on wheat, potato, citrus, mango and dates has benefitted the industry and significantly improved the life and livelihood of farmers and rural communities.

**Dr. Muhammad Sarwar Khan** (Pakistan) is Dean of the Faculty of Agriculture, University of Agriculture, Faisalabad, Pakistan. He holds a PhD from the University of Cambridge, UK. He established a pioneer group on chloroplast molecular biology, and Biotech Interdisciplinary Division (BID). He is a member of the panels of referees and editorial boards of several international journals. **Awards**: Fellow of the Royal Society of Biology (2022); Cambridge Commonwealth Society (1994); Fellow of the Cambridge Philosophical Society (1995); Rockefeller Foundation Postdoctoral Fellowship under the Rice Biotechnology Program.

**Dr. Bai-Jian Lin** (China) is a PhD candidate at the College of Agronomy and Biotechnology, China Agriculture University. His research mainly focuses on the synergistic mechanism between soil organic carbon sequestration and crop yield under agricultural measures.

**Prof. Dr. Magdy Madkour** (Egypt) is Emeritus Professor of Biotechnology at the Arid Lands Agricultural Research Institute, Ain Shams University, Cairo. He was appointed Chairman of the Technical Committee of the Agricultural Research for Development Fund (ARDF), Egypt; Member, Technical Advisory Committee of the African Biotechnology Network of Expertise ABNE (NEPAD). He holds a PhD from the University of Wales, Aberystwyth, UK. In 1990 he established the Agricultural Genetic Engineering Research Institute (AGERI). He was Visiting Professor, University of California, Davis, USA (1989–1990); President of the Agricultural Research Center, Egypt (2001–2004); Member, Technical Advisory Committee of the CGIAR (TAC) (1996–2000); Member, Board of Trustees of ICARDA and IPGRI (Biodiversity); Editor-in-Chief of the *Journal of Agriculture and Food Security* until 2019. Three patents for novel biological insecticides were issued in his name which led to the commercialization of AGERIN© an environmentally friendly biological insecticide applied successfully throughout the last decade in controlling cotton and potato pests (average 300,000 acres annually). **Awards**: Egypt's Presidential Gold Medal and the State Award of Appreciation in Advanced Agricultural Technologies; World Academy for Science (TWAS) Arab Regional Prize (2012); Fellow of the World Academy of Science for the Developing World (TWAS) (2014). **Publications**: 142 peer-reviewed scientific publications; contributed chapters in 7 books.

**Ms. Dena Merriam** (USA) is founder of the Global Peace Initiative of Women (GPIW), an organization chaired by a multi-faith group of women spiritual leaders, with the mission to enable women to facilitate healing and reconciliation in areas of conflict and post-conflict, and to bring spiritual resources to help address critical global problems. Since its founding, GPIW has organized dialogues in Israel/Palestine, Iraq, Sudan, Afghanistan, Cambodia and Kashmir, India. For several years after its founding, GPIW also developed, in partnership with the UN, a global leadership programme for young community leaders. GPIW's work in the area of peace building has expanded to include fostering new models of development, inclusive and sustainable, mobilizing faith communities to address the climate crisis, and to changing attitudes toward the environment, regaining the sense of awe, respect and reverence for the Earth and its life systems. Her work at GPIW has been devoted to creating a global platform for women religious and spiritual leaders, and to engaging the Hindu and Buddhist leadership more actively on the world stage and ensuring East-West balance, enabling participation from the Eastern religious traditions equal to that of the Abrahamic traditions. She is one of the founding members of the Contemplative Alliance, an initiative to bring together contemplative practitioners from across traditions to demonstrate how the mainstreaming of contemplative practice is changing the American religious landscape and can ultimately contribute to the positive transformation of American society. For over 35 years, she has been a student of the Hindu Master Paramahansa Yogananda and a practitioner of Kriya Yoga meditation. She is also a long-time student of the great texts of the Vedic tradition. She continues to serve on the boards of The All-India Movement (AIM) for Seva, the Manitou Foundation, is an advisor to the board of Dharma Drum Mountain Buddhist Association and is on the advisory council of The Gross National Happiness Center in Bhutan. She holds a Master's degree from Columbia University and has served on the boards of the Harvard University Center for the

Study of World Religions, the International Center for Religion and Diplomacy, The Interfaith Center of New York, and Seven Pillars. **Awards**: Niwano Peace Prize by the Niwano Peace Foundation in Japan for her many years of commitment to interfaith dialogue (2014). **Publications**: *My Journey Through Time: A Spiritual Memoir of Life, Death and Rebirth*, *The Untold Story of Sita*, and *When the Bright Moon Rises: Awakening of Ancient Memories*.

**Dr. Sharon Mijares** (USA) is a psychologist and leads the faculty at the California Institute for Human Science. She is a professor at the National University. **Publications**: author or co-author of seven books focused on psychological and spiritual development, including *The Revelation of the Breath: A Tribute to Its Wisdom, Power, and Beauty* (2009). Her 2013 edited book, *A Force Such as the World Has Never Known: Women Creating Change*, brings women together from many cultures. *The Power of the Feminine: Facing Shadow/Evoking Light*, is co-authored with Aliaa Rafea, an Egyptian Sufi leader, Dhardon Sharling, a Tibetan feminist, Joshua Amponsem, an environmentalist from Ghana, and a Mary Mallory, a US professor. She has presented workshops in Costa Rica, Ecuador, Egypt, India, Mexico, Scotland, Uganda, USA and Venezuela.

**Prof. Dr. Adel M. Aboul Naga** (Egypt) is Eminent Senior Researcher, Animal Production Research Institute, ARC, Egypt. He was Advisor to H.E. Minister of Agriculture and Land Reclamation, Egypt; Supervisor of Animal Production Research Institute; Senior Advisor to ICARDA DG, Research for Development; FAO Representative to Pakistan; FAO Contact to Afghanistan Program; Agricultural Counselor and Deputy Representative of Egypt to FAO, WFP and IFAD, Rome; Chairman Group 77, UN Chapter in Rome; Under Secretary for Animal Production Sector, Ministry of Agriculture, Egypt; Director of TT, National Agriculture Research Project (NARP); President, International Goat Association; Board member, European Association Animal Production. **Publications**: 185 scientific publications in international journals and periodicals.

**Prof. Dr. Vinay Nangia** (India) is the Research Program Leader – Soil, Water and Agronomy at ICARDA; visiting professor, Tottori University in Japan; adjunct faculty at Texas A&M University, USA. He serves on FAO WASAG, UNCCD IWG on Drought, UNFCCC focal point for ICARDA and other high-level expert bodies and is an advisory Board member, GIWEH (a Geneva-based water think tank). He holds a PhD in Water Resources Science and an MSc in Biosystems Agricultural Engineering and in Geographic Information Science from the University of Minnesota. His leadership and team-building skills have helped solve complex challenges to water resource management posed by climate change.

**Olatunde Pelumi Oladele** (Nigeria) holds a Master's in resource use and plant protection and is a PhD candidate at the China Agricultural University, focusing on crop cultivation, soil microbiome and nutrient cycling. His research combines field trials, lab work and metagenomics analysis, aiming to advance sustainable agriculture and environmental preservation.

**Dr. Steven Were Omamo** (Kenya) is Director of Development Strategy and Governance with the International Food Policy Research Institute (IFPRI). Prior to this, he was CEO of New Growth International, a development consulting firm he founded in Kenya in 2007. He holds PhD and MA degrees from Stanford University, an MSc from the University of Connecticut, and a BSc from California State University, Fresno. He was Representative and Country Director for the UN World Food Programme (WFP) in Ethiopia; Deputy Director of Policy and Programme, WFP; Coordinator of Food Systems Strategy, Policy and Support; Representative to the African Union and UN Economic Commission for Africa in Addis Ababa; Deputy Director of Policy Planning and Strategy; Chief of Food Security and Safety Nets; Chief of Social Protection and Livelihoods; Director of Global Engagement and Research with the International Fund for Agricultural Development (IFAD); Director of Policy and Advocacy with the Alliance for a Green Revolution in Africa (AGRA); Senior Research Fellow and Coordinator of the Eastern Africa Food Policy Network with the International Food Policy Research Institute (IFPRI); Research Officer with the International Service for National Agricultural Research (ISNAR); Research Scientist with the International Livestock Research Institute (ILRI); Technical Advisor with the Kenya Agricultural Research Institute (KARI); Technical Advisor with the Association for Strengthening Agricultural Research in Eastern and Central Africa (ASARECA); Lecturer, Egerton University, Kenya. He has published extensively on topics related to agricultural and rural growth and transformation, economic development, social protection, and emergency preparedness and response.

**Dr. Theib Y. Oweis** (Jordan) is a world-leading specialist in water harvesting and management for agriculture, especially under conditions of water scarcity. He has PhD and MSc degrees in Agriculture and Irrigation Engineering from Utah State University and has spent over 35 years in researching, developing, teaching and publishing in water harvesting, supplemental irrigation, deficit irrigation, water productivity, salinity, and other areas associated with improving the productivity of water use in agriculture. He was Director of the Integrated Water and Land Management Program at the International Center for Agricultural Research in the Dry Areas (ICARDA). **Publications**: over 200 refereed journals, books/book chapters and conference proceedings.

**Dr. V. Padmaja-Karanam** is a Senior Soil and Environmental Researcher at the International Crops Research Institute for the Semi-Arid Tropics. She holds a Master's in Agriculture-Soil Science and Agricultural Chemistry from PJTSAU, ICAR, and a PhD in Agriculture with specialization in Environmental Sciences from the Indian Agricultural Research Institute, New Delhi, ICAR. She has over 20 years' experience in the areas of agricultural research for development and environmental sustainability. She was selected for international programs with scholarship on Sustainable Environmental Management in the USA; and for the Science, Research, and Innovation Program in the UK. Her work areas encompass climate change adaptation, mitigation studies, carbon sequestration, carbon farming, and soil health studies to enhance agricultural productivity. She focuses on integrating innovative and interdisciplinary approaches to address agricultural food and nutritional security and related environmental challenges.

**Dr. Raj S. Paroda** (India) is Chairman, Trust for Advancement of Agricultural Sciences (TAAS), New Delhi; Member, Board of Directors of International Fertilizer Development Center (IFDC), USA; Foreign Fellow of the Australian Academy of Technological Science and Engineering, Third World Academy of Sciences, Academy of Agricultural Sciences of Russia, Georgia, Armenia, Tajikistan; Honorary Member, Crop Science Society and Agronomy Society of America. He was Director General, Indian Council of Agricultural Research, and Secretary DARE, Government of India; he has modernized the Indian national agricultural research system and created more than 30 new institutions including the National Gene Bank; President of Indian Science Congress and the National Academy of Agricultural Sciences; founder Chairman, Global Forum on Agricultural Research (GFAR); Chairman, ICRISAT Board of Trustees; Executive Secretary of Asia-Pacific Association of Agricultural Research Institutions (APAARI); Member, ACIAR, CABI WMO and CGIAR-SIMEC Advisory Committees. **Awards**: recipient of many national and international awards, including Padma Bhushan; 21 DSc. degrees, including Ohio State University.

**Dr. Juan Andrés Quincke** (Uruguay) graduated as agronomist from the College of Agriculture at Udelar (Uruguay) and obtained his MSc and PhD degrees at the University of Nebraska-Lincoln (USA); Principal Researcher at the National Agricultural Research Institute (INIA, Uruguay) in the field of soil fertility and management; Member, INIA's long-term experiments team (Plataforma Agroambiental); leading scientist of the old Crop-Pasture Rotations Experiment at INIA La Estanzuela. Specific research topics include soil carbon dynamics under agricultural management practices, soil tests for phosphorus management, and soil erosion.

**Prof. Dr. Aliaa R. Rafea** (Egypt) is Professor of Anthropology at Ain Shams University, Women's College. She has worked as an expert for an Arab League Project. She established the Human Foundation (an NGO), and co-established women's activities within the Egyptian Society for Spiritual and Cultural Research (ESSCR) more than 25 years ago. She was Visiting Professor at Randolph-Macon Women. She teaches Analytical Psychology (Jungian psychology), training trainers for spiritual guiding, serving multiple purposes. **Awards**: prize from All Ladies League (2016) for her contribution to Human Unity. **Publications**: author of *The Root of All Evil: An Exposition to Prejudice, Fundamentalism and Gender Bias* (2007); numerous chapters and articles, including *A Force Such as the World Has Never Known: Women Changing the World* (2013). She participated in a cultural project initiated by Master Ali Rafea and co-authored with him and Aisha Rafea several books. She translated and commented on C.J. Jung's *Memories, Dreams and*

*Reflections* (2015), and wrote about her spiritual development in an autobiography *Insight* (2016). Many articles in *Nahdet Misr* newspaper.

**Dr. Cherukumalli Srinivasa Rao** (India) is Director, ICAR-National Academy of Agricultural Research Management (NAARM), Hyderabad (2017– ). His major research interests are climate change, soil carbon sequestration, dryland systems, contingency planning, research management and HRD. He was Director, ICAR-Central Research Institute for Dryland Agriculture (CRIDA) (2014–2017); Member, International Dryland Development Commission (IDDC) Board; Asian Long Term Experimental Network for Agriculture (ALTENA), Japan; Expert Committee of National Biodiversity Authority, India. He represented India in UNFCCC CoP Meetings held in France, Germany and Spain, and as Global Climate Change Negotiator and IPCC Meeting at Japan and APN Meeting at Bangkok. **Awards**: 25 national and international awards from ICAR, NAAS, ISCA, ICRISAT, IARI, DST and FAI.

**Prof. Dr. João Carlos de Moraes Sá** (Brazil) has been a Senior Associate Researcher at the Lal Carbon Center at The Ohio State University since July 2023 and has served as President of the Scientific Committee of the Brazilian No-till System Federation since 2018. From 1995 to 2019, he held the position of Associate Professor of Soil Science at the State University of Ponta Grossa. He holds a PhD in soil science from São Paulo University, School of Luiz de Queiroz, and Ohio State University. He was Visiting Professor, Ohio State University, School of Natural Resources (2015); Associate Researcher, CIRAD, France (2007). He held the position of visiting professor at the universities of Santiago de Compostela (Spain) and Évora (Portugal). He has coordinated over a dozen research projects examining the potential for carbon sequestration by no-till system soils in Brazil.

**Dr. Ismail Serageldin** (Egypt) is the Emeritus Librarian of Alexandria, and was the Founding Director of the Bibliotheca Alexandrina, the New Library of Alexandria in Egypt (2001–2017). He is a member of many Academies, and currently serves on many boards and advisory committees for academic, research and NGO institutions. He is a Patron of the International Science Council (ISC) and co-Chair of the Board of the Nizami Ganjavi International Center (NGIC). He was Vice-President of the World Bank (1993–2000) in charge of Environmentally Sustainable Development (ESD), whose mandate included oversight over all World Bank financed projects in Agriculture, Infrastructure, and Environment; Chairman of the Consultative Group for International Agricultural Research (CGIAR); Founding Chairman of the Global Water Partnership (GWP) and the Consultative Group to Assist the Poorest: A Micro-finance Program (CGAP); Co-founding member, World Water Council (WWC); Chair of the UN Commission on Water in the 21st Century (1999–2000); Distinguished Professor in Wageningen University and the College de France. **Awards**: Order of the Rising Sun of Japan; Knight of the French Legion d'Honneur; Commander of Arts and Letters of the French Republic; 40 honorary doctorates from all over the world. **Publications**: over 100 books and 500 articles.

**Prof. Dr. Abdel-Aziz S. Sheta** (Egypt) works on urbanization monitoring and crop area estimation and monitoring of land reclamation activities in Egypt using Remote Sensing and GIS. He holds a PhD in Agricultural Sciences from Ain Shams University, Cairo, under a joint PhD programme with University of Guelph, Ontario. He was Chair of the Soils Department of the Faculty of Agriculture of Ain Shams University, Egypt; has specialized in the fields of soil pedology, soil survey and evaluation, and arid land management. His research includes the use of natural clays for water conservation in arid lands and soil amendments; soil survey, classification and evaluation for irrigated agriculture; soil and water management in irrigated lands; improvement of sandy soils through the use of zeolite and bentonite; and application of GIS and remote sensing for soil survey and land evaluation. He was Chairman of Urban Zoning Committee of the Ministry of Agriculture and Land Reclamation, Egypt; Chairman of the Technical Committee for Sustainable Land Use in Egypt.

**Dr. Rajbir Singh** (India) is an Assistant Director General (Agronomy, Agroforestry & Climate Change) in the Natural Resource Management Division of ICAR, New Delhi, where he is instrumental in planning, coordinating and monitoring R&D programmes for sustainable agricultural production and resource conservation under the changing scenario of climate change. His major research interests are Conservation Agriculture, Technology Application and Climate Smart Agriculture. He was Director, ICAR-Agricultural Technology Application Research Institute, Ludhiana, for 8 years. **Awards**: Fellow of NAAS, ISA, HIS. **Publications**: *c.*125 research papers and 10 books.

**Dr. Ren-Hua Sun** (China) is Senior Agronomist, Rural Energy and Environment Agency, Ministry of Agriculture and Rural Affairs. Her research centres on comprehensive utilization of crop straw and other waste, and ecological agriculture models. She has published *c.*20 papers.

**Dr. Yuan-Feng Sun** (China) is an Agronomist, Rural Energy and Environment Agency, Ministry of Agriculture and Rural Affairs. Her research mainly focuses on technical specifications of crop residue, carbon sequestration and agro-ecology. She has published *c.*10 papers.

**Dr. Kuldeep Tripathi** (India) is a Scientist at ICAR-National Bureau of Plant Genetic Resources (NBPGR), New Delhi, since 2015; manages four external projects; contributes to teaching at the Graduate School of ICAR-IARI. His work is instrumental in advancing pulse crop research and genetic enhancement. His expertise lies in managing plant genetic resources, particularly in pre-breeding and genetic enhancement of pulse crops. His research has led to the development of Indian pulse core sets of lentil, pea, and cowpea, derived from the extensive collections within the Indian National Genebank. He has registered 26 legume genetic stocks, and deposited gene sequences in the NCBI database. **Publications**: over 70 research papers.

**Prof. Dr. Syed Aftab Wajid** (Pakistan) is a crop modeller and climate change specialist. He specializes in understanding the intricate relationship between agricultural systems and changing climate. His expertise lies at the intersection of crop modelling and climate change, allowing him to analyse, predict, mitigate and adapt the impact of climate change on crop production and food security. He utilizes advanced mathematical and statistical techniques to develop models to simulate the growth, development, and yield of various crops under different environmental conditions. He has collaborated with an international group of scientists in the field of climate, crop and economic modelling called AgMIP (Agricultural Inter-comparison and Improvement Project) under the DFID Program. His work with AgMIP involves assessing potential risks during early- (2010–2039), mid- (2040–2069) and late- (2070–2099) century scenarios for wheat, rice and cotton crops, considering climate change and mitigating strategies.

**Dr. Xing Wang** (China) is an Associate Professor at the College of Agriculture, Northwest A&F University (NWAFU). His research focuses on water conservation, carbon sequestration and emission reduction in agricultural systems and agro-ecology, especially in the interaction mechanism between water conservation and carbon sequestration in arid areas. He is the Associate Editor of *Heliyon* and guest editor of *Agronomy*.

**Dr. Yu-Xin Wei** (China) is a PhD candidate at the College of Agronomy and Biotechnology, China Agricultural University. His research focuses on the correlation between soil organic carbon and ecological stoichiometry ratio under conservation tillage.

**Dr. Meryl Williams** (Australia) has worked for over 40 years in international fisheries and aquaculture, aquatic resource conservation and development. Currently, she is focusing on research and advocacy on women and gender equality in the fisheries, aquaculture and food value chains, and science for fair, inclusive, and responsible fish production for food security and nutrition. She was a member of the Governing Board of the International Crop Research Center for the Semi-Arid Tropics; Chair of the ACIAR Board of Management and Policy Advisory Council (2004–2007); Chair of ACIAR Commission (2007–2010); Director General of WorldFish Center (1994–2004); Director of Australian Institute of Marine Science; Industry Fellow, University of Technology, Sydney. **Awards**: Honorary Life Member, AFS; Australian Centenary Medal, and Crawford Medal (Crawford Fund, Australia) for contributions to international agricultural research; Outstanding Alumnus of James Cook University; FAO Margarita Lizarraga Medal, International Institute of Fisheries, Economics and Trade Distinguished Service Award.

**Dr. Jian-Fu Xue** (China) works at the College of Agriculture, Shanxi Agricultural University. His research centers on dryland farming systems, soil organic carbon sequestration, and the key technology of green production of high yield and high efficiency. He has published more than 30 papers.

**Prof. Dr. Mohamed K. Yousef** (USA) is Emeritus Distinguished Professor, University of Nevada, Las Vegas, USA (2005– ); President, Academic Consulting Enterprise, Las Vegas (1995– ); Deputy Vice Chancellor for Academic Affairs, UAE University (1994– ). He was Dean, College of Science, UAE University (1974–1990); Visiting Professor, Department of Physiology, University Occupational & Environmental Health, Japan. (1985–1986); Visiting Professor, Department of Physiology, University of New England, Australia.

**Prof. Dr. Hai-Lin Zhang** (China) is Associate Dean of the College of Agronomy and Biotechnology, China Agriculture University (CAU); Associate Chair, Chinese Farming System Division of China Association of Agricultural Science Societies; Associate Editor of *Heliyon*, *Agronomy*, *Carbon Management* and *Anthropocene Science*. His research centres on conservation agriculture, farming systems, climate change, residue management, carbon sequestration and agro-ecology. He and other colleagues launched a new Golden Open Access journal *Farming System*, published by Elsevier.

**Dr. Xin Zhao** (China) is Associate Professor of Agronomy, China Agricultural University. His research interests include conducting meta-analysis in agronomy, establishing conservation agriculture in China, and soil organic carbon sequestration and greenhouse gases emission from agricultural practices. He has published *c*.40 papers.

# List of Contributors

Abousabaa, Aly: a.abousabaa@cgiar.org
Agrawal, Anuradha: Anuradha.Agrawal@icar.gov.in; anuagrawal1@yahoo.co.in
Allegri, Mario: mario.alleri@gmail.com
Bai, Wei: libai200008@126.com
Baum, Michael: m.baum@cgiar.org
Bista, Prakriti: pbista@nmsu.edu
Chen, Zhe: chenzhe990327@163.com
Ehui, Simeon: simeonehui@yahoo.com; S.Ehui@cgiar.org
El-Beltagy, Adel S.: elbeltagy@drylanddevelop.org
El-Gindy, Abdel-Ghany: abdelghany.elgindy@ksiu.edu.eg; Elgindy47@gmail.com
Falk, Jim: jimfalk2@gmail.com
Ghanem, Hafez: hafez.ghanem@gmail.com
Ghimire, Rajan: rghimire@nmsu.edu
Hadid, Ayman Abou: aabouhadid@yahoo.com; ayman_abouhadeed@agr.asu.edu.eg
He, Cong: hecong@cau.edu.cn
Jat, M.L.: mangilal.jat@icrisat.org
Jenane, Chakib: cjenane@worldbank.org
Khan, Iqrar Ahmad: iqrar@uaf.edu.pk; iqrarahmadkhan2008@gmail.com
Khan, Muhammad Sarwar: sarwarkhan_40@hotmail.com
Karanam, Padmaja: padmaja.karanam@icrisat.org
Kennel, Charlie: ckennel@ucsd.edu
Lal, Rattan: Lal.1@osu.edu
Lin, Bai-Jian: linbaijian@cau.edu.cn
Madkour, Magdy: madkour.magdy@gmail.com
Malik, Kauser: kausermalik@fccollege.edu.pk; Kauser45@gmail.com
Mariam, Dena: dena.merriam@finnpartners.com; merriamd@gpiw.org
Mijares, Sharon: sharon.mijares12@gmail.com
Naga, Adel M. Aboul: adelmaboulnaga@gmail.com
Nangia, Vinay: V.Nangia@cgiar.org
Oladele, Olatunde Pelumi: ola.ola@cau.edu.cn
Omamo, Steven Were: swomamo@gmail.com
Oweis, Theib: theib.y.oweis@gmail.com
Paroda, Raj: raj.paroda@gmail.com

Rao, Ch. Srinivasa: cherukumalli2011@gmail.com
Quincke, Juan Andrés: aquincke@inia.org.uy
Rafea, Aliaa: aliaa.readah.rafea@women.asu.edu.eg; aliaa.rafea@yahoo.com
Sá, João Carlos de Moraes: jcmoraessa@yahoo.com.br
Serageldin, Ismail: Ismail@serageldin.com
Sheta, Abdel-Aziz: sheta11us@gmail.com
Sun, Ren-Hua: sunrh_abc@163.com
Sun, Yuan-Feng: sunyuanfeng2022@126.com
Tripathi, Kuldeep: Kuldeep.Tripathi@icar.gov.in
Wajid, Syed Aftab: aftab.wajid@uaf.edu.pk
Wang, Xing: wangxing0103@nwafu.edu.cn
Yu-Xin Wei: weiyuxin@cau.edu.cn
Williams, Meryl: meryljwilliams@gmail.com
Xue, Jian-Fu: fudange95@163.com
Youssef, Mohamed: mkyousef@hotmail.com
Zhang, Hai-Lin: hailin@cau.edu.cn
Zhao, Xin: zhaox@cau.edu.cn

# Foreword: Climate Change and Sustainable Agro-ecology in Global Drylands

Despite growing political momentum, the world remains deep in the triple planetary crisis of climate change, nature and biodiversity loss, and pollution and waste. This triple crisis threatens the well-being of current and future generations. Food systems are at the heart of this crisis, as both a driver and victim of environmental change. Climate variability and extreme weather have already slowed global agricultural productivity growth, with further disruption to come.

In this light, *Climate Change and Sustainable Agro-ecology in Global Drylands* is a timely work; it highlights the importance of pursuing innovative pathways that will make agriculture in drylands sustainable. These ecosystems cover about 40% of the land area and may expand to cover 50% of the land area by 2100 as climate change worsens. Building local adaptation in drylands is therefore critical. This book presents important insights and analysis from a senior and distinguished group of scientists, highlighting experiences and outlining an important set of innovative adaptation strategies to support the 2.5 billion people who inhabit dryland ecosystems.

The authors cast their net far and wide, looking at practices in Africa, the Americas, China, India and Pakistan. The articles look at the potential of increasing carbon stocks in dryland biomass and soils to slow climate change and create income for farmers through payments. They look at using smart phones and AI to modernize agriculture and make it more efficient, using plant genetic resources to boost climate adaptation in drylands, how to deploy water-smart irrigation and rainwater harvesting, and so much more.

Such practices safeguard people, biodiversity, and crucial ecosystem services, especially in environments that are highly vulnerable to climate change. The transition towards sustainable food systems and a climate-adapted future will depend on the widespread adoption of such innovative net-positive agricultural practices and technologies.

This book is a testament to the scientists and local stakeholders who have dedicated much of their lives to this cause and can serve as a source of knowledge and a catalyst for change. By following their lead, we can, together, protect and nurture the planet that sustains us all.

**Inger Andersen**
*Under-Secretary-General of the United Nations and Executive Director of the United Nations Environment Programme*

# Preface: The Day after COP28 – The Road Ahead: Climate Change and Sustainable Agro-ecology in Global Drylands

This book deals with an essential topic: climate change and sustainable agro-ecology in global drylands. It is an exceptionally large effort by an eminent group of scientists and development specialists to address the challenges of saving the lives and livelihoods of the poor and vulnerable, especially in the drylands of the world, where fragile ecosystems are particularly exposed to the increasingly visible consequences of climate change. These drylands cover about 40% of the Earth's surface and are home to about 40% of the world's population, of whom some 80% are the poorest and most vulnerable of the human family.

An introductory and a concluding chapter bracket the contents of the book and are contributed by the editorial board that guided the production of the materials in these diverse chapters. The concluding chapter is appropriately titled 'The Day after COP28', not only referring to the time the book is being issued, but also to this moment in humanity's struggle to respond adequately to the challenge of climate change.

The day after COP28 is indeed a proper moment to reflect on how far we have come in the last three decades, and how much remains to be done. It reminds us all that climate change is a global problem that does not recognize political borders or societal boundaries, and that it requires a truly global collective effort by all the nations of the world.

**We cannot overstate how serious the situation is. We are facing an existential challenge.** The IPCC report (AR6) of 2021 shows that:

> ...we've currently reached 1.1°C above the average temperatures from 1850–1900. Not in 125,000 years has earth seen a time period with such a high average temperature for such a prolonged period of time, indicating that this is more than just a natural heat spike...This is coupled with reporting that atmospheric $CO_2$ is at a 2,000,000-year peak and that other greenhouse gasses like methane and nitrous oxide are at 800,000-year peaks.

Everyone now recognizes that global warming is due to the accumulation of greenhouse gases (GHG) in the atmosphere, mainly $CO_2$ but also methane and nitrous oxide. That means that we must stop adding to these GHGs, and then seek their reduction if we are to keep the world's average surface temperature to no more than 1.5 degrees warmer than pre-industrial levels by 2100. That implies reaching net-zero emissions by 2050. That, in turn, requires that we reduce annual emissions by about 45% come 2030. Keeping global warming below 1.5 degrees is the commitment that 195 countries took in the Paris Agreement of 2015, and yet we are continuing to increase our global emissions every year. Indeed, in 2022, global energy-related $CO_2$ emissions grew by 0.9%, or 321 million tons, reaching a new high of more than 36.8 billion tons. So, we are not on track to fulfil our 2015 Paris Agreement pledge, and mitigation of GHG emissions is certainly not on track.

But beyond mitigation, we must also focus on adaptation and resilience, especially in the global south, where so many of the poor and vulnerable members of the human family live, and who are suffering the already visible consequences of climate change. This requires support to implement policies and programmes to promote adaptation and nurture resilience to the devastation of climate change. That is especially true for the drylands that cover 40% of the land area on the planet and are inhabited by some 40% of the world's population, but where we find some 80% of the poorest and most vulnerable of the human family. They are the ones who are suffering most from the consequences of this global existential challenge, even though they contributed least to the creation of the problem. We made a major step in that direction in COP27 when we adopted the idea of a 'Loss and Damage Fund' to support the necessary funding for adaptation and resilience in the global south. This book covers many of the aspects of both the challenges that these poor vulnerable populations face, and what can be done to respond to them.

## Where are we as we emerge from COP28?

It is clear that we are not on track to fulfilling our obligations as defined in the Paris Agreement of 2015, and the Nationally Determined Commitments (NDCs) that followed it. These NDCs themselves, even if fully implemented, would not have been enough to reach our Paris Agreement pledge. We are falling short.

Fossil fuels continue to be the prime cause of the growing emissions of GHGs. However, for the first time, the countries agreed to a statement about fossil fuels, accepting the wording 'transition away from fossil fuels', a much weaker statement than 'phasing out fossil fuels', which was the wording advocated by many of the climate activists present. In addition, the adopted wording came without any milestones or time horizons.

Another positive development at COP28 has been the attention given to the food and agriculture sector. For the first time, COP28 allocated an entire day to discussing the transformation of food systems, and it was only the second time the conference included a pavilion dedicated to food systems. The food and agriculture sector does contribute to the world's total emissions. But we believe that the transformation of the food and agriculture sector, from being a contributor to GHGs to being part of the solution, can emerge as a major force for carbon capture and sequestration (CCS). After all, plants are nature's system of CCS! A transformed sector, reducing or abolishing the enormous wastage that now exists as we move towards science-based precise agriculture and climate-smart agriculture (CSA), will also noticeably increase the efficiency and effectiveness of the sector to move towards the goal of abolishing hunger in the world.

Also positive in this domain is that, following this progress, countries including Brazil, Norway, Sierra Leone, Cambodia and Rwanda established the Alliance of Champions for Food Systems Transformation. This alliance aims to ensure that nations are responsible for reporting developments in this sector. Supporting these efforts, the UN's Food and Agriculture Organization (FAO) unveiled its strategy, designed to offer a worldwide perspective on transformation; more specific regional analyses, financial projections, and national strategies are now expected in future conferences.

The contents of this book cover the intense and urgent collective action required to tackle the dual challenges of the influence of climate change on food security as well as the impact of our food systems on climate change. It is particularly timely to be coming out after COP28.

Methane, significantly more potent than carbon dioxide in warming the atmosphere during its initial decade, contributes to around one third of current global warming, as per the IPCC. However, it dissipates in a few decades as opposed to $CO_2$ which takes centuries to dissipate. Thus, urgent and decisive action to eliminate methane emissions would have a relatively fast impact on the emissions problem. COP28 recognized the importance of stopping methane emissions but did not really add meaningfully to the timing of implementing the necessary measures.

Given methane's short lifespan, yet high impact, immediate action to reduce emissions is critical for achieving global climate objectives and mitigating both immediate and long-term effects of climate change. While COP28 saw encouraging developments regarding methane, the fundamental problem of dependence on pollutant fuels persists, and mere efficiency improvements, though beneficial, are insufficient.

Human health, and the intricate links between it and the health of the planet, were also broached in COP28. Indeed, COP28 dedicated an entire day to exploring how climate change affects human health. It also brought together health ministers who participated with their environmental counterparts for the first time. Finally, over 120 countries endorsed the Climate and Health Declaration, underscoring the need for all countries to look at the links between climate change and human health in their national plans.

COP28 also looked at financing. Another positive outcome, was the confirmation of the creation of the Loss and Damage Fund, aimed at providing financial support to low- and middle-income countries (LMICs) grappling with the effects of climate change, such as severe weather events and rising sea levels, and even the start of some modest funding for that new facility. However, the funding provided is still way short of what is required and placing it within the World Bank was not what a number of the LMICs wanted, as they consider that institution too closely associated with the G7 countries.

In fact, the $700 million provided for the fund is a mere 0.2% of what experts estimate is necessary to cover all damages caused by climate change.

That, of course, highlights that approaching the financing issue in a new and innovative approach is desperately needed. We estimate that the total financing required will be around 5–7 trillion dollars, annually, for the coming decade or more. This sum cannot be reached by the conventional approaches when total Official Development Assistance (ODA) from all OECD countries for all sectors is a mere 200 billion dollars annually. Therefore, finance for climate change, covering emissions, adaptation, and resilience, remains high on the agenda for any future discussion of climate change.

Much of the material in this book is very timely as we prepare for COP29 in Baku, Azerbaijan, in late 2024. May the collective experience of all the specialists who contributed to the multifaceted discussions reported in this volume take us a step closer to help humanity curb the negative orientations and advance the caring and nurturing approaches for the poor and the vulnerable. May it help us to develop the necessary science-based policies and evidence-based regulations that the world must apply to rise to the challenge of climate change and return to the path of sustainable development.

**Ismail Serageldin**

Honorary President of the IDDC. Former VP of the World Bank. Former Chair of the CGIAR.

# Acknowledgments


This book is the result of collaborative efforts from a diverse group of authors spanning various domains and agro-ecological zones within dry areas worldwide. The collective contributions of 53 authors, representing more than 51 institutions across 16 nationalities, have significantly enriched the content of this book, and we extend our sincere gratitude to each of them.

We express our appreciation to Dr. Inger Andersen, Under-Secretary-General of the United Nations and Executive Director of the United Nations Environment Programme (UNEP), for graciously contributing the Foreword to this publication. Additionally, our heartfelt thanks go to Dr. Ismail Serageldin, Honorary President of the IDDC, Former Vice President of the World Bank, and Former Chair of the CGIAR, for contributing the Preface.

A special acknowledgment is extended to Dr. Mohan Saxena, Executive Secretary of the International Dryland Development Commission (IDDC), for his invaluable contribution to this project.

We wish to recognize the pivotal role played by Eng. Aida Ghazi, Former Under-Secretary of Extension & Training, Ministry of Agriculture and Land Reclamation, Egypt, who skillfully coordinated the task leading to the completion of this book. Our sincere gratitude also goes to Ms Dina Taha, illustrator, UX/UI designer and graphic designer, for her significant input and contribution during the revision process. Dr. Hanaa Adly is acknowledged for her valuable insights and assistance in processing materials, while Dr. Sameh Hassanein is thanked for his technical expertise.

The collective efforts of these individuals have played an integral role in shaping and refining the content of this book, and we are truly grateful for their dedication and expertise.

The editors would also like to thank the team at CABI, including David Hemming, Emma McCann, Shankari Wilford and Marion Doy.

The Editorial Board

# List of Abbreviations and Acronyms

| | |
|---|---|
| % | - Per cent |
| >= | - Greater than or equal to |
| °C | - Degrees Celsius |
| °E | - Degress East |
| °N | - Degrees North |
| 1 Faddan | = 1.04 acres |
| 13 T | - Cotton, genetically modified |
| $^{13}C$ | - Carbon-13 |
| $^{230}Th$ | - Thorium-230 |
| $^{234}U$ | - Uranium-234 |
| $^{238}U$ | - Uranium-238 |
| ABA | - Abscisic acid |
| ABS | - Access and benefit sharing (for genetic resources) |
| ACC | - Anthropogenic climate change |
| ACIAR | - Australian Centre for International Agricultural Research |
| ACs | - Agriculture cooperatives |
| AE | - Agro-ecological |
| AEZs | - Agro-ecological zones |
| AF | - Adaptation fund |
| AFD | - Agence Françoise de Developpment |
| AGERI | - Agriculture Genetic Engineering Research Institute |
| AGMERRA | - AgMIP dataset by Modern-Era Retrospective Analysis for Research & Applications |
| AgMIP | - Agricultural Model Intercomparison and Improvement Project |
| AI | - Artificial intelligence |
| AIB | - Allied Irish Bank |
| ALRC | - Arid Land Research Center, Japan |
| ALYM | - African League of Young Masters |
| AMV | - Alfalfa mosaic virus |
| API | - Aquaculture and photo-voltaic integration |
| APSIM | - Agricultural Production Systems Simulator |
| AR | - Assessment Report of IPCC |
| AR4D | - Agricultural research for development |
| ARC | - Agricultural Research Center |

ARDC - Agricultural Research & Development Council
B - Billion
BD - Bulk density
BGM - Botrytis grey mold
BWYV - Beet Western Yellow Virus
C - Carbon
Ca - Calcium
CA - Conservation agriculture
$Ca^{2+}$ - Calcium ion
CAADP - Comprehensive Africa Agriculture Development Program
CAS - Conservation agriculture systems
CASI - Conservation agriculture-based sustainable intensification
CBA - Community-based adaptation
CBD - Convention on Biological Diversity
CC - Climate change
CCAFS, CCFS - Climate change agriculture and food security
CCS - Carbon capture and sequestration
CECA - Centre of Excellence for Conservation Agriculture
CF - Carbon farming
CFAES - College of Food, Agricultural, and Environmental Sciences
CFS - Committee on World Food Security
CGIAR - Consultative Group for International Agricultural Research
CGR - Crop genetic resources
CGRFA - Commission on Genetic Resources for Food and Agriculture
$CH_4$ - Methane
CI - Concentration index
CIMMYT - International Maize and Wheat Improvement Center
CIS - Climate information system
CLAC - The Central Laboratory for Information Center
cm - Centimeters
$CO_2$ - Carbon dioxide
COP - Conference of the Parties (to the UN Framework Convention on Climate Change)
COP27 - 27th Conference of the Parties
CoPs - Community of Practitioners
CpCSV - Chickpea Chlorotic Stunt Virus
CRA - Climate-resilient agriculture
CRISPR - Clustered regularly interspaced short palindromic repeats
CRISPR/Cas9 - The fundamental technology for genome editing
CRV - Climate-resilient village
CSA - Climate-smart agriculture
CSDs - Civil society organizations
CSE - Center for Science and Environment
CT - Conventional tillage
CTCN - Climate Technology Centre and Network
CWANA - Central and West Asia & North Africa
CWR - Crop wild relatives
DACP - District Agricultural Contingency Plan
DAFW - Department of Agriculture & Farmers Welfare
DARE - Department of Agricultural Research & Education
DHA - Dehydrogenase activity

| | |
|---|---|
| DIIVA-PR | - Dissemination of interspecific ICARDA varieties and elites through participatory research |
| DNA | - Deoxyribonucleic acid |
| DREP | - Dehydration responsive element binding |
| DRR | - Disaster risk reduction |
| DSR | - Direct seeded rice |
| DSS | - Decision support systems |
| DSSAT | - Decision support system for agro-technology transfer |
| EEZ | - Exclusive Economic Zone |
| EGP | - Eastern Gangetic Plains |
| ESCWA | - Economic and Social Commission for Western Asia of the United Nations |
| ESs | - Essential ecosystem services |
| EST | - Expressed sequence tag |
| EU | - European Union |
| FAO | - Food and Agriculture Organization of the United Nations |
| FAS | - Food and agriculture sector |
| Fig. | - Figure |
| FIGS | - Focused identification of the germplasm strategy |
| FMNR | - Farmer-managed natural regeneration |
| FP | - Flat planting |
| FSFPS | - Food security and food production systems |
| FSN | - Food security and nutrition |
| FW | - Fusarium wilt |
| FYM | - Farmyard manure |
| g | - Grams |
| G20 | - Group of Twenty |
| GBF | - Global Biodiversity Framework |
| GCC | - Global carbon cycle |
| GCF | - Green Climate Fund |
| GCM | - General circulation model |
| GCM | - Global climate models |
| GCSAR | - General Commission for Scientific and Agricultural Research |
| GDLs | - Global drylands |
| GDP | - Gross domestic product |
| GERD | - Grand Ethiopian Renaissance Dam |
| GEWEX | - Global energy and water exchanges |
| GFAR | - Global Forum on Agricultural Research |
| GGP | - Global Gene Bank Partnership |
| GHG | - Greenhouse gas |
| GIM | - Green India Mission |
| GIS | - Geographic information system |
| GM | - Genetically modified |
| Govt | - Government |
| GPP | - Gross primary production |
| GPS | - Global positioning system |
| GRDC | - Grains Research and Development Corporation |
| GS | - Genomic selection |
| Gt C $yr^{-1}$ | - Gigatonnes of carbon per year |
| GVA | - Gross value added |
| GW | - Global warming |
| GWAS | - Genome-wide association study |
| GWP | - Global warming potential |

| | |
|---|---|
| $H_2O$ | - Water |
| ha | - Hectare |
| HEIS | - High-efficiency irrigation systems |
| HFCs | - Hydrofluorocarbons |
| HKH | - Himalayan-Karakoram-Hindukush |
| HLI | - Heat Load Index |
| HLPE | - High-level panel of experts |
| HS | - Heat stress |
| IAC | - International Agrobiodiversity Congress |
| IBPES | - Intergovernmental Science-Policy Platform on Biodiversity and Ecosystem Services |
| ICAR | - Indian Council of Agricultural Research |
| ICARDA | - International Center for Agricultural Research in the Dry Areas |
| ICC | - India Climate Collaborative |
| ICCIP | - International Climate Change Information Program |
| ICLARM | - International Center for Living Aquatic Resources Management |
| ICRISAT | - International Crops Research Institute for the Semi-Arid Tropics |
| ICT | - Information and communication technology |
| IDDC | - International Dryland Development Commission |
| IEA | - International Energy Agency |
| IFAD | - International Fund for Agriculture Development |
| IGP | - Indo-Gangetic Plain |
| IICA | - Inter-American Institute for Cooperation on Agriculture |
| IITA | - International Institute of Tropical Agriculture |
| ILO | - International Labor Organization |
| INCCA | - Indian Network for Climate Change Assessment |
| INIA | - National Agricultural Research Institute |
| INM | - Integrated nutrient management |
| INR | - Indian rupee |
| IoT | - Internet of Things |
| IPCC | - Intergovernmental Panel on Climate Change |
| IPM | - Integrated pest management |
| IPPC | - International Plant Protection Convention |
| ISAAA | - International Service for the Acquisition of Agri-biotech Applications |
| ISRO | - Indian Space Research Organization |
| ISSM | - Integrated soil-crop system management |
| IT | - Information technology |
| ITPGRFA | - International Treaty on Plant Genetic Resources for Food and Agriculture |
| IUCN | - International Union for the Conservation of Nature |
| IWM | - Integrated water management |
| K | - Potassium |
| $K^+$ | - Potassium ion |
| kg | - Kilograms |
| km | - Kilometers |
| KUSUM | - Kisan Urja Suraksha evam Utthaan Mahaabhiyan |
| LDCs | - Least-developed countries |
| LICs | - Low-income countries |
| LLL | - Laser land leveling |
| LMIC | - Low- and middle-income countries |
| m | - Meter |
| M | - Million |
| MALR | - Ministry of Agriculture & Land Reclamation |

MENA - Middle East and North Africa region
MERRA - Modern-Era Retrospective Analysis for Research & Applications
Mg - Magnesium
Mg - Megagram
mg - milligrams
$Mg^{2+}$ - Magnesium ion
mha - Million hectares
MIDH - Mission for Integrated Development of Horticulture
mm - Millimeters
MoEFCC - Ministry of Environment, Forest, and Climate Change
MRB - Mechanized raised bed
MT - Million tonnes
MW - Maize-wheat
N - Nitrogen
$N_0$ - Neutron
$N_{135}$ - Nitrogen-135
$N_{180}$ - Nitrogen-180
$N_2O$ - Nitrous oxide
$N_{45}$ - Nitrogen-45
$N_{90}$ - Nitrogen-90
$Na^+$ - Adsorb sodium
NAARM - National Academy of Agricultural Research Management
NABARD - National Bank for Agriculture and Rural Development
NAFCC - National Adaptation Fund for Climate Change
NAPCC - National Action Plan on Climate Change
NARES - National Agricultural Research and Education System
NARS - National Agricultural Research System
NASA - National Aeronautics and Space Administration
NBAIR - National Bureau of Agricultural Insect Resources
NBIAM - National Bureau of Agriculturally Important Microorganisms
NBM - National Bamboo Mission
NCCS - National Climate Change Strategy
NDCs - Nationally determined contributions
NEI - Not elsewhere included
NEP - Net ecosystem productivity
NFSM - National Food Security Mission
NFT - Non-fungible token
NGB - National Gene Bank
NGIC - Nizami Ganjavi International Center
NGO - Non-governmental organization
$NH_3$ - Ammonia
$NH_4^+$ - Ammonia ion
NICRA - National innovations on climate-resilient agriculture
NIDIS - National Integrated Drought Information System
NIE - National implementing entity
NLM - National Livestock Mission
NMSA - National Mission for Sustainable Agriculture
NOAA - National Oceanic & Atmospheric Administration
NPV - Net return value
NT - No-till
NTC - No-till cropland
NTS - No-till system

OECD - Organization for Economic Co-operation & Development
OFWM - On-farm water management
OR-N - Organic and synthetic nitrogen fertilizers
P - Phosphorus
PA - Precision agriculture
PCR - Polymerase chain reaction
PDMC - Per Drop More Crop
PFCs - Perfluorocarbons
Pg - Petagram
Pg C $year^{-1}$ - Petagram of carbon per year
PGR - Plant genetic resources
PIC - Pedogenic inorganic carbonates
PKVY - Paramparagat Krishi Vikas Yojana
PMFBY - Pradhan Mantri Fasal Bima Yojana
PMKSY - Pradhan Mantri Krishi Sinchayee Yojana
QTL - Quantitative trait loci
R4D - Research for development
RACC - Regional Action on Climate Change
RAP - Representative agricultural pathways
RAPD - Random Amplified Polymorphic DNA
RAS - Recirculating aquaculture system
RCM - Regional climate model
RCP - Representative concentration pathway (IPCC)
RDF - Recommended dose of fertilizers
RH - Relative humidity
RKVY - Rashtriya Krishi Vikas Yojana
RMP - Recommended management practice
RUE - Rain-use efficiency
RW - Rice-wheat
RWH - Rainwater harvesting
RWP - Rainwater productivity
S - Sulfur
SADS - Sustainable Agriculture Development Strategy
SAPCC - State Action Plan on Climate Change
SDGs - Sustainable Development Goals
SDI - Sub-surface drip-irrigation
SDS - Sand dust storms
SF6 - Sulfur hexafluoride
SFDRR - Sendai Framework for Disaster Risk Reduction
SFS - Sustainable food systems
SH - Soil health
SHM - Soil-health management
SI - Supplemental irrigation
SIC - Soil inorganic carbon
SICLU - Sustainable intensification of cultivated land use
SIDS - Small island developing states
SLM - Sustainable land management
SLR - Sea-level rise
SMBC - Soil microbial biomass carbon
SNP - Single nucleotide polymorphisms
SOC - Soil organic carbon
SON - Soil organic nitrogen

| | |
|---|---|
| SPM | - Summary for policymakers of IPCC |
| SQI | - Soil Quality Index |
| SRB | - Surface radiation budget |
| SRFSI | - Sustainable and resilient farming systems intensification |
| SRI | - System of rice intensification |
| SSA | - Sub-Saharan Africa |
| SSF | - Small-scale fisheries/fishing |
| SSNM | - Site-specific nutrient management |
| SSR | - Simple sequence repeats |
| STC | - Strip-tilled cropland |
| STS | - Science & Technology for Society Forum |
| STS | - Socio-technical systems |
| T Sh. | - Tanzania shilling |
| TAAS | - Trust for Advancement of Agricultural Sciences |
| TAAT | - Technology for Africa Agriculture Transformation |
| TEC | - Technology Executive Committee |
| Tg | - Teragram (= million ton) |
| THI | - Temperature humidity index |
| TVET | - Technical & vocational education & training |
| TYLCV | - Tomato Yellow Leaf Curl Virus |
| US | - United States |
| USA | - United States of America |
| UGL | - Ungrazed native grassland |
| UN | - United Nations |
| UNCCD | - United Nations Convention to Combat Desertification |
| UNDP | - United Nations Development Program |
| UNEP | - United Nations Environment Program |
| UNESCO | - United Nations Educational, Scientific and Cultural Organization |
| UNFCCC | - United Nations Framework Convention on Climate Change |
| UNICEF | - United Nations International Children's Emergency Fund (now officially United Nations Children's Fund) |
| UNIDO | - United Nations Industrial Development Organization |
| UN-SDG | - United Nations Sustainable Development Goal |
| UNU | - United Nations University |
| US $ | - United States dollar |
| USAID | - United States Agency for International Development |
| USDA | - United States Department of Agriculture |
| USGCRP | - United Nations Global Change Research Program |
| VCRMC | - Village-level climate risk management committee |
| WANA | - West Asia and North Africa |
| WAPDA | - Water and Power Development Authority |
| WBT | - Wet bulb temperature |
| WEF | - World Economic Forum |
| WG | - Working group |
| WHO | - World Health Organization |
| WMO | - World Meteorological Organization |
| WP | - Water productivity |
| WPI | - Water Productivity Index |
| WRKY | - Transcription factor 'worky' |
| WUE | - Water-use efficiency |
| Yr | - Year |
| ZT | - Zero tillage |

# 1 Introduction

**Adel S. El-Beltagy[1,2], Rattan Lal[3], and Kauser Malik[4]**
[1]*Chair, International Dryland Development Commission (IDDC);* [2]*Arid Land Agricultural Graduate Studies & Research Institute (ALARI), Ain Shams University, Egypt;* [3]*Rattan Lal Center for Carbon Sequestration and Management, The Ohio State University, USA;* [4]*Forman Christian College (A Chartered University), Lahore, Pakistan*

Drylands are home to nearly 2.5 billion people with 80% of the Earth's poorest people living in dryland developing countries. The inhabitants of drylands are facing unprecedented challenges to their sustainability because of climate change.

Global drylands are highly vulnerable to anthropogenic climate change, and land area under drylands may expand from 41%, presently, to 50% of the Earth's surface by the year 2100. Drylands include 45% of agricultural land, which will be affected by climate change in aggravating prevalence of undernutrition and malnutrition (El-Beltagy *et al.*, 2005; Sivakumar *et al.*, 2013) because of adverse effects on the quantity and quality of food production in these regions. The world population is currently 8.1 billion and is projected to reach 9.8 billion by 2050. A radical transformation of the world's food systems will be needed. Seventy per cent of the increase in world population between now and 2050 will be in 24 of the poorest countries of the world. Meanwhile, the projected increase in temperature beyond 1.5°C, which may reach 3°C, will reduce food production by 20–50%.

The objective of this book is to adopt a holistic approach to sustainable management of drylands to make agriculture drought-resilient and to sustain the livelihood of the vulnerable inhabitants (Traylor *et al.*, 1999). Eminent scientists from around the world have been invited to share their knowledge and experiences for adaptation and mitigation of anthropogenic climate change through innovation in sustainable management of natural resources of water, soil, and biodiversity (crops, livestock, fisheries).

Time is of the essence as the Earth's average surface temperature in 2022 was the fifth warmest on record. Continuing the planet's warming trend, global temperature in 2022 was 1.06°C above the average for NASA's baseline period (1951–1980) and confirmed by NOAA Record (NOAA, 2023).

In addition, NASA clocks stated that July 2023 was the hottest month on record since 1880 (Wong, 2023). Tollefson (2023) stated that Earth's hottest month happened in July and that heatwave projections could see a 1.5°C rise, surpassing it in a matter of years at around 2°C of warming. This is likely even if governments meet their climate commitments. Based on current policies alone, climate scientists estimate that temperatures could rise nearly 3°C in this century. In each of these scenarios, models project that many places on Earth could experience 30–40 extra days each year above 35°C, affecting public health and ecosystems globally (Tollefson, 2023).

Launched in 2001 and involving more than 1300 leading scientists, the Millennium Ecosystems Assessment is a study of how humans have altered ecosystems and the impact on human well-being. Integrating findings at local, regional, and global scales, it is the first truly comprehensive picture of the health of the planet (Millennium Ecosystem Assessment, 2005).


DOI: 10.1079/9781800624870.0001

The anticipation that climate change will have major impacts on agro-ecosystems requires continuous dynamic assessments, globally, regionally, and at the local level where the major action of adaptation would have to occur. The assessment will require international cooperation, national capacity building, and a solid early-warning system.

This book presents a cross-section of the global, regional, and national impact of climate change on dryland agro-ecology. The authors detail not only constraints but also lay out the different pathways to achieve a more climate-resilient society.

The book emphasizes the importance of utilizing the best scientific findings including smart and precision agriculture, conservation agriculture, and innovation and new technology as tools for adaptation and mitigation.

Several chapters are devoted to the human dimensions and policy considerations with emphasis on enhancing coping and adaptive capacity. The book addresses events after COP27 including loss and damage, governance, and finance. The code of conduct of mankind in relation to nature should be addressed and emphasized globally with regards to the following:

**1.** promoting sustainable management of drylands as a solution to environmental and food security issues by adopting a holistic approach for sustainable development of these fragile eco-regions and translating science into action for greening global drylands;
**2.** undertaking dynamic assessment of the impact of climate change on agro-ecological zones to provide guidelines for sustainable management of global drylands; and
**3.** following up on declarations of COP27 and COP28, it is pertinent to address loss and damage, coping with uncertainties, enhancing adaptive capacity, governance, policies, and finance.

## Main Features of the Book

A unique feature of the book is a holistic approach by authors from around the world to address the following aspects: innovations in agriculture; climate-smart agriculture/e-agriculture; agriculture 4.0; precision agriculture; conservation agriculture; carbon farming; utilization of genetic resources (crop, livestock, fisheries) through new modes of genetic engineering/gene editing etc.; continuous assessment of the impact of climate change (CC) on agro-ecological zones; translating science into action; enhancing coping and adaptive capacity; and policy interventions.

It is important to strengthen cooperation with the local community including cooperative private sector research for development (R4D) continuum.

The book makes special reference to the dynamic change of the boundaries of agro-ecology impacted by climate change and the need for continuous assessment of the dynamic shift, which will require massive changes in cropping systems. There is a need for enhancing coping and adaptive capacity and policy aspects, including finance.

The book addresses the various convergence zones which were presented globally and during four webinars during the four pre-COP meetings organized by the International Dryland Development Commission (IDDC); Regional Action for Climate Change (RACC)/Science and Technology for Society (STS) forum; the Nizami Ganjavi International Center (NGIC); and the African League of Young Masters (ALYM), hosted by Ain Shams University (2022) and Alexandria University (2023) (IDDC, 2022, 2023) and STS/RACC in Kyoto, Japan (2022, 2023) (STS, 2023; Falk *et al.*, 2023, 2024). These meetings are projecting efforts needed to bridge the gap, especially after the COP27 breakthrough agreement on the new Loss and Damage Fund for vulnerable countries.

## Main Elements in the Book's Chapters

### Global chapters

*Ch. 2: Navigation through Uncertainties*

**Part A: Regional Action on Climate Change**
This part of the chapter states that our epoch of unprecedented human planetary impact demands an urgent collective response. Since the Anthropocene's conception in 2000, assessments confirm extensive disruption of climate, ecology, and environment, quantified by concepts like post-WWII's Great Acceleration. Crossing of planetary

boundaries nears catastrophe, defying one-size-fits-all solutions. At peak population, Wilson predicted that ecosystems surviving ecological bottlenecks will be irrevocably altered. Despite freshwater security linking diverse regions, integrated action lags. Recent African and G20 declarations (African Union, 2023; G20, 2023) signal a commitment to climate-smart agriculture, nature conservation, and biodiversity protection. However, averting biodiversity annihilation requires international campaigns leveraging gene banks and international and national organizations along with sufficient funding to introduce new genetic resources for the dynamic adaptation of agriculture. With crises converging, we do not have the luxury of time. Avoiding climate injustice demands equitable resilient responses. While daunting, humanity retains the urgency to act decisively, steering away from collapse towards sustainability. This pathway exists but it is narrowing. The stakes could not be higher.

**Part B: Agro-ecosystems Affected by Dynamic Impact of Climate Change**

The main elements impacting the vulnerability of agro-ecologies in the dry areas are elaborated with emphasis on the impact of water, biodiversity, land, livelihood of the population, and, consequently, the process of desertification. The danger of outmigration and its impact are discussed on global, regional, and national socio-political stability. The main outcomes of international agreements and conclusions in different COP meetings, including the Sendai framework for disaster risk reduction, the COP27 loss & damage fund and Justice are vividly discussed. It also elaborates on early-warning systems linked to the dynamic impact of climate change on agro-ecological zones. Adaptation as a priority for global climate action is one of the notable achievements of COP27. Aspects related to the adaptation were discussed in detail with emphasis on international cooperation, especially in relation to science and technology and the need to create a knowledge action network.

This part also addresses the shift in agro-ecological zones and the attendant changes in habitats which affect crops, livestock, and fisheries. Collaboration among global gene banks, both national and international (like CGIAR and Svalbard), is essential to ensure that collections give sufficient emphasis to stress-adapted genetic material that can contribute to adaptation to climate change, and ensure that policies facilitate international exchange of the genetic resources needed to adapt farming systems to shifts in the distribution of climatic agro-ecological zones. It is critical to cope with uncertainty, recognizing that there will be shifting baselines as further assessment of climate change is done.

Funding is necessary to enhance gene bank passports with molecular characteristics and continue collection missions from dry areas, establish a global breeding program aimed at the development of new genetic make-up resistance to biotic and abiotic stressors, and develop an early-warning system for climate change impact on the shift of local agro-ecological zones. It is also important to support the transformation and modernization of agro-management techniques.

This chapter strongly recommends that a solid consortium be established which includes the national, regional, and international gene banks with international UN institutions, i.e. UNEP, FAO, IFAD, CGIAR, UNDP, IUCN, and Crop Diversity Trust, to activate the global gene bank partnership (GGP) platform and ensure investment for a research and information system which will serve the mission to address the shifting baseline in agro-ecology by providing suitable crops (animal, fisheries), and to continuously introduce new genetic resources for ever-shifting agro-ecology.

Emphasis is needed on human resources development and knowledge sharing at regional and international level. There is a need for establishment of the Loss and Damage Fund with clear indicators for its operation.

This chapter states that resilience knowledge must be seen as a universal asset. The gap between the 'haves' and the 'have-nots' must be bridged to ensure peace, stability, and prosperity. A shift in consciousness among the 'haves' and the 'have-nots' can lead to a new understanding of humanity's shared destiny.

### *Ch. 3: Carbon Farming in Global Drylands*

The focus of this chapter is on sustainable management of global drylands (GDLs) for farming carbon (C), which implies growing C in land-based sinks to generate another income stream for farmers. Total C stock in GDLs comprises soil organic C (SOC) and soil inorganic C (SIC), and, in general, stock of SIC may be more than that of SOC.

GDLs play a significant role in regulating the global C cycle (GCC) and can be a sizeable sink for atmospheric $CO_2$ due to their large surface area and degraded soil. If decertified and degraded GDLs are restored, they have the potential to be a large sink for atmospheric $CO_2$. Yet scientific understanding of the controls of GDL-$CO_2$ exchange is poor. Thus, this chapter focuses on C farming as a strategy to address these issues.

Enhancing SOM content is a critical adaptation strategy in global drylands through improvement in water retention and nutrient cycling. Despite sparse vegetation but a large land area (> 40% of Earth's surface), GDLs have a large influence on global biogeochemical cycles and dominate the long-term trends in the GCC.

Some techniques discussed in this chapter to grow C stock in soil include the following:

**1.** Integration of livestock with crop and trees can increase diversification. Indeed, livestock play an important role in sustainable management of drylands at the farm, household and territorial levels with specific reference to food security, well-being, value-chain performance, food supply and maintenance of common grazing resources.
**2.** Afforestation of marginal drylands may strengthen several ecosystem services especially in temperate regions. However, there may be conflicts between additional forest area and water consumption with adverse effects on water yield.
**3.** Agroforestry systems are recommended for management of GDLs based on the production of indigenous crops. An agroforestry-based approach to mitigating climate change (and other issues including hunger, poverty, wildlife habitat etc.) is appropriate for sustainable management of GDLs.
**4.** Conservation agriculture (CA) in combination with integrated soil fertility management and complex crop rotations can be a conservation and soil-enhancing system for GDLs. Residue retention is a critical component of CA without which crop yield may decline and degrade soil functions. In addition to retention of crop residue mulch, effectiveness of CA also depends on adopting the strategy of integrated nutrient management (INM).
**5.** Soil-water conservation and management is the key to achieving food security and improving the livelihood of the large population of GDLs. Thus, conserving water in the root zone, increasing water-use efficiency (WUE) and achieving more crops per drop of water are the major goals and principal challenges of agriculture in GDLs.

Sustained C sequestration in GDLs is a major challenge addressed in this chapter because of multiple resource limitations of resource-poor farmers. In addition to water, low nutrient availability may be one factor that inhibits C sequestration under dryland conditions. Judicious use of N fertilizer is critical to enhancing SOC stock but also moderating fluxes of $CO_2$, $CH_4$, and $N_2O$.

The chapter concludes with recommendations for a way forward to farming carbon in GDLs:

**1.** Protection of native biomes, restoration of degraded soils, and sustainable management of soils of GDLs offer an opportunity to reconcile the need to produce more food with the necessity of improving the environment.
**2.** Whereas the rate of soil C sequestration is low, GDLs cover a large area of 6–7 billion ha of land and can sequester a large amount of C as SOC and SIC.
**3.** Sequestration of SIC, through formation of secondary carbonates and leaching of bicarbonates, is an important component which needs to be addressed in the GDLs.
**4.** A system should be developed for payments to farmers for ecosystem services strengthened through sequestration of C in GDLs.

### *Ch. 4: Plant Genetic Resources for Adaptation to Climate Change in Drylands*

This chapter, prepared under the lead authorship of Dr. R.S. Paroda, former DG of ICAR and Director of DARE, vividly explains the role of plant genetic resources (PGR) in the future and the adverse effects of climate change in the drylands of India. The chapter covers the conservation, utilization, and management strategies of PGR towards climate-resilient agriculture to ensure both agricultural sustainability and future food security. The authors deliberate that genetic resources could contribute greatly to cope with climate change, which is already affecting natural ecosystems and food production systems globally. The chapter focuses specifically on the effects of climate change on drylands in general and those of India in particular. The chapter begins with an explanation of the impacts of climate change on agriculture, especially on agronomic yield under the conditions of rising temperatures and increase in frequency and intensity of droughts among increases in biotic stresses.

**Significance of PGRs**

With this background, the chapter focuses on the significance of PGRs under changing climate. In this context, therefore, the authors state that genetic diversity is a valuable resource for breeding programs, enabling the development of improved crop varieties with enhanced tolerance to abiotic stresses (e.g. drought, heat, salinity) and biotic stresses (e.g. pests, diseases). They further argue that PGRs also serve as a source of nutritionally rich and culturally important crops, promoting dietary diversity and food sovereignty, which is specifically relevant to countries where large populations are prone to hidden hunger and malnutrition.

With the goal of adaptation and mitigation of climate change in dryland agriculture, the authors explain that traits that would be needed would include short life cycles, deep root systems, waxy leaves, water storage capabilities in roots and trunks, dormancy during droughts, and tolerance to high temperatures. Among the three main components of climate change, temperature has large effects, both in terms of lethality at high and low temperatures and in affecting both phenology and distribution of crops, and PGRs can play a significant role in addressing these issues.

**Case Study on PGR Diversity in Indian Perspective**

The authors appropriately elaborate on the unique success of Indian agriculture in achieving the Green, White, Yellow, and Blue Revolutions. Yet, they also highlight the impact on degradation of natural resources which are particularly severe in the Green Revolution corridors of India. Therefore, they focus on:

1. **Dryland agriculture in India**
   Drylands occupy 80 mha of cropland where are also located 50% of the total rural workforce and 60% of India's livestock. In addition to horticultural crops, the forage genetic resources (e.g. lucerne, cowpea, range grasses, legumes, trees, and shrubs) can play a significant role in improving adaptability to harsh environmental conditions.
2. **PGR diversity in dryland**
   The chapter explains the rich crop diversity in the Indian arid zone in terms of species, genotypes, and ecotypes, etc. The authors highlight that agrobiodiversity contributes to resilience through the protection and restoration of ecosystems, the sustainable use of soil and water resources, agroforestry, diversification of farming systems, various adjustments in cultivation practices and the use of stress-tolerant crops and crop improvement.
3. **PGR management and climate-resilient agriculture**
   The authors propose the development of a Red Data Book devoted to arid region biodiversity and suggest that incentives and rewards be provided to farmers who actively conserve the agrobiodiversity. Germplasm conservation can be either *in situ* protectionism or *ex situ* conservationism. Indian gene bank conserves germplasm *ex situ*. Authors also propose an integrated or *trans situ* approach based on ecology, biology, diversity and utility of the targeted plants. The chapter also describes the importance of traditional knowledge and community involvement and provides specific examples from Rajasthan, Haryana, and Punjab, and that followed by the Bishnoi community.
4. **Utilization of PGR**
   Among examples of utilization of PGR, the authors describe options of:
   1. breeding for climate resilience;
   2. use of biotechnology and molecular breeding;
   3. identification of desirable traits and reviewing mitigation strategies;
   4. strategies for climate-smart agriculture;
   5. conservation agriculture;
   6. crop diversification; and
   7. numerous other approaches for reshaping and evolving production systems to cope with changing climate.
5. **Global forums and international treaties**
   Finally, the chapter describes examples of international treaties such as those of the UNFCCC, UNCCD, CBD and the Commission on Genetic Resources for Food and Agriculture (CGRFA) under the auspices of FAO, along with an International Treaty on Plant Genetic Resources for Food and Agriculture (ITPGRFA). This section also lists many global

online portals which serve as a source CWR inventories.

The chapter concludes with a set of 14 specific recommendations to cope with problems in agriculture arising due to climate change in dryland ecosystems.

### *Ch. 5: Agro-management Techniques and Precision Agriculture/Agriculture 4.0 in Arid Lands*

This chapter on precision agriculture (PA) or e-agriculture is authored by a team of researchers from ICARDA, from Salman University in Sinai, and Ain Shams University. PA, or site-specific crop management, is defined as agriculture that increases the number of (correct) decisions per unit of area of land per unit time with associated net benefits. It is an approach to agriculture that uses data sensors, connected devices, remote controls, and other advanced technologies to give farmers more control over the field. More specifically, PA links agricultural management systems with modern technology that contributes to improving and sustaining the use of natural resources, especially soil and water, increasing the quantity of production and improving its quality.

The chapter outlines specific soil functions which can be enhanced by the use of PA including remote-sensing technology, GIS, data management and the Internet of Things (IoT) for an accurate measurement of differences in the field at the right time, allowing the right decision to be made at the right time and maximizing the return from available natural resources. Authors vividly explain that PA is a management method aimed at reaching the highest production using the lowest inputs while maintaining the integrity of the surrounding environment and managing fields based on soil types, soil units in different areas, and dynamic soil quality indicators.

Some applications of PA in the context of arid agriculture lands discussed in the chapter include the following:

1. **Applications of IoT**
   The chapter explains that PA is an approach used in farm management and crop control through information and communication technology, remote sensors, remote control systems, and autonomous machines to obtain accurate data and invest these data in directing agriculture accurately towards greater production at a lower cost, and the production of high-quality crops. Other important applications discussed in the chapter include the use of drones to monitor and evaluate crops, photograph, and map agricultural lands, measure air components, spray crops with pesticides quickly and safely, send data instantly to software that analyzes them and instructs farmers to implement the best procedures.
2. **Integration of PA techniques to achieve agricultural sustainability**
   In this section, authors list examples of some sensors used in PA. Prominent among these are:
   **1.** sensors capable of aerial photography such as cameras carried on drones;
   **2.** soil quality sensors and the condition of the elements in the plant;
   **3.** electrochemical sensors; and
   **4.** smart temperature sensors.
3. **Revitalizing dryland rainfed systems under changing climate**
   The chapter describes the use of PA systems in revitalizing the dryland rainfed systems with specific focus on water resources as a critical factor for all production systems in relation to agriculture in Egypt and elsewhere in global drylands. The authors emphasize that by 2025, as many as 29 countries in the world are expected to experience water scarcity, including 18 countries in the Middle East and northern Africa classified as water-scarce. Therefore, many countries have recognized the urgent need to secure, rationalize, and utilize new irrigation technologies and techniques such as SMART irrigation management to sustain minimum resources base, as well as modify the traditional irrigation systems. The low annual crop yields increase in many Arab regions (1.5% per year) and can be improved by adopting PA systems for sustainable management of water resources. The authors argue that potential loss in rainfed areas in some MENA countries due to climate change may lead to a 20% reduction in annual precipitation.

However, there is a considerable potential for vertical productivity increases in most of the crops. Rainfed grain yields, on average, around

1.0 t/ha compared with about 3.0 t/ha in irrigated agriculture. Thus, the authors argue that there is a large untapped potential in rainfed agriculture in the drylands which can be realized by the adoption of PA.

**Constraints to improved dryland farming and the need for change of strategies**

The chapter outlines three specific constraints to improving dryland farming. These include:

1. climate constraints due to temporal and special variability in rainfall;
2. soil constraints due to poor soil water infiltration and storage capacity, poor soil fertility, and nutrient deficiency; and
3. adoption constraints due to a lack of knowledge and low financial capacity.

The authors explain that these constraints can be addressed by adopting strategies such as: enhancing water availability, increasing crop water uptake, improving rainwater productivity, and adopting climate-smart agriculture under rainfed conditions.

**Key practices and approaches**

Examples of key practices outlined in the chapter include supplemental irrigation, rainwater harvesting, agronomic and conservation practices, and adoption of digital rainfed agriculture. However, specific focus is given to PA and smart irrigation systems along with explanation of devices and tools for operating smart on-farm irrigation systems. The authors also list devices required for the automatic operation of smart irrigation systems including soil moisture sensors along with measuring and control devices.

**Energy in PA system, agricultural mechanization, and post-harvest services**

In these sections the authors explain renewable energy such as solar, wind, and green hydrogen energies. The section is devoted to agricultural mechanization with a focus on modern agricultural power and machinery.

The chapter concludes by stating that on-farm irrigation systems and their management are one of the most important agricultural sectors in applying the concept of precision and digital sustainable agriculture. Water saving, a critical factor in the success of farming in dryland, can be achieved by precisely computing the amount of water that plants require and applying it at the right time and by the right methods, which enhance use efficiency of the most precious but finite resource.

### *Ch. 6: Impact of Climate Change on Inland and Ocean Fisheries and Aquaculture*

This chapter provides an in-depth analysis of the multifaceted effects of climate change on these critical sectors. It begins by emphasizing the global significance of both inland and ocean fisheries and aquaculture, highlighting their roles in providing nutrition, employment, and economic benefits worldwide. The chapter then delves into the historical evolution of these sectors, noting their growth and increasing contribution to global fishing activities.

It also explores various adaptation strategies to mitigate the impacts of climate change on fisheries and aquaculture. These include adopting new technologies, modifying fishing practices, and implementing policy changes to support sustainable fishing and aquaculture practices. Emphasizing the need for localized solutions, the chapter suggests integrating traditional knowledge with scientific research to develop effective adaptation strategies. It calls for collaborative efforts between governments, research institutions, and communities to develop and implement policies that ensure the long-term sustainability of these sectors in the face of climate change.

It explains how changing climatic conditions are altering fish habitats, thus leading to shifts in fish stock abundance and distribution. This phenomenon is particularly evident in both inland and ocean fisheries, where species are migrating to cooler waters, disrupting traditional fishing patterns and practices. The chapter also discusses the increased frequency and intensity of extreme weather events like storms and heatwaves, which have direct and indirect impacts on fisheries and aquaculture operations. These changes pose significant challenges to the sustainability of these sectors and the livelihoods they support.

Another critical aspect covered in the chapter is the role of genomics in managing and conserving the genetic resources of fish species. The advancement in genomic technologies offers potential solutions to the challenges posed by climate change. By enhancing the understanding of the genetic make-up of fish species, these technologies aid in developing more resilient fish breeds and conservation strategies. The chapter explains how genomics is being integrated into breeding programs to develop species that can better withstand changing environmental conditions.

In addition, the chapter also stresses current operations, and emphasizes that the emerging and pending stresses of climate change have pushed better-endowed fishing and farming enterprises, research and development institutes and entrepreneurs to strive for more cutting-edge technological fixes.

In aquaculture, the technological fixes address current bottlenecks such as the supply and cost of feeds in aquaculture and/or the desire for much greater control over the aquaculture process. In marine fisheries, fisheries management and fish stock assessment technologies are of paramount importance.

The chapter also addresses the unique challenges faced by the impact of climate change on dry areas in developing countries. The impacts of climate change are often more acute, with increased vulnerability due to limited resources and adaptive capacities. The chapter discusses how changes in rainfall patterns and increased temperatures are affecting inland fisheries, especially in arid and semi-arid regions. This situation exacerbates the challenges for communities that rely heavily on these resources for food and livelihoods. In developing countries, the reliance on fisheries and aquaculture is often a critical component of food security and economic stability. The chapter highlights the need for targeted strategies in these areas, focusing on sustainable practices, capacity building and the development of climate-resilient species through advanced genomics and breeding programs. It emphasizes the importance of international cooperation and support to help these regions adapt to the changing climate, ensuring the sustainability and resilience of fisheries and aquaculture in some of the world's most vulnerable areas.

### *Ch. 7: Impact of Climate Change on Livestock in Drylands*

This chapter offers a comprehensive analysis of how changing climatic conditions affect livestock and the adverse interrelationship between them. It begins by highlighting the critical role of livestock as a vital asset in dryland regions, both for their economic value and cultural importance. The chapter delves into the direct impacts of climate change, such as temperature fluctuations, water scarcity, and prevalence of diseases, along with indirect consequences such as changes in feed crop yields and quality.

A significant portion of the chapter is dedicated to discussing the contribution to greenhouse gas (GHG) emissions, drawing attention to the need for strategies to mitigate these effects. The chapter also emphasizes the adaptive capabilities of different livestock species to environmental stressors, underscoring the importance of genetic diversity and the development of heat-tolerant animal breeds.

Mitigation strategies proposed in this chapter include improving animal management practices, enhancing nutrition, and advancing selective breeding and genetic engineering to develop resilient livestock breeds. The authors argue for an accurate measurement of heat stress in animals and the development of early-warning systems as essential tools in managing climate-induced challenges.

The chapter concludes by stressing the urgency of addressing the interplay between climate change and livestock farming. It calls for more research, policy interventions, and community engagement to ensure the sustainable management of livestock in dryland regions, highlighting the need for coordinated efforts in this direction. The holistic approach presented in the chapter underscores the complexity and interdependence of environmental, economic, and social factors in managing livestock under the looming threat of climate change.

## Regional chapters

### *Ch. 8: Sub-Saharan Africa*

This chapter is contributed by competent scientists experienced in a wide range of issues. Urgency to address Africa's food and climate security are highlighted by the fact that the continent is prone to extreme weather events: heatwaves and wildfires in the north; rising temperatures, droughts and destructive floods in western and central regions; and cyclones and catastrophic flooding in the southern regions. Thus, the continent continues to suffer from vulnerability of hunger and malnutrition.

The authors state that many ecoregions of Africa are fragile and prone to drastic effects. For example, drylands cover 19.6 million $km^2$ and are home to 40% of Africa's population or 525 million people, and are growing at 3.0% per year. Further, the regional warming in drylands may be up to twice the global increase in temperature. The problem of food insecurity is aggravated by the prevalence of drought and by the war in eastern Europe. Therefore, the authors highlight the need for African countries to strengthen social safety nets, invest more in research and development, avoid trade restrictions, and implement initiatives such as the Comprehensive Africa Agriculture Development Program (CAADP).

The chapter examines emerging and potential impacts on food security on the continent, considering each dimension of food security: availability, access, utilization, and food system stability. Climate-related food insecurity challenges in two especially threatened areas are considered in detail: the Sahel and the Horn of Africa. Specific focus is given to trends and projects of climate change, droughts, floods, cyclones, storms, heatwaves and desertification. The authors discuss the effects of these climatic factors on all interrelated aspects affecting food security: namely food availability, food access and food utilization.

The chapter also focuses on the stability of Africa's food systems in the context of climate and economic shocks. The authors highlight that hardly 1% of arable land in Africa is equipped for irrigation. Food insecurity is also aggravated by food loss because of poor storage capacity of households and warehouses resulting in significant food loss, averaging 9% each year in Africa. Food insecurity in Africa is also aggravated by increases in food prices, decrease in family income and poor distribution to the affected areas.

The authors highlight the food insecurity hotspots which comprised 16 countries during 2022, out of the world's total of 24 countries. These food insecurity hotspots are attributed to chronic poverty and vulnerability, low and erratic rainfall, uneven access to feed in depleted rangelands, flooding, pest attacks, domestic food price inflation, and conflict and displacements. Because of these factors, food insecurity in the Sahel region has increased by 50% since 2021. Consequently, 18 million people are affected in the Sahel and 24 million in the Horn of Africa.

The chapter also focuses on possible solutions to this problem that has affected Africa since the 1960s. The authors emphasize that Africans are responsible for their own food security and need to carry out the necessary reforms and investments to assure food security through development and implementation of policies and investments in African countries to help translate science into action. Priority actions with regards to investment highlighted in the chapter include the following:

**1.** enhance the financial, technical, and organizational capacities of national agricultural research and extension systems;
**2.** strengthen regional and international agricultural research and development organizations and consortiums (i.e. CGIAR);
**3.** strengthen the structure and functioning of national and regional farmer and trader organizations; and
**4.** expand agricultural market and value-chain infrastructure and services, prioritizing climate-smart common assets for value addition.

The chapter emphasizes the importance of investment in water resources and environment, with a focus on reducing the emission of greenhouse gases by restoring degraded and decertified ecosystems and on cross-sectoral actions (e.g. energy).

Among the principal recommendations highlighted by the authors are:

**1.** progress towards a pan-African climate adaptation and food security initiative;
**2.** establishment of the African Council of Agricultural Research and the Forum for Agricultural Research in Africa (FARA), which will guide research and provide funding support;
**3.** funding support from the IMF and other international organizations at $50 billion per annum for adaptation, and another $190 billion per annum for mitigation efforts on global warming in Africa;
**4.** the creation of a Green Bank at about $2 trillion per annum; and
**5.** translating science into action, urgently needed for Africa through creation of the Technologies

for African Agriculture Transformation (TAAT) program to replicate the success stories such as cassava in Nigeria, maize in Kenya, rice in Sierra Leone, and sorghum in Mali, among others.

The chapter focuses on identification of pathways to achieve food security in Africa and concludes by stating that Africa is the biggest victim of climate change caused mostly by developed economies. The authors emphasize that more investment is needed in agriculture, water, and environment, along with those in agricultural research and education. The importance of continental free trade is highlighted as an opportunity to increase regional trade in food and agricultural products. Two options recommended are:

**1.** creating a Green Bank to finance private sector green projects; and
**2.** debt for climate swaps to provide African governments with adequate fiscal space to finance climate investments.

## *Ch. 9: Case Studies in America on Carbon Management and Sequestration in Agriculture*

This chapter focuses on case studies from Latin America. The authors have long experiences in Brazil, Uruguay, and other countries. It focuses on biomass C-based practices as the primary pathway to develop resilient, regenerative, profitable, and environmentally friendly production systems. It includes case studies on C balance with biomass-C inputs across production systems in South and North America. It also highlights the role of the Inter-American Institute for Cooperation on Agriculture (IICA), which plays a strategic role by partnering with national scientific and technological institutions, and promotes Living Soils of the Americas project, implemented in cooperation with the CFAES Rattan Lal Center for Carbon Management and Sequestration.

Three case studies presented in the chapter are:

**1.** biomass-C input interaction and C balance across production systems under no-till (NT) in Brazil;
**2.** diverse crop and soil management practices from the western USA; and
**3.** evidence from different crop rotations in a long-term experiment in Uruguay.

### Case study from Brazil

This case study focuses on the use of no-till in Brazil for about 50 years. The no-till system (NTS) was used in Paraná state, followed by Rio Grande do Sul (southern region). These were the ecoregions where the NTS was initially adopted to control soil erosion. In the long-term, the NTS proved to be one of the most important technologies that supported upscaling of Brazilian agriculture. Brazil currently has 36.8 million ha under NTS, but only 10–15% have successfully adopted NTS based on the three pillars (absence of soil disturbance, permanent soil cover, and diverse crop systems).

The data on SOC sequestration rates for tropical (−0.03–2.15 Mg ha$^{-1}$ yr$^{-1}$) and subtropical (−0.07–1.4 Mg ha$^{-1}$ yr$^{-1}$) regions of Brazil vary widely because of differences in the amount of biomass-C input and the agronomic potential of agro-ecosystems. Large variability in SOC sequestration rates may be attributed to a high diversity of cropping systems, the amount and frequency of biomass-C input, and the difference in inherent soil properties. The authors argue that making Brazilian agriculture emission-negative can be achieved by maximizing the soil C sink by increasing the $CO_2$ fixed by crops during photosynthesis and its accumulation in soil profile upon land application of biomass-C.

The authors propose that the key component to reach a positive C balance in farming production systems is to manage biomass-C input in association with NTS envisaging to accumulate and stabilize C by three mechanisms:

**1.** stimulate formation of macro-aggregation through continuous soil cover;
**2.** control the rate of SOC oxidation; and
**3.** ensure stabilization of C in intra-aggregate and the associated organic C.

### Case study in the western USA

This study presents the evidence from studies comparing tillage systems, crop rotation, cover cropping, and perennial cropping systems in the semi-arid eastern New Mexico and other studies from across the western USA. The study highlights the importance of:

**1.** reducing soil disturbance;
**2.** intensifying and diversifying cropping systems;
**3.** using perennial cropping systems and grassland restoration; and

**4.** exploring dual opportunities of carbon and water management.

The authors of this study argue that improved farming practices such as cropping system intensification and diversification, cover cropping, crop rotation, mixed-cropping, and conservation tillage can improve water-use efficiency and enhance biomass-C input for enhancing SOC sequestration and soil-water conservation benefits. Furthermore, integrating reduced and no-till with cropping systems intensification, diversification, and enhanced residue inputs could increase SOC in arid and semi-arid regions and also improve soil-water conservation.

**Case study from long-term experiments in Uruguay**

This case study is focused on different crop or crop-pasture rotations studied for 60 years in the oldest long-term experiment in Uruguay, initiated in 1963 and located at the INIA Experimental Station 'La Estanzuela' (34°20′S, 57°41′W). The data from long-term study showed the following:

**1.** the SOC decreased nearly constantly in continuous cropping without N and P fertilization;
**2.** SOC declined at a lesser rate and became stable at SOC level of 1.5% under continuous cropping but with N and P fertilization;
**3.** crop-pasture rotation with red clover (*Trifolium pratense*) resulted in a relatively steady SOC content over the long term compared to the starting condition;
**4.** with less intensive crop-pasture rotations that had 50% crops and 50% pastures, SOC slightly increased over the long term; and
**5.** SOC content increased steadily in a crop-pasture rotation that presented only two years of crops rotating with two types of pastures, one lasting one year and the other lasting three years. A pattern of decrease in SOC during cropping phases and increase during pastures was recognized for several crop-pasture rotations within the rotation cycles of this long-term experiment.

### *Ch. 10: Central and West Asia and North Africa (CWANA)*

The region is experiencing an increase in the events of climate change resulting in droughts, floods, and high temperatures compounding events with significant consequences for society and the environment. To overcome the negative effects of climate change, the International Centre for Agricultural Research in Dry Areas (ICARDA), alongside its national and international partners and donors, promotes diversified cropping systems and delivers its solutions as integrated packages including innovative technologies, capacity-building programs, instruments for better market access, and the current best agronomic approaches. These include sustainable, water-smart irrigation, cultivation of drought-tolerant crop varieties, water management approaches such as rainwater harvesting and deficit irrigation, precision farming technologies, and low-water-footprint cropping systems, raised-bed technology, and others.

One of the important contributions of ICARDA has been the collection of germplasm, its characterization and distribution to all member states of the region.

ICARDA in collaboration with the Pavlov Institute in Russia and GRDC Australia has developed FIGS (The focused identification of the germplasm strategy), which is a trait-based approach and assists gene bank managers in identifying desired gene material with high probability of having the sought-after trait. FIGS is based on the premise that the environment strongly influences natural selection and, consequently, the geographical distribution of organisms. FIGS has been successfully used to identify sources of resistance for several useful traits for breeding globally such as Sunn pest in wheat in Syria, Russian wheat aphid in bread wheat, abiotic stresses, such as drought adaptation in *Vicia faba* L., resistance to stem rust in bread and durum wheat, and stem rust and stripe rust in accessions of wheat landraces.

**Heat- and drought-resistant variety development by ICARDA**

With this information, together with the new genotyping technologies, breeders have been able to identify the genetic mechanisms behind superior tolerance to heat and drought. Cropping season 2021–2022 was one of the harshest of the last 50 years in Morocco.

It is worth mentioning that, until now, ICARDA has released more than 880 new varieties for cultivation – and generated benefits worth US$850 million every year. Technology 'packages' have been developed – crop management, irrigation, pest control – to maximize yields from the new varieties. In the last five years, more than 120 cereal and legume varieties of ICARDA origin have been released by the national programs in more than 20 countries.

In addition to providing various genotypes, ICARDA is actively involved in water conservation technologies and different agronomic practices for mitigating climate change effects. This has been demonstrated through different case studies of intervention made in different regions/countries.

The chapter concludes that restoration of soil health should be based on the following:

**1.** increasing the biomass-C input according to the three pillars that support conservation agriculture, adding the amount at the rate above that of the dynamic equilibrium, guaranteeing a controlled soil C oxidation rate, and promoting the continuous C and N flow to reorganize the aggregates and the new structure;
**2.** stimulating macro-aggregation and enhancing C protection within intra-aggregates and stabilizing it over time and restoring soil functionality by improving soil's chemical, physical and biological attributes; and
**3.** managing the rate of input of biomass-C as a key strategy to decrease C turnover time and leading to the restoration of natural capital and ecosystem services to society.

## National chapters

### *Ch. 11: Innovations, Strategies, and Policies for Building Resilience of India's Dryland Farming to Climate Change*

This chapter is contributed by authors from ICRISAT, ICAR-NAARM and ICAR Natural Resources Management Division, and is aimed at building resilience of India's dryland farming to climate change. The chapter outlines India's new programs and innovative strategies for climate proofing of agriculture through National Action Plan on Climate Change (NAPCC). It deliberates major initiatives such as National Mission on Sustainable Agriculture (NMSA), National Innovations on Climate Resilient Agriculture (NICRA) and international collaborative programs, i.e. conservation agriculture (CA), landscape approach, etc.

The chapter focuses on climate risks in Indian agriculture and then outlines strategies to address these risks. Important climate risks are erratic monsoons, extreme weather events, heatwaves and extreme temperatures, water scarcity and depletion of groundwater resources, pests and diseases, shifts in crop suitability zones, and livestock vulnerability and adaptation challenges. Thereafter, the authors describe the need for climate-resilient agriculture (CRA), which encompasses a combination of sustainable and regenerative agricultural techniques such as conservation agriculture, use of climate-resilient crop varieties, agroforestry, measures to manage water resources effectively during droughts and floods, crop rotation and diversification, and precision farming.

#### National initiatives on climate proofing in Indian agriculture

The chapter outlines various initiatives undertaken in India such as the National Action Plan on Climate Change (NAPCC) in 2008. Other initiatives discussed include:

**1.** National Innovations in Climate Resilient Agriculture (NICRA) in February 2011. It is operated with the objectives to:
**(i)** undertake strategic and applied research on climate change adaptation and mitigation;
**(ii)** validate, demonstrate, and assess the impact of climate-resilient technologies on farmers' fields; and
**(iii)** strengthen the capacity of scientists, farmers, and other stakeholders on climate-resilient agriculture.
**2.** National Mission for Sustainable Agriculture (NMSA), which focuses on promoting climate-resilient agriculture through the adoption of climate-resilient practices, sustainable water management, and integrated farming systems.
**3.** Green India Mission (GIM) with the objective to safeguard the biological resources of India and associated livelihoods against the peril of adverse climate change and to recognize the vital impact of forestry on ecological sustainability, biodiversity conservation and food-, water- and livelihood-security. Climate finance under which the National Adaptation Fund for Climate Change (NAFCC) was established in 2015 to meet the cost of adaptation to climate change for the state and union territories of India that are particularly vulnerable to the adverse effects of climate change. Also discussed are the National

Food Security Mission, the Mission for the Integrated Development of Horticulture, the National Bamboo Mission, and the National Livestock Mission.

The authors discuss international initiatives and programs on CRA in India including the CGIAR Research Program on Climate Change, the Climate Change Agriculture and Food Security (CCAFS), Sustainable and Resilient Farming Systems Intensification (SRFSI) in the eastern Gangetic Plains, landscape approach for scaling climate-resilient, regenerative and carbon-neutral agricultural systems, and the conservation agriculture (CA) initiative.

The chapter focuses on CA and states that science-based CA management has the potential to produce more food (10–15%) from less water (20–75%) and energy (20–45%) while increasing farmers' income (25–50%) in an environmentally responsible manner through lowering the C footprint by 25–30%. Thus, the chapter includes a table of references citing evidence of climate-resilient agriculture practices in different regions of India for site-specific farming systems. The authors outline some policy needs on climate proofing both at national and state levels.

**Community of practitioners (CoPs) and networks on climate proofing in India**

The chapter outlines several communities of practitioners (CoPs) and networks in India focusing on climate proofing and addressing climate change challenges. These initiatives are aimed to foster collaboration and knowledge sharing among various stakeholders, including government agencies, non-governmental organizations (NGOs), academic institutions, and local communities. The chapter also lists some examples of potential CoPs and networks on climate proofing in India.

**Potential opportunities for scaling of climate-resilient agriculture**

The authors claim that CRA is the most cost-effective and scalable solution. It offsets or reduces the net $CO_2$ emissions and is the key to mitigating the major factor of global warming and climate change. The chapter outlines some potential opportunities for scaling climate-resilient agriculture such as the need for policy and financial support, knowledge sharing and capacity development, climate-resilient technologies, climate finance mechanisms, carbon credits, public-private partnerships, farmer cooperative networks, climate-resilient certification labelling, etc.

The concluding section outlines some key steps to move forward with CRA in India. Important among these are:

**1.** scaling up successful initiatives;
**2.** enhancing climate research and data;
**3.** mainstreaming climate into development plans;
**4.** empowering local communities;
**5.** private sector engagement;
**6.** capacity building and training;
**7.** incentivizing climate-resilient agriculture through carbon credits;
**8.** innovative financing mechanisms;
**9.** public awareness behavioral change;
**10.** collaboration and international cooperation; and
**11.** review and adaptation.

The chapter concludes with the statement that addressing climate change is a complex and multifaceted challenge and building resilience to climate change is a long-term endeavor that requires sustained commitment from all stakeholders.

### *Ch. 12: Climate Change and Sustainable Agro-ecology in China's Dryland: A Case Study*

This chapter reviews the problems and challenges in climate-smart agricultural practices of China's dryland farming. The chapter is based on the premise that application of CSA management practices in China's dryland can increase the amount of soil organic carbon (SOC) and maintain or increase crop yield, with the hope of also reducing the emission of GHGs. Of the total amount of arable land area in China, at 130–171.3 million ha, approximately 50% is drylands. The focus of this chapter is adoption and upscaling of CSA technologies in China's drylands. Important among these technologies are improved systems of nutrient management and the use of water-saving agricultural technology, conservation agriculture, crop diversification, plastic film management, and integrated management practices that facilitate soil fertility. Based on the review of the literature, the authors emphasize that conservation and organic agriculture can improve soil nutrients, reduce environmental costs, and maintain soil health, but the benefits for yields are controversial. While irrigation can reduce GHG emissions and improve crop productivity in dryland agriculture, the site-specific effects depend on irrigation methods: flooding and furrow or sprinkler irrigation.

The authors emphasize that most studies on CSA management practices focus primarily on a single technology or the entire farmland system, but few studies have systematically reviewed the progress and challenges of CSA in drylands. Therefore, this chapter is an attempt to provide a deeper understanding of the methods of dryland farm management in China and proposes new strategies to advance CSA. This review is specifically focused on the following CSA practices:

1. **Conservation tillage**
It is estimated that conservation tillage is practiced on some 8.6 mha of cropland in China. However, the responses of crop yields to conservation tillage are controversial and debatable in China. Therefore, the authors emphasize the need for appropriate strategies that alleviate yield limitations and harness the benefits of conservation tillage to improve production. The need for investigation of the mechanisms of SOC sequestration is highlighted. Research is also needed on environmental factors (e.g. temperature, moisture, and pH) and soil C and N components as the key factors that affect the emissions of $N_2O$ from soil managed by conservation tillage. With proper management package, conservation tillage could be a preferred 'climate-resilient' technology to lower the carbon footprint and enhance the ecosystem services of climate regulation for the winter wheat-summer maize system in the north China plain.
2. **Optimizing chemical fertilizer input**
Integrated nutrient management is a strategy to maintain optimal ecosystem productivity and sustainability for fertilizer use in dryland systems of China. It involves combined use of organic and mineral fertilizers, combined with patterns of crop nutrient demands to improve the nutrient use efficiency. A judicious combination of organic and synthetic nitrogen fertilizers offers a nutrient-rich source that complements intermediate nutrients and micronutrients. A 26-year study showed that combined application of organic and mineral fertilizers increased the total concentration of organic carbon, organic carbon stocks, and labile fractions of organic C. However, applications of inorganic fertilizers often produce contradictory effects on the concentrations of SOC and its fractions.
3. **Water-saving agriculture**
Some options to save water discussed in the chapter for semi-arid regions of China include conservation tillage, limited irrigation, drought-resistant chemical control, and the use of water-saving varieties over and above the use of drip irrigation or micro-irrigation. In the semi-arid areas of the Loess Plateau, the yield-enhancing effects of plastic film mulching are reported in this chapter to be somewhat limited and even leading to reduced yields and production. Use of full plastic film mulching may negatively affect the storage of soil carbon and control of GHG emissions. Yet, the application of highly absorbent polymers can promote wheat growth, improve root vigor, increase wheat dry-matter weight and tiller number, enhance soil physicochemical properties and water-use efficiency, and increase agronomic yield. Site-specific effects of these water-saving options, i.e. drip- and micro-irrigation are discussed.
4. **Biochar**
Despite numerous reports on the use of biochar, the authors also highlight debate and disagreement in the results. The chapter elaborates on the available data and states that the efficacy of biochar as a soil enrichment is based on appropriate biochar technology. In other words, field management practices must fully account for soil type, basic fertility, and the type and amount of biochar used as a feedstock. The chapter discusses that biochar may be more effective in fertilizing degraded land with poor soil fertility than on a fertile soil with high productivity. Also discussed are saline soils and those with other constraints to crop production.
5. **Plastic film mulch**
Because of its widespread use in China, pros and cons of plastic film mulching are elaborated in this chapter, especially with regard to net radiation input in arid regions where the warming effect follows a U-shaped trend throughout the crop reproductive period. While plastic film mulching can increase the

soil-water content of the cultivated layer by 1–4%, in years with less rainfall, plastic film mulching can reduce the subsoil water content, with no difference in soil-water storage between 0 m and 2 m depth. Nonetheless, the authors discuss that plastic mulching has an impact on early sprouting, seedling emergence, and the growth of stems with leaves, specifically the root system, and on the effectiveness of soil nutrients uptake by crops. They highlight that extended and irrational use of mulch over the longer term can nullify the initial gains in yield by reductions in yield over the longer period when the residues would persist for a long time and aggravate plastic pollution.

The chapter concludes with a detailed case study from north-east China. The case study is based on long-term research involving technologies such as intercropping, dryland autumn mulching, and monopoly film sowing over the straw to improve the efficiency of water and fertilizer use in dryland crops. It highlights challenges and opportunities for future development in the region, which is the breadbasket of China. The chapter concludes that CSA management faces significant challenges in balancing the relationship between organic carbon sequestration, food security, and GHG emissions under site-specific conditions. Thus, identifying and prioritizing appropriate CSA practices are necessary to help establish sustainable agricultural production systems in arid regions.

### *Ch. 13: Dryland Agriculture in Pakistan under Changing Climate: A Case Study*

Pakistan has a predominantly arid to semi-arid climate, where 5 mha of the cultivable land is rainfed. Droughts have severe repercussions in the country, including food shortage, animal losses, crop failure, and water scarcity. This situation makes Pakistan one of the most vulnerable countries to climate change. In order to overcome these conditions, Pakistan has adopted various mitigation strategies for conserving water through better water reservoir management, lining irrigation channels. In addition, high-efficiency irrigation systems (HEIS) are being adopted to conserve water.

**Adaptation measures for agriculture**

Climate change adaptation strategies for the agricultural sector are crucial. Adaptation measures for agriculture can help farmers cope with the challenges of climate change and build resilience in their farming systems. However, the success of these measures depends on a range of factors, including access to finance, technology, supportive policy frameworks, and effective partnerships between farmers, governments, and other stakeholders.

Here are some examples of adaptation measures that can be implemented in agriculture:

i. **Crop diversification.** Farmers can diversify their crops by growing a variety of crops that are better suited to changing climate conditions, crop rotation, and intercropping.
ii. **Water management.** Using efficient irrigation systems, rainwater harvesting, and agriculture practices for water conservation can help to increase water-use efficiency, reduce soil erosion because of runoff, and increase soil moisture to improve soil health.
iii. **Improved soil health.** Improving soil health with practices like conservation agriculture, cover cropping, and agroforestry can help to enhance soil organic matter, lessen soil erosion, and promote soil fertility, thus improving crop productivity and resilience.
iv. **Use of climate-smart technologies.** Farmers can use climate-smart technologies such as solar-powered irrigation systems, improved seeds, and precision farming technologies to increase productivity and reduce vulnerability to climate impacts.
v. **Early warning systems.** Implementing early-warning systems for extreme weather events, pests, and diseases can help farmers to take appropriate measures to reduce damage to their crops and lessen their susceptibility to the effects of the climate.
vi. **Improved seeds and technologies.** Selecting drought-tolerant crops together with modern agriculture practices such as precision farming (e.g. drip-irrigation and mechanization) can significantly increase dryland productivity.
vii. **Training and education.** Providing farmers and stakeholders with training and education on climate-smart agriculture practices can help them to adopt water-efficient irrigation, crop management, and soil conservation, which are critical to improving dryland farming.

The impacts of climate change, such as droughts, floods, and changing weather, are already being felt by farmers nationwide, particularly in the dryland regions. To address these challenges and build more resilient and sustainable farming systems, there is a need to adapt scientific interventions that consider the local context and the needs of farmers. Adaptation measures such as developing climate-change-resilient genotypes, crop diversification, conservation agriculture, and water management can help farmers cope with the impacts of climate change and improve their livelihoods. Though the scientific community is playing its role, more focused research needs to be initiated. Policy decisions and legislation for adopting new technologies must be sensitized and revised. Likewise, re-zoning is also equally important to uplift agriculture in the drylands. Certain initiatives taken by the government are encouraging water conservation and sustainable agriculture practices, thus helping to mitigate these challenges. Hence, there is a dire need to devise short-, medium-, and long-term policies to uplift dryland agriculture in the country.

### *Ch. 14: Resilience and Adaptation of Egyptian Agriculture*

This chapter provides a comprehensive analysis of the challenges faced by Egyptian agriculture in the context of climate change. The primary focus is on the impacts of limited resources, sea-level rise, dwindling water supplies, and population growth. The chapter underscores the need for a significant transformation in agricultural practices to combat the effects of climate change. It highlights the 'Sustainable Agriculture Development Strategy towards 2030', which incorporates advances in science and technology, including smart agriculture, and emphasizes the importance of human resources development, institutional support, policies, and legislation.

The chapter provides a detailed examination of the vulnerabilities of Egyptian agriculture to climate change, including issues related to sea-level rise, soil salinity, and desertification. It emphasizes the serious implications of these challenges for the economy, food security, outmigration patterns, public health, and livelihoods. The chapter also addresses the vulnerability of Egypt to rising sea levels, particularly in the Nile Delta, Alexandria, and North Coast, and the associated risks to agriculture, infrastructure, and population displacement.

In addressing these challenges, the chapter proposes several strategies and interventions. These include risk aversion strategies to counter the impacts of sea-level rise and initiatives to mitigate the threat of diminishing water supply from the Nile. The chapter emphasizes the importance of integrating scientific advancements and sustainable practices into agricultural strategies to build a resilient agricultural sector. It also highlights the role of human resource development, institutional support, and robust legislation in this transformative journey.

Early-warning systems are highlighted as a crucial tool in managing the risks associated with climate variability. These systems provide timely information that allows for better planning and response to potential threats, thereby minimizing losses in agriculture. The chapter discusses the integration of these systems into the national agricultural framework to enhance preparedness and resilience.

Adaptation strategies are a key focus, emphasizing the need for Egyptian agriculture to evolve in response to climate change impacts. This involves developing and implementing practices that are more resilient to climatic extremes, such as droughts and heatwaves. The chapter describes various adaptive techniques being employed, including crop diversification, improved irrigation methods, and the adoption of drought-resistant crop varieties.

The optimization of water-use efficiency is another significant aspect of this chapter. Given Egypt's reliance on the Nile for its water needs and the increasing pressure from the Grand New Dam, Ethiopia, and population growth and climate change, efficient water management is vital. The chapter delves into modern irrigation techniques, water-saving technologies, and policies aimed at maximizing the utility of available water resources.

Agro-management techniques, particularly precision agriculture and AI, are discussed as innovative approaches to enhance productivity and sustainability. Precision agriculture involves the use of advanced technologies such as satellite imagery, GPS, and sensors to monitor and manage crops more efficiently. AI is employed to analyze data and make informed decisions, optimizing resource use and crop yields.

The chapter explores the role of genetic resources in bolstering the resilience of Egyptian agriculture. It discusses the development and use of genetically modified crops and livestock that are resistant to pests, diseases, and environmental stresses. This approach is seen as critical to ensuring food security and sustainable agricultural development in the face of changing climatic conditions.

The chapter concludes by emphasizing the critical juncture faced by Egyptian agriculture, pivotal to the country's economy and food security, under the strain of climate change, limited natural resources, and rapid population growth. It calls for resilience, innovation, and adaptability in the agricultural sector, leveraging science and technology, fostering institutional reforms, and engaging in comprehensive policy development to secure a sustainable and prosperous future for Egypt's agricultural sector and its people.

The chapter also investigates the impact of climate change on Sustainable Development Goals (SDGs) and explores the opportunities arising from the outcomes of COP27, particularly in the context of loss and damage. It discusses various modalities including green transition, the adaptation fund, carbon tax, and other mechanisms, and their impact on the livelihood of Egyptian inhabitants.

Overall, the chapter provides a comprehensive view of the multifaceted efforts required to transform Egyptian agriculture into a more resilient, efficient, and sustainable system in the face of unprecedented climatic challenges.

### *Ch. 15: Humanity and Nature*

This chapter delves into the complex relationship between humanity and nature. It becomes increasingly evident that the authors' diverse perspectives converge on several significant themes. The undeniable consensus that emerges from the four presentations is the profound interconnectedness of humans with the natural world, with the Industrial Revolution marking a pivotal moment that significantly contributed to the shaping of climate change. Collectively, these authors underscore the urgent call for shared responsibility, emphasizing that those who bear greater responsibility for the harm inflicted upon nature must also shoulder the responsibility of protecting vulnerable communities.

Nevertheless, while common ground exists in many aspects, each author brings a unique perspective, highlighting distinct nuances. Dr. Serageldin and Dr. Sharon Mijares draw our attention to the Holocene epoch as the starting point for human-induced alterations to nature, setting in motion an irreversible trajectory. Aliaa Rafea differentiates between naturally occurring climate variations and those exacerbated by human actions.

Furthermore, Dr. Serageldin, Dr. Mijares, and Dr. Rafea all stress the importance of international collaboration in addressing the intricate challenges of climate change. Dr. Serageldin and Dr. Rafea share a common belief in the indispensable role of science in confronting and navigating the complexities of our changing climate.

Spirituality emerges as a recurring theme, with Dr. Mijares, Honorable Dena Merriam, and Dr. Rafea incorporating it into their narratives, albeit through different lenses. Dr. Mijares perceives spirituality as intrinsically linked to the feminine essence of the universe. Honorable Dena Merriam shares her profound spiritual connection with a whale, encouraging humanity to listen and resonate with the diverse voices of nature genuinely. Dr. Rafea centers on introspection, proposing that the path to self-realization is rooted in a spiritual quest that inevitably deepens our bond with the natural world.

While each author contributes a unique tapestry of thoughts, their collective aspiration remains abundantly clear: preserving our planet. In all its magnificence, nature implores us to safeguard and heal its wounds. A radical shift in our collective consciousness is imperative. We must unite in thought and action, ensuring that future generations are spared from the devastating impact of nature's wrath.

### *Ch. 16: Protecting Vulnerable Lives and Livelihoods in Times of Climate Change*

Authored by Ismail Serageldin, the chapter emphasizes the dire situation of over 2 billion people living in drylands who are highly susceptible to the impacts of climate change.

The chapter covers several key areas:

1. **Climate change impact.** This section deals with the undeniable reality of climate change

and its diverse manifestations, including extreme weather events and their impact on biodiversity and sea levels.

**2. Capacity building for adaptation and resilience.** The chapter emphasizes the importance of enhancing adaptation and resilience, especially in poorer regions, and discusses the moral responsibility of richer countries to assist in these efforts.

**3. Vulnerabilities in drylands.** It focuses on the unique challenges faced by populations in drylands, such as water scarcity, soil degradation, and limited economic opportunities.

**4. Human and institutional capacity.** The importance of investing in human resources, health, education, and institutional development as a means to sustainable development in poor countries is highlighted.

**5. Migration issues.** The chapter addresses various facets of migration, including internal and international migration, and the phenomenon of 'brain drain'.

**6. Gender dimensions.** It touches on the differential impacts of climate change on women, highlighting their crucial role in adaptation and resilience.

**7. Technological and scientific solutions.** The potential of science and technology, including ICT and new biological techniques, in addressing climate change issues is explored.

**8. Financing and funding.** The need for a significant shift in traditional mechanisms of aid and international finance is highlighted. The authors suggests that current financial structures are inadequate to address the challenges posed by climate change. The chapter proposes more effective and sustainable funding solutions, particularly for vulnerable regions like drylands, stressing the need for developed countries to play a more substantial role in providing financial assistance. This section delves into the complexities of funding climate change initiatives and underscores the urgency of rethinking financial strategies to combat environmental challenges effectively.

### *Ch. 17: The Day after COP28*

This chapter addresses and discusses what is required to employ the best of science and technology, policy, enhancing coping and adaptive capacity, and ethical consideration to save mankind and humanity and cater for the most vulnerable populations on Earth.

The chapter sheds light on the major outcomes of COP27 and COP28 and major international meetings: G20 New Delhi Leaders' Declaration, 9–10 September 2023 and African Leaders' Nairobi Declaration on Climate Change, 4–6 September 2023 (African Union, 2023; G20, 2023); in addition to pre-COP27 and 28 international webinars and workshops focusing on the impact of climate change in dry areas with emphasis on coping and adaptive capacity.

It concludes by stressing major elements related to action required (policy, funding, and legislation). It calls for a new world order to facilitate the improvement of the people's livelihoods and resilience in the face of climate change.

## References


African Union (2023) The African leaders' Nairobi declaration on climate change and call to action. Available at: https://au.int/sites/default/files/decisions/43124-Nairobi_Declaration_06092023.pdf (accessed 17 June 2024).

El-Beltagy, A., Valkoun, J., Amri, A., Thomas, R. and Shideed, K. (2005) Biodiversity, land degradation and poverty alleviation in dry and sub-humid lands. In: *Success Stories in Implementation of the Programme of Work in Dry Lands and the Global Taxonomy Initiative*. 11th Meeting of the Subsidiary Body on Scientific, Technical and Technological Advice of the Convention on Biological Diversity, Montreal, Canada, 21, 13–20. ISBN 92-9225-035-3.

G20 (2023) G20 New Delhi Leaders' Declaration, G20 summit 2023, New Delhi. Available at: https://www.mea.gov.in/Images/CPV/G20-New-Delhi-Leaders-Declaration.pdf (accessed 17 June 2024).

IDDC (International Dryland Development Commission) (2022) Impact of climate change on food production in the dry areas. Background note. RACC/IDDC/ICARDA/ALARI hybrid webinar. Available at: http://www.drylanddevelop.org/uploads/6/1/7/8/61785389/background_note__as_of_30_august_2022_.pdf (accessed 17 June 2024).

IDDC (International Dryland Development Commission) (2023) The Pre-COP28 Conference, Alexandria, 11–12 September. Statement. Available at: http://www.drylanddevelop.org/uploads/6/1/7/8/61785389/pre-cop28_statement.pdf (accessed 17 June 2024).

Falk, J., Colwell, R.R., Behera, S.K., El-Beltagy, A.S., Gleick, P.H. *et al.* (2023) An urgent need for COP27: confronting converging crises. *Sustainable Science* 18, 1059–1063. DOI: 10.1007/s11625-022-01253-5

Falk, J., Gleick, P.H., Asayama, S., Attig-Bahar, A., Behera, S. *et al.* (2024) Critical hydrologic impacts from climate change: addressing an urgent global need. *Sustainable Science* 19, 241–244. DOI: 10.1007/s11625-023-01428-8

Millennium Ecosystem Assessment (2005) *Ecosystems and Human Well-being: Synthesis*. Island Press, Washington, DC. Available at: www.millenniumassessment.org/documents/document.356.aspx.pdf (accessed 17 June 2024).

NOAA (National Centers of Environmental Information) (2023) 2022 was world's 6th-warmest year on record. Available at: https://www.noaa.gov/news/2022-was-worlds-6th-warmest-year-on-record (accessed 17 June 2024).

STS (Science and Technology in Society Forum) (2023) RACC 2023. Available at: https://www.stsforum.org/racc2023/ (accessed 17 June 2024).

Sivakumar, M.V.K., Lal, R., Selvaraju, R. and Hamdan, I. (2013) *Climate Change and Food Security in West Asia and North Africa*. Springer, Dordrecht, The Netherlands. DOI: 10.1007/978-94-007-6751-5

Tollefson, J. (2023) Earth's hottest month: these charts show what happened in July and what comes next. *Nature* 620, 703–704. DOI: 10.1038/d41586-023-02552-2

Traylor, I.R., Dregne, H. and Mathis, K. (eds) (1999) Proceedings of the Fifth International Conference on Desert Development: *Desert Development: The Endless Frontier*. Volume 1. The International Center for Arid and Semiarid Land Studies of the Office of International Affairs, Texas Tech University, Lubbock, Texas.

Wong, C. (2023) Earth just had its hottest year on record: climate change is to blame. *Nature*. DOI: 10.1038/d41586-023-03523-3

# 2 Navigation through Uncertainties

## Part A: Regional Action on Climate Change

**Charles Kennel[1,2*], Adel S. El-Beltagy[3,4], and Jim Falk[5]**

*[1]Scripps Institution of Oceanography, University of California – San Diego, USA; [2]Centre for Science and Policy, University of Cambridge, UK; [3]Chair, International Dryland Development Commission (IDDC); [4]Arid Land Agricultural Graduate Studies & Research Institute (ALARI), Ain Shams University, Egypt; [5]University of Wollongong, Australia*

**Abstract**

Our epoch of unprecedented human planetary impact demands an urgent collective response. Since the Anthropocene's inception in 2000, assessments confirm extensive disruption of climate, ecology and environment, quantified by concepts like post-WWII Great Acceleration. Crossing of planetary boundaries nears catastrophe, defying one-size-fits-all solutions. At peak population, Wilson predicted ecosystems surviving an ecological bottleneck will be irrevocably altered. Despite freshwater security linking diverse regions, integrated action lags. Recent African and G20 declarations signal commitment to climate-smart agriculture, nature conservation, and biodiversity protection. However, averting biodiversity annihilation requires international campaigns leveraging gene banks and international and national organizations, along with sufficient funding to introduce new genetic resources for dynamic adaptation of agriculture. With crises converging, time is not on our side. Avoiding climate injustice demands equitable resilient responses. While daunting, humanity retains the urgency to act decisively, steering away from collapse towards sustainability. This pathway exists but it is narrowing. The stakes could not be higher.

### The Anthropocene Era

Since World War II, the world's population has more than doubled, and the global economy has grown more than seven times in real terms. Humans burned a major fraction of the hydrocarbons produced biologically and sequestered geologically over hundreds of millions of years. The product of this combustion, emissions of carbon dioxide, is changing the climate. Humans also increased their consumption of living biological resources at an unprecedented rate, multiplying the species extinction rate by 100–1000 times that in other geological eras. Camilleri and Falk (2009) and Dasgupta (2021) concluded that, shaped by emerging social and physical crises, an epochal transition to a new era of social organization is in progress, but to an unclear destination. Nobel Laureate atmospheric scientist Paul Crutzen and limnologist Eugene Stoermer imagined that future geologists digging down in the Earth would find a layer dominated by human impacts, in the same way as earlier paleontologists had identified evolutionary epochs to which they gave now-familiar

*Corresponding author: ckennel@ucsd.edu

 *Climate Change and Sustainable Agro-ecology in Global Drylands* (A.S. El-Beltagy, R. Lal and K. Malik eds.)
DOI: 10.1079/9781800624870.0002

names like Cretaceous, Jurassic and Triassic. Crutzen and Stoermer proposed that our present epoch would merit a name of its own, *Anthropogenic* (Crutzen and Stoermer, 2000), to emphasize the human responsibility for the transition to a new geological and evolutionary era.

There is much debate about when we should consider that the Anthropocene era emerged. Some argue that the transition to the new epoch should be taken to be when 'the great acceleration' of energy and natural resource use after World War II (Steffen *et al.*, 2015) pushed several natural systems close to the planetary boundaries. Transformation of the environment, however, did not start then, because small-scale human societies had altered landscapes and exploited local ecologies before the Agricultural Revolution (Ellis *et al.*, 2021). Humans may have gained a decisive advantage in exploiting environmental and ecological resources when cooperative learning societies formed. If so, one could argue that the origin of Anthropocene behavior lies in the evolutionary history of the human mind. Joseph Henrich, in *The Secret of Our Success* (Henrich, 2016), argues that humans have inherited a neuro-psychological predisposition to act collectively on culturally transmitted knowledge.

For Crutzen and Stoermer (2000), the important point was that human society's actions equal past geological cataclysms in producing species extinctions (Wilson, 1984). Wilson's attitude in 2001 was: 'Why should the human species differ from those that perished in other geobiological transformations?' (Wilson, 2002). No matter when the Anthropocene era started, 21st-century society finds itself in a planetary scale transition with little agreement on where and when it began and little sense of how and when it will end. Divining the future by extrapolating current trends has obvious limitations. Wilson expressed his fear to Charles Kennel during a visit to the Scripps Institution of Oceanography to receive the inaugural Nierenberg Prize for Science in the Public Interest (2001).

Nonetheless, today's convergence of climatic, environmental and ecological trends, individually already alarming, threatens to lead to a comprehensive crisis, which it seems natural to call 'the Anthropocene Crisis' (Kennel, 2021).

Population projections open the possibility that the adverse climatic, environmental and ecological trends experienced today might generally subside (with locally distressing exceptions) after the peak population era. But whether the Anthropocene Crisis subsides or continues to increase thereafter will not prevent the crisis from undermining – conceivably destroying – the cohesion of global society, quality of life and much of the planet's ecology. Global society in its present form, let alone one that responds to the aspirations of its bottom 3 billion (Ramanathan, 2022), cannot be considered sustainable until it builds the resilience to survive the Anthropocene Crisis with acceptable losses. Societies may be unable to escape facing up to a 'life-or-death' proposition – achieve sustainability or perish (Rees, 2003).

If the origins of Anthropocene behavior are obscured in debate, it is possible to see when Anthropocene issues began their move to the center of the diplomatic and international science agendas. The concept of sustainable development appeared in 1972 in the declaration following the Stockholm Conference. The same year saw *The Limits to Growth*, with the use of a computer world model, give a clear if controversial message that globally human environmental activity was heading towards a major crisis (Meadows *et al.*, 1972). In contrast, it called for a sustainable future, which is said to be the first modern use of 'sustainability' (Caradonna, 2020). Birch, writing in 1993, developed in more detail what might be needed for a 'sustainable society'. The Report of a UN commission, *Our Common Future*, published in 1987, played a foundational role in defining the concept of 'sustainable development', emphasizing that the requirements for a sustainable future needed also to provide for a progressive reduction globally in economic injustice (Brundtland *et al.*, 1987).

The Intergovernmental Panel on Climate Change was established in 1988, and the Berlin Wall came down in 1989. Those were years when the dominant security challenge involving science was switching from averting global military disaster to avoiding global eco-environmental disaster (Kennel, 2023). Thirty-two years later, in 2021, two major reports characterized the state of Anthropocene issues. The Intergovernmental Panel on Climate Change in 2021 served a grave and urgent warning: climate change is impacting the terrestrial environment in ways unprecedented in thousands of years, and some of the changes will prove irreversible. And the UK Treasury's *Dasgupta Review* (Dasgupta, 2021) began the process of integrating biodiversity into quantitative economics and, hence, into national and international decision making.

## Wilson's Bottleneck and the demographic transition

National populations, while growing in many, if not most, countries earlier, are now in decline in China, Japan, South Korea and parts of Europe (World Population Review, 2023). The planet is in the middle stages of a human demographic transition. In the absence of effective action, ecological and environmental vulnerability to climatic (and other) disasters may continue to grow at least until the demographic transition has run its course and the population declines globally. The peak global population period of the demographic transition threatens to be one not only of intense worldwide resource consumption but also of population stress on ecologies and environments. Climatic extreme events could interact with ecologies at their most vulnerable, precisely when the number of humans at risk is the largest it will have ever been.

The great ecologist, Edward O. Wilson (1929–2021) feared the demographic maximum would be an ecological bottleneck (Wilson, 2002). A 'bottleneck' was the way Wilson found to communicate his concern to a biologically naïve physicist, Charles F. Kennel (1998–2006), who was burdened with an institutional responsibility for marine conservation as the Director of the Scripps Institution of Oceanography, in his private remarks in 2001, during his visit to La Jolla to receive the Inaugural William A. Nierenberg Prize from that Institution: those ecosystems that survive it set initial conditions for the post-Anthropocene Crisis world of nature.

Wilson considered the mass extinctions he was a witness to existential crimes of the highest order; as he had written earlier (Wilson, 1984):

> The worst thing that will probably happen – in fact is already well underway – is not energy depletion, economic collapse, conventional war, or the expansion of totalitarian governments. As terrible as these catastrophes would be for us, they can be repaired in a few generations. The one process now going on that will take millions of years to correct is loss of genetic and species diversity by the destruction of natural habitats. This is the folly our descendants are least likely to forgive us.

In Wilson's time, then, the complexity of the ways climate change is altering ecological habitats was not as widely appreciated, nor was it as widely perceived that the combination of climatic and eco-environmental stresses acts in a habitat-by-habitat manner; it is now better understood that it is wrong-minded to seek 'one-size-fits-all' solutions to biodiversity threats, either geographically or temporally; solutions effective in one place and time do not automatically translate. It was simpler for Wilson in 2001 to say to a non-ecologist that when the number of humans on Earth is maximum, so also will be the global threat to biodiversity.

More than two decades have passed; the environmental, ecological and ecological sciences have progressed; and it is more widely understood that taking the spatio-temporal complexity of ecological systems into account is the essential precursor to taking action. A 2023 review article in *Science* (Pörtner *et al.*, 2023) provides an up-to-date summary of this approach. This article and those referenced therein make clear that it is unrealistic to hope that future projections of climate change and its eco-environmental impacts will be any more uniform across regions and nations. (The projections of population growth are also unclear, with different expert groups arriving at different dates; see Adam, 2021). How far will global population rise? Researchers cannot agree. What seems certain is that demographic transitions are in progress in many countries, and that the UN estimate of a global demographic maximum arrival near the end of the present century is the furthest off all the expert estimates.

Furthermore, there is emerging evidence that crises once thought causally independent may be on track to coalesce into more comprehensive forms (Falk *et al.*, 2022). Whether coalescence occurs is context-dependent.

Expert communities of knowledge and practice, sometimes called knowledge-action networks (Falk *et al.*, 2021), have formed around many anticipated climate change impacts. Since the goal is making global knowledge actionable using local understanding, the networks focus on local eco-environmental vulnerabilities whose action strategies draw on global knowledge and experience. The present effort on drylands ecology and agriculture takes this approach. The anticipation that climate change will have major impacts on agro-ecosystems and land degradation requires continuous dynamic assessments, regionally and at the local level (El-Beltagy, 2001; Millennium Ecosystem Assessment, 2005; El-Beltagy and Madkour, 2012).

In a recent declaration by the leaders of Africa on climate change (African Union, 2023) they recognized and committed to:

- climate-smart and restorative agricultural practices and essential protection and enhancement of nature and biodiversity;
- actions to halt and reserve biodiversity loss, deforestation and desertification;
- restore degraded lands to achieve land degradation neutrality;
- finalize and implement the African Union biodiversity strategy and action plan with

**Fig. 2.1.** Executive Board of Crop Diversity Trust for 2008 visiting Svalbard Global Seed Vault. *Left to right:* Wangari Maathai, Nobel Prize winner 2004; Margaret Catley-Carlson, former Chair, now Patron of Global Partnership; Emile Frison, former Director General of Biodiversity international; Adel El-Beltagy, former Global Forum of Agricultural Research (GFAR); Mangala Rai, former Secretary of the Government of India's Department of Agricultural Research and Education (DARE) – Oct 2007. (Svalbard Global seed image is CC by 2.0).

the view to realizing the 2050 vision of living in harmony with nature;

- integrate climate, biodiversity and ocean agendas into national development plans and processes to increase resilience of local communities and nation economies.

This was followed by the leaders of G20 declaration (New Delhi, September, 2023) (G20, 2023). They stated that:

> ...it is with the philosophy of living in harmony with our surrounding ecosystem that we commit to concrete action to address global challenges. We will urgently accelerate our actions to address development and climate challenges, promote Lifestyles for sustainable development (LIFE), and conserve biodiversity, forests, and oceans. Commit to the swift, full, and effective implementation of the Kunming-Montreal global biodiversity Framework (GBF), and encourage others to do the same, and encourage actions to halt and reverse biodiversity loss by 2030. We also call for enhanced financial support from all resources. To this end, we welcome the recent establishment of the global biodiversity framework fund within the global environment facility (GEF).

To avoid Wilson's Bottleneck, we need to initiate a global campaign to sustain biodiversity (collection conservation and utilization) using a network of gene banks international, regional and national supported by FAO, UNEP, IFAD, UNESCO, WMO and the Consultative Group on International Agricultural Research (CGIAR), the International Treaty on Plant Genetic Resources for Food and Agriculture, and benefit-sharing fund Crop Trust (dedicated to conserving and making crop diversity available for use globally). The Crop Trust played a central role in the establishment and funding of Svalbard Global Seed Vault with the Norwegian Government and Nordic Genetic resources Center (NorGen) (Fig. 2.1).

This global initiative is an essential element for introducing new genetical make-up suitable for the ever-changing agro-ecological zones resulting from climate change, and hence securing food and maintaining the balance between food production and nature.

We, the authors of these introductory remarks, come from different regions – El-Beltagy from Egypt, Falk from Australia, and Kennel from California, USA, where freshwater availability has been the primary prerequisite for human settlement. Images of the Empty Quarter on the Arabian Peninsula illustrate to us what could happen worldwide if every region does not deal effectively with its own version of Wilson's Bottleneck.

## References


Adam, D. (2021) How far will global population rise? Researchers can't agree. *Nature* 597(7877), 462–465. DOI: 10.1038/d41586-021-02522-6

African Union (2023) The African Leaders Nairobi Declaration on Climate Change and Call to Action. Available at: https://au.int/sites/default/files/decisions/43124-Nairobi_Declaration_06092023.pdf

Birch, C. (1993) *Regaining Compassion for Humanity and Nature*. Kensington, New South Wales University Press, Sydney, Australia.

Brundtland, G., Khalid, M., Agnelli, S., Al-Athel, S., Chidzero, B. *et al.* (1987) *Our Common Future*. Report of the World Commission on Environment and Development. Available at: http://www.bne-portal.de/fileadmin/unesco/de/Downloads/Hintergrundmaterial_international/Brundtlandbericht.File.pdf?linklisted=2812 (accessed 25 June 2024).

Camilleri, J. and Falk, J. (2009) *Worlds in Transition: Evolving Governance across a Stressed Planet*. Edward Elgar Publishing, Cheltenham, UK.

Caradonna, J.L. (2020) *Sustainability: A History*. Oxford University Press, New York.

Crutzen, P.J. and Stoermer, E.F. (2000) The Anthropocene. *Global Change Newsletter* 41, 17–18. DOI: 10.12987/9780300188479-041

Dasgupta, P. (2021) *The Economics of Biodiversity: The Dasgupta Review*. HM Treasury, London. Available at: https://bibliotecadigital.ciren.cl/handle/20.500.13082/32687 (accessed 25 June 2024).

El-Beltagy, A. (2001) Land degradation: a global and regional problem. In: Van Ginkel, H. (ed.) *Human Development and the Environment: Challenges for the United Nations in the New Millennium*. United Nations University Press, Tokyo, New York, Paris, pp. 245–263.

El-Beltagy, A. and Madkour, M. (2012) Impact of climate change on arid lands agriculture. *Agriculture & Food Security* 1(3). DOI: 10.1186/2048-7010-1-3

Ellis, E.C., Gauthier, N., Klein Goldewijk, K., Bliege Bird, R., Boivin, N. *et al.* (2021) People have shaped most of terrestrial nature for at least 12,000 years. *Proceedings of the National Academy of Sciences* 118(17), e2023483118. DOI: 10.1073/pnas.2023483118

Falk, J., Attig-Bahar, F., Colwell, R.R., Behra, S.K., El-Beltagy, A. *et al.* (2022) Addressing our planetary crisis. *Sustainability Science* 17, 5–7. DOI: 10.1007/s11625-021-01059-x

Falk, J., Colwell, R.R., Kennel, C.F. and Murray, C.A. (2021) Link knowledge and action networks to tackle disasters. *Nature* 599, 372. DOI: 10.1038/d41586-021-03419-0

G20 (2023) G20 New Delhi Leaders' Declaration, G20 Summit, New Delhi. Available at: https://www.mea.gov.in/Images/CPV/G20-New-Delhi-Leaders-Declaration.pdf (accessed 25 June 2024).

Henrich, J. (2016) *The Secret of Our Success: How Culture Is Driving Human Evolution, Domesticating Our Species, and Making Us Smarter*. Princeton University Press. Princeton, New Jersey. DOI: 10.2307/j.ctvc77f0d

Kennel, C.F. (2021) The gathering anthropogenic crisis. *The Anthropocene Review* 8(1), 83–95. DOI: 10.1177/2053019620957355

Kennel, C.F. (2023) *From the Cold War to Global Warming: A Scientific Odyssey*. California Digital Library, University of California, David, California. Available at: https://escholarship.org/uc/item/6pj6c3rf (accessed 26 June 2024).

Meadows, D.H., Meadows, D.L., Randers, J. and Behrens, W.W. (1972) *The Limits to Growth: A Report for the Club of Rome's Project on the Predicament of Mankind*. Universe Books, New York. DOI: 10.1349/ddlp.1

Millennium Ecosystem Assessment (2005) *Ecosystems and Human Well-being: Synthesis*. Island Press, Washington, DC. Available at: www.millenniumassessment.org/documents/document.356.aspx.pdf (accessed 26 June 2024).

Pörtner, H.O., Scholes, R.J., Arneth, A., Barnes, D.K.A., Burrows, M.T. *et al.* (2023) Overcoming the coupled climate and biodiversity crises and their societal impacts. *Science* 380(6642). DOI: 10.1126/science.abl4881

Ramanathan, V. (2022) Climate resilience for the poorest three billion. YouTube Google Corporations, 10 May. Available at: https://www.youtube.com/watch?v=OO6pVI80TTs (accessed 26 June 2024).

Rees, M.J. (2003) *Our Final Hour: A Scientist's Warning: How Terror, Error, and Environmental Disaster Threaten Humankind's Future in this Century: On Earth and Beyond*. Basic Books, UK.

Steffen, W., Broadgate, W., Deutsch, L., Gaffney, O. and Ludwig, C. (2015) The trajectory of the Anthropocene: the great acceleration. *The Anthropocene Review* 2(1), 81–98. DOI: 10.1177/2053019614564785

Wilson, E.O. (1984) *Biophilia*. Harvard University Press. DOI: 10.2307/j.ctvk12s6h

Wilson, E.O. (2002) The bottleneck. *Scientific American* 286(2), 82–91. Available at: http://www.jstor.org/stable/26059564 (accessed 26 June 2024).

World Population Review (2023) Countries with declining population. Available at: https://worldpopulationreview.com/country-rankings/countries-with-declining-population (accessed 26 January 2023).

# Part B: Agro-ecosystems Affected by Dynamic Impact of Climate Change

**Adel S. El-Beltagy**[1,2*]
[1]*Chair, International Dryland Development Commission (IDDC);* [2]*Arid Land Agricultural Graduate Studies & Research Institute (ALARI), Ain Shams University, Egypt*

**Abstract**

To address the multifaceted challenges of climate change, a comprehensive strategy is essential. Central to this is a unified approach that combines scientific and indigenous knowledge to create an early-warning system, focusing on the local impact assessment of climate change, especially where livelihoods are threatened. Modernizing agricultural practices is crucial, utilizing state-of-the-art technologies such as Artificial Intelligence (AI), and emphasizing precision and regenerative techniques. The changing agro-ecological zones necessitate collaboration among global and national gene banks to initiate breeding programs for resilient crops. Advanced AI-driven Decision Support Systems are vital for effective crop management in these evolving zones. Human resource development and institutional restructuring will amplify resilience. It is also pivotal to gauge climate-induced loss and damage. In embracing the principle of inclusivity, international collaboration is paramount to mitigate climate change's disproportionate burdens on vulnerable populations. This initiative will necessitate the cooperation between UN organizations and regional and global financial institutions.

## Introduction

Home to over 2 billion people, dryland areas account for 40% of the world's population (Fig. 2.2). A significant 55% of these individuals reside in rural regions, while more than 90% live in developing countries. Drylands are where approximately half of the world's impoverished population can be found, and they are characterized by the highest population growth rates.

Drylands occupy 41% of Earth's surface and include 45% of global agricultural land (Mirzabaev *et al.*, 2019) (Fig. 2.3). These ecosystems are among the most vulnerable to anthropogenic climate change and land-use shifts, with desertification as an impending threat (El-Beltagy, 2002; Lickley, 2018; Adam, 2021).

The current global population stands at 8.1 billion and is projected to reach approximately 10 billion by 2050. This expected growth would increase global food demand by nearly 75%. To meet this challenge, developing countries will need to nearly double their agricultural production, all while grappling with climate change induced by anthropogenic factors.

In an alarming contrast, one third of the food produced globally goes to waste every year. Meanwhile, around 805 million people suffer from hunger annually. A 4°C temperature rise is

*Address for correspondence: elbeltagy@drylanddevelop.org

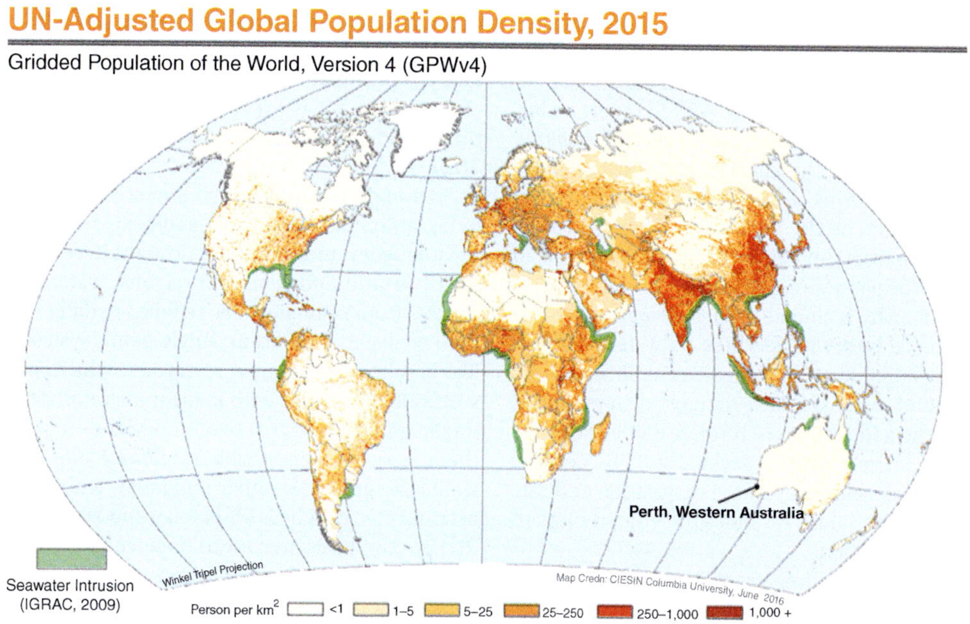


**Fig. 2.2.** World population density map derived from gridded population of the world. (version 4. CC by 4.0) (Data source: Center for International Earth Science Information Network – CIESIN – Columbia University 2018). Gridded population of the world version 4(GP Wv4): Population count, revision 11. Palisades, NY: NASA Socioeconomic Data and Applications Center (SEDAC). Available at: https://doi.org/10.7927/H4JW8BX5. (https://sedac.ciesin.columbia.edu/data/collection/gpw-v4/maps/gallery/search accessed 22 June 2024).

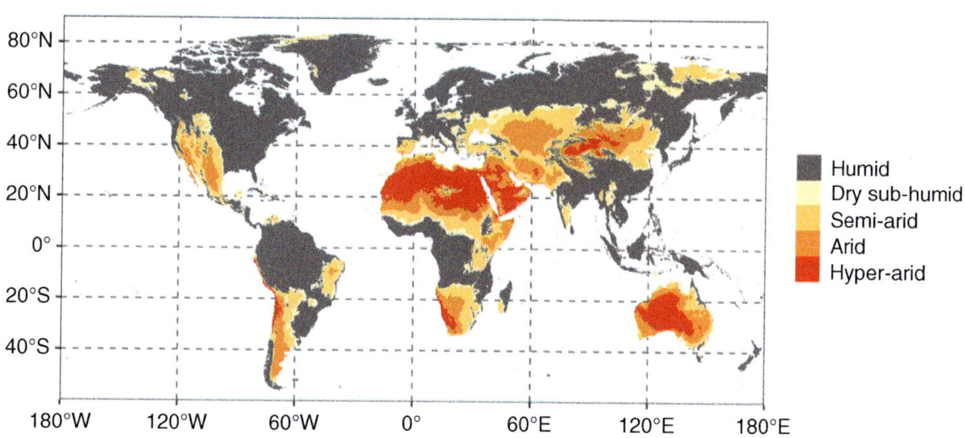


**Fig. 2.3.** Geographical distribution of drylands, delimited based on the aridity index (AI). The classification of AI is: Humid AI > 0.65; Dry sub-humid 0.50 < AI ≤ 0.65; Semi-arid 0.20 < AI ≤ 0.50; Arid 0.05 < AI ≤ 0.20; Hyper-arid AI < 0.05. Data: TerraClimate precipitation and potential evapotranspiration (1980–2015). *Climate Change and Land*: an IPCC special report; Cambridge University Press; reused with permission. (Mirzabaev *et al.*, 2019) Available at: https://www.cambridge.org/core/books/climate-change-and-land/reviewers-of-the-ipcc-special-report-on-climate-change-and-land/A25F2E06D1E4C401DDF6B3F184AB0BF1 (accessed 22 June 2024).

projected to reduce food production (encompassing crops, fish and livestock) by 50% by the end of the century.

According to the IPCC's 2022 report on impacts (IPCC, 2022), *Adaptation, and Vulnerability*, unprecedented climatic changes have already affected all forms of life across the globe. A staggering 70% of the world's population growth between 2010 and 2050 is expected to occur in 24 of the poorest countries.

So why is climate change of particular significance for drylands? These ecosystems already face a multitude of challenges: increasing resource demands, unsustainable management practices, and the added stress of human-induced climate change. The fragility of dryland systems, coupled with their sensitivity to the magnitude and rate of climate change, escalates the vulnerability of local populations.

The direct impact of climate change in these areas will be reflected in changes in temperature, rainfall, length of the growing season, and timing of extreme and critical threshold events that can disrupt the food chain and affect crop development. The Paris goal in 2015 to retain temperature rise at 1.5°C seemingly is not attainable and it is expected that the temperature could go up 3.0°C by year 2060 (IPCC, 2021) (Fig. 2.4).

Drought is a common characteristic of drylands, and climate change is predicted to amplify the frequency and severity of drought in certain areas. In arid and semi-arid regions, water resources from rainfall will be reduced in different agro-ecologies and from single-point systems like boreholes or isolated reservoirs, which are susceptible to failure with inadequate rains. The fragile natural resource base of drylands makes them especially susceptible to climate change, leading to greater weather anomalies with devastating consequences (El-Beltagy and Madkour, 2012). Amid this precarious balance of despair and hope, dryland communities require more support to use their natural resources sustainably and maintain their livelihoods in the face of changing climate.

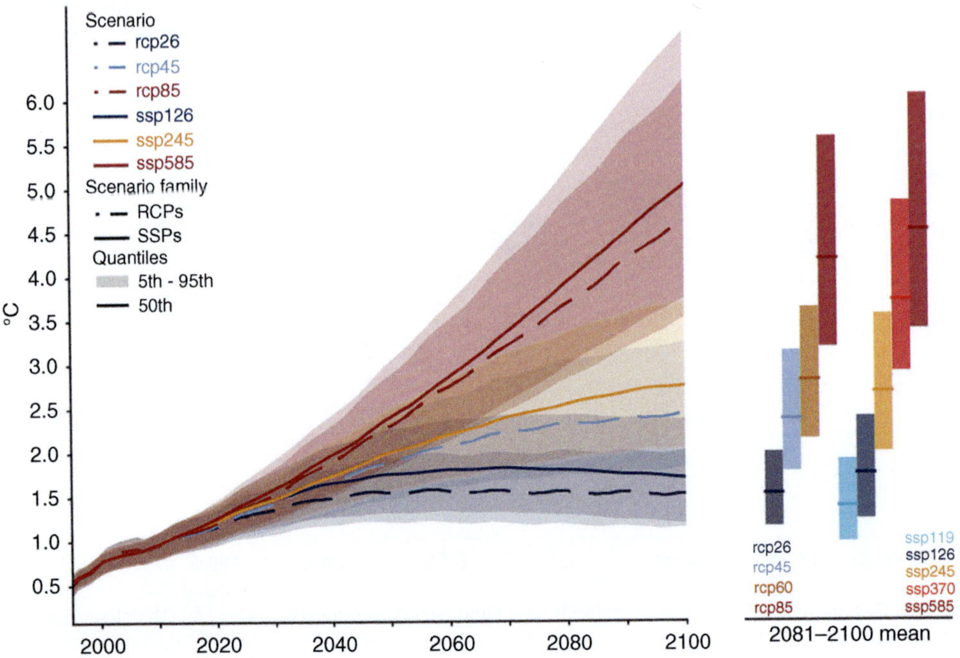


**Fig. 2.4.** Air surface temperature change. Adapted from IPCC Climate Change 2021 (IPCC, 2021), Cambridge University Press, reused with permission. Available at: https://www.cambridge.org/core/books/climate-change-2021-the-physical-science-basis/415F29233B8BD19FB55F65E3DC67272B (accessed 22 June 2024).

The effects of this vulnerability depend on the initial conditions of the water supply system, population growth, demand changes, available technology, and socio-economic and legislative conditions. Given the high spatial variability of dry areas, the generalized information provided by global climate change modeling is often too coarse to capture regional and local climate processes.

Increases in temperature will reduce agriculture production (Agnolucci *et al.*, 2020; Zhu *et al.*, 2022). Notably, yields will decrease in the drylands of the tropics and sub-tropics, where crops are already at their maximum temperature tolerance level. It is clear, therefore, that dry areas are expected to suffer the most from climate change (Fig. 2.5).

This will put increased pressure on natural resources such as land, water and biodiversity due to population growth, further exacerbated by climate change. Consequently, the process of desertification (Fig. 2.6) is likely to accelerate, with its damaging impacts on natural resources and human well-being (El-Beltagy, 1999, 2001; Burrell *et al.*, 2020). A review of articles on soil viral diversity, ecology and climate change concluded that there is a need for more research to understand how different aspects of climate change influence soil viruses, their ecological rules and their potential to be pathogens (soil quality and functionality as a blueprint for provision of human health and reducing future pandemics) (Jansson and Wu, 2023).

We can expect the loss of genetic resources, degradation of soil's physical, chemical and biological properties, deterioration of water quality, and an increase in salinity effects.

Improving resilience and capacity to cope with the adverse effects of climate change necessitates a holistic approach. This includes applying scientific and technological advances and providing socio-economic, policy and institutional support. Rising anthropogenic pressures, particularly in developing countries due to population boom, are straining natural resources, which are further being devastated by conflicts and wars. Therefore, proactive measures are needed to halt this destructive process and, where possible, reverse this destructive trajectory (Fig. 2.7).

## Outmigration

The potential for mass migration resulting from climate change is an issue of paramount importance.

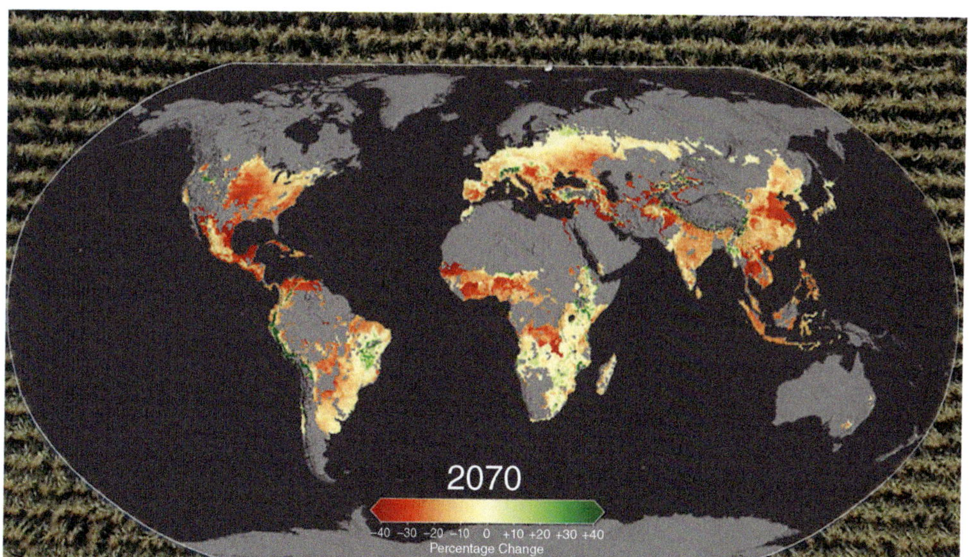


**Fig. 2.5.** Map of the world showing, in red, where decreases in corn yields are projected to occur in 2071: parts of North America, South America, West Africa, Central Europe, India, China. NASA/Katy Mersmann; https://climate.nasa.gov/news/3124/global-climate-change-impact-on-crops-expected-within-10-years-nasa-study-finds/ (accessed 22 June 2024).

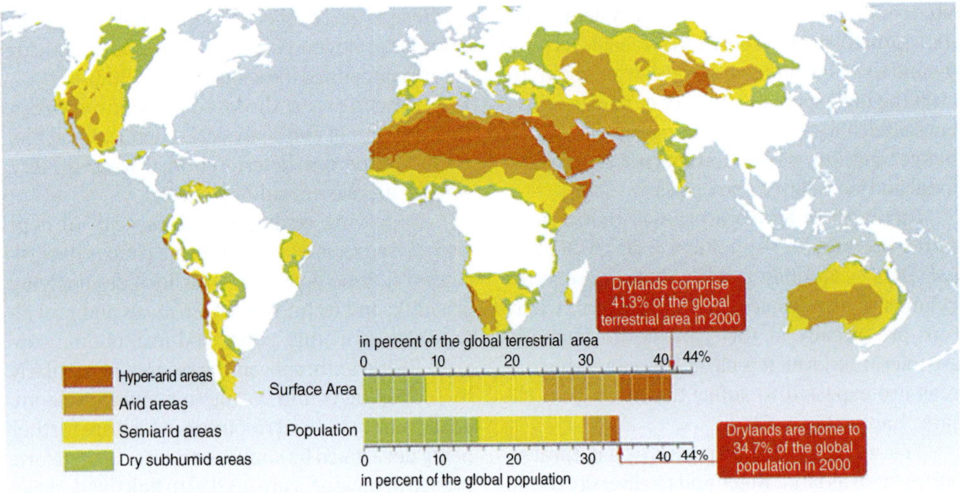


**Fig. 2.6.** Drylands vulnerable to desertification and their categories. From Millennium Ecosystem Assessment (Adeel *et al*., 2005); public domain. Available at: https://www.millenniumassessment.org/en/index.html (accessed 22 June 2024).

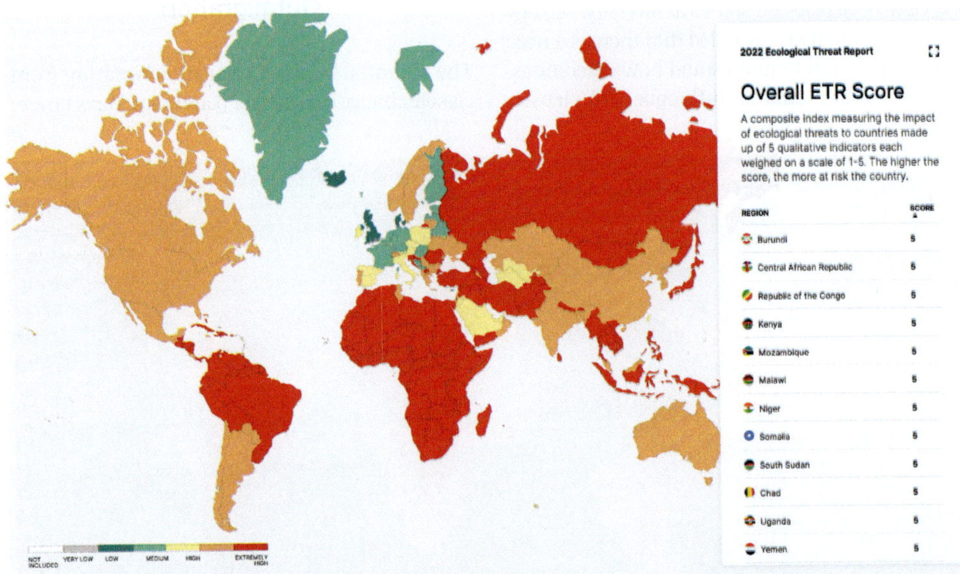


**Fig. 2.7.** Projections to 2050 show that much of sub-Saharan Africa will be unsustainable, with 95% population growth projected. From Vision of Humanity; public domain. Available at: https://www.visionofhumanity.org/maps/ecological-threat-report/#/ (accessed 22 June 2024).

As global temperatures rise, sea levels increase, and extreme weather events become more frequent, several regions will be rendered uninhabitable. Consequently, an unprecedented number of individuals will face the harrowing decision to abandon their homes.

The Stern Review on the Economics of Climate Change (Stern, 2007) conservatively projects that by mid-century, climate change could displace upwards of 150 million individuals. However, some assessments present even graver figures: Christian Aid (2008) estimates that this number could escalate to a staggering 1 billion.

Such immense migration flows can catalyse substantial socio-political upheavals. As these climate migrants traverse borders and settle in regions where resources are already in short supply, host communities may confront resource-based conflicts, infrastructure overload, and economic destabilization. Furthermore, the mass displacement will undoubtedly precipitate humanitarian emergencies. Many of these migrants may find themselves in dire conditions, devoid of essential life-sustaining resources like food, clean water, and shelter.

The evidence presented underscores the pressing need to devise and execute effective strategies that both mitigate the impacts of climate change and bolster community adaptation to these changes. A multifaceted approach is vital:

1. **Emissions reduction**: a concerted effort to reduce greenhouse gas emissions can decelerate the trajectory of climate change.
2. **Resilient infrastructure**: investments should be channeled into fortifying infrastructure, ensuring that it can withstand the challenges posed by extreme weather events.
3. **Economic and social support**: comprehensive economic and social policies are indispensable to assist displaced individuals and communities, ensuring that they can rebuild their lives post-migration.

Furthermore, the complexities of climate-induced migration necessitate robust international cooperation. Given that climate change transcends national boundaries, its ripple effects on migration will similarly breach borders. This global challenge demands a collective response. The international community must unite in its efforts, ensuring that support systems and solutions are in place to aid those most vulnerable to the effects of our changing climate (Fig. 2.8).

Finally, a pivotal area of focus should be enhancing the resilience of communities, particularly in developing countries that frequently bear the brunt of climate change's effects. Concrete steps include:

1. **Sustainable agriculture**: promoting agricultural practices that are environmentally sustainable and resilient to changing climate patterns will ensure food security and livelihoods for many.

**Fig. 2.8.** The impact of disasters and crises on agriculture and food security. From Food and Agriculture Organization of the United Nations. Available at: https://www.fao.org/3/I8656EN/i8656en.pdf (accessed 23 June 2024).

**2. Effective water management**: proper water management techniques can mitigate the impacts of droughts and floods, preserving this vital resource.
**3. Renewable energy**: investment in renewable-energy technologies not only combats the greenhouse gas emissions driving climate change but also offers a sustainable energy source for these communities.

By prioritizing these measures, we can potentially alleviate the compulsion for many to migrate. This dual-benefit approach not only enhances the quality of life for residents but also champions the sustainable use of natural resources, paving the way for a more resilient and sustainable future.

## The Evolution of International Cooperation

In recent decades, the global community has progressively recognized the urgent need for cooperation in confronting the multifaceted challenges of climate change:

- COP15 (2009) in Copenhagen: a focus on adaptation and technical funds;
- COP16 (2009) Cancun Adaptation Framework;
- Pivotal agreements of 2015: this year marked a watershed moment in international consensus on sustainable development and climate change mitigation. Three monumental agreements were forged:
    - Sustainable Development Goals (SDGs);
    - COP21 Paris Agreement; and
    - Sendai Framework for Disaster Risk Reduction;
- COP27 (2022) Sharm El-Sheikh: a focus on loss, damage, and justice.

### COP15 in Copenhagen: a focus on adaptation and technical funds

COP15 took place in December 2009. While the conference has been frequently remembered for the controversies and challenges in achieving a legally binding treaty, it did set some important precedents, particularly in the areas of adaptation and financial support:

1. Copenhagen Accord:
    - The conference led to the drafting of the Copenhagen Accord, a political agreement that, while not legally binding, signified the commitment of major economies to take immediate action on climate change.
    - Under the accord, countries recognized the need to limit global temperature rises to less than 2°C above pre-industrial levels.
2. Commitment to financial support:
    - One of the major outcomes of COP15 was the pledge by developed countries to mobilize financial resources to assist developing countries in their climate change mitigation and adaptation efforts.
    - Under the Copenhagen Accord, developed countries committed to a goal of mobilizing jointly $100 billion annually by 2020 to address the needs of developing countries. This funding was to come from a mix of public and private sources.
3. Establishment of the Green Climate Fund (GCF):
    - Though the Green Climate Fund would be officially established during COP16 in Cancun, the seeds for its inception were sown during the discussions in Copenhagen.
    - The GCF aims to be a major channel for funding adaptation and mitigation activities in developing countries.
4. Emphasis on adaptation:
    - The accord recognized the significance of enhancing protective actions against the adverse impacts of climate change. This highlighted the importance of strengthening resilience and adaptive capabilities, especially in the most vulnerable countries, such as Small Island Developing States (SIDS) and African nations.
5. Technical cooperation:
    - The Copenhagen Accord also underscored the importance of technology development and transfer to developing countries as a way to enhance their actions on mitigation and adaptation. This led to discussions about establishing mechanisms to facilitate better technological cooperation between nations.

The pledges and discussions from COP15 laid essential foundations for subsequent climate negotiations and frameworks.

### COP16 in Cancun: stepping forward on adaptation and technical funds

The 16th Conference of the Parties (COP16) took place in December 2010. The Cancun conference was crucial in restoring faith in the multilateral process and saw tangible progress, particularly regarding adaptation and technical support mechanisms (UNFCC, 2011). The key outcomes of COP16 were:

1. Cancun Agreements:
   - These created a comprehensive framework and structure to advance global efforts on climate change.
   - They acknowledged the deepened commitment to act on climate change by all parties, emphasizing both mitigation and adaptation.
2. Adaptation Framework:
   - The Cancun Adaptation Framework was established during COP16 to enhance global adaptation efforts.
   - The Framework recognized the need to strengthen international cooperation on adaptation to reduce vulnerability and increase resilience in developing countries, particularly in those that are most vulnerable, like the SIDS, Least Developed Countries (LDCs), and African nations.
3. Green Climate Fund (GCF):
   - COP16 formalized the creation of the GCF to support projects, programs, policies, and other activities in developing countries using thematic funding windows.
   - The GCF was set as a central mechanism to channel the financial commitment made in Copenhagen to mobilize $100 billion annually by 2020 for climate change initiatives in developing countries.
4. Technology mechanism:
   - This mechanism was intended to boost technology development and transfer in support of mitigation and adaptation actions.
   - The mechanism consists of the Technology Executive Committee (TEC) and the Climate Technology Centre and Network (CTCN). Their combined role is to advise on technological needs and facilitate the dissemination and deployment of technology to developing countries.
5. Fast-start finance:
   - Developed countries, under the Cancun Agreements, were committed to provide $30 billion in new and additional fast-start finance between 2010 and 2012 for mitigation, adaptation, technology development and transfer, and capacity building in developing countries.
6. Long-term finance:
   - The agreements also saw a reiteration of the commitment from developed countries to mobilize the aforementioned $100 billion annually by 2020 to address the needs of developing countries.

In conclusion, COP16 solidified many of the concepts introduced during COP15. The establishment of structured mechanisms, especially around adaptation and technology, provided a clear pathway to enhance global climate action. The Cancun conference was successful in detailing and operationalizing many of the pledges made in Copenhagen, ensuring progress in both adaptation and technical funding efforts.

### Sustainable Development Goals (SDGs)

At the UN Sustainable Development Summit held 25–27 September 2015 in New York, the United Nations introduced the 2030 Agenda for Sustainable Development, a transformative plan of action that consists of 17 'Global Goals' with 169 targets in total (UN, 2015). This landmark initiative, officially known as 'Transforming Our World: The 2030 Agenda for Sustainable Development', was forged through a comprehensive deliberative process that engaged all 193 Member States of the United Nations, in conjunction with a broad spectrum of global civil society.

While every goal has implications for the environment, society, and the economy, it is worth noting that out of the 17 SDGs, ten, directly or indirectly, tie into disaster risk reduction (DRR). This inter-linkage underscores the significance of DRR as an integral component of sustainable development.

Two goals that particularly stand out in the context of climate change are:

- Goal 13 (Climate Action): it mandates urgent measures to combat climate change and its impacts, underscoring the global consensus on adaptation.
- Goal 15 (Life on Land): it emphasizes sustainable forest management, combating desertification, and halting land degradation and biodiversity loss. These actions inherently incorporate climate change adaptation strategies.

In evaluating the effectiveness of our actions towards these goals and the broader context of climate change, the Intergovernmental Panel on Climate Change (IPCC), operating under the auspices of WMO (World Meteorological Organization) and UNEP (United Nations Environment Programme), plays a pivotal role. The IPCC has published six comprehensive assessment reports since its inception in 1990 (1990, 1995, 2001, 2007, 2014, 2023).

The IPCC functions through three specialized working groups:

- WG I concentrates on the scientific aspects of climate systems and the shifts observed in climate patterns.
- WG II evaluates the susceptibility of socioeconomic and natural systems to climate change. This group scrutinizes both the adverse and advantageous outcomes of climate change, offering adaptation strategies.
- WG III focuses on possible solutions to abate climate change. This involves studying means to curtail greenhouse gas emissions and identifying actions to enhance the removal of these gases.

Notably, several predictions laid out in the IPCC's assessment reports have been realized, with many manifesting even sooner than anticipated. This underlines the precision of the panel's research and the accelerating pace of climate change challenges.

## Paris Agreement

Adopted on 12 December 2015, the Paris Agreement stands as a landmark in the global climate struggle. It aims to limit global warming to well below 2°C above pre-industrial levels, striving for 1.5°C. A significant component of the Agreement concerns adaptation:

- The Agreement recognizes that adaptation is a global challenge faced by all, with local, sub-national, and national dimensions.
- It emphasizes the importance of 'enhancing adaptive capacity, strengthening resilience, and reducing vulnerability to climate change'.
- The Agreement mandates regular 'global stocktakes' to assess collective progress towards achieving its purpose, which includes adaptation efforts.

## The Sendai Framework for Disaster Risk Reduction (SFDRR) 2015–2030

In March 2015, global stakeholders converged at the Third United Nations World Conference on Disaster Risk Reduction in Sendai City, Miyagi, Japan. The culmination of this gathering was the unanimous adoption of the Sendai Framework for Disaster Risk Reduction 2015–2030 (UNDRR, 2015).

This pivotal framework charted a new course in disaster risk reduction by delineating its scope through a global outcome, a singular goal, 13 guiding principles, and four priorities of action intended to be implemented at local, national, regional, and global tiers. Central to the framework are seven global targets designed to gauge progress and effectiveness. Some implications of the framework include:

- Holistic approach: the Sendai Framework adopts a more inclusive approach to disaster risk, acknowledging that climate change is a driver of risk and reinforcing the need for climate change adaptation to be integrated into disaster risk strategies.
- Shift from disaster management to risk management: while earlier frameworks primarily focused on disaster response, Sendai emphasizes understanding disaster risk, its reduction, and the strengthening of resilience. This approach aligns with climate change adaptation principles.
- Local-level actions: recognizing that disasters and the impacts of climate change are often felt most acutely at the local level, the

Sendai Framework underscores the importance of local risk assessments and strategies, intertwining closely with localized climate change adaptation measures.
- Integration with Development Goals: the framework encourages countries to align their disaster risk reduction policies with other relevant frameworks, including the SDGs and the Paris Agreement. This integrated approach ensures that climate change adaptation is not siloed but is considered within a larger developmental context.

One of the stand-out aspects of the Sendai Framework is its explicit monitoring mechanism, detailed comprehensively to nullify any potential ambiguities (Fig. 2.9). This approach ensures transparent oversight and accountability in its application.

If the commitments undertaken by the Member States within the framework are faithfully executed, the result would be a tangible enhancement in the resilience of communities to climate change. The progress achieved on various levels – local, national, and regional – would be meticulously documented in subsequent reports. These reports will serve as periodic checkpoints to assess strides made and address challenges, ensuring consistent dialogue and introspection at designated intervals (UNISDR/GE/2015 – ICLUX EN5000 1st edition).

Together, the SDGs, the Paris Agreement, and the Sendai Framework provide a holistic approach to understanding and addressing the multifaceted challenges posed by climate change, emphasizing both mitigation and adaptation strategies.

The milestone agreements of Paris and Sendai are only partially legally binding, leaving much room for voluntary action. Therefore, a vigorous and proactive campaign is essential to establish an internationally binding legal framework. Though a commitment was made in Cancun to establish a technology box and a fund to support developing countries in enhancing their resilience and reducing their vulnerability to climate change, its implementation has been ineffective. The promise of making 100 billion dollars available annually until 2020 to help developing countries adapt to climate change has not been fulfilled. The voluntary Nationally Determined Contributions (NDCs) expressed by each country reflect the pillars of implementation: (a) political will and effective governance; (b) long-term mitigation strategies; (c) integrated adaptation planning and implementation programs; (d) financial framework for coping with climate change; and (e) the establishment of new institutions, credits and legislation. It is crucial to optimize and enhance synergies between international, regional and national research systems.

## Major outcomes of COP27: a focus on loss, damage, and justice

The 27th Conference of the Parties (COP27) held in Sharm El-Sheikh, Egypt, in November

**Fig. 2.9.** Monitoring of action on disaster risk reduction at global, regional and national levels under the Sendai Framework. Image credit: UN Office for Disaster Risk Reduction (2015). CC BY 4.0. Available at: https://www.undrr.org/media/48692/download?startDownload=true, https://gar.undrr.org/infographics.html (accessed 23 June 2024).

2022, signaled a turning point in the ongoing global efforts to combat the catastrophic effects of climate change. Below are the critical outcomes and statements from the UN Secretary-General Antonio Guterres at its conclusion.

1. Establishment of a loss and damage fund:
   - The decision to establish a loss and damage fund was one of the highlights of the COP27. This fund is vital because it recognizes the irreparable damage and irreversible losses that countries, especially the most vulnerable, are already experiencing due to climate change.
   - The operationalization of this fund in the forthcoming period will provide a mechanism to support nations grappling with the consequences of climatic events and will provide a concrete means of redressing the balance.
2. Rebuilding trust between nations:
   - A significant emphasis was placed on rebuilding the shattered trust between developed and developing countries. This commitment signifies the collective realization that only through mutual trust and cooperation can the world effectively tackle the challenges of climate change.
3. Revamping the financial models:
   - The Secretary-General underscored the necessity to alter the business models of multilateral development banks and international financial institutions. This change aims to ensure that these entities are more aligned with the global climate objectives and are more responsive to the needs of countries most affected by climate change.
4. Ending the assault on nature:
   - In his pointed remarks, the Secretary-General implored the international community to cease the 'suicidal war on nature'. The rampant destruction of ecosystems, accelerating species extinction, and the exacerbating climate crisis necessitate an immediate re-evaluation of our actions and priorities.
5. Early-warning systems for vulnerable regions:
   - Recognizing the need for preparedness, especially for countries most susceptible to the immediate impacts of climate change, the Secretary-General highlighted the imperative to establish an early-warning system.
   - To this end, Antonio Guterres announced a landmark US$3.1 billion plan, aiming to ensure that everyone across the globe is shielded by early warning systems within the next half-decade (https://news.un.org/en/story/2022/11/1130277; accessed 23/6/24).
   - Such systems will be pivotal in offering countries, especially those that are most vulnerable, a fighting chance against the sudden onslaught of climatic events.

In essence, the outcomes of COP27, particularly in the realms of loss, damage, and justice, underscore the global community's renewed commitment to both corrective and preventive measures. Governments also agreed to establish a 'transitional committee' to make recommendations on how to operationalize both the new funding arrangements and the fund at COP28. The first meeting of the transitional committee is expected to take place before COP28. The emphasis on trust-building, reimagined financial models, protection of ecosystems, and preparedness indicates a holistic approach to tackling the multifaceted challenges of the climate crisis.

### Impact on ecosystems

IPCC projections are at odds with the food production goals of increasing intensification, diversification, and productivity (Kaur *et al.*, 2017). Climate-induced system transitions are expected to be unprecedented in scale and will intersect different ecosystems (Hillel and Rosenzweig, 2010). Land-use transitions under a 1.5°C temperature increase will be similar to those projected under a 2°C-rise model. Such substantial transitions present formidable challenges for the sustainable management of land to meet human settlement, food, livestock feed, fiber, bioenergy, carbon storage, biodiversity, and other ecosystem service demands (Fig. 2.10). Overcoming socio-economic, institutional, technological, financing, and environmental barriers is crucial for implementing land-based mitigation and adaptation options (IPCC, 2018). These barriers vary across regions, making the formulation of a

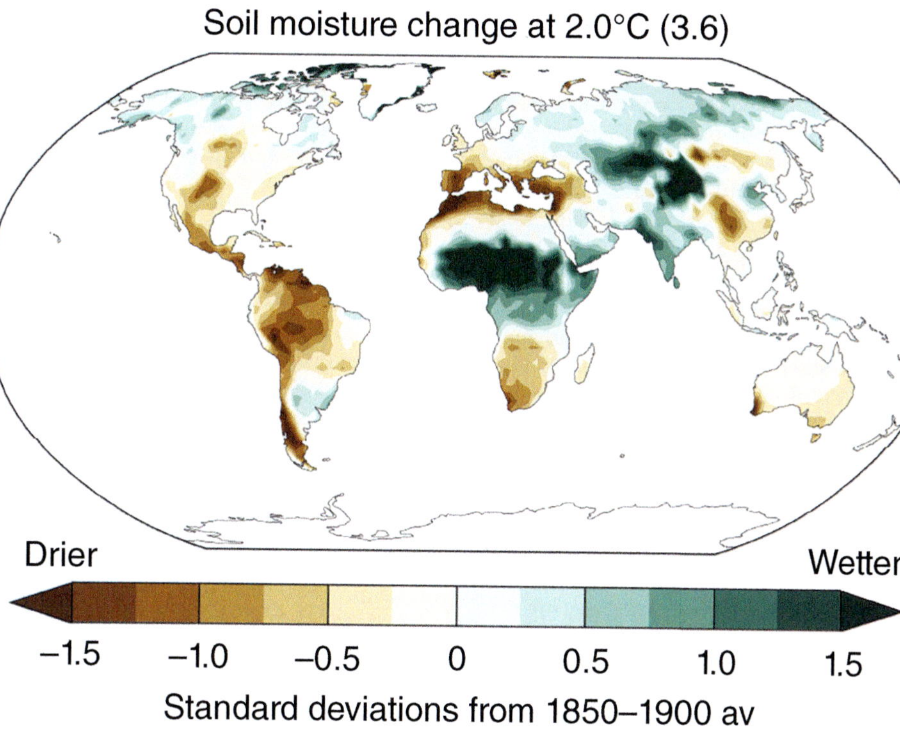


**Fig. 2.10.** The sixth IPCC Assessment Report projects changes in average soil moisture that can disrupt agriculture and ecosystems. A reduction in soil moisture by one standard deviation means that average soil moisture will approximately match the ninth driest year between 1850 and 1900 at that location. CC by 4.0. Available at: https://commons.wikimedia.org/wiki/File:Soil_moisture_and_climate_change.svg (accessed 23 June 2024).

one-size-fits-all strategy untenable. Context-specific solutions, designed keeping in mind regional challenges and assets, will be vital in ensuring that land remains a sustainable resource for future generations.

The primary challenge we need to address for a deeper understanding of potential environmental changes is that of cumulative impacts. The study of cumulative environmental impacts is inherently intricate. Unlike isolated events, cumulative effects evolve over time and can have ripple effects across multiple ecosystems. For instance, a combination of deforestation, urbanization, and climate change can together result in more significant and unpredictable impacts than each factor would produce individually. Addressing them in isolation may lead to superficial solutions that would not stand the test of time. To find long-lasting and effective solutions, there is a need for:

- holistic approaches: recognize and address the interconnectedness of environmental challenges;
- integrated management: instead of addressing one issue at a time, strategies should incorporate various factors and their potential cumulative impacts; and
- predictive modeling: use data and technology to predict potential cumulative impacts, thus enabling proactive solutions.

Sometimes, the science of cumulative impact is straightforward, such as establishing migration corridors in response to sea-level rise induced by climate change. However, even clear-cut cases necessitate extra effort, broader partnerships, and more time. Addressing cumulative issues is a priority if we are to find viable solutions to the imminent environmental crises of the coming decades (McNutt, 2013; Kefford *et al.*, 2023; Wang and Lei, 2023).

The establishment of a loss and damage fund at COP27 underscores the urgency to intricately understand and evaluate climate vulnerabilities at regional and local levels. The need for developing countries to organize themselves for more granular financial commitments, including: more aid for low-income and vulnerable countries; compensation for loss and damage from climate change; a new mandate and larger ambition for multilateral development banks; and a serious effort to mobilize private financing and drive down its cost by smart use of official guarantees and other de-risking instruments, was suggested in Brookings paper (Bhattacharya *et al.*, 2023). The European Environment Agency's study 'Just Resilience for Europe' (Lager *et al.*, 2023), emphasizes this by introducing a nuanced approach that combines traditional vulnerability assessments with considerations of equity and justice. By embedding these principles, the fund can be utilized more effectively and fairly, ensuring that adaptation strategies holistically address the multifaceted challenges of climate change and societal disparities.

## Adaptation: A Priority for Global Climate Action

The intensifying effects of climate change – be it heatwaves, flooding or forest fires – are no longer sporadic occurrences; they have established themselves as regular facets of our global environment. COP26 reinforced this reality, leading to world leaders, governments, and convention

**Box 2.1.** European Environment Agency (EEA, 2023)

In the face of impending climate challenges, Europe is actively exploring ways to refine its methodologies for vulnerability assessment, with the aim of sharing these advancements with developing nations. The goal is to provide these countries with concrete, result-oriented tools to gauge both vulnerability and injustice. Critical areas identified in this endeavor include:

- **Vulnerability.** The IPCC portrays vulnerability as the predisposition of communities to be adversely impacted by socio-economic inequalities. This concept extends beyond superficial factors like age or income and delves deep into an individual's capability to cope with, and adapt to, climate-related challenges. EU guidelines categorize vulnerable groups as the elderly, disabled, displaced, socially marginalized, economically disadvantaged, and certain genders and minorities.
- **Transboundary and cascading effects.** Climate-change consequences often transcend borders, potentially jeopardizing global stability. Individuals in precarious situations are particularly at risk. Nevertheless, some adaptation strategies by the EU may inadvertently harm external entities. In recognizing this, several EU nations are restructuring their developmental aid, focusing on aiding countries least responsible for, yet most affected by, climate change.
- **Monitoring frameworks for just resilience.** Europe acknowledges the need for monitoring systems that can evaluate just resilience at a national level. Such systems would assess the disproportionate climate burden, evaluate the accessibility of funds for adaptive strategies, and ensure the participation of vulnerable groups in significant decisions.
- **Enhancement of Knowledge and Adaptation Efforts.** To optimize adaptive methodologies, it is vital to share best practices and discern overlaps between policy sectors. There are notable parallels in justice elements, whether it is addressing energy poverty in urban regions or protecting small businesses in agriculture.
- **Reduction of inequality and injustice.** Although adaptation policies increasingly recognize the imperativeness of social justice, there exists a palpable gap in monitoring frameworks to measure just resilience. The existing policy landscape in the EU portrays a disconnect, with certain strategies addressing justice on a global scale but overlooking national or sectoral priorities. To bridge this gap, the EU aims to infuse justice into diverse policy domains, ensuring that funds meant for a just transition achieve their intended purpose. Climate adaptation, in this context, is not merely a reactive strategy but an opportunity to rectify entrenched inequalities.

The EU's intent is clear: harness the lessons of adaptation, promote a culture of shared knowledge, and champion policies that, while addressing the climate crisis, simultaneously rectify systemic injustices.

state parties elevating their dedication to combat these challenges through Adaptation at COP27. One of the most notable achievements of COP27 was the establishment of the Global Goal on Adaptation. As we progress towards COP28, it is imperative that the momentum is not only maintained but intensified. All participating entities must display unwavering political will to monitor, evaluate, and augment their strides towards fortifying resilience and supporting the world's most susceptible populations.

However, our journey should not halt with merely establishing a global goal. COP28 must further consolidate the global adaptation agenda, upholding the commitments established in Paris and elaborated upon in the Glasgow pact. The essence of this global strategy lies in proactively adjusting agro-management techniques and harnessing genetic resources apt for the fluidly shifting agro-ecological landscapes (Fig. 2.11).

In acknowledgment of the pressing need for immediate and effective adaptation strategies, the International Climate Change Information Program (ICCIP) has partnered with Springer to produce an essential reference work on the subject, as detailed by Walter L. Filho in 2015. This work underpins the collective global commitment to adapt, evolve, and confront the challenges climate change thrusts upon us (Box 2.2).

Local communities are on the frontline of climate change impacts, making them vital stakeholders in the development of effective adaptation strategies. Embracing locally led adaptation assessments involves:

**1.** collaborative decision making: working closely with communities to identify pivotal issues and determine the factors that influence their vulnerabilities to climate change;
**2.** defining success: understanding from the communities themselves what they perceive as effective adaptation and the outcomes they desire;
**3.** risk mapping and data collection: deploying tools like risk mapping to grasp the unique challenges posed by the changing climate in specific localities; and
**4.** merging local and scientific knowledge: fusing traditional local wisdom with contemporary scientific findings can provide a more holistic view of adaptation measures. This blend can create strategies that are culturally appropriate and scientifically robust.

In addition, understanding how climate change affects biodiversity is essential. The capacity of species and ecosystems to adapt – or their inability to do so – has significant ramifications for both nature and people (El-Beltagy *et al.*, 2005). For instance, changes in crop yields, livestock, and fisheries can severely threaten food security. The interplay between food systems and climate adaptation is highlighted by entities like the Committee on World Food Security and the

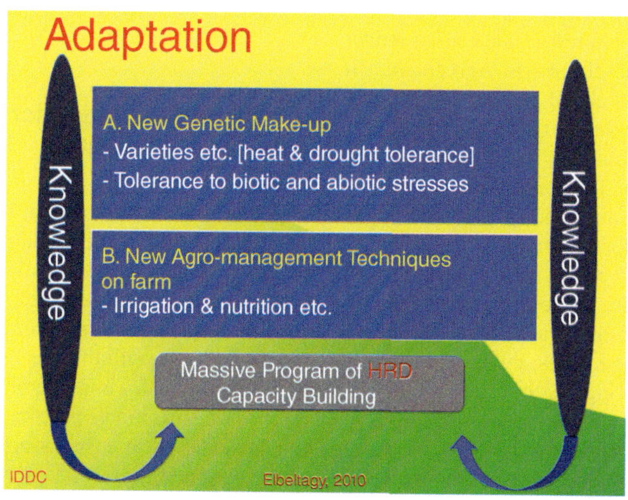


**Fig. 2.11.** Major elements for dynamic adaptations with continuous feed of new knowledge driven from science and technology innovation. (From El-Beltagy, 2010)

**Box 2.2.** *Comprehensive Handbooks on Climate Change Adaptation*

The handbooks curated by Walter Leal Filho (2015) are an exhaustive resource that delves into the multifaceted aspects of climate change adaptation. These handbooks tackle:

- climate change impact and management: this delves into understanding the tangible and intangible repercussions of climate change and strategizing effective management techniques;
- policy: crafting and implementing robust policies is vital in spearheading efforts to combat and adapt to climate change impacts;
- adaptation, agriculture, and water security: given the significant impacts of climate change on agriculture and water resources, this section focuses on sustainable adaptation strategies that ensure food and water security; and
- climate change adaptation technologies and cross-cutting issues: this segment not only discusses technological interventions for adaptation but also delves into intertwined societal concerns such as socio-economics, resilience, trade, justice, poverty, health, population security, and education.

For these strategies and interventions to be effective at grassroots level, there is a palpable need for precise vulnerability assessments, incorporating metrics like exposure, sensitivity, and adaptive capacities. Such evaluations are indispensable in tailoring interventions that cater to the unique needs of local communities.

Adaptation, as elucidated by Moser and Ekstrom (2010), encompasses alterations in socio-ecological systems in the face of the tangible impacts of climate change and potential future implications. Such adaptation measures span from immediate short-term coping mechanisms to profound, long-term transformations. Notably, these strategies are not exclusively aimed at addressing climate change but also strive to tackle broader socio-ecological challenges. However, their efficacy in mitigating detrimental impacts or leveraging potential opportunities remains variable.

The concept of 'fragility', as defined by the OECD (2016), amalgamates the two phenomena of risk exposure and the limited ability (of states, systems or communities) to manage, absorb or mitigate these risks.

Furthermore, engaging vulnerable populations in the adaptation process extends beyond merely avoiding harm. It envisions climate adaptation as a unique opportunity to systematically and sustainably address and mitigate deep-rooted inequalities and injustices that exacerbate vulnerabilities in the first place.

High-Level Panel of Experts on Food Security and Nutrition (HLPE-CFS).

Recognizing the unavoidable consequences of climate change, the Paris Agreement, in 2015, established a global goal for adaptation. This initiative underscores that mitigation alone is not sufficient; countries must also invest in adaptive measures. This sentiment is particularly relevant for vulnerable nations that might lack the resources for adaptation.

The Paris Agreement was a landmark in global climate action. While it saw commitments from major emitters like Brazil, China, and India to curtail greenhouse gases, vulnerability maps spotlight the sobering reality: developing nations are poised to experience the harshest impacts of climate change (Fig. 2.12).

Their susceptibility arises from a mix of geographical, economic, and infrastructural factors that constrain their adaptive capacities. Moving forward, these vulnerabilities must be at the forefront of global climate discussions and actions (Fig. 2.13).

## Holistic Adaptation for a Resilient Future

Improvement in infrastructure and policy are crucial elements of these adaptation strategies. Better rural infrastructure, such as roads, market buildings, storage facilities, and institutional support like extension programs, credit, and input markets, can enhance resilience and adaptation capacity. Policy improvements are needed to incentivize sustainable practices and internalize the environmental costs associated with different economic activities.

On the point of population growth, it is true that a growing population could exacerbate the impacts of climate change, particularly in regions

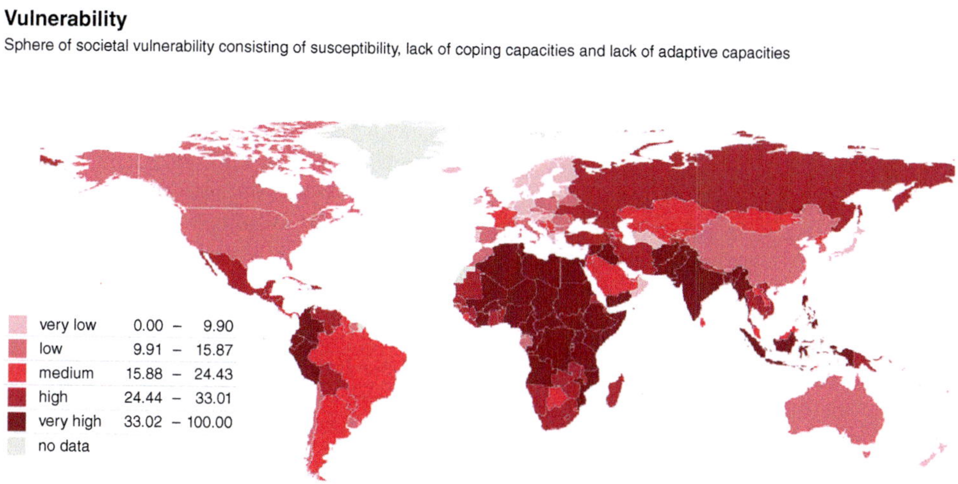


**Fig. 2.12.** Vulnerability map, published by Bündnis Entwicklung Hilft e.V. & Institute for International Law of Peace and Armed Conflict. CC BY 4.0. World Risk Report, 2022. Available at: https://en.wikipedia.org/wiki/File:WorldRiskIndex_2022-en.svg (accessed 24 June 2024)

**Fig. 2.13.** The two core aspects of just resilience: dealing with unequal burdens (climate change impacts and risk) and leaving no one behind (inclusion in and effects of adaptation action) (Lager *et al.*, 2023). CC by 4.0 international; available at: https://www.eionet.europa.eu/etcs/etc-ca/products/just-resilience-for-europe-towards-measuring-justice-in-climate-change-adaptation (accessed 24 June 2024)

already facing significant climate risks. These factors underline the urgency of implementing comprehensive and effective measures to mitigate and adapt to climate change.

Yet, addressing climate change cannot be viewed in isolation. It is intricately linked to myriad pivotal concerns – from food security and water availability to social equity and robust governance. Any effective climate strategy, therefore, should be a tapestry of interventions spanning these domains. This holistic vision is enshrined in the UN's Sustainable Development Goals, which spotlight the inextricable ties between environmental action and broader developmental imperatives. The future lies in integrated solutions, where climate action is harmoniously woven into the broader fabric of sustainable development.

## Agro-management techniques

Dryland agriculture is particularly susceptible to the brunt of climate extremes and the consequences of unsustainable management practices. With the urgency to ensure food security, alleviate poverty, and address hunger and malnutrition, there is a pressing need to understand the implications of climate change on these fragile systems. By conducting comprehensive reviews, significant gaps can be identified, paving the way for technologically driven interventions.

Technological collaboration at regional levels, encompassing universities, agricultural institutions, growers, and corporate entities, offers potential remedies. These encompass water conservation techniques, supplemental irrigation, innovative foliar sprays, integrated nutrient management, resilient crop cultivars, and the harnessing of artificial intelligence in precision agriculture. Process-based models, applied globally, can predict climate change impacts, allowing for optimized management of dryland cropping systems. Such models tackle various elements – from water, nutrient, pesticides, and herbicides management to crop diversification. Such models lead to optimized recommendation of resource usage by artificial intelligence (Talaviya *et al.*, 2020).

Highlighting the disparities in global agricultural research, it is noted that while the majority of the world's farms are smaller than 2 ha, the bulk of research prioritizes larger farms. This needs reconsideration, especially as dryland farming systems, already vulnerable, face escalating threats from climate-induced changes. Efficient resource use, coupled with strategic soil management, can potentially counteract these negative trajectories.

One approach to counter the setbacks of climate change is through process-based models, simulating 'what if' scenarios. By such methods, the effects of various management practices can be anticipated, both in the short and long term. These insights can further inform strategic interventions, from the integration of legumes in cropping systems and development of resilient crop varieties to precision farming and carbon sequestration.

A notable paradigm addressing these challenges is climate-smart agriculture (CSA). Envisioned as a transformative pathway for the future of agriculture in a changing climate, CSA prioritizes resilience and resource efficiency. By encompassing a multi-stakeholder approach, CSA seeks to enhance adaptive capacities, marry climate and agricultural policies, and innovate in financing mechanisms. Distinctly different from traditional approaches, CSA's strength lies in its adaptability, offering context-driven solutions bolstered by innovative policies and financing (Lipper *et al.*, 2014).

Lastly, the doctrine of Sustainable Intensification of Cultivated Land Use (SICLU) underscores the balance between burgeoning food demand and the limits of resources and the environment. Rather than merely maximizing ecological, economic, and social benefits, the aim is a dynamic equilibrium among these facets, ensuring sustained benefits from cultivated land use (Lyu *et al.*, 2022).

Rudel (2020) argues that there is no single optimal strategy; rather, optimal strategies vary across socio-ecological context, i.e. (i) in preurban agriculture; (ii) on large farms situated on prime agriculture land; and (iii) in smallholder-dominated agriculture districts. Government intervention to promote sustainable intensification should recognize and build on these distinct, place-based economic and agro-ecological dynamics.

## Agro-ecology

A critical question is whether agro-ecology can promote climate change mitigation and adaptation outcomes without compromising food security. A study by Dittmer *et al.* (2023) evaluated smallholder agricultural methods in low- and middle-income countries (LMICs) based on 35 indicators pertaining to mitigation, adaptation, and yield. This evaluation encompassed 50 articles which presented 77 instances of agro-ecological interventions as compared to traditional practices. Notably, in 63% of the cases that reported on yields, the crop outputs were higher.

There is a need to increase the yields through ecological intensification of agricultural systems as well as to identify the needs in terms of research innovation, training, collective organizations and public polices required to achieve this purpose (Kremen, 2020).

Local and scientific know-how should be jointly mobilized to encourage the dynamics of ecological intensification, regenerative agriculture (Khangura *et al.*, 2023), and climate smart agriculture (Zhao et al., 2023). Therefore, result-based choices should be made in utilizing the best options for different agro-ecologies, which will maintain the balance of nature and agriculture productivity in facing the risk of climate change.

A report presented by the High Level Panel of Experts on Food Security and Nutrition (HLPE, 2019) describes several innovative approaches to sustainable food systems (SFS) and clusters them into two main categories: (i) sustainable intensification of production systems and related approaches (including climate-smart agriculture, nutrition-sensitive agriculture and sustainable food value chains) that generally involve incremental transitions towards SFS; and (ii) agro-ecological and related approaches (including organic agriculture, agroforestry and permaculture) that some stakeholders consider to be more transformative. While the former category starts from a premise that, to address future challenges, productivity per unit of land needs to increase in a sustainable manner – which is what is meant by sustainable intensification – the latter emphasizes reducing inputs and fostering diversity alongside social and political transformation focused on improving ecological and human health and addressing issues of equity and governance.

The report highlights the points of convergence and divergence existing among these different innovative approaches, building its comparative analysis upon the following nine characteristics:

1. regenerative production, recycling, and efficiency;
2. biodiversity, synergy, and integration;
3. economic diversification versus specialization;
4. climate change adaptation and mitigation;
5. knowledge generation and dissemination;
6. equity;
7. labor versus capital intensification;
8. connectivity versus globalization; and
9. governance and participation.

Each characteristic is described in a dynamic way, as a spectrum of various possible positions lying between two opposite poles.

The Committee on World Food Security (CFS) can serve as a model of inclusive civil society and private sector involvement and a starting point for implementing transitions towards food security and nutrition (FSN). Strategies and planning for implementing agro-ecological approaches at different scales (local, territorial, national, regional and global) can help achieve this fundamental transformation of food systems by: setting long-term goals; ensuring policy coherence

**Box 2.3.** Multiple definitions of agro-ecology.

As a science, agro-ecology is: (i) the integrative study of the ecology of the entire food system, encompassing ecological, economic and social dimensions, or, in brief, the ecology of the food system (Francis *et al.*, 2003); (ii) the application of ecological concepts and principles to the design and management of sustainable food systems (Gliessman, 2007); and, more recently, (iii) the integration of research, education, action and change that brings sustainability to all parts of the food system: ecological, economic and social (Gliessman, 2018).

Agro-ecological practices aim at improving agro-ecosystems by harnessing natural processes, creating beneficial biological interactions and synergies among their components (Gliessman, 1990), and using, in the best way, ecological processes and ecosystem services for the development and implementation of practices (Wezel *et al.*, 2014). As a social movement, agro-ecology is seen as a solution to current challenges such as climate change and malnutrition, contrasting with the so-called 'industrial' model and transforming it to build locally relevant food systems that strengthen the economic viability of rural areas based on short marketing chains, and fair and safe food production. It supports diverse forms of smallholder food production and family farming, farmers and rural communities, food sovereignty, local knowledge, social justice, local identity and culture, and indigenous rights for seeds and breeds (Altieri and Toledo, 2011; Rosset *et al.*, 2011; Nyéléni, 2015). This dimension of agro-ecology as a political movement is becoming increasingly prominent (Gonzalez de Molina, 2013; Toledo and Barrera-Bassols, 2017).

(Source: Agroecology Europe, 2017; FAO Agroecology Knowledge Hub, 2023)

across sectors (agriculture, trade, health, gender, education, energy, environment); and involving all relevant actors through consultative multi-stakeholder processes.

In conclusion, while the task ahead is significant and the risks are high, the efforts taken so far, such as the Paris Agreement and the focus on adaptation measures, are encouraging. We can work towards a more resilient and sustainable future with continued international cooperation, investments in sustainable development, and a commitment to follow through on climate commitments.

There is no 'one-size-fits-all' solution to realizing the transformation of food systems globally required to achieve FSN; it will require supporting a diversity of transitions from different starting points, along different pathways, adapted to the local conditions and challenges faced in different places by different people. The following recommendations, distilled from the deliberations of this report, aim to help decision makers develop concrete actions that will encourage and support the innovation required at local, territorial, national, regional and global scales to follow appropriate transition pathways towards SFS that enhance FSN (HLPE, 2019).

## Enhancing Coping Capabilities through Adaptation

Climate change poses a substantial risk to our planet's resilience, potentially resulting in grave ecological crises. Customized integrated natural resource management strategies for different ecosystems are necessary to meet the needs of the reliant communities. Such a strategy necessitates the development of extensive knowledge and understanding of coping mechanisms for drought risk, management and restoration of ecological functions; sustainable biodiversity; and diversification of production systems and livelihoods. This technological knowledge needs to be disseminated globally, regionally and nationally. This will be supplemented by supportive policy and institutional mechanisms, forming the foundation of sustainable development.

Enhancing the coping capacities of the communities will require appropriate adaptation mechanisms. Developing these measures would require new tools of science and technology in the fields of remote sensing and GIS/GPS; biotechnology and genetic engineering; functional breeding; simulation modeling; information technology; AI (machine learning, expert systems, bioinformatic, neural networks, simulation modeling, IOT, image processing etc.); renewable energy; new energy-saving techniques for desalination and transportation of seawater; nanotechnology; and molecular machines and devices (Rafea and El-Beltagy, 1987; Rafea, 2002; El-Beltagy and Rafea, 2021).

Gene-editing techniques in crops and livestock aim to enhance diverse traits, many of which could be advantageous for climate change adaptation. Work has been done to highlight specific gene-editing applications and research directed towards bolstering the resilience of

**Box 2.4.** Adaptation will require intensive knowledge in the following fields:

**A.** New genetic make-up: genetic engineering/genome editing, e.g. for developing:

- C4 rice;
- nitrogen-fixing wheat;
- enhanced photosynthetic pathways;
- biotic- and abiotic-resistant crop varieties through functional breeding.

**B.** New agro-management techniques related to on-farm irrigation and nutrition management; integrated pest management; CSA; precision agriculture; conservation agriculture; dynamics of ecological intensification; regenerative agriculture:

- moisture sensors (to conserve water);
- optimizing use of fertilizer nutrients;
- remote sensing;
- simulation modeling artificial intelligence;
- urban agriculture hydroponic, NFT, etc.; and
- cloud computing.

crops and livestock against climatic shifts. By considering the technological boundaries and potentialities, several studies indicate the imminent promise of these techniques in addressing climate-related issues (Karavolias *et al.*, 2021). The recent developments in different reverse genetic approaches have made it possible to identify the function of genes controlling nutritional, biochemical and metabolic traits imparting drought, heat, cold, salinity tolerance as well as diseases and insect pests (Kumar *et al.*, 2022).

Upscaling (or extrapolation) and downscaling effectively take place within similar agro-ecological environments. However, scaling-out, or transfer and adaptation, may be conceived as the movement of a technology or management option from one environment to a different environment to address a problem across environments and systems. For this, additional tools are needed. If upscaling mainly utilizes GIS techniques, then transfer and adaptation need to combine GIS much closer with other modeling techniques and disciplines, combining data from numerous disciplines and sources to arrive at estimates of the differing environmental effects on the utility and performance of alternative options (Fig. 2.14).

## Science and Technology

Climate change undeniably poses significant challenges to agriculture. To address these, a considerable increase in investment within the realm of agricultural science and technology is imperative. A specific emphasis should be laid on advanced crop-breeding techniques, incorporating biotechnology approaches that address both abiotic and biotic stresses, fostering biological resilience. Efficient water management, optimized for usage, is equally paramount.

The application of agro-management techniques should be discerningly chosen, tailored to the unique conditions of each agro-ecological zone, ensuring the best outcomes. Addressing the knowledge disparities concerning the local and regional impacts of climate change necessitates a concerted global effort and collaboration. The onus is on developed nations to perceive their knowledge and financial contributions not merely as aid but as pivotal investments in safeguarding the well-being of their citizens. With projections suggesting the potential displacement of up to 200 million people by 2050 due to climate-related issues, the ramifications of inaction become even more pronounced. Such predictions underscore the dire need for holistic support and collaboration. To protect and sustain our Earth, a united front is indispensable.

For inhabitants of dry regions to effectively fortify their resilience in the face of climate change, it is imperative to integrate the following key considerations:

**1.** comprehensive climate-change scenario analysis: undertaking detailed assessments at

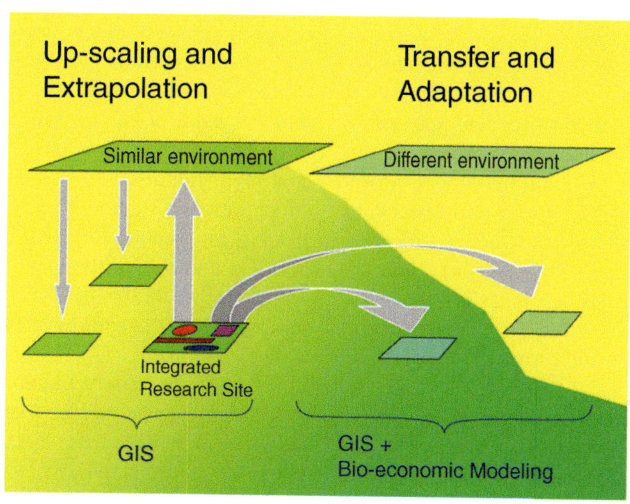


**Fig. 2.14.** Upscaling and extrapolation. (From El-Beltagy, 2019)

local, national, and regional scales is essential for strategic navigation and effective decision-making;

**2.** embracing inter-sectoral approaches swiftly: it is crucial to establish robust institutions and chart out policies that promote responsible and sustainable agricultural practices, inclusive of aquaculture and fisheries;

**3.** adapting to the fluidity of baselines: as new data emerge and the landscape of climate change understanding evolves, it is important to be adaptive, acknowledging that previously established baselines may change;

**4.** continuous assessment of shifting agro-ecological zones: as the impact of climate change becomes manifest, local communities are affected, thwarting previous crop rotation and agro-management systems. Continuous monitoring and assessment of these shifts will be essential to develop adaptive strategies and maintain agricultural productivity;

**5.** elevating research and information system investments: ensuring substantial investments in research and advanced information systems is a fundamental step towards informed and proactive action.

It is worth mentioning that the impact of climate change will force a dynamic navigation through scales (global, regional, and local) and it will have a prime impact on shifting agro-ecological zones and hence will require a dynamic shift of suitable crops, livestock, and fisheries to match this change (Laurence *et al.*, 2023; Nainggolan *et al.*, 2023).

Shifts in agro-ecological zones due to climate change will have profound effects on local fauna and biota. Such changes will necessitate not just modifications in cropping rotations but also adjustments in the local microflora. Moreover, there will be repercussions on livestock, as well as marine and inland fisheries, affecting their productivity and suitability for specific regions. Addressing these changes will be paramount for ensuring sustainable agricultural and fisheries practices in the future (Barange *et al.*, 2018).

In a study entitled 'Impact of future climate change on shifting patterns of Agro-ecological zones in China', it was found that future climate change will lead to significant local damage of agro-ecological zones in China in each of the four periods 2011–2020, 2021–2030, 2031–2040, and 2041–2050 (Lin *et al.*, 2013).

Climate variation has been responsible for shifting observed green biomass boundaries 250 km between wetter and drier years. Assessment priorities require a strong national monitoring program and reliable data, and effective planning, and therefore it calls for international cooperation to ensure a high degree of coordination among scientists (Dregne, 1987).

Collaboration among communities, policy-makers, and scientists is essential. Consequently, a strong global alliance must be forged for food security, peace, and prosperity of people living in dry areas under a changing climate.

Developing effective coping strategies necessitates a comprehensive understanding of the system, which is characterized by significant spatial and temporal variations. Global climate change impact assessments fail to reveal the actual changes occurring at regional and local levels and their effects on the production system and livelihoods. Therefore, more precise local assessments are required. Models and methodologies that can assist in better understanding the impacts under different ecological conditions will need to be developed for this purpose, enabling the development of targeted adaptation strategies.

Since its inception in 1978, the International Dryland Development Commission (IDDC) has committed itself to enhancing synergies between international, regional and national systems within the context of global research and technology (www.drylanddevelop.org). Professor Hans van Ginkel, Rector of the United Nations University, Tokyo (1997–2007) and President of the International Association of Universities (2000–2004), pioneered the concept of Regional Centers of Expertise (RCE) in education for sustainable development as a strategy for meeting the goals of the United Nations Decade of Education for Sustainable Development (UN DESD) by translating its global objectives into the context of local communities (Glasser, 2008).

It is of the utmost importance to establish and synergize knowledge networks, regional and national, i.e. future harvest and Central West Asia and North Africa+ (CWANA+) (Van Ginkel *et al.*, 2007) and the Global Forum for Agriculture Research (GFAR). Developing countries with large dryland populations should start preparing now, establishing connections with the international scientific community working on climatology for appropriate assessments. The development of coping mechanisms to enhance resilience and reduce vulnerability will depend on these assessments.

The consensus from international deliberations on climate change is that the cost of inaction will be enormous. Actions, both on adaptation and mitigation, must commence immediately. Coping mechanisms that are easy to implement for adapting to the changes that are already happening must be developed. However, this will only be possible when we clearly understand the impacts at the local level, rather than the vague regional or global levels. Global classifications mask regional differences, regional assessments mask sub-regional differences, and national classifications mask different agro-ecology. Therefore, navigating across scales is necessary (OECD, 2009; Gross *et al.*, 2016;) (Fig. 2.15).

The 'nested' regional climate modeling strategy offers a promising approach (Randall *et al.*, 2007). Global climate models (GCMs) operate at a resolution that is too coarse to capture regional and local climate processes effectively. Therefore, a regional climate model (RCM) is 'nested' within a GCM to increase model resolution locally. The RCM simulates the effect of sub-GCM-grid scale forcing and provides detailed regional information. Refining similar methodologies for local assessments will be necessary. Efforts in this direction have already begun, with the state of California leading the way. National programs will require training and technology transfer to undertake precise local assessments. The World Meteorological Organization (WMO) has been assisting 20 national organizations in conducting local and regional impact assessments. The agreements reached in Cancun will facilitate this process (Lori *et al.*, 2020; Lamba *et al.*, 2023; Von Jeetze *et al.*, 2023).

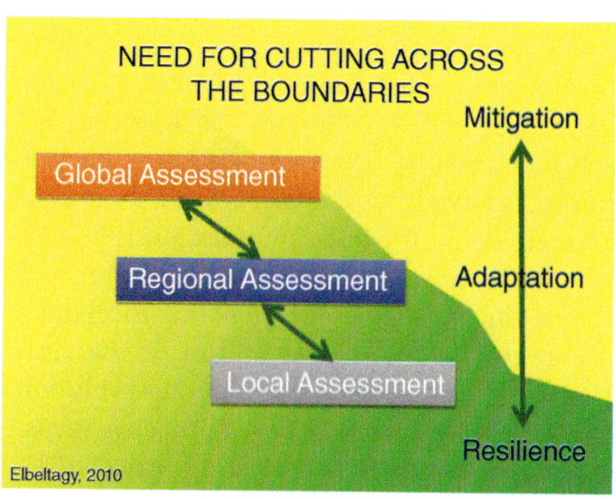


**Fig. 2.15.** Navigation across scales. (Author's own figure.)

Furthermore, localized evaluations are pivotal because they account for region-specific vulnerabilities, resource availabilities, and socio-economic conditions. By anchoring these assessments, national strategies can be tailor-made, ensuring that interventions are not just broadly applicable but also contextually relevant.

Inter-ministerial collaboration is paramount to ensure both financial and strategic objectives are allocated effectively to execute these interventions. Specialized agencies and research institutions bring in expertise in understanding and predicting climate-related changes and events. Meanwhile, institutions for disaster prevention play a crucial role in averting and managing crises stemming from climatic extremes.

## Action-oriented Knowledge Networks for Dealing with Climate Change

There is an urgent need for establishing an action-oriented knowledge network to accelerate and enhance the effort to develop local assessment of the impact of climate change and facilitate more precise prediction of agro-ecosystem sustainability and future change. An attempt to create knowledge networks is being noticed in the USA, Europe, Asia and other regions. A group of concerned scientists and experts started a Regional Action in Climate Change (RACC) in 2009, which has been established within the context of the Science and Technology for Society (STS) Forum. The group created an adjunct session to the STS Forum in Kyoto, Japan, to discuss the challenges climate change poses for governments, organizations and regions as they develop adaptation strategies (STS Forum, 2018).

RACC is based on 'knowledge action network' to connect the global science, technology and policy communities to realize locally applicable solutions. These are sponsored social networks connecting the generators of pertinent knowledge with local actors and decision makers. Every region has knowledge leaders who can forge relationships with local decision makers, but often there are not enough of them. The critical mass sufficient to characterize the multiple impacts of climate change and communicate them to decision makers is often lacking. The objectives of the RACC are: (i) learning from each other's experiences, issues, problems and solution approaches, and from the integration of local and traditional knowledge with newest advances in science and technology; (ii) creating a growing community of individuals across societal institutions and disciplines who talk the same language; (iii) developing templates based on local cases for successful solutions in climate change adaptation; and (iv) building capacities around the globe for dealing efficiently with local challenges for successful climate change adaptation. RACC, in its tenth meeting in October 2018, agreed to mobilizing the effort to harness the synergy between the UN, FAO, UNDP, WMO, UNESCO, UNEP, CGIAR, GFAR, World Bank, regional banks, and newly established networks.

Furthermore, these efforts cannot function in isolation; a consolidated and synergistic approach is vital. To that end, stakeholder involvement extends beyond governmental and specialized entities. It is about engaging local communities, NGOs, businesses, academia, and the general public. Their perspectives, expertise, and on-the-ground experiences enrich the adaptation strategies, ensuring they are holistic, inclusive, and, most importantly, actionable.

A national strategy can help to: (i) provide a framework for coordinating adaptation activities; (ii) create a mid- to long-term vision for adaptation; (iii) enable informed decision making based on vulnerabilities, impacts, and adaptation options; (iv) raise awareness in all affected sectors; (v) mobilize support nationally and internationally; and (vi) prepare the ground for appropriate institutional structures for adaptation.

Developing countries must conduct studies using the Climate Change Integrated Assessment Methodology. This method, developed for assessing cross-sectoral adaptation and vulnerability hotspots in Europe, uses an 'integrated assessment program'. The European Commission's Seventh Framework Programme funded the development of this methodology.

The conclusion from the assessment methodology exercise was that this approach to vulnerability assessment was replicable and transferable. It allowed the integration of capital and coping-capacity concepts with stakeholder-derived scenarios, creating patterns of coping capacity expected for different socio-economic scenarios. The hotspot maps created using this methodology for different parts of Europe suggested that human well-being was most at risk from southern Europe's water stress and biodiversity loss, and from lack of food production and land-use diversity in northern Europe (Lager, 2023).

However, implementing such integrated assessments in developing countries at local, sub-regional, and regional levels would necessitate infrastructure development and trained human resources. Support from industrialized countries would be crucial for this purpose.

## Conclusion

Developing the ability to cope with and adapt to climate change necessitates harnessing knowledge and formulating appropriate policies, institutional frameworks, legislation, and financial structures. A key aspect of a dynamic response is establishing a comprehensive knowledge network to tackle the far-reaching implications of climate change. It is essential to raise awareness among local communities about the risks they face and to encourage preventive and mitigative action. This necessitates efficient communication and information dissemination on global, regional, and local scales.

Several United Nations affiliates, such as FAO, UNDP, WMO, UNEP, and UNESCO, have established existing networks tackling climate change, alongside various regional organizations. However, the pressing question is how to enhance these networks to optimize coping and adaptation capabilities in response to climate change. Given the urgent nature of the situation, we cannot afford to wait for this evolution to occur organically. Significant proactive measures should be initiated, beginning with COP28, supported by a solid business plan that builds on the progress achieved in COP27 and earlier meetings.

The fact that rising sea levels and temperatures will impact all life forms on Earth underscores the urgency of the situation. In order to bolster the knowledge network, regional alliances should be formed, incorporating regional banks, UN organizations, universities, research centers, and other stakeholders. Financial support for these initiatives could be drawn from adaptation funds, risk management mechanisms, international and regional banks, private sectors, and civil societies.

Legislative bodies and parliaments play a crucial role in streamlining and enabling the operation of the knowledge network at national and local levels. Recognizing the pivotal role of knowledge institutions in developing countries and leveraging their research outputs, innovations, and technologies for climate change adaptation mitigation is fundamental. Given that climate change's impact predominantly affects the most vulnerable local communities, mainly due to the actions of industrialized nations, it is incumbent upon policymakers in these countries to comprehend that aiding adaptive mechanisms is not mere charity but a crucial strategy to prevent socio-economic devastation in their own countries and globally.

The following are the main components of a global campaign for enhancing adaptive and coping capacities in drylands (Fig. 2.16):

1. **Holistic integration.** A unified approach is paramount. This involves bridging existing scientific and indigenous knowledge to build an early-warning system. Such a system should provide insights into the granular impacts of climate change, especially in regions where livelihoods are at risk.
2. **Incorporating advanced technology in agriculture.** Agro-management techniques must evolve by integrating precision agriculture, climate-smart agriculture, regenerative agriculture, and intensive agro-ecology. The fusion of innovative technologies – like artificial intelligence, Big Data, neural networks, data mining, expert systems, drones, sensors, robots, and more – can optimize resources such as water, biodiversity, and soil.
3. **Addressing the shift in agro-ecological zones.** Changes in habitats affect crops, livestock,

**Fig. 2.16.** Enhancing coping and adaptive capabilities through optimization and enhancing synergies between international, regional, and national systems. (From El-Beltagy, 2018)

and fisheries. Collaboration among global gene banks, both national and international (like CGIAR and Svalbard), is essential to ensure that collections give sufficient emphasis to stress-adapted genetic material that can contribute to adaptation to climate change; and ensure that policies facilitate international exchange of the genetic resources needed to adapt farming systems to shifts in the distribution of climatic agro-ecological zones. Coping with uncertainty is needed, recognizing that there will be shifting baselines as further assessment of climate change is done.

Funding is necessary for:

(a) enhancing gene bank passports with molecular characteristics alongside geo-referencing and morphological descriptions (Volk and Richard, 2011);
(b) continuation of collection missions; this effort is crucial for programs aiming to harness genetic resources adapted to climate-changed agro-ecological zones, factoring in altered temperature, soil conditions, moisture levels, wind patterns, and precipitation;
(c) establishing global breeding programs to introduce new crop varieties resistant to both biotic and abiotic stressors; strengthen policies to promote dynamic seed systems, including the establishment of community-based seed enterprises at strategic locations within the country;
(d) an early-warning system (assessment) for the impact of climate change on shift of local agro-ecological zones which will require continuous effort to match new genetical resources suitable for this changing agro-ecology;
(e) decision support systems for agronomy: as new genetic compositions are introduced to changing agro-ecological zones, advanced agronomic decision systems based on AI tools will be essential to manage them effectively. These advanced systems can analyze vast amounts of data, encompassing genetic information, soil conditions, weather patterns, and more. With the inevitable shifts in agro-ecological zones due to climate change, these AI-powered DSS can provide real-time insights and recommendations for optimal crop management. This ensures that newly introduced genetic compositions are cultivated in a manner that maximizes yield and sustainability in their new environments. It is highly recommended that a solid consortium will be established which includes the national, and international gene banks with international institutions, i.e. UNEP, FAO, IFAD, CGIAR, UNDP, IUCN. Crop diversity to activate the global gene bank partnership (GGP) platform and ensure investment for research and information systems which will serve this mission to address the shifting baseline in agro-ecology by providing suitable crops (animal, fisheries) and to continuously introduce new genetic resources for ever-shifting agro-ecology.

**4. Human resource development.** All these global initiatives necessitate comprehensive human resource development. This aligns with the ideas discussed in international forums like Copenhagen and Cancun, especially concerning the technology fund.

**5. Institutional restructuring.** Existing institutions should be restructured and interconnected at regional and international levels to boost resilience and coping capabilities.

**6. Assessment of loss and damage.** Implement a results-oriented program to gauge the loss and damage using precise indicators tied to the vulnerability of distinct agro-ecological zones.

**7. Promotion of global solidarity.** To truly honor the principle of 'leaving no one behind', nations must engage in significant collaborative efforts. These partnerships are vital to counterbalance the uneven burdens of climate change, especially in safeguarding the livelihoods of the most vulnerable populations in developing countries.

As an international collective, we must collaborate to enhance our resilience and adaptive capabilities. A global research system will optimize efforts and bolster synergies between international and regional research frameworks and national systems in developing countries. Resilient knowledge must be seen as a universal asset. The gap between the 'haves' and 'have-nots' must be bridged to ensure peace, stability, and prosperity. A shift in consciousness among the 'haves and 'have-nots', leading to a new understanding of humanity's shared destiny, is crucial to promptly achieve our sustainable development goals (Fig. 2.17).

**Fig. 2.17.** The way forward; new understanding of the oneness of humanity. (From El-Beltagy, 2018)

## References

Adeel, Z., Safriel, U., Niemeijer, D., White, R., de Kalbermatten, G. *et al.* (2005) Dry lands. In: Sarukhan, J. and Whyte, A. (eds) *Ecosystems and Human Well-being: Desertification Synthesis*. Millennium Ecosystem Assessment. World Resources Institute. Washington, DC. Available at: https://www.millenniumassessment.org/en/Condition.html#download (accessed 25 June 2024).

Agnolucci, P., Rapti, C., Alexander, P., De Lipsis, V., Holland, R. *et al.* (2020) Impacts of rising temperatures and farm management practices on global yields of 18 crops. *Nature Food* 1, 562–571. DOI: 10.1038/s43016-020-00148-x

Agroecology Europe (2017) Our understanding of agroecology. Available at: http://www.agroecology-europe.org/our-approach/our-understanding-of-agroecology/ (accessed 25 July 2024).

Altieri, M.A. and Toledo, V.M. (2011) The agroecological revolution in Latin America: rescuing nature, ensuring food sovereignty and empowering peasants. *Journal of Peasant Studies* 38(3), 587–612.

Barange, M., Bahri, T., Beveridge, M.C.M., Cochrane, K.L., Funge-Smith, S. *et al.* (2018) *Impacts of Climate Change on Fisheries and Aquaculture: Synthesis of Current Knowledge, Adaptation and Mitigation Options*. FAO Fisheries and Aquaculture Technical Paper No. 627. FAO, Rome. Available at: https://openknowledge.fao.org/server/api/core/bitstreams/9aeb8ade-a623-4954-8adf-204daae3b5de/content (accessed 25 July 2024).

Bhattacharya, A., Kharas, H. and W. McArthur, J.W. (2023) *Keys to Climate Action. Chapter 1: Overview: How developing countries could drive global success and local prosperity*. Center for Sustainable Development, Brookings Institution. Working Paper 180.1. Available at: https://www.brookings.edu/articles/keys-to-climate-action-how-developing-countries-could-drive-global-success-and-local-prosperity/ (accessed 25 June 2024).

Burrell, A.L., Evans, J.P. and De Kauwe, M.G. (2020) Anthropogenic climate change has driven over a million $km^2$ of drylands towards desertification. *Nature Communications* 11, 3853. DOI: 10.1038/s41467-020-17710-7

Christian Aid (2008) *Christian Aid Annual Report 2007-08*. Available at: https://www.christianaid.org.uk/sites/default/files/2022-08/christian-aid-annual-report-2007-2008-september-2008.pdf (accessed 11 September 2024).

Dittmer, K.M., Rose, S., Snapp, S.S., Kebede, Y., Brickman, S. *et al.* (2023) Agroecology can promote climate change adaptation outcomes without compromising yield in smallholder systems. *Environmental Management* 72, 333–342. DOI: 10.1007/s00267-023-01816-x

Dregne, H.E. (1987) Desert expansion: monitoring and assessment. In: Bishay, A. and Dregne, D. (eds) Proceedings of the 2nd International Desert Development Conference. Part 1: Desert Agriculture and Ecology, Cairo. Harwood Academic Publishers, London, Paris, New York.

EEA (European Environment Agency) (2023) Towards 'just resilience': leaving no one behind when adapting to climate change. Available at: https://www.eea.europa.eu/publications/just-resilience-leaving-no-one-behind/towards-just-resilience-leaving-no (accessed 18 September 2023).

El-Beltagy, A. (1999) Can desertification trends be reversed in West Asia and North Africa? In: *New Technologies to Combat Desertification: Proceedings of an International Symposium*. UNU/Ministry of Agriculture, Tehran, pp. 65–78. Available at: https://archive.unu.edu/env/workshops/iran-1/index.htm (accessed 25 June 2024).

El-Beltagy, A. (2002) Reversing desertification: the potential for research. John, R. (ed.) *Proceedings of 'Desert and Dryland Development: Challenges and Potential in the New Millennium'*. International Dryland Development Commission. ICARDA. ISBN: 92-9127-130-0. www.drylanddevelop.org (accessed 25 June 2024).

El-Beltagy, A. (2010) Unpublished presentation for the IDDC.

El-Beltagy, A. (2017) Achieving Sustainable Development Goals (SDGs) in dry areas – a myth or a real possibility. *Proceedings of the Twelfth International Dryland Development Conference*, 21–24 August, The Bibliotheca Alexandrina, Alexandria, Egypt, pp. 15–28.

El-Beltagy, A. (2019) Navigating through uncertainties: agro-ecosystems affected by dynamic impact of climate change. In: El-Beltagy, A., Saxena, M.C., Yadav, O.P. and Panwar, N.R. (eds) *Converting Dryland Areas from Grey into Green*. Proceedings of 13th International Dryland Development Conference, 11–14 February, Jodhpur, India. International Dryland Development Commission.

El-Beltagy, A. and Rafea, A. (2021) Modernization of Egyptian agriculture. In: El-Beltagy (ed.) *Sustainable Agricultural Development 4.0: Smart Agriculture/Precision Agriculture/E.Agriculture*. Specialized Scientific Councils Academy of Scientific Research and Technology, Egypt.

El-Beltagy, A., Valkoun, J., Amri, A., Thomas, R. and Shideed, K. (2005) Biodiversity, land degradation and poverty alleviation in dry and sub-humid lands. In: *Success in Implementation of the Programme of Work in Dry Lands and the Global Taxonomy Initiative*. 11th Meeting of the Subsidary Body on Scientific, Technical and Technological Advice of the Convention on Biological Diversity. Technical Series no. 21. Secretariat of the Convention on the Biological Diversity, Montreal, Canada.

FAO Agroecology Knowledge Hub (2023). Overview: What Is Agroecology? Available at: https://www.fao.org/agroecology/overview/en/ (accessed 18 September 2023).

Francis, C., Lieblein, G., Gliessman, S., Breland, T.A., Creamer, N. *et al.* (2003) Agroecology: the ecology of food systems. *Journal of Sustainable Agriculture* 22(3), 99–118. DOI: 10.1300/J064v22n03_10

Glasser, H. (2008) Hans van Ginkel: On the vision, history and status of the regional centres of expertise in ESD programme. *Journal of Education for Sustainable Development* 2(2), 109–117. DOI: 10.1177/097340820800200208

Gliessman, S.R. (1990) *Agroecology: Researching the Basis for Sustainable Agriculture*. Springer, New York.

Gliessman, S.R. (2007) *Agroecology: The Ecology of Sustainable Food Systems*, 2nd edn. CRC Press, Boca Raton, Florida.

Gliessman, S.R. (2018) Defining agroecology. *Agroecology and Sustainable Food Systems* 42(6), 599–600.

Gonzalez de Molina, M. (2013) Agroecology and politics. How to get sustainability? About the necessity for apolitical agroecology. *Agroecology and Sustainable Food Systems* 37(1), 45–59.

Gross, J.E., Watson, J., Welling, L.A. and Woodley, S. (2016) *Adapting to Climate Change: Guidance for Protected Area Managers and Planners: Best Practice*. Protected Area Guidelines Series No. 24. IUCN, Gland, Switzerland. DOI: 10.2305/IUCN.CH.2017.PAG.24.en

Hillel, D. and Rosenzweig, C. (2010) Introduction: climate change and agroecosystems: key issues. In: Hillel, D. and Rosenzweig, C. (eds) *Handbook of Climate Change and Agroecosystems: Impacts, Adaptation, and Mitigation*, vol. 1. Imperial College Press, London. DOI: 10.1142/9781848166561_0001

HLPE (High-Level Panel of Experts on Food Security and Nutrition of the Committee) (2019) Agroecological and other innovative approaches for sustainable agriculture and food systems that enhance food security and nutrition. Committee on World Food Security. Available at: https://www.fao.org/family-farming/detail/en/c/1200043/ (accessed 26 June 2024).

IPCC (Intergovernmental Panel on Climate Change) (2018) *Global Warming of 1.5°C. An IPCC Special Report on the impacts of global warming of 1.5°C above pre-industrial levels and related global greenhouse gas emission pathways, in the context of strengthening the global response to the threat of climate change, sustainable development, and efforts to eradicate poverty* (eds: Masson-Delmotte, V., Zhai, P., Pörtner, H.-O., Roberts, D., Skea, J., *et al.*). Cambridge University Press, Cambridge, UK and New York, DOI: 10.1017/9781009157940

IPCC (Intergovernmental Panel on Climate Change) (2021) Future global climate: scenario-based projections and near-term information (lead authors: Lee, J.-Y. and Marotzke, J.). In: *Climate Change 2021:*

*The Physical Science Basis*. Contribution of Working Group I to the Sixth Assessment Report of the Intergovernmental Panel on Climate Change. Cambridge University Press, Cambridge, UK and New York, pp. 553–672. DOI: 10.1017/9781009157896.006

IPCC (Intergovernmental Panel on Climate Change) (2022) *Climate Change 2022: Impacts, Adaptation, and Vulnerability*. Contribution of Working Group II to the Sixth Assessment Report of the Intergovernmental Panel on Climate Change. Cambridge University Press, Cambridge, UK and New York. DOI: 10.1017/9781009325844

Jansson, J.K. and Wu, R. (2023) Soil viral diversity, ecology and climate change. *Nature Reviews Microbiology* 21, 296–311. DOI: 10.1038/s41579-022-00811-z

Karavolias, N.G., Horner, W., Abugu, M. and Evanega, S. (2021) Application of gene editing for climate change in agriculture. *Frontiers in Sustainable Food Systems* 5. DOI: 10.3389/fsufs.2021.685801

Kaur, A., Scarborough, P. and Rayner, M. (2017) A systematic review, and meta-analyses, of the impact of health-related claims on dietary choices. *International Journal of Behavioral Nutrition and Physical Activity* 14(1), 93. DOI: 10.1186/s12966-017-0548-1

Kefford, B.J., Nicholas, S. and Duncan, R. (2023) The cumulative impacts of anthropogenic stressors vary markedly along environmental gradients. *Global Change Biology* 29, 590–602. DOI: 10.1111/gcb.16435

Khangura, R., Ferris, D., Wagg, C. and Bowyer, J. (2023) Regenerative agriculture: a literature review on the practices and mechanisms used to improve soil health. *Sustainability* 15(3), 1–41. Available at: https://ideas.repec.org/a/gam/jsusta/v15y2023i3p2338-d1048464.html (accessed 26 June 2024).

Kremen, C. (2020) Ecological intensification and diversification approaches to maintain biodiversity, ecosystem services and food production in a changing world. *Emerging Topics in Life Sciences* 4(2), 229–240. DOI: 10.1042/ETLS20190205

Kumar, J., Kumar, A., Sen Gupta, D., Kumar, S. and DePauw, R.M. (2022) Reverse genetic approaches for breeding nutrient-rich and climate-resilient cereal and food legume crops. *Heredity* 128, 473–496. DOI: 10.1038/s41437-022-00513-5

Lager, F., Coninx, I., Beril, M., Bakhtaoui, I., Pedersen, A.B. *et al.* (2023) Just resilience for Europe: towards measuring justice in climate change adaptation. *ETC CA*. DOI: 10.25424/CMCC-BATP-3M95

Lamba, A., Teo, H.C., Sreekar, R., Zeng, Y., Carrasco, L.R. *et al.* (2023) Climate co-benefits of tiger conservation. *Nature Ecology & Evolution* 7, 1104–1113. DOI: 10.1038/s41559-023-02069-x

Laurence, T.J., Vilbig, J., Kangogo, G., Fèvre, E.M, Deem, S. *et al.* (2023) Shifting climate zones and expanding tropical and arid climate regions across Kenya (1980–2020). *Regional Environmental Change* 23, 59. DOI: 10.1007/s10113-023-02055-w

Lickley, M. and Solomon, S. (2018) Drivers, timing and some impacts of global aridity change. *Environment Research Letters* 13(10). DOI: 10.1088/1748-9326/aae013.

Lin, Y., Liu, A., Ma, E. and Zhang, F. (2013) Impact of future climate change on shifting patterns of agro-ecological zones in China. *Advances in Meteorology* 2013. Article ID 163248. DOI: 10.1155/2013/163248

Lipper, L., Thornton, P., Campbell, B.M., Baedeker, T., Braimoh, A. *et al.* (2014) Climate-smart agriculture for food security. *Nature Climate Change* 4(12), 1068–1072. Available at: https://www.nature.com/nclimate/journal/v4/n12/full/nclimate2437.html (accessed 26 June 2024).

Lori, M., Piton, G., Symanczik, S., Legay, N., Brussaard, L. *et al.* (2020) Compared to conventional, ecological intensive management promotes beneficial proteolytic soil microbial communities for agro-ecosystem functioning under climate change-induced rain regimes. *Scientific Reports* 10, 7296. DOI: 10.1038/s41598-020-64279-8

Lyu, X., Peng, W., Niu, S., Qu, Y. and Xin, Z. (2022) Evaluation of sustainable intensification of cultivated land use according to farming households' livelihood types. *Ecological Indicators* 138. DOI: 10.1016/j.ecolind.2022.108848

McNutt, M. (2013) Climate change impact. *Science* 341(6145). DOI: 10.1126/science.124356

Mirzabaev, A., Wu, J., Evans, J., García-Oliva, F., Hussein, I. *et al.* (2019) Desertification. In: Shukla, P., Skea, J., Calvo Buendia, E., Masson-Delmotte, V., Pörtner, H.-O. *et al.* (eds) *Climate Change and Land: An IPCC Special Report on Climate Change, Desertification, Land Degradation, Sustainable Land Management, Food Security, and Greenhouse Gas Fluxes in Terrestrial Ecosystems*. DOI: 10.1017/9781009157988.005

Moser, S.C. and Ekstrom, J.A. (2010) A framework to diagnose barriers to climate change adaptation. *Proceedings of the National Academy Sciences USA* 107, 22026–22031. DOI: 10.1073/pnas.1007887107

Nainggolan, D., Abay, A.T., Christensen, J.H. and Termansen, M. (2023) The impact of climate change on crop mix shift in the Nordic region. *Scientific Reports* 13. DOI: 10.1038/s41598-023-29249-w

Nyéléni (2015) International Forum for Agroecology. Nyéléni Center, Sélingué, Mali. 24–27 February. Available at: https://www.foodsovereignty.org/wp-content/uploads/2023/02/NYELENI-2015-ENGLISH-FINAL-WEB.pdf (accessed 25 July 2024).

OECD (2009) *Integrating Climate Change Adaptation into Development Co-operation: Policy Guidance*. OECD Publishing, Paris. ISBN: 9789264054769. DOI: 10.1787/9789264054950-en

OECD (2016) *States of Fragility 2016: Understanding Violence*. OECD Publishing, Paris. DOI: 10.1787/9789264267213-en

Rafea, A.A. (2002) Expert systems for agriculture development. In: John, R. (ed.) *Proceedings for Desert and Dryland Development: Challenges and Potential in the New Millennium*. Organized by the International Dryland Development Commission. ICARDA. ISBN: 92-9127-130-0.

Rafea, A.A. and El-Beltagy, A.S. (1987) Expert system as an appropriate computer technique to speed up agricultural desert development in Egypt. In: Bishay, A. and Dregne, H. (eds) *Proceedings of the International Desert Development Conference, Part 1: Desert agriculture, Ecology*. Harwood Academic Publishers (London, Paris, New York).

Randall, D.A., Wood, R.A, Boney, S., Colman, R., Fichefet, T. *et al.* (2007) Climate models and their evaluation. In: *Climate Change 2007: The Physical Science Basis*. Contribution of Working Group I to the Fourth Assessment Report of the Intergovernmental Panel on Climate Change. Cambridge University Press, Cambridge, UK and New York.

Rosset, P.M., Sosa, B.M., Jaime, A.M.R. and Lozano, D.R.A. (2011) The Campesino-to-Campesino agroecology movement of ANAP in Cuba: social process methodology in the construction of sustainable peasant agriculture and food sovereignty. *The Journal of Peasant Studies* 38(1), 161–191. DOI:10.1080/03066150.2010.538584

Rudel, T.S.K. (2020) The variable path to sustainable intensification in agriculture. *Regional Environmental Change* 20, 126. DOI: 10.1007/s10113-020-01720-8

Stern, N. (2007) *The Economics of Climate Change: The Stern Review*. Cambridge University Press, UK. DOI: 10.1017/CBO9780511817434

STS Forum (2018) Annual meeting 2018 I NPO STS forum. Available at: https://www.stsforum.org/kyoto2018/ (accessed 25 July 2024).

Talaviya, T., Shah, D., Patel, N., Yagnik, H. and Shah, M. (2020) Implementation of artificial intelligence in agriculture for optimization of irrigation and application of pesticides and herbicides. *Artificial Intelligence in Agriculture* 4, 58–73. DOI: https://doi.org/10.1016/j.aiia.2020.04.002

Toledo, V.M. and Barrera-Bassols, N. (2017) Political agroecology in Mexico: a path toward sustainability. *Sustainability* 9(2), 268. DOI:10.3390/su9020268

UNDRR (United Nations Office for Disaster Risk Reduction) (2015) *Sendai Framework for Disaster Risk Reduction 2015–2030*. DOI: 10.1163/2210-7975_hrd-9813-2015016

UN (2015) Summit charts new era of sustainable development. UN News, 24 September. Available at: https://www.un.org/sustainabledevelopment/blog/2015/09/summit-charts-new-era-of-sustainable-development-world-leaders-to-gavel-universal-agenda-to-transform-our-world-for-people-and-planet/ (accessed 25 July 2024).

UNFCC (United Nations Framework Convention on Climate Change) (2011) Cancún Climate Change Conference, November 2010. Available at: https://unfccc.int/conference/cancun-climate-change-conference-november-2010 (accessed 25 July 2024).

Van Ginkel, H., El-Beltagy, A. and Sohl, M. (2007) Living with deserts: CWANA-plus partnership, human and nature – working together for sustainable development of drylands. In: El-Beltagy, A., Saxena, C. and Wang, T. (eds) *Proceedings of the Eighth International Conference on Development of Drylands*. Beijing.

Volk, G. and Richard, C. (2011) Integration of georeferencing, habitat, sampling, and genetic data for documentation of wild plant genetic resources. *HortScience* 46(11), 1446–1449. DOI: 10.21273/HORTSCI.46.11.1446

Von Jeetze, P.J., Weindl, I., Johnson, J.A., Borrelli, P., Panagos, P. *et al.* (2023) Projected landscape-scale repercussions of global action for climate and biodiversity protection. *Nature Communications* 14, 2515. DOI: 10.1038/s41467-023-38043-1

Wang, M. and Lei, G. (2023) Relative and cumulative effects of climate and land use change on hydrological ecosystem services in Northeast China. *Land* 12(7), 1298. DOI: 10.3390/land12071298

Wezel, A., Casagrande, M., Celette, F., Vian, J.F., Ferrer, A. *et al.* (2014) Agroecological practices for sustainable agriculture: a review. *Agronomy for Sustainable Development* 34(1), 1–20.

World Risk Report (2022) Bündnis Entwicklung Hilft, Ruhr University Bochum – Institute for International Law of Peace and Conflict. Available at: https://weltrisikobericht.de/weltrisikobericht-2022-e (accessed 26 June 2024).

Zhao, J., Lui, D. and Huang, R. (2023) A review of climate-smart agriculture: recent advancements, challenges, and future directions. *Sustainability* 15, 3404. DOI: 10.3390/su15043404

Zhu, P., Burney, J., Chang, J., Jin, Z., Mueller, N.D. *et al.* (2022) Warming reduces global agricultural production by decreasing cropping frequency and yields. *Natural Climate Change* 12, 1016–1023. DOI: 10.1038/s41558-022-01492-5

# 3 Carbon Farming in Global Drylands

**Rattan Lal***
*Rattan Lal Center for Carbon Sequestration and Management,*
*The Ohio State University, USA*

**Abstract**

Global drylands, the largest biome among terrestrial ecoregions, covering about 40% of the Earth's land area, are expanding because of climate change and may cover 50% of the Earth's surface by 2100. Dryland agriculture is vulnerable to extreme events driven by anthropogenic climate change. Global drylands are a large reservoir of carbon in soil and vegetation and impact the global carbon cycle. Soils of drylands contain relatively more soil inorganic carbon (SIC) than soil organic carbon (SOC). The SIC comprises of primary or lithogenic carbonates and secondary carbonates formed through biotic and abiotic processes. Drylands may be a net carbon sink through nocturnal sequestration by abiotic processes. Upscaling of recommended land use and management practices can enhance terrestrial C sequestration and render agriculture a solution to environmental issues. Carbon farming can create another income stream for farmers through payments for ecosystem services and advance Sustainable Development Goals.

## Global Drylands: Area, Extent, Population

Food demands of the growing and increasingly affluent population may necessitate additional food-grain production by 1 billion Mg and meat production by 200 million Mg (Stewart and Lal, 2018), for which global drylands (GDLs) can play an important role.

Estimates of GDL, as a percentage of the Earth's surface area, vary widely and range from 40% (Smith *et al.*, 2019) to 41% (Prăvălie, 2016; Yao *et al.*, 2020) to 43% (Squires, 1998) to 45.36% (Kou *et al.*, 2018; Lal, 2019) to 47% (Inanaga *et al.*, 2005) or about 60–66.7 M $km^2$. GDL supports 38% of the global human population (García-Palacios *et al.*, 2019) and half of the world's livestock (Galvin, 2021), even with harsh climate and poor soils for agriculture. Africa and Asia each have 23 M $km^2$ or 15% of the global land area under drylands (Prăvălie, 2016). Drylands in Central Asia cover 5.2 M $km^2$ and represent 80% of global total temperate desert area (Zhu *et al.*, 2019). In general, GDL is highly vulnerable to climate change and environmental perturbations. Such ecological disturbances are affecting 46 countries in Africa and 38 in Asia and must be addressed to minimize adverse effects on human well-being and nature conservancy. Drastic environment change can aggravate human migration and lead to political unrest (Neumann *et al.*, 2015).

*Address for correspondence: Lal.1@osu.edu


DOI: 10.1079/9781800624870.0003

Because of their total area and large carbon (C) stocks, GDLs are the largest sources of inter-annual variability in the global C sink, and GDLs may play a major role in key aspects of Earth's C cycle (Choi *et al.*, 2022). With large land area and large C stock in soil (both organic and inorganic), GDLs can be an important factor in moderating future climate change by being a source or sink of atmospheric $CO_2$ and affecting the global C cycle (GCC) (Lal, 2019).

With adoption of restorative land use and site-specific recommended management practices (RMPs), GDLs can be a sizeable sink for atmospheric $CO_2$ due to their large area and degraded soils (Robertson *et al.*, 2018). If deserted and degraded GDLs are restored, they have a potential to sequester atmospheric $CO_2$ by 1.0 Pg C/yr (Squires, 1998; Lal, 2000, 2003). Yet, scientific understanding of the controls of GDL–$CO_2$ exchange is poor. Wertin *et al.* (2018) observed that abiotic factors (i.e. soil moisture regime) strongly moderate the response of soil respiration to biotic factors and soil C and N pools in dryland communities, and especially in dry years. In addition to climate factors, future food needs in GDLs are also likely to be affected by reduced natural and social capitals, food-energy-water insecurities, resource-driven political instability, and increased risks of soil degradation. Thus, compared to the historic successes of the Green Revolution strategies implemented under favorable soil and climate regions, future food security must be based on sustainable management of natural resources of GDL.

Therefore, the objective of this chapter is to describe the potential and challenges of sustainable management of GDL with specific focus on soil carbon sequestration and carbon farming so that farmers can avail another income stream while advancing SDGs of the United Nations. This chapter also collates and synthesizes information on processes and factors which affect soil C stocks in GDL and considers the potential and challenges of upscaling proven RMPs, which may lead to re-carbonization of GDL.

## Carbon Farming

Carbon footprint (CFP) is the fastest-growing part of humanity's overall footprint. Thus, protection, restoration, and sustainable management of soil health of agro-ecosystems can reduce CFP and management of GDL can be of critical importance. Soil health refers to:

> ...soil's capacity, as a dynamic and biologically active entity, within natural and managed landscapes, to sustain multiple ecosystem services including net primary productivity, food and nutritional security, biodiversity, water purification and renewability, carbon sequestration, air quality and atmospheric chemistry and elemental cycling for human wellbeing and nature conservancy.
>
> (Lal, 2016)

Sustainable management of soil health is intricately interconnected with the emerging concept of carbon farming.

Carbon farming (CF) implies management of carbon pools, flows, and greenhouse gas (GHG) fluxes at farm level as a strategy to make agriculture environmentally friendly and transform it from being a problem to a solution while also creating another income stream for land managers. CF specifically refers to growing C on land (soil and vegetation) as a farm commodity that generates income for farmers as farm produce. In this context, CF includes management of all GHG emitted from farm operations including $CO_2$, $CH_4$ and $N_2O$. The private sector (especially the agro-industry) can play an important role in commodification of terrestrial C grown on farmland. The private sector, in cooperation with scientists and policymakers, can promote C farming by upscaling nature-positive agriculture, increasing access to farm inputs, and enhancing investments in agricultural research and development.

## Climate Change and Global Drylands

GDLs are characterized by low annual rainfall of approximately 300–500 mm, most of which is received during spring or early summer. The rainfall is both inadequate and irregular and a major challenge to sustained agriculture (Inanaga *et al.*, 2005). In addition to harsh environments, GDLs are also vulnerable to anthropogenic climate change (ACC; Fig. 3.1), and the environmental conditions may get harsher with ACC. It is estimated that surface area of GDLs

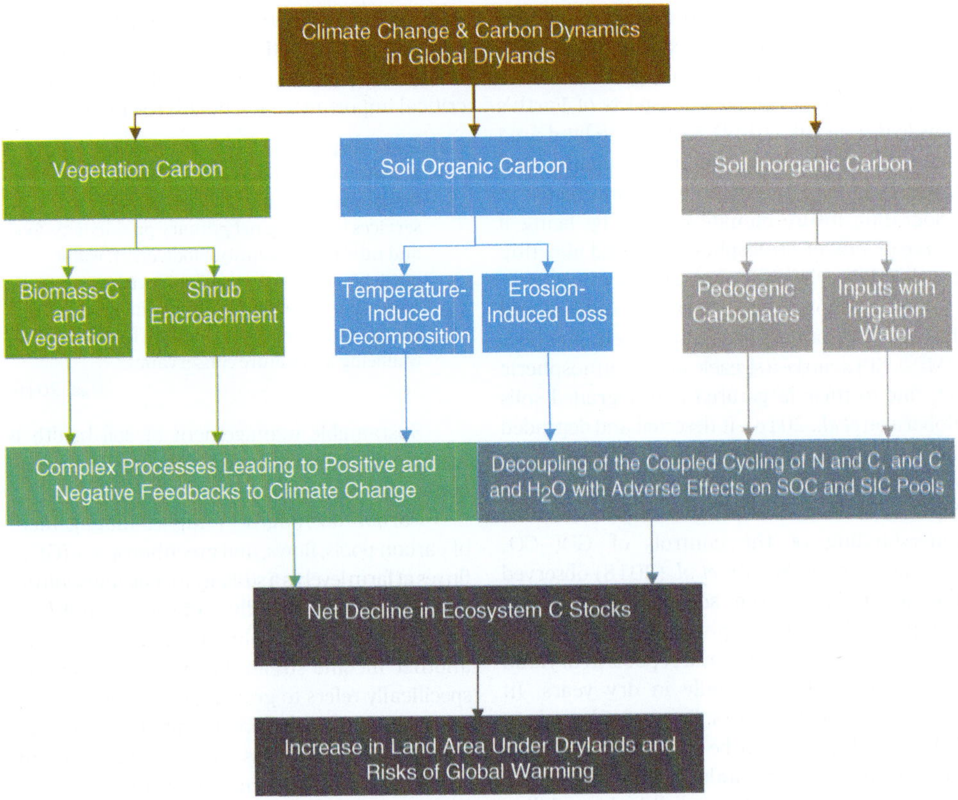


**Fig. 3.1.** Vulnerability of global drylands to anthropogenic climate change and positive feedback to global warming.

has expanded over the 60 years from 1948 to 2008 and the projected area may increase by 10% by *c.*2100 compared with the baseline of 1961–1990 (Feng and Fu, 2013). Yao *et al.* (2020) hypothesized that by 2100 gross primary production (GPP) of GDLs will increase by 12% ± 3% relative to the 2000–2014 baseline. Furthermore, ACC is expected to cause progressively increased frequency and severity of extreme events with adverse effects on crop growth and agronomic yields (Li *et al.*, 2020). ACC may also lead to loss of soil organic carbon (SOC) from the terrestrial biomes. In tropical drylands of Mexico, Abbruzzini *et al.* (2021) observed that increasing aridity with global warming would disrupt the biogeochemical cycles and homogenize soil biodiversity. The latter greatly increases the vulnerability of agricultural ecosystems to drought with severe negative consequences to rainfed agriculture in drylands of Mexico. In Kalahari savannahs, Dintwe and Okin (2018) observed that the SOC loss due to anthropogenic climate change may be as much as 1.1 Tg C/yr. Upscaling these data to global savannahs indicated the SOC loss of 28.4 Tg C/yr or even more. These losses may lead to positive-feedback mechanisms between climate change and processes moderating SOC.

In the Sonora Desert and adjacent regions of California, Hantson *et al.* (2021) reported a large decrease in vegetation cover between 1984 and 2017. Hantson and colleagues concluded that ACC has already significantly impacted these drylands and there are growing risks for dryland ecosystems in other regions. Extreme heatwaves can also bring about alterations in vegetation patterns. For example, the heatwave in Australia during the summer of 2019 strongly affected the vegetation. Qiu *et al.* (2020) observed that vegetation growth was reduced and that the heatwave had different effects on the vegetation during the morning and the afternoon. Dintwe and Okin (2018) used the Century model to assess the impact of ACC on soil C dynamics of the Kalchan Desert at 1.1 Tg C/yr to 2.0 Tg C/yr until the end of the century. Upscaling these results to global savannah regions, SOC loss may range from 28.4 Tg C/yr to 64.1 Tg C/yr, leading to positive feedback to the already accelerating ACC. Effects of ACC to feedback in Australian dryland was studied by Martins *et al.* (2016). They observed that dryland soils respond to ACC with an offset under interactive climate treatments. Thus, future studies must consider both biotic and abiotic soil characteristics for the development of future mitigation studies.

In Central Asia, the GDL area of 5.6 M $km^2$ represents 80–90% of the world's temperate deserts. Li *et al.* (2015) reported that C stock was 31.3–34.2 Pg in the top 1m of soil and another 10.4–11.4 Pg in 1–3 m soil of the temperate deserts. This stock represents 18–24% of the global C stock in deserts and dry shrublands. But 90% of Central Asia's C stock was stored in soil which has a relatively high SOC density compared to the hot deserts of the world, and such a high SOC density is vulnerable to ACC. Between 1979 and 2011, for example, the dryland of Central Asia lost approximately 0.46 Pg C, and the largest losses were found in Kazakhstan where annual precipitation declined at a rate of 90 mm/decade.

The predicted increase in aridization during the 21st century may threaten the balance between coupled cycles of different elements ($H_2O$, C, N, P, S) and adversely affect the availability of essential elements. Delgado-Baquerizo *et al.* (2013) reported negative effects of aridity on the concentration of SOC and N but positive effects on the concentration of P. Because of its negative effect on plant cover, ACC may aggravate physical processes (i.e. rock weathering) and reduce the concentration of N and C in GDL, but increase that of P leading to uncoupling of the cycles of C, N and P with the attendant negative effects on essential ecosystem services or ESs (Delgado-Baquerizo *et al.*, 2013).

Despite sparse vegetation but large land, GDL has a large influence on global biogeochemical cycles and dominates the long-term trends in the global C cycle (GCC). Thus, reliable estimates of global land C sink depend on accurate representation of dryland C cycle processes (MacBean *et al.*, 2021). Indeed, the strategy of CF in GDL must account for the adverse effects of historic and future global warming on the C stocks and fluxes in GDL and consider the effects of anthropogenic perturbations on the decoupling of the cycles of different elements. The biogeochemical cycles of C, N, P and sulfur (S) are closely linked with that of water ($H_2O$). However, coupled cycles could get uncoupled due to rapid ACC, which are specifically relevant in GDLs because water limitation is the critical factor (Delgado-Baquerizo *et al.*, 2013).

In this context, understanding of the biogeochemical cycling of C in GDL under diverse land use and soil/crop/livestock management is of critical importance.

## Biogeochemical cycling of carbon in drylands and effects of land use and land-use change

Dryland ecosystems, being fragile and under harsh conditions, are sensitive to small changes in resource availability because most resources are at low levels (Choi *et al.*, 2022). Thus, in some cases, conversion of natural drylands (deserts and shrublands), which have low SOC stocks, to managed croplands, can enhance SOC stocks. Innovative strategies to enhance SOC stock in GDLs are outlined in Fig. 3.2 and discussed below. A meta-analysis by Wang *et al.* (2023) indicated that SOC stocks were increased to 1 m depth in managed croplands by 278.9%. However, SOC stocks to 1 m depth decreased by 24.1% and 10.7% in clearing forestland and grassland for cropland, respectively. A study in Sonora Desert of Mexico by Castellanos *et al.* (2022) indicated that annual net ecosystem productivity (NEP) in the shrubland was a C-source in drier years and a sink in others. Castellanos and colleagues observed that discrepancies in annual water use, and C-use efficiencies, indicated the importance of ecohydrological feedback.

In the rice-wheat system of India, Dutta *et al.* (2022) reported that soil C sequestration rate ranged between 0.81 and 1.23 Mg C/ha yr in treatments receiving mineral fertilizers only. Similarly, Choi *et al.* (2022) observed that N addition alone results in no changes in soil C cycling, but when N was applied in conjunction

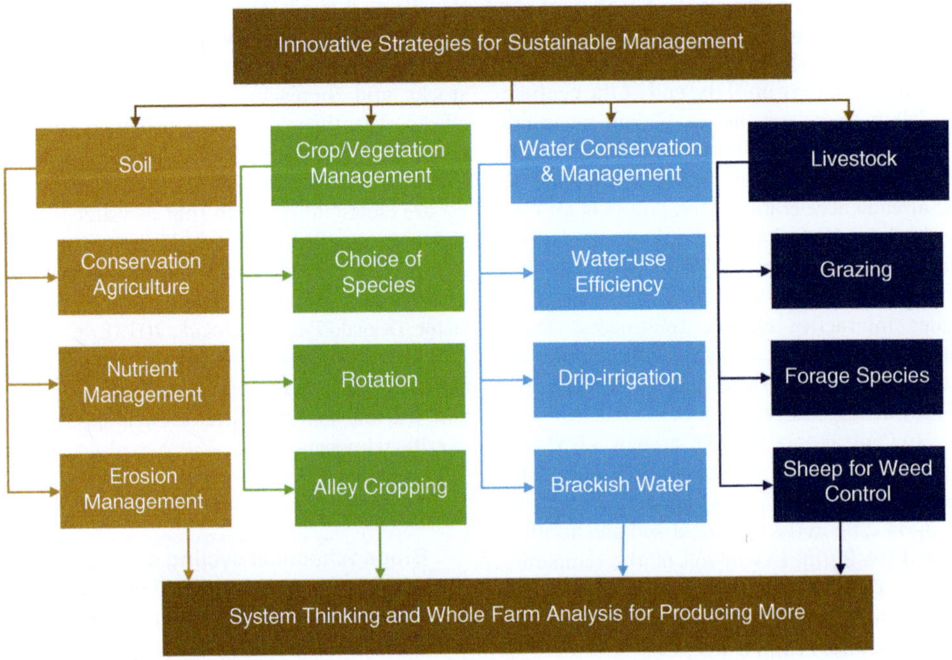


**Fig. 3.2.** Innovative strategies of enhancing eco-efficiency under harsh conditions of multiple resource limitation.

with water and C, N greatly increased soil C cycling rates relative to addition of water and C without N. These trends supported the Transient Maxims Hypothesis of ecology that multiple resource limitation explains ecosystem response better than water limitation alone (Choi *et al.*, 2022).

Enhancing SOM content is a critical adaptation strategy in GDL through improvement in water retention (Lal, 2020) and nutrient cycling. However, land managers face the so-called 'soil carbon dilemma' (Janzen, 2006): using it or storing it? Van der Pol *et al.* (2022) concluded that including legumes in rotation could allow for 10% greater SOC gains over time compared to wheat-fallow system. Because legumes also increase crop yield, Van der Pol and colleagues concluded that growers could solve the 'soil C dilemma' by simultaneously increasing the quantity of C and N contents of soil input through continuous cropping with both grains and legumes.

Root-derived C, rather than the root biomass, has a significant impact on SOC sequestration and dynamics. Yang *et al.* (2021) used the $^{13}C$ method to quantify net root C input in a study conducted in the Loess Plateau of China. They observed that root-derived C in soil represented a considerable contribution to below-ground net primary production (57–81%), and that to SOC was 3% or similar to 5% during one-year growth. Yang *et al.* concluded that fencing increases the contribution of root-derived C to SOC mainly by increasing root C content.

A two-year study using $^{13}C$-labelled maize straw conducted by Mo *et al.* (2021) in the Loess Plateau region of China showed that 49–56% of straw $^{13}C$ was sequestered as $CO_2$-C, 35–43% was sequestered as SOC pool, and 7–10% remained undecomposed at the end of the two-year study period. Mo and colleagues also observed that plastic mulching significantly decreased the straw-derived $CO_2$-C emissions by 15% due to increased incorporation of straw-C into the SOC pool. Mo and colleagues concluded that crop straw return combined with plastic mulching could enhance SOC sequestration in dryland cropping systems. Dacal *et al.* (2020) evaluated SOC losses through respiration by experimental warming in a dryland ecosystem and observed that warming-induced increases in soil temperature were the main drivers of the short-term soil-respiration responses. Based on a long-term study on CA in dryland farming systems in eastern Washington state, Wachter *et al.* (2019) observed that profile SOC showed significant negative trends in conventional tillage (–3.1 Mg C/ha yr) and in mixed crop-livestock (–4.1 Mg C/ha yr) but not in organic mixed crop-livestock and in organic hay. Wachter and colleagues recommended that integrating perennial crops (e.g. alfalfa and forage grasses, into organic farming systems) can build soil quality and supply nutrients and water to succeeding crops.

Woody plant encroachment into grasslands can have strong adverse impacts on soil C stock. Puttock *et al.* (2014) observed that woody encroachment leads to changes in both the structure and function of dryland ecosystems, and accelerated soil erosion and the attendant loss of plant nutrients. Yet, scientific information about the relationship between erosion and soil C dynamics is scanty and uncertain. Based on results from a study in the southwestern USA, Puttock and colleagues showed that woody encroachment aggravated loss of SOC via increase in erosion and by destabilizing the legacy C under grass cover.

These and other studies indicate the importance of long-term research information under site-specific conditions on the soil C dynamics to understand the C budget in soils of GDL.

## Carbon Budget in Global Drylands

Total C stock in GDLs comprises SOC and SIC (Fig. 3.3), and in general, the stock of SIC may be more than that of SOC (Lal, 2019). The current and projected global warming may have a strong impact on soil C stock in GDLs, especially decline of the SOC stock may be aggravated with adverse impacts on soil functions and ESs. Despite strong impacts of SOC/SIC dynamics on soil health and functions, research information for diverse land use and management is scanty for GDLs. Whatever information exists, it is for the surface/ plow layer only. However, there is a strong need

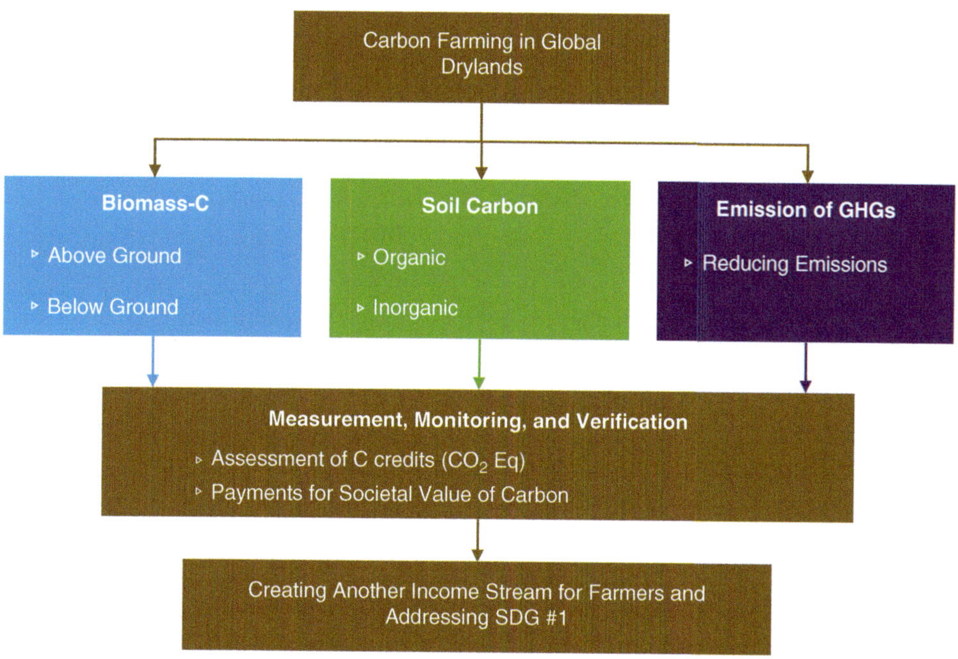


**Fig. 3.3.** Carbon farming in global drylands by managing biomass and soil carbon, and moderating gaseous emisisons. Author's own figure.

to understand SOC/SIC dynamics for sub-soil layers to at least 1-m depth and preferably for deeper layers. There is a wide range of innovative strategies for enhancing eco-efficiency under harsh conditions of multiple resource limitation (Fig. 3.3). Some available information on the effects of land use and management on soil C stock is discussed below.

In China's Loess Plateau, Song *et al.* (2020) evaluated C and N stocks to 5-m depth as affected by conversion of cropland to woodland. The concentration of SIC ranged from 12.9 g/kg to 20.2 g/kg for Fuefeng site and 0.18 g/kg to 0.36 g/kg for Yangshou site. Conversion of cropland to woodland increased SOC content to 5-m depth at both sites but the effect was variable on SIC content. The concentration of SIC and soil total C increased in Yangshou but deceased in Fuefeng. Based on a meta-analysis to identify global patterns of how warming impacts soil C dynamics, Yan *et al.* (2020) observed that warming reduced SOC by 4.96% and stimulated soil microbial biomass C, soil respiration, and heterotrophic respiration by 6.30%, 14.56%, and 8.42%, respectively. Yan, and colleagues concluded that warming induces accelerated transition of soils from C sink to C source and exacerbates the positive feedback loop in terrestrial ecosystems.

Pedogenic or secondary carbonates play an important role in soil C dynamics of GDLs (Lal *et al.*, 1999; Monger *et al.*, 2015). Nyachoti *et al.* (2019) determined accumulation rates of pedogenic carbonates in southwestern USA by using U series ($^{238}U$-$^{234}U$-$^{230}Th$) disequilibrium dating technique to investigate Ca sources, ages, and formation rate of pedogenic carbonates in irrigated southwestern USA. Nyachoti and colleagues observed the presence of much younger carbonates at the irrigated sites compared to natural sites because of influxes from dissolved Ca and inorganic C in irrigation water. This study highlights the importance of agriculture expansion and land-use change and irrigation in drylands to increase the rate of pedogenic carbonate accumulation.

Shrub encroachment in GDLs may also increase plant biomass and change SOC/SIC stocks. In Mongolian grasslands, Liu *et al.* (2020) assessed the impact of shrub encroachment on SIC dynamics and observed decline in SIC density with an increase in shrubs in the upper 100 cm (3.85 vs. 4.74 kg C/m$^2$). Liu and colleagues also observed that the lost SIC was primarily released into the atmosphere but not redistributed into the sub-soil layers. The SIC emission into the atmosphere was attributed to soil acidification by shrub encroachment. In Central Asia, Zhu *et al.* (2019) conducted a model study for the 1980–2014 period to assess C stocks of drylands in response to ACC. Over the study period, total stocks of C were 2.8 ± 0.05 Pg in vegetation, 45.2 ± 0.01 Pg as SOC stock, and 0.03 ± 0.004 Pg C in litter biomass. In the meanwhile, total C declined by 0.7 Pg. The ACC declined in total C by 1–3 Pg.

Soil C dynamics is also influenced by erosional processes (Lal, 2003, 2022). However, dynamics of SIC, which forms the predominant component of total soil C stock in GDLs, has not been widely studied. Wang (2019) studied selective transport of SOC and SIC as affected by different raindrop sizes and inflow rates. The data showed that SOC content was sensitive to the presence of fine particles while the SIC was sensitive to coarse particles. Wang and colleagues also observed that SOC in sediment tended to be depleted under intensive rainfall on steep slopes while the SIC was slightly enriched. However, both SOC and SIC were selectively transported by water erosion.

Jia *et al.* (2019) observed that afforestation of the Loess Plateau increased SOC content in all sites with accumulation rates of 0.09–0.34 g C/kg yr for different aggregates and particle size fractions. However, the response to SIC differed among sites and was negatively correlated with the change in SOC.

Deng *et al.* (2016) synthesized data from 103 publications including 160 sites in 29 countries and concluded that:

**i)** Land-use conversions reduced soil C stocks by 0.39 Mg/ha yr;
**ii)** SOC stocks increased after conversion from farmland to grassland by 0.30 Mg C/ha yr;
**iii)** SOC stocks increased by conversion of forest to grassland at 0.68 Mg C/ha yr;
**iv)** SOC stocks significantly decreased after conversion from grassland to farmland by 0.89 Mg C/ha yr;
**v)** SOC stocks decreased by 1.74 Mg C/ha yr by change of forest to farmland; and
**vi)** SOC stocks decreased by 0.63 Mg C/ha yr by change of forest to forestland/plantations.

Deng and colleagues observed that SOC did not change upon conversion from farmland to forest and grassland to forest.

Wang *et al.* (2016) studied SIC dynamics at different successional stages along a 250-year natural vegetation restoration chrono-sequence. The results showed that SIC storage in the top 10-cm layer decreased by 0.092 Mg C/ha yr along the vegetation restoration sequence. They concluded that variation in SOC induced by vegetation restoration was the main driving force for SIC changes in the topsoil.

Based on a study in the Yanqi Basin of China, Wang *et al.* (2015) observed that SOC had a significant relationship with SIC and pedogenic inorganic carbonates (PIC). Further, converting shrubland to cropland increased PIC stock by 5.2 kg C/m$^2$ which was 3.6 times that of the SIC stock. The data from a long-term study showed greater accumulation of PIC (21.49 g C/m$^2$ yr.) than SOC (10–39 g C/m$^2$ yr.) over 0–20 cm depth. Wang and colleagues concluded that intensive cropping in the arid and semi-arid regions leads to an increase in both SOC and PIC stocks.

Based on a study in eastern Central Asia from 1981, in Xinjiang, China, covering an area of 1.6 M km$^2$, Li *et al.* (2013) reported that this region acted as a C sink of 138 Tg because of the increased vegetation C due to the $CO_2$ fertilization effect estimated at 124 Tg C over this period. Whereas the temperature change resulted in 12 Tg of C loss, precipitation change resulted in 24 Tg C sequestration. Rising temperatures increased heterotrophic respiration causing 17 Tg of C loss from soil.

## Sand Dust Storms and Aeolian Processes

Sand dust storms (SDS) are common in many of GDLs including the Sakura, the Middle East, Central and Eastern Asia, and parts of Australia (Middleton, 2017). Such aeolian processes affect the biosphere in GDLs by altering landform evolution, biogeochemical cycles, regional climate, human health, and desertification (Ravi *et al.*, 2011). Ravi and colleagues observed that aeolian dust emissions are driven by ACC, soil properties, soil moisture, vegetation and roughness, and land use/management.

In addition to the natural factors, SDS are also caused by anthropogenic factors such as those related to overgrazing, deforestation, and misuse of water resources. Zucca *et al.* (2022) observed that SDS in the Thar Desert of India are caused by chronic and extensive historical land degradation (or desertification) under arid climatic conditions. Under some cases, SDS may also be caused by extreme weather conditions (drought, windstorms), soil properties (e.g. erodibility), and site-specific factors (e.g. bare soil after crop harvest). Thus, SDS may be considered indicators of increasing risks of desertification and land vulnerability to ACC (Zucca *et al.*, 2022). Bullard and McTainsh (2003) focused on the rise of aeolian–fluvial interaction in dryland regions, and suggested that landscape sensitivity, or the effectiveness of links between the process systems, spatial environmental transitions and temporal environmental change are the three primary driving forces which determine the geomorphological significance of aeolian/fluvial interactions.

## Livestock-based Systems

The global trend of segregating livestock from cropping systems, leading to shortening of crop rotations and simplification of farming systems, has increased food production but aggravated environmental costs. Thus, integration of livestock with crops and trees can increase diversification. Indeed, livestock play an important role in sustainable management of drylands at the farm, household, and territorial levels with specific reference to food security, well-being, value-chain performance, food supply and maintenance of common grazing resources (Alary *et al.*, 2022). However, inter-annual variation in rainfall patterns can strongly affect economics, forage production, livestock reproduction, and mortality. Thus, there is a need to identify risk management strategies related to climate variability (Espeland *et al.*, 2020).

Livestock-based systems also facilitate use of farmyard manure (FYM) as an integral component of an integrated nutrient management (INM) system. Bado *et al.* (2021) observed that the highest grain yield of millet was obtained with NPK fertilizer used in combination with FYM as INM with regard to risk management related to climate variability in rangeland-based livestock production. Espeland *et al.* (2020) outlined the following strategies:

**i)** dynamic management of forage supply by choice of the judicious land use;
**ii)** dynamic management of animal demand by herd manipulation; and
**iii)** diversification and use of social networks by combination of land use and herd manipulation.

Additional diversification within these areas would further reduce climate-related risks.

Use of cacti (*Opuntia* and *Nopalia*) as forage for livestock may be an advantageous strategy for GDLs. Dubeux Jr. *et al.* (2013) reported that potential dry mass production of 20 Mg/ha yr is achievable in drylands such as those in the Brazilian semi-arid regions. Different strategies for forage production from cacti include scattered planting in marginal lands (< 200 mm/yr rainfall), alley cropping with forage legumes or cash crops, intensive cactus production in rainfed systems, and drip irrigation with minimal amount of 2–5 mm/week (Dubeux Jr. *et al.*, 2013).

Sheep grazing is pertinent to GDL ecosystems with benefits of weed control. Barsotti *et al.* (2013) observed that sheep grazing on a cropping sequence containing fallow may not reduce net GHG emissions compared with herbicide use for weed control in continuous cropping.

One long-term experiment in northwest Syria by Larbi *et al.* (2009) indicated that integration of shrubs into the smallholder crop-rangeland-livestock farming systems in non-tropical dry areas could reduce risks of feed gaps, rangeland degradation and desertification, and sequester C. Pertinent shrubs included *Atriplex halimus*, *A. canescens*, *A. polycarpa*, *A. lentiformis*, *A. herba-alba*, and *A. nummularia*, which were found to have a greater potential for development of dryland agro-technologies. These shrubs would provide fodder, fuelwood, shade, and medicine; stabilize sand dunes and sequester C in soil and vegetation (Larbi *et al.*, 2009).

## Agroforestry and Forest-based Systems

Afforestation of marginal GDLs may strengthen several ESs, especially in temperate regions. However, there may be conflicts between additional forest area and water consumption with adverse effects on water yield. For example, Liu *et al.* (2022) argued that additional tree cover can amplify water consumption through increase in evapotranspiration but can also strengthen other ESs critical to sustainable development of drylands under ACC.

Rather than afforestation *per se*, Leakey *et al.* (2022) suggested agroforestry systems for Africa based on the production of indigenous crops for African agriculture. Leakey and colleagues advocated a holistic approach to mitigating ACC (and other issues including hunger, poverty, wildlife habitat, etc.). In addition to five types of natural capital (financial, natural, produced, human, and social), 'political will' may be the missing sixth capital especially in the context of sub-Saharan Africa (Leakey *et al.*, 2022). Alley cropping is a common form of agroforestry promoted by the International Institute of Tropical Agriculture (IITA), Ibadan, Nigeria. In the dryland region of Dodoma, Tanzania, Swamila *et al.* (2021) recommended the use of alley cropping for addressing issues of declining fertility and climate extremes in sub-Saharan Africa. Swamila and colleagues reported that inter-cropping of maize with *Gliricidia sepium* (Jacq.) exhibited a higher net return value (NPV) (T Sh. 19.3M) and benefit:cost ratio (4.27) than that with unfertilized sole maize. In the West African Sahel, Bado *et al.* (2021) suggested that no-till with *Ziziphus* (80 plants/ha) is an affordable and a sustainable agroforestry system for cultivation of millet.

## Conservation Agriculture

Mulch-based conservation agriculture (CA) in combination with integrated soil fertility management and complex crop rotations (Lal, 2015), can be a conservation and soil-enhancing system for dryland farming. Residue retention is a critical component of CA without which no-till *per se* and by itself may decline crop yield and degrade soil functions (Xiao *et al.*, 2021). In addition to retention of crop residue mulch, effectiveness of CA also depends on adopting the strategy of INM. Based on a study with pearl millet in northwestern India, Chandana *et al.* (2021) reported that replacing 25% of N with

FYM and other organic amendments increased $CO_2$ emissions by 9– 16% and $N_2O$ emissions by 48–56% compared to full recommended dose of fertilizer through neem-coated urea treatment. Chandana and colleagues also observed that agroforestry would reduce the overall GHG emissions. Based on a study in South Africa, Strauss *et al.* (2021) observed that the holistic principles of CA are applicable not only to farming of grains but also other commodities. A four-year study conducted in the West African Sahel by Bado *et al.* (2021) showed that no-till with and without *Ziziphus* decreased SOC by 15% and 21%, and K by 105% and 110%, compared with the original soil. However, soil N content was decreased only in systems without *Ziziphus*. Bado and colleagues concluded that including *Ziziphus* trees at the density of 80/ha improves productivity and farm income.

In the Shaanxi province of China, Yu *et al.* (2020) conducted a ten-year study and concluded that sub-soil tillage with plowing is a uitable agricultural tillage scenario for ameliorating bulk density, porosity, water-stable aggregation, water use and storage, and yields and profits that meet the variable climate of dryland farming ecosystems of the future.

In South Asia, Dixon *et al.* (2020) observed that CA-based sustainable intensification is an appropriate strategy for small landholder farmers practicing dryland farming. Dixon and colleagues proposed that CA-based sustainable intensification could also be adopted in the Middle East and Africa. Sainju (2020) observed that crop yields under CA in North America may be valuable in irrigated cropping systems, but agronomic yields are similar or greater with CA in dryland cropping systems. Thus, SOC content with CA in dryland farming is also similar to or greater than that under conventional systems. Whereas the emission of $N_2O$ may be more under CA, the overall GHG intensity can be lower under CA (Sainju, 2020).

Based on the synthesis of almost 20 years data from CA plots in the Chinese Loess Plateau, Kuhn *et al.* (2016) observed differences in total SOC stocks between CA and conventional till plots. While crop yields increased by 20% under CA, the topsoil achieved maximum SOC stocks after about ten years of CA.

In India, Pratibha *et al.* (2015) observed that adoption of CA (zero tillage) in pigeon pea and castor reduced GHG emission by 21–23% due to lower fuel consumption but with no reduction in crop yield. Based on a study in eastern Montana from 2006 to 2011, Sainju (2014) observed that surface residue C and SOC were related with grain yield. Sainju reported that the greater amounts of crop residue returned to the soil and turnover rate may have increased residue C, soil C storage, and crop yields.

In Washington State, Papendick and Parr (1997) observed that continuous CA (no-till) is the most effective and practical approach for restoring and improving soil quality, which is an essential prerequisite for both sustained food production and healthy environment. Papendick and Parr concluded that continuous use of CA enhances SOC content, improves soil structure, controls soil erosion, and sustains crop yield because of improved water relations and nutrient availability. However, there may be yield reduction during the transition year or the first two years.

## Soil-water Conservation and Management

Lack of adequate water, especially in the soil's root zone, is an important challenge to improving agronomic productivity of drylands. Therefore, drought management is the key to achieving food security and improving the livelihood of the large population living on the drylands (Fig. 3.4) by: (i) being prepared to cope with drought; (ii) developing more resilient ecosystems; (iii) improving resilience to recover from drought; and (iv) mitigating the impacts of droughts (Solh and Van Ginkel, 2014). Pertinent among preparedness strategies are: (i) choosing agricultural systems according to geographical locations; (ii) increasing climate resilience of rainfed systems; (iii) improving water-use efficiency (WUE) of rainfed systems; and (iv) expanding the land area equipped for irrigation (Solh and Van Ginkel, 2014). Thus, conserving water in the root zone, increasing WUE, and achieving more crops per drop of water are the major goals and principal challenges of dryland agriculture. In addition to adopting drought-tolerant species and developing/breeding improved varieties through morphological, physiological, and molecular modifications

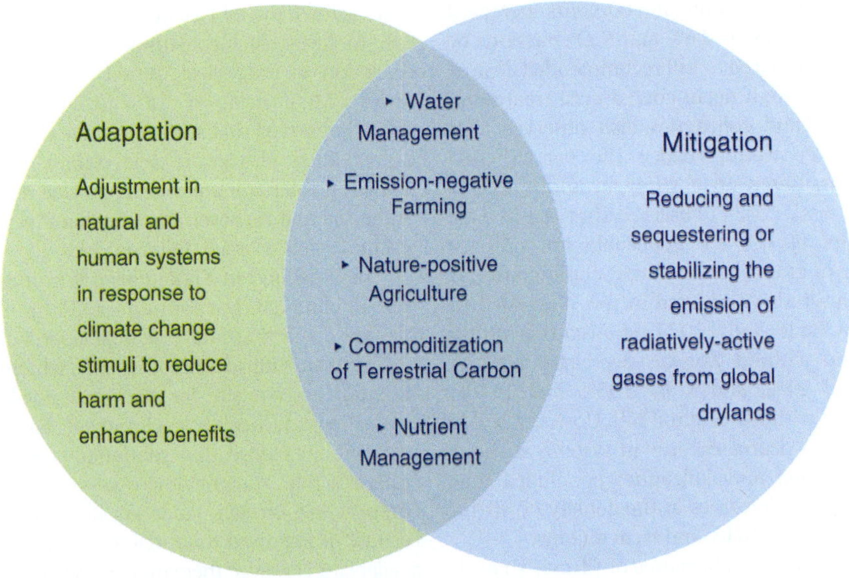


**Fig. 3.4.** Protecting, restoring, and sustainably managing soil and water resources of global drylands through innovative farming systems which enhances both adaptation to and mitigation of anthropogenic climate change.

(Li *et al.*, 2020), it is critical to identify and adopt soil/crop livestock management practices which conserve rainfall and enhance WUE. Such strategies would include biological engineering (Li *et al.*, 2020) and ecological engineering (e.g. conservation agriculture).

Similar to the concept of INM, the strategy of integrated water management (IWM) is extremely pertinent to advancing food security in GDLs. In the Loess Plateau, Liang *et al.* (2018) observed that mulching with maize straw significantly increased maize yield and net income by 20% in dryland agriculture. Improving WUE of agro-ecosystems in GDLs entails 'system thinking' to farming innovations involving all stakeholders (Solh and Van Ginkel, 2014).

Use of drip-irrigation (now used on 1% of total cultivated land) in crops and plantations of GDL may save water and produce more crops per unit drop. Friedlander *et al.* (2013) observed that drip irrigation can increase yield and improve WUE, especially in Africa's drylands. Use of drip-irrigation is constrained by water storage problems and destructive wildlife, poor maintenance, and repair, etc.

Rain-use efficiency (RUE), defined as the ratio of above-ground net primary productivity to annual precipitation (Fensholt *et al.*, 2013), is an appropriate indicator for assessing land degradation. Fensholt and colleagues calculated the RUE for the West African Sahel for the period 1982–2010 and observed positive trends indicating that non-precipitation-related degradation is not a widespread phenomenon, and estimated annual rates of C sequestration within the range of 0.08–0.17 Mg C/ha yr averaged over the next 50 years. The most effective practices were those that maintained the input of organic matter such as FYM (up to 0.09 Mg C/ha yr), established trees (0.15 Mg C/ha yr), and adopted CA (0.04 Mg C/ha yr).

A study in the Old Peanut Basin of Senegal by Tschakert *et al.* (2004) indicated that average soil and woody biomass C contents were 11.3 Mg/ha and 6.3 Mg/ha, respectively. Simulated practices based on the Century model resulted in C dynamics ranging from –0.13 Mg C/ha yr from a worst-case millet-sorghum rotation, to 0.43 Mg C/ha yr on intensively managed agricultural fields. Agroforestry innovations based on *Faidherbia* were important to sustaining productivity. These rates of C sequestration can be achieved through adoption of site-specific best management practices based on efficient use of water, plant stands pertinent to the soil-water supply, practices which conserve water and minimize soil erosion, managing soil fertility and increasing SOC stock, and choice of suitable species and drought-tolerant varieties of crops. Inanaga *et al.* (2005) recommended that these principles could be integrated with indigenous knowledge to accomplish successful transformation of agro-ecosystems in GDLs. Addressing the concerns about nutrient depletion in soils managed by small landholders of Africa (Nigeria, Senegal, Niger), Mortimore and Harris (2005) found evidence of farmers' achievements in terms of sustainable production and investments in soil-fertility maintenance. Rather than focusing on fertilizer *per se*, it is important for rural livelihoods and the private sector to invest in sustainable management of natural resources (Mortimore and Harris, 2005).

## Soil Nutrient Management

In addition to enhancing productivity of agro-ecosystems, plant nutrients (N, P, K, Ca, Mg, S) are also critical to mitigating GHG emissions through innovative management of agro-ecosystems. Judicious use of N fertilizer is critical to moderating fluxes of $CO_2$, $CH_4$, and $N_2O$. In eastern Montana, Sainju *et al.* (2012) observed that total $CH_4$ uptake in no-till continuous barley system was greater with 80 than with 0 kg N/ha. Based on several long-term studies in the West Asia/North Africa (WANA) region, Ryan *et al.* (2012) identified fertilizer best-management practices which embrace 4 Rs (right source, right place, right time, and right method of application). Ryan and colleagues observed that N-use efficiency increases with increasing rainfall and is influenced by crop rotation, and recommended balanced fertilization especially involving K and micronutrients. Judicious use of N fertilizer can also increase SOC stocks. In China's Loess Plateau, Guo *et al.* (2012) observed that each kg of fertilizer N application increased SOC by 0.51 kg in the topsoil from 1984 to 2009. Using the Roth C model, Guo and colleagues calculated that annual inputs of C to the soil from all biomasses were 0.62, 0.71, 0.86, and 0.97 Mg C/ha yr for $N_0$, $N_{45}$, $N_{90}$, $N_{135}$, and $N_{180}$ kg/ha, respectively. These inputs enhanced SOC sequestration at the rate of 48, 48, 68, and 100 kg/ha yr for N rate of 45, 90, 135, and 180 kg/ha, primarily due to an increase in input of root biomass to the soil.

Based on 135 studies of 114 long-term experiments located at 180 sites throughout the world, Ladha *et al.* (2011) observed a decline of 16% in SOC and 4–11% in SON with no use of N amendments. With use of synthetic N, the average increases were 8% and 12% for SOC and SON, respectively.

## Sustained Carbon Sequestration

Historical land use has induced SOC depletion from agro-ecosystems. Thus, conversion to restorative land use and best management practices can reverse the trend and transform soil into a C sink. Furthermore, GDLs are severely and extensively degraded, and the farmers are resource-poor with little capacity to invest in soil restoration. Protecting, restoring and sustainably managing soil and water resources of GDLs through innovative farming systems can enhance adaptation to and mitigation of anthropogenic climate change (Fig. 3.4). Some examples of rate of SOC sequestration in relation to land use in GDLs are outlined in Table 3.1. Synthesis of the available data from GDLs shown indicates that rates of SOC sequestration in drylands are low and in the range of 100–400 kg C/ha yr. The rate of sequestration can be negative in some cases (such as conversion of grassland and forest land to cropland), in cotton-based systems and in some soils with low crop yield. With irrigation and sustainable intensification, such as for the rice-wheat system in the Indo-Gangetic Plains, the rate of SOC sequestration can be 800–1200 kg C/ha yr.

The strategy is to create positive soil/ecosystem C and nutrients (NPK and micronutrients) through INM and IWM, and conservation of water in the root zone by adopting a range of site-specific farming systems as outlined in Table 3.2. These technologies include livestock-based systems, agroforestry-based systems, conservation agriculture, soil and water conservation and management, and soil fertility management practices based on judicious combination of organic and inorganic sources including biofertilization,

**Table 3.1.** Some examples of rates of sequestration of soil organic carbon in global drylands.

| Country/ region | Site | Land use/management | Rate (kg C/ha yr) | Reference |
|---|---|---|---|---|
| India | Northwestern | Rice-wheat system | 810–1230 | Dutta *et al.* (2022) |
| Global | All regions | Farmland to grassland | 300 | Deng *et al.* (2016) |
| | | Forest to grassland | 680 | Deng *et al.* (2016) |
| | | Grassland to farmland | –890 | Deng *et al.* (2016) |
| | | Forest to farmland | –1740 | Deng *et al.* (2016) |
| West Africa | Sahel | Use of farmyard manure | 90 | Fensholt *et al.* (2013) |
| | | Afforestation | 150 | Fensholt *et al.* (2013) |
| | | Conversation agriculture | 40 | Fensholt *et al.* (2013) |
| | | Long-term average for 50 years | 80–170 | Fensholt *et al.* (2013) |
| Senegal | Old Peanut Basin | Millet-sorghum rotation | 130–430 | Tschakert *et al.* (2004) |
| Australia | Western | Tagasaste treatment | 900 | Wocheslӓnder *et al.* (2016) |
| Tropics | Nigeria, Sudan, Argentina | *Leucaena leucocephala* | 120–220 | Farage *et al.* (2007) |
| China | Loess Plateau | N use | 48–100 | Guo *et al.* (2012) |

**Table 3.2.** Land-use and farming systems which enhance soil carbon stock.

| Family system | Management | Reference |
|---|---|---|
| Livestock-based systems | Integrated nutrient management | Espeland *et al.* (2020); Chandana *et al.* (2021) |
| | Use of cacti as fodder | Dubeux Jr. *et al.* (2013) |
| | Sheep grazing for weed control | Barsotti *et al.* (2013) |
| | Integration of shrubs into the smallholder crop-rangeland-livestock farming systems | Larbi *et al.* (2009) |
| Agroforestry-based systems | Agroforestry systems for Africa based on indigenous crops | Leakey *et al.* (2022) |
| | Alley cropping in Tanzania | Swamila *et al.* (2021) |
| | No-till with *Ziziphus* in West African Sahel | Bado *et al.* (2021) |
| Conservation agriculture | Holistic approach to CA in South Africa | Strauss *et al.* (2021) |
| | CA-based sustainable intensification in South Asia | Dixon *et al.* (2020) |
| | CA in Chinese Loess Plateau | Kuhn *et al.* (2016) |
| | CA for pigeon pea and castor-based system in India | Pratibha *et al.* (2015) |
| | CA in Western USA | Papendick and Parr (1997) |
| Soil-water conservation and management | Maize straw mulching in China's drylands | Liang *et al.* (2018) |
| | System thinking in GDL management | Solh and Van Ginkel (2014) |
| | Drip-irrigation | Friedlander *et al.* (2013) |
| | Improving rain-use efficiency in West African Sahel | Fensholt *et al.* (2013) |
| Soil fertility management | N use in China's Loess Plateau | Guo *et al.* (2012) |

biological N fixation and use of mycorrhizae (Table 3.2). Thus, the major challenge lies in achieving more sustainable produce per unit of land area, protection and restoration of degraded soils and the environment, identification and adoption of site-specific best management practices, cost-effective and accessible innovations, and favorable policies that are pro-nature and pro-farmer (Inanaga *et al.*, 2005). Translation of science into action requires cooperation between researchers, land managers, private sector, and policymakers.

Soil C sequestration and carbon farming are a major challenge in GDL because of multiple resource limitations and resource-poor farms.

Experiments conducted on Australian Vertisols by Hulugalle *et al.* (2017) showed that ideal stoichiometric ratio C:N:P:S of 10,000: 833:200: 143 is required for sustained C sequestration. Data from experiments between 1993 and 2012 observed that soil C sequestered in cotton was mostly negative. Low nutrient availability may not be the cause of soil losses under irrigation but may be one factor that inhibits C sequestration under dryland conditions.

In addition to water and nutrient management through adoption of CA and mulch farming, establishment of perennial vegetation is a viable option to enhance C sequestration in GDLs. Wocheslӓnder *et al.* (2016) measured above- and below-ground biomass-C and SOC to 2 m depth in a 22-year experiment in Western Australia (rainfall of 498 mm/yr). They observed that C sequestration over the 22-year lifespan was 2.5 Mg/ha in tagasaste treatment along with a change of 0.9 Mg/ha yr in SOC and 1.6 Mg C/ha yr in biomass. Tagasaste plots contained significantly more SOC stocks compared with control plots to 0.9 m depth.

Farage *et al.* (2007) used models (Roth-C-263 and Century) to predict C sequestration potential for soils of Nigeria, Sudan, and Argentina, and observed that it would be possible to adopt best management practices and to convert these soils from C sources to net sinks. Farage and colleagues observed that *Leucaena leucocephala* resulted in a C gain of 0.22 and 0.12 Mg C/ha yr, respectively. Tschakert (2004) also presented a farmer-centered *ex ante* cost-benefit analysis of 15 beneficiaries of C-offset projects in the Old Peanut Basin of Senegal. Tschakert observed significant differences in anticipated net benefit for various management options ranging from \$1400 to \$9600/Mg C. Thus, a carefully designed cost-sharing mechanism would be necessary to achieve equitable and efficient local participation in C sequestration projects.

## Sequestration of Soil Inorganic Carbon in GDLs

SIC is an important component of the C budget of GDLs (Lal and Kimble, 2000). For example, SIC pool in soils of China is 55.3 ± 10.7 Pg with a current average SIC content of 6.3 ± 1.2 kg C/m$^2$, and representing 5.8% of the global SIC pool (Wu *et al.*, 2009). While 51% of the total cultivated soils in China have experienced C loss, Wu and colleagues have observed that SIC in drylands has increased in irrigated soils in the northwestern regions. Yet, total loss of SIC in all cropland soils decreased by 1.6 Pg C due to extensive human activities.

Drylands also experience nocturnal soil $CO_2$ uptake and its relationship to sub-surface soil and ecosystem C flux. In a study based on Chihuahuan Desert shrubland, Hamerlynck *et al.* (2013) observed negative nocturnal flux of $CO_2$ (flux from the atmosphere into soil). This negative flux, also observed in other deserts, depended on air-soil temperature gradient and was more frequent and stronger in inter-canopy soils. Further sub-surface profiles associated with negative C flux indicate sustained carbonate dissolution lowered shallow soil $CO_2$ below atmosphere levels. Hamerlynck and colleagues concluded that abiotic nocturnal $CO_2$ uptake is an important factor that modulates dry land ecosystem-level carbon dynamics, and these dynamics may be affected by future higher atmospheric $CO_2$ levels.

## Way Forward to Carbon Farming in Global Drylands

Three widely recognized approaches to mitigating environmental degradation and reducing risks of anthropogenic climate change in GDL are: (i) using energy more efficiently; (ii) using no-C or low-C energy resources; and (iii) capturing and storing C in stable reservoirs with long mean residence time. Thus, protection of native biomes, restoration of degraded soils, and sustainable management of soils of agro-ecosystems offer an opportunity to reconcile the need to produce more food with the necessity of improving the environment. Whereas the rates of soil C sequestration in GDLs are low (Table 3.2), drylands cover a large area of 6–7 bha (in 50% of the world's land area). The WANA region has a land area of 1.7 bha and has SOC content of < 0.5% (Lal, 2002), and soil C sink capacity of 3.7 Pg over about 50 years. Past estimates indicate technical potential of soil C sequestration of GDLs at 1.1 Pg C/yr (Squires, 1998; Lal,

2000). Thus, even a small incremental increase in soil C sequestration in GDLs can have a large drawdown of atmospheric $CO_2$. Then, there are four potential land-based sinks: two each in biomass (above-ground and below-ground) and soil (soil organic C and soil inorganic C). Furthermore, a large proportion of GDLs are degraded and deserted and have a large C sink capacity. The time horizon over which the depleted soil C reserves of GDLs can be restored may take as long as 50 years (Mermut, 2003).

Calculations and upscaling of data in Table 3.1 and Table 3.2 through upscaling of proven technologies (Table 3.3) can make agriculture a solution to climate change and GDLs a major sink of atmospheric $CO_2$. Realizing the vast potential of GDLs is a win-win-win option to mitigate climate change, end poverty, and advance food and nutritional security.

## Payments to Farmers for Ecosystem Services to Operationalize Carbon Farming

Farmers and land managers must be rewarded for upscaling of RMPs and growing C as a farm commodity (Fig. 3.5). The strategy is to encourage farmers to adopt site-specific RMPs which lead to eco-intensification of agro-ecosystems in GDLs and produce more from less – less land, less water use, less fertilizers and pesticides, less emission of GHGs, and less use of energy. Simply put, land

**Table 3.3.** Some examples of rate of sequestration of soil inorganic carbon in global drylands. (Mermut *et al.*, 2000 and Ryskov *et al.*, 2000 are adapted from Lal and Kimble, 2000)

| Type | Region | Rate | Units | Reference |
|---|---|---|---|---|
| I. Bicarbonate flux downward in groundwater | Steppes of Russia | 2.1–7.4 | gC/m² yr | Lapenis *et al.* (2008) |
| | Irrigated land in California, USA | 0.9–7.15 | kgC/m² yr | Eshel *et al.* (2007) |
| | Global flux | 0.2–0.36 | $\frac{PgC}{m}$ as $HCO_3$ | Monger *et al.* (2015) |
| II. Pedogenic carbonates | Drylands of USA | 0.06–1.5 | gC/m² yr | Mermut *et al.* (2000) |
| | Chernozems, Russia | 3.2–5.6 | gC/m² yr | Ryskov *et al.* (2000) |
| | Chernozems, Russia | ⁻9.8–26 | gC/m² yr | Ryskov *et al.* (2000) |

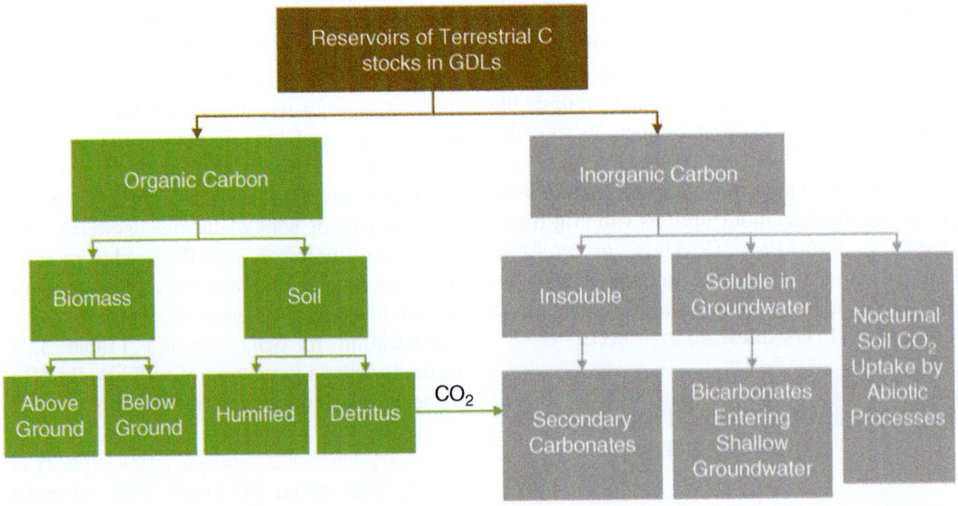


**Fig. 3.5.** Components of the terrestrial carbon stock in global drylands involved in carbon farming.

managers are encouraged to promote the Green Revolution, which is soil-based (soil resilience), ecosystem-based (eco-efficiency), and knowledge-based (combination of traditional and scientific innovations). This would involve increasing agronomic productivity from existing land, restoring degraded lands, enhancing biological nitrogen fixation by legumes, and converting some marginal agricultural land for nature conservancy without any conversion of natural land to agro-ecosystems through eco-intensification. In this context, the focus is on managing soil health through judicious use of CNPK rather than NPK, because improvement in soil C stock would increase eco-efficiency by minimizing losses of inputs (i.e. fertilizers, pesticides, irrigation, energy).

Farmers should be paid according to the societal value of C, which may be as much as \$130 per metric tonne of C or \$30–35 per credit (Lal, 2014). However, the price may be \$150–200 per metric tonne of C in 2024–2025 or \$45–55 per credit of C. Rather than intensive monitoring of pool and fluxes of C on farmland, payment can be made on the basis of land use and farming/cropping systems and the available research information for the ecoregion. Simplification of the protocol and direct payments to farmers may be the best option so that farmers receive the most benefit of their hard work and commitments to restoration and sustainable management of soil health.

## Conclusion

GDLs comprise the largest terrestrial biome, which is also vulnerable to global warming and is projected to increase in area from about 40% in 1960s to 50% by the end of 21st century.

GDLs have a large terrestrial C stock in both vegetation and soil. The soil C stock consists of SOC and SIC and SIC is larger in stock than SOC. Both SOC and SIC stocks are prone to climate change, land use and farming systems.

Through judicious land use and innovative management, global drylands can transform agriculture into a solution to environmental issues of the 21st century while also achieving food/nutritional security, adapting/mitigating to anthropogenic climate change, and advancing Sustainable Development Goals of the United Nations.

Recommended management practices for increasing soil carbon stock include system-based conservation agriculture, agroforestry, water conservation and management, drip sub-fertigation, integrated water management to enhance water use efficiency and produce more crop per drop of water, and judicious grazing.

The rate of soil carbon sequestration in GDLs may range from 0 to 250 kg C /ha yr for SOC and 5 to 100 kg C/ha yr for SIC as secondary carbonates. Sequestration of C at these rates in large and degraded land areas can make GDL a major sink for atmospheric $CO_2$ and render agriculture in GDLs a solution to global issues including putting SDGs on track.

Payments to farmers for ecosystem services at the societal value of C (US \$30 per credit in 2010s and \$50 per credit in 2020s) can promote adoption of recommended management practices and bring about a rapid transformation in agricultural productivity, soil quality, human wellbeing, and nature restoration.

Carbon farming is an important strategy to promote translation of science into action. Indeed, carbon farming is the game changing solution which can advance food and nutrition security, alleviate poverty, and restore degraded soils and ecosystems.

Widespread adoption of carbon farming in GDLs necessitates development and implementation of policies which are pro-nature, pro-farmer, and pro-agriculture.

The strategy of carbon farming in GDLs is a win-win-win option. It is the nature-based solution to adaptation and mitigation of climate change, advancement of SDGs, and restoration of degraded soils and desertified ecosystems.

## References

Abbruzzini, T.F., Avitia, M., Carrasco-Espinosa, K., Peña, V., Barrón-Sandoval, A. *et al.* (2021) Precipitation controls on soil biogeochemical and microbial community composition in rainfed agricultural systems in tropical drylands. *Sustainability* 13(21), 11848. DOI: 10.3390/su132111848

Alary, V., Lasseur, J., Frija, A. and Gautier, D. (2022) Assessing the sustainability of livestock socio-ecosystems in the drylands through a set of indicators. *Agricultural Systems* 198, 103389. DOI: 10.1016/j.agsy.2022.103389

Bado, B.V., Whitbread, A. and Manzo, M.L.S. (2021) Improving agricultural productivity using agroforestry systems: performance of millet, cowpea, and ziziphus-based cropping systems in West Africa Sahel. *Agriculture, Ecosystems & Environment* 305, 107175. DOI: 10.1016/j.agee.2020.107175

Barsotti, J.L., Sainju, U.M., Lenssen, A.W., Montagne, C. and Hatfield, P.G. (2013) Net greenhouse gas emissions affected by sheep grazing in dryland cropping systems. *Soil Science Society of America Journal* 77(3), 1012–1025. DOI: 10.2136/sssaj2012.0386

Bullard, J.E. and McTainsh, G.H. (2003) Aeolian-fluvial interactions in dryland environments: examples, concepts and Australia case study. *Progress in Physical Geography* 27(4), 471–501. DOI: 10.1191/0309133303pp386ra

Castellanos, A.E., Hinojo-Hinojo, C., Rodriguez, J.C., Romo-Leon, J.R., Wilcox, B.P. *et al.* (2022) Plant functional diversity influences water and carbon fluxes and their use efficiencies in native and disturbed dryland ecosystems. *Ecohydrology* 15(5), e2415. DOI: 10.1002/eco.2415

Chandana, P., Lata, A.M., Prasad, J.V.N.S. and Khan, M.A. (2021) Greenhouse gas emissions from integrated nutrient management practices in pearl millet plus *Melia dubia* agri-silvi system. *Current Science* 121(5), 704. DOI: 10.18520/cs/v121/i5/704-708

Choi, R.T., Reed, S.C. and Tucker, C.L. (2022) Multiple resource limitation of dryland soil microbial carbon cycling on the Colorado Plateau. *Ecology* 103(6), e3671. DOI: 10.1002/ecy.3671

Dacal, M., García-Palacios, P., Asensio, S., Cano-Díaz, C., Gozalo, B. *et al.* (2020) Contrasting mechanisms underlie short- and longer-term soil respiration responses to experimental warming in a dryland ecosystem. *Global Change Biology* 26(9), 5254–5266. DOI: 10.1111/gcb.15209

Delgado-Baquerizo, M., Maestre, F.T., Gallardo, A., Bowker, M.A., Wallenstein, M.D. *et al.* (2013) Decoupling of soil nutrient cycles as a function of aridity in global drylands. *Nature* 502(7473), 672–676. DOI: 10.1038/nature12670

Deng, L., Zhu, G.Y., Tang, Z.S. and Shangguan, Z.P. (2016) Global patterns of the effects of land-use changes on soil carbon stocks. *Global Ecology and Conservation* 5, 127–138. DOI: 10.1016/j.gecco.2015.12.004

Dintwe, K. and Okin, G.S. (2018) Soil organic carbon in savannas decreases with anthropogenic climate change. *Geoderma* 309, 17–16. DOI: 10.1016/j.geoderma.2017.08.030

Dixon, J., Rola-Rubzen, M.F., Timsina, J., Cummins, J. and Tiwari, T.P. (2020) Socioeconomic impacts of conservation agriculture based sustainable intensification (CASI) with particular reference to South Asia. In: Dang, Y., Dalal, R. and Menzies, N. (eds) *No-till Farming Systems for Sustainable Agriculture*. Springer, Cham, Switzerland, pp. 377–394. DOI: 10.1007/978-3-030-46409-7_22

Dubeux Jr. J.C.B., Dos Santos, M.V.F., De Mello, A.C.L., Vieira da Cunha, M., de A. Ferreira, M. *et al.* (2013) Forage potential of cacti on drylands. In: *VIII International Congress on Cactus Pear and Cochineal*, Vol. 1067. pp. 181–186. DOI: 10.17660/ActaHortic.2015.1067.24

Dutta, D., Singh, V.K., Upadhyay, P.K., Meena, A.L., Kumar, A. *et al.* (2022) Long-term impact of organic and inorganic fertilizers on soil organic carbon dynamics in a rice-wheat system. *Land Degradation & Development* 33(11), 1862–1877. DOI: 10.1002/ldr.4267

Eshel, G., Fine, P. and Singer, M.J. (2007) Total soil carbon and water quality: an implication for carbon sequestration. *Soil Science Society of America Journal* 71(2), 397–405. DOI: 10.2136/sssaj2006.0061

Espeland, E.K., Schreeg, L. and Porensky, L.M. (2020) Managing risks related to climate variability in rangeland-based livestock production: What producer driven strategies are shared and prevalent across diverse dryland geographies? *Journal of Environmental Management* 255, 109889. DOI: 10.1016/j.jenvman.2019.109889

Farage, P.K., Ardö, J., Olsson, L., Rienzi, E.A., Ball, A.S. *et al.* (2007) The potential for soil carbon sequestration in three tropical dryland farming systems of Africa and Latin America: a modelling approach. *Soil and Tillage Research* 94(2), 457–472. DOI: 10.1016/j.still.2006.09.006

Feng, S. and Fu, Q. (2013) Expansion of global drylands under warming climate. *Atmospheric Chemistry and Physics* 13, 10081–10094. DOI: 10.5194/acp-13-10081-2013

Fensholt, R., Rasmussen, K., Kaspersen, P., Huber, S., Horion, S. *et al.* (2013) Assessing land degradation/recovery in the African Sahel from long-term earth observation based primary productivity and precipitation relationships. *Remote Sensing* 5(2), 664–686. DOI: 10.3390/rs5020664

Friedlander, L., Tal, A. and Lazarovitch, N. (2013) Technical considerations affecting adoption of drip irrigation in sub-Saharan Africa. *Agricultural Water Management* 126, 125–132. DOI: 10.1016/j.agwat.2013.04.014

Galvin, K.A. (2021) Transformational adaptation in drylands. *Current Opinion in Environmental Sustainability* 50, 64–71. DOI: 10.1016/j.cosust.2021.03.003

García-Palacios, P., Alarcón, M.R., Tenorio, J.L. and Moreno, S.S. (2019) Ecological intensification of agriculture in drylands. *Journal of Arid Environments* 167, 101–105. DOI: 10.1016/j.jaridenv.2019.04.014

Guo, S.L., Wu, J., Coleman, K., Zhu, H., Li, Y. and Liu, W. (2012) Soil organic carbon dynamics in a dryland cereal cropping system of the Loess Plateau under long-term nitrogen fertilizer applications. *Plant and Soil* 353, 321–332. DOI: 10.1007/s11104-011-1034-1

Hamerlynck, E.P., Scott, R.L., Sánchez-Cañete, E.P. and Barron-Gafford, G.A. (2013) Nocturnal soil $CO_2$ uptake and its relationship to subsurface soil and ecosystem carbon fluxes in a Chihuahuan Desert shrubland. *Journal of Geophysical Research: Biogeosciences* 118(4), 1593–1603. DOI: 10.1002/2013JG002495

Hantson, S., Huxman, T.E., Kimball, S., Randerson, J.T. and Goulden, M.L. (2021) Warming as a driver of vegetation loss in the Sonoran Desert of California. *Journal of Geophysical Research: Biogeosciences* 126(6), e2020JG005942. DOI: 10.1029/2020JG005942

Hulugalle, N.R., Strong, C., McPherson, K. and Nachimuthu, G. (2017) Carbon, nitrogen and phosphorus stoichiometric ratios under cotton cropping systems in Australian Vertisols: a meta-analysis of seven experiments. *Nutrient Cycling in Agroecosystems* 107, 357–367. DOI: 10.1007/s10705-017-9837-y

Inanaga, S., Eneji, A.E., An, P. and Shimizu, H. (2005) A recipe for sustainable agriculture in drylands. In: Omasa, K., Nouchi, I. and De Kok, L.J. (eds) *Plant Responses to Air Pollution and Global Change*, pp. 285–293. Springer, Tokyo. DOI: 10.1007/4-431-31014-2_32

Janzen, H.H. (2006) The soil carbon dilemma: Shall we hoard it or use it? *Soil Biology and Biochemistry* 38(3), 419–424. DOI: 10.1016/j.soilbio.2005.10.008

Jia, X.X., Wang, X., Hou, L., Wei, X., Zhang, Y. *et al.* (2019) Variable response of inorganic carbon and consistent increase of organic carbon as a consequence of afforestation in areas with semiarid soils. *Land Degradation & Development* 30(11), 1345–1356. DOI: 10.1002/ldr.3320

Kou, D., Ma, W., Ding, J., Zhang, B., Fang, K. *et al.* (2018) Dryland soils in northern China sequester carbon during the early 2000s warming hiatus period. *Functional Ecology* 32(6), 1620–1630. DOI: 10.1111/1365-2435.13088

Kuhn, N.J., Hu, Y., Bloemertz, L., He, J., Li, H. *et al.* (2016) Conservation tillage and sustainable intensification of agriculture: regional vs. global benefit analysis. *Agriculture, Ecosystems & Environment* 216, 155–165. DOI: 10.1016/j.agee.2015.10.001

Ladha, J.K., Reddy, C.K., Padre, A.T. and van Kessel, C. (2011) Role of nitrogen fertilization in sustaining organic matter in cultivated soils. *Journal of Environmental Quality* 40(6), 1756–1766. DOI: 10.2134/jeq2011.0064

Lal, R. (2000) Carbon sequestration in drylands. *Annals of Arid Zone* 39(1), 1–10. Available at: https://epubs.icar.org.in/index.php/AAZ/article/view/65880 (accessed 3 July 2024).

Lal, R. (2002) Carbon sequestration in dryland ecosystems of West Asia and North Africa. *Land Degradation & Development* 13(1), 45–59. DOI: 10.1002/ldr.477

Lal, R. (2003) Soil erosion and the global carbon budget. *Environment International* 29(4), 437–450. DOI: 10.1016/S0160-4120(02)00192-7

Lal, R. (2014) Societal value of soil carbon. *Journal of Soil and Water Conservation* 69(6), 186A–192A. DOI: 10.2489/jswc.69.6.186A

Lal, R. (2015) A system approach to conservation agriculture. *Journal of Soil and Water Conservation* 70(4), 82A–88A. DOI: 10.2489/jswc.70.4.82A

Lal, R. (2016) Soil health and carbon management. *Food and Energy Security* 5(4), 212–222. DOI: 10.1002/fes3.96

Lal, R. (2019) Carbon cycling in global drylands. *Current Climate Change Reports* 5(3), 221–232. DOI: 10.1007/s40641-019-00132-z

Lal, R. (2020) Soil organic matter and water retention. *Agronomy Journal* 112(5), 3265–3277. DOI: 10.1002/agj2.20282

Lal, R. (2022) Biophysical controls that make erosion-transported soil carbon a source of greenhouse gases. *Applied Sciences* 12(16), 8372. DOI: 10.3390/app12168372

Lal, R. and Kimble, J.M. (2000) Pedogenic carbonates and the global carbon cycle. In: Lal, R., Kimble, J.M., Stewart, B.A. and Eswaran, H. (eds) *Global Climate Change and Pedogenic Carbonates*. Lewis Publishers, Boca Raton, Florida, pp. 1–14.

Lal, R., Hassan, H.M. and Dumanski, J. (1999) Desertification control to sequester C and mitigate the greenhouse effect. In: Rosenberg, N.J., Izzauraulde, R.C. and Malone, E.L. (eds) *Carbon Sequestration in Soils: Science, Monitoring and Beyond*. Batelle Press, Columbia, Ohio, pp. 83–107.

Lapenis, A.G., Lawrence, G.B., Bailey, S.W., Aparin, B.F., Shiklomanov, A.I. *et al.* (2008) Climatically driven loss of calcium in steppe soil as a sink for atmospheric carbon. *Global Biogeochemical Cycles* 22(2). DOI: 10.1029/2007GB003077

Larbi, A., Khatib-Salkini, A., Bolus Jamal, P. and Iniguez, L. (2009) Shrub yield and fodder quality variations in a non-tropical dryland environment in West Asia. *Agroforestry Systems* 75, 147–155. DOI: 10.1007/s10457-008-9156-z

Leakey, R.R., Tientcheu Avana, M.L., Awazi, N.P., Assogbadjo, A.E., Mabhaudhi, T. *et al.* (2022) The future of food: domestication and commercialization of indigenous food crops in Africa over the third decade (2012–2021). *Sustainability* 14(4), 2355. DOI: 10.3390/su14042355

Li, C.F., Zhang, C., Luo, G. and Chen, X. (2013) Modeling the carbon dynamics of the dryland ecosystems in Xinjiang, China from 1981 to 2007: the spatiotemporal patterns and climate controls. *Ecological Modelling* 267, 148–157. DOI: 10.1016/j.ecolmodel.2013.06.007

Li, C.F., Zhang, C., Luo, G., Chen, X., Maisupova, B. *et al.* (2015) Carbon stock and its responses to climate change in Central Asia. *Global Change Biology* 21(5), 1951–1967. DOI: 10.1111/gcb.12846

Li, G.X., Xu, B.C., Yin, L.N., Wang, S.W., Zhang, S.Q. *et al.* (2020) Dryland agricultural environment and sustainable productivity. *Plant Biotechnology Reports* 14, 169–176. DOI: 10.1007/s11816-020-00613-w

Liang, G.M., Wang, Y.G., Chi, B.L., Li, N.N., Chen, W.L. *et al.* (2018) Exploring optimal soil mulching to enhance maize yield and water use efficiency in dryland areas in China. *Acta Agriculturae Scandinavica, Section B – Soil & Plant Science* 68(3), 273–282. DOI: 10.1080/09064710.2017.1394486

Liu, H., Xu, C., Allen, C.D., Hartmann, H., Wei, X. *et al.* (2022) Nature-based framework for sustainable afforestation in global drylands under changing climate. *Global Change Biology* 28(7), 2202–2220. DOI: 10.1111/gcb.16059

Liu, S.S., Zhou, L., Li, H., Zhao, X., Yang, Y. *et al.* (2020) Shrub encroachment decreases soil inorganic carbon stocks in Mongolian grasslands. *Journal of Ecology* 108(2), 678–686. DOI: 10.1111/1365-2745.13298

MacBean, N., Scott, R.L., Biederman, J.A., Peylin, P., Kolb, T. *et al.* (2021) Dynamic global vegetation models underestimate net CO2 flux mean and inter-annual variability in dryland ecosystems. *Environmental Research Letters* 16(9), 094023. DOI: 10.1088/1748-9326/ac1a38

Martins, C.S., Macdonald, C.A., Anderson, I.C. and Singh, B.K. (2016) Feedback responses of soil greenhouse gas emissions to climate change are modulated by soil characteristics in dryland ecosystems. *Soil Biology and Biochemistry* 100, 21–32. DOI: 10.1016/j.soilbio.2016.05.007

Mermut, A.R. (2003) Carbon sequestration for mitigation of environmental degradation. In: Lynch, J.M., Schepers, J.S. and Unver, I. (eds) *Innovative Soil-plant Systems for Sustainable Agricultural Practices. Biological Resource Management in Agriculture*. OECD Publishing, Paris, pp. 105–118. DOI: 10.1787/9789264099722-en

Mermut, A.R., Amundson, R. and Cerling, T.E. (2000) The use of stable isotopes in studying carbonates in soils. In: Lal, R., Kimble, J.M., Eswaran, H. and Stewart, B.A. (eds) *Global Climate Change and Pedogenic Carbonates*. Lewis Publishers, Boca Raton, Florida, pp. 65–85.

Middleton, N.J. (2017) Desert dust hazards: a global review. *Aeolian Research* 24, 53–63. DOI: 10.1016/j.aeolia.2016.12.001

Mo, F., Zhu, Y., Wang, Z.Y., Deng, H.L., Li, P.F. *et al.* (2021) Polyethylene film mulching enhances the microbial carbon-use efficiency, physical and chemical protection of straw-derived carbon in an Entisol of the Loess Plateau. *Science of the Total Environment* 792, 148357. DOI: 10.1016/j.scitotenv.2021.148357

Monger, H.C., Kraimer, R.A., Khresat, S.E., Cole, D.R., Wang, X. *et al.* (2015) Sequestration of inorganic carbon in soil and groundwater. *Geology* 43(5), 375–378. DOI: 10.1130/G36449.1

Mortimore, M. and Harris, F. (2005) Do small farmers' achievements contradict the nutrient depletion scenarios for Africa? *Land Use Policy* 22(1), 43–56. DOI: 10.1016/j.landusepol.2003.06.003

Neumann, K., Sietz, D., Hilderink, H., Janssen, P., Kok, M. *et al.* (2015) Environmental drivers of human migration in drylands – a spatial picture. *Applied Geography* 56, 116–126. DOI: 10.1016/j.apgeog.2014.11.021

Nyachoti, S., Jin, L., Tweedie, C.E. and Ma, L. (2019) Insight into factors controlling formation rates of pedogenic carbonates: a combined geochemical and isotopic approach in dryland soils of the US Southwest. *Chemical Geology* 527, 118503. DOI: 10.1016/j.chemgeo.2017.10.014

Papendick, R.I. and Parr, J.F. (1997) No-till farming: the way of the future for a sustainable dryland agriculture. *Annals of Arid Zone* 36(3), 193–208.

Pratibha, G., Srinivas, I., Rao, K.A., Raju, B.M.K., Thyagaraj, C.R. *et al.* (2015) Impact of conservation agriculture practices on energy use efficiency and global warming potential in rainfed pigeonpea–castor systems. *European Journal of Agronomy* 66, 30–40. DOI: 10.1016/j.eja.2015.02.001

Prăvălie, R. (2016) Drylands extent and environmental issues: a global approach. *Earth-Science Reviews* 161, 259–278. DOI: 10.1016/j.earscirev.2016.08.003

Puttock, A., Dungait, J.A., Macleod, C.J., Bol, R. and Brazier, R.E. (2014) Woody plant encroachment into grasslands leads to accelerated erosion of previously stable organic carbon from dryland soils. *Journal of Geophysical Research: Biogeosciences* 119(12), 2345–2357. DOI: 10.1002/2014JG002635

Qiu, B., Ge, J., Guo, W., Pitman, A.J. and Mu, M. (2020) Responses of Australian dryland vegetation to the 2019 heat wave at a subdaily scale. *Geophysical Research Letters* 47(4), e2019GL086569. DOI: 10.1029/2019GL086569

Ravi, S., D'Odorico, P., Breshears, D.D., Field, J.P., Goudie, A.S. *et al.* (2011) Aeolian processes and the biosphere. *Reviews of Geophysics* 49(3). DOI: 10.1029/2010RG000328

Robertson, A.D., Zhang, Y., Sherrod, L.A., Rosenzweig, S.T., Ma, L. *et al.* (2018) Climate change impacts on yields and soil carbon in row crop dryland agriculture. *Journal of Environmental Quality* 47(4), 684–694. DOI: 10.2134/jeq2017.08.0309

Ryan, J., Sommer, R. and Ibrikci, H. (2012) Fertilizer best management practices: a perspective from the dryland West Asia–North Africa region. *Journal of Agronomy and Crop Science* 198(1), 57–67. DOI: 10.1111/j.1439-037X.2011.00488.x

Sainju, U.M. (2014) Cropping sequence and nitrogen fertilization impact on surface residue, soil carbon sequestration, and crop yields. *Agronomy Journal* 106(4), 1231–1242. DOI: 10.2134/agronj14.0026

Sainju, U.M. (2020) No-till farming systems in North America. In: Dang, Y., Dalal, R. and Menzies, N. (eds) *No-till Farming Systems for Sustainable Agriculture*. Springer, Cham, Switzerland, pp. 587–599. DOI: 10.1007/978-3-030-46409-7_32

Sainju, U.M., Caesar-TonThat, T., Lenssen, A.W. and Barsotti, J.L. (2012) Dryland soil greenhouse gas emissions affected by cropping sequence and nitrogen fertilization. *Soil Science Society of America Journal* 76(5), 1741–1757. DOI: 10.2136/sssaj2012.0076

Smith, W.K., Dannenberg, M.P., Yan, D., Herrmann, S., Barnes, M.L. *et al.* (2019) Remote sensing of dryland ecosystem structure and function: progress, challenges, and opportunities. *Remote Sensing of Environment* 233, 111401. DOI: 10.1016/j.rse.2019.111401

Solh, M. and van Ginkel, M. (2014) Drought preparedness and drought mitigation in the developing world's drylands. *Weather and Climate Extremes* 3, 62–66. DOI: 10.1016/j.wace.2014.03.003

Song, Y., Yao, Y., Qin, X., Wei, X., Jia, X. *et al.* (2020) Response of carbon and nitrogen to afforestation from 0 to 5 m depth on two semiarid cropland soils with contrasting inorganic carbon concentrations. *Geoderma* 357, 113940. DOI: 10.1016/j.geoderma.2019.113940

Squires, V.R. (1998) Dryland soils: their potential as a sink for carbon and as an agent in mitigating climate change. In: Blume, H.P., Eger, H., Fleischhauer, E., Hebel, A., Reij, C. and Steiner, K.G. (eds) *Advances in GeoEcology*, Vol. 31, pp. 209–216.

Stewart, B.A. and Lal, R. (2018) Increasing world average yields of cereal crops: it's all about water. *Advances in Agronomy* 151, 1–44. DOI: 10.1016/bs.agron.2018.05.001

Strauss, J.A., Swanepoel, P.A., Laker, M.C. and Smith, H.J. (2021) Conservation agriculture in rainfed annual crop production in South Africa. *South African Journal of Plant and Soil* 38(3), 217–230. DOI: 10.1080/02571862.2021.1891472

Swamila, M., Philip, D., Akyoo, A.M., Manda, J., Mwinuka, L. *et al.* (2021) Profitability of gliricidia-maize system in selected dryland areas of Dodoma region, Tanzania. *Sustainability* 14(1), 53. DOI: 10.3390/su14010053

Tschakert, P. (2004) The costs of soil carbon sequestration: an economic analysis for small-scale farming systems in Senegal. *Agricultural Systems* 81(3), 227–253. DOI: 10.1016/j.agsy.2003.11.004

Tschakert, P., Khouma, M. and Sene, M. (2004) Biophysical potential for soil carbon sequestration in agricultural systems of the Old Peanut Basin of Senegal. *Journal of Arid Environments* 59(3), 511–533. DOI: 10.1016/j.jaridenv.2004.03.026

Van der Pol, L.K., Robertson, A., Schipanski, M., Calderon, F.J., Wallenstein, M.D. *et al.* (2022) Addressing the soil carbon dilemma: legumes in intensified rotations regenerate soil carbon while maintaining yields in semi-arid dryland wheat farms. *Agriculture, Ecosystems & Environment* 330, 107906. DOI: 10.1016/j.agee.2022.107906

Wachter, J.M., Painter, K.M., Carpenter-Boggs, L.A., Huggins, D.R. and Reganold, J.P. (2019) Productivity, economic performance, and soil quality of conventional, mixed, and organic dryland farming systems in eastern Washington State. *Agriculture, Ecosystems & Environment* 286, 106665. DOI: 10.1016/j.agee.2019.106665

Wang, K.B., Ren, Z.P., Deng, L., Zhou, Z.C., Shangguan, Z.P. *et al.* (2016) Profile distributions and controls of soil inorganic carbon along a 150-year natural vegetation restoration chronosequence. *Soil Science Society of America Journal* 80(1), 193–202. DOI: 10.2136/sssaj2015.08.0296

Wang, L., Huang, X., Fang, N.F., Niu, Y.H., Wang, T.W. *et al.* (2019) Selective transport of soil organic and inorganic carbon in eroded sediment in response to raindrop sizes and inflow rates in rainstorms. *Journal of Hydrology* 575, 42–53. DOI: 10.1016/j.jhydrol.2019.05.033

Wang, X.J., Wang, J., Xu, M., Zhang, W., Fan, T. *et al.* (2015) Carbon accumulation in arid croplands of northwest China: pedogenic carbonate exceeding organic carbon. *Scientific Reports* 5(1), 11439. DOI: 10.1038/srep11439

Wang, Y.G., Luo, G., Li, C., Ye, H., Shi, H. *et al.* (2023) Effects of land clearing for agriculture on soil organic carbon stocks in drylands: a meta-analysis. *Global Change Biology* 29(2), 547–562. DOI: 10.1111/gcb.16481

Wertin, T.M., Young, K. and Reed, S.C. (2018) Spatially explicit patterns in a dryland's soil respiration and relationships with climate, whole plant photosynthesis and soil fertility. *Oikos* 127(9), 1280–1290. DOI: 10.1111/oik.04935

Wocheslӓnder, R., Harper, R.J., Sochacki, S.R., Ward, P.R. and Revell, C. (2016) Tagasaste (*Cytisus proliferus* Link.) reforestation as an option for carbon mitigation in dryland farming systems. *Ecological Engineering* 97, 610–618. DOI: 10.1016/j.ecoleng.2016.10.039

Wu, H.B., Guo, Z., Gao, Q. and Peng, C. (2009) Distribution of soil inorganic carbon storage and its changes due to agricultural land use activity in China. *Agriculture, Ecosystems & Environment* 129(4), 413–421. DOI: 10.1016/j.agee.2008.10.020

Xiao, L.C., Kuhn, N.J., Zhao, R. and Cao, L. (2021) Net effects of conservation agriculture principles on sustainable land use: a synthesis. *Global Change Biology* 27(24), 6321–6330. DOI: 10.1111/gcb.15906

Yan, C., Yuan, Z., Shi, X., Lock, T.R. and Kallenbach, R.L. (2020) A global synthesis reveals more response sensitivity of soil carbon flux than pool to warming. *Journal of Soils and Sediments* 20, 1208–1221. DOI: 10.1007/s11368-019-02513-1

Yang, X., Wang, B. and An, S. (2021) Root derived C rather than root biomass contributes to the soil organic carbon sequestration in grassland soils with different fencing years. *Plant and Soil* 469, 161–172. DOI: 10.1007/s11104-021-05144-z

Yao, J., Liu, H., Huang, J., Gao, Z., Wang, G. *et al.* (2020) Accelerated dryland expansion regulates future variability in dryland gross primary production. *Nature Communications* 11(1), 1665. DOI: 10.1038/s41467-020-15515-2

Yu, Q., Wang, H., Wen, P., Wang, S., Li, J. *et al.* (2020) A suitable rotational conservation tillage system ameliorates soil physical properties and wheat yield: an 11-year in-situ study in a semi-arid agroecosystem. *Soil and Tillage Research* 199, 104600. DOI: 10.1016/j.still.2020.104600

Zhu, S.H., Li, C., Shao, H., Ju, W. and Lv, N. (2019) The response of carbon stocks of drylands in Central Asia to changes of CO2 and climate during past 35 years. *Science of the Total Environment* 687, 330–340. DOI: 10.1016/j.scitotenv.2019.06.089

Zucca, C., Fleiner, R., Bonaiuti, E. and Kang, U. (2022) Land degradation drivers of anthropogenic sand and dust storms. *Catena* 219, 106575.

# 4 Plant Genetic Resources for Adaptation to Climate Change in Drylands

**Raj Paroda[1*], Anuradha Agrawal[2], and Kuldeep Tripathi[3]**
*[1]TAAS, New Delhi, India; [2]ICAR–Directorate of Knowledge Management in Agriculture, New Delhi, India; [3]ICAR–National Bureau of Plant Genetic Resources, New Delhi, India*

**Abstract**

Dryland areas, covering approximately 47% of the Earth and supporting the livelihood of millions of people, face challenges such as prolonged droughts, increased temperatures, deteriorating soil health, and desertification. As global warming intensifies, these regions, specifically, face a major challenge of sustaining agricultural production. The problems confronting dryland agriculture, global food security, and the sustainable management of natural resources are enormous and interconnected. Ensuring crop production and food security in the coming years would, therefore, require enhancing agricultural resilience and adaptation to climate change. This, in turn, has detrimental consequences for agrobiodiversity, particularly crops in drylands that have undergone a continuous evolutionary process over millennia, shaped by the occurrence of mutations, genetic adaptation, and natural selection. These processes have resulted in the development of specific traits that enable plants and animals to adapt well to such challenging environments. Consequently, drylands also provide ample opportunities for addressing the adverse impacts of climate change.

This chapter enumerates the role of plant genetic resources (PGR) in the future and the adverse effects of climate change in drylands, with specific reference to India. It broadly covers the conservation, utilization, and management strategies of PGR towards climate-resilient agriculture to ensure both agricultural sustainability and future food security.

## Introduction

### Background

Plant genetic resources (PGR) for food and agriculture play a crucial role in food security, nutrition and livelihoods, and in providing environmental services. They are key components of sustainability, resilience, and adaptability in production systems. Climate change is expanding and has grown more severe in the last few decades, accelerated by anthropogenic activities, and if current trends continue, the global temperature may rise by 1.5°C by 2040 and reach up to 3.5°C by 2100 (IPCC, 2021). The world, including India, is witnessing shifts in seasons, increased variability and intensity of rainfall, rising sea levels, and more frequent and intense extreme weather events such as cyclones, heatwaves, heavy rains, floods, and droughts (MoEFCC, 2021). Tackling climate change is central to achieving a sustainable future in terms of food security for the world's growing population. Balancing food security with the reduction of greenhouse gas

*Corresponding author: raj.paroda@gmail.com


DOI: 10.1079/9781800624870.0004

emissions, poses a significant challenge for agriculture globally (Pathak, 2023). Despite notable emission reductions over the last decades, there is a need to transform production and consumption systems to achieve climate neutrality (Pathak, 2023). Genetic resources could contribute greatly to cope with climate change, but in many cases, the magnitude and speed of climate change may surpass our ability to identify, select, breed, and eventually use these resources in the field. Climate change is already affecting natural ecosystems and food production systems (Pathak, 2023).

Dryland regions encompassing arid, semi-arid, and dry sub-humid regions, exhibit distinct features, including limited and unpredictable rainfall, elevated temperatures, high evaporation rates and sparse vegetation. Drylands are important ecosystems that occupy a significant portion of the Earth's surface (~ 47%), provide net primary productivity of around 40%, and are home to approximately 39% of the global population, including some of the world's poorest communities, underscoring their significance in terms of both human sustenance and ecological functions (Koutroulis, 2019; Nath *et al.*, 2021). Dryland ecosystems (grasslands, shrublands, and savannahs) found across diverse continents, encompassing sections of Africa, Asia, Australia, the Middle East, and the Americas, are presently experiencing desertification and degradation on a widespread scale due to the combined effects of global warming, rapid economic development, population growth, and urbanization (Huang *et al.*, 2016). By 2050, it is anticipated that major crop yields in drought-prone regions will undergo a reduction exceeding 50%, which is estimated to increase to nearly 90% by 2100 (Malhi *et al.*, 2021).

The primary livelihoods of human populations residing in drylands revolve around agriculture and animal husbandry. They have long been inhabited by indigenous communities with distinct traditional agricultural methods of selecting and preserving a rich diversity of crops and livestock breeds tailored to their niche-specific cropping and pastoral systems (Agrawal *et al.*, 2019). Local inhabitants rely on the agrobiodiversity found in the drylands for various purposes such as food production, animal feed, wool production, dairy products, and medicinal uses. However, ensuring food security in drylands poses a significant challenge due to the fragile nature of the natural resource base. The looming threat of climate change exacerbates this challenge, further placing the livelihoods of people in these areas at considerable risk. The threat of massive outmigration adds to the vulnerability of these communities and their livelihoods. However, the extensive agrobiodiversity maintained *in situ* by these native populations, as well as those held *ex situ* in gene banks, increases the chances of successful adaptation to these stressors (Paroda, 2019b).

The objective of this narrative is to highlight the pivotal role of PGR in combating climate change impacts in dryland areas. By examining their significance, evaluating adaptation strategies, assessing conservation efforts, and identifying policy recommendations, the chapter aims to contribute to the development of effective strategies for climate resilience and sustainable agriculture in these vulnerable regions.

## Impact of Climate Change on Agriculture

By definition, climate is the average of weather over a long time (IPCC, 2014). Climate change refers to the long-term modification in the characteristics of the climate system, which can be identified through statistical analysis by observing changes in the average and/or variability of climate properties. These alterations persist over an extended period, indicating a shift in the state of the climate. As per reports, climate change may be due to natural internal processes or external forces such as modulations of the solar cycles, volcanic eruptions, and anthropogenic changes in the composition of the atmosphere or land use (Campbell *et al.*, 2011; IPCC, 2014).

Meteorological data and model projections suggest that significant climate changes are ongoing and are likely to continue for the next century, requiring important land management and crop adaptations to sustain food supply to growing human populations (Hansen *et al.*, 2010). Global observations on surface air temperature show that most parts of the Earth's surface have already experienced notable warming over the last century, with the longest observational dataset indicating that global temperatures have risen by 0.72°C–0.85°C in the period from 1850–1900 to 2003–2012 (Morice *et al.*, 2012).

Certain types of gases in the atmosphere have the property of absorbing part of the solar radiation reflected by the Earth's surface. These gases are called greenhouse gases (GHGs), which prevent radiation from being reflected into space and cause warming of the atmosphere. GHGs emitted mostly due to industrialization and other activities include carbon dioxide ($CO_2$), methane ($CH_4$), nitrous oxide ($N_2O$), hydrofluorocarbons (HFCs), perfluorocarbons (PFCs) and sulfur hexafluoride ($SF_6$) (IPCC, 2001). Over the years, their concentrations in the atmosphere have greatly increased due to industrial activities and the associated burning of fossil fuels and deforestation (IPCC, 2021). The projected rise in global temperatures has been largely attributed to increasing atmospheric concentrations of GHGs, which absorb energy emitted by the Earth's surface (long-wave radiation) and retain a fraction of this energy in the atmosphere. Faster accumulation (than removal) of these gases in the atmosphere allows progressively more energy to be retained near the Earth's surface, causing an increase in temperature (IPCC, 2013).

Twenty Earth-system modeling groups employed a set of climate models to assess potential changes in future climate with increasing GHGs (IPCC, 2013). These studies suggest that the changes in an intensification of hydroclimatic patterns will be a prominent characteristic of 21st-century global warming (Huang *et al.*, 2016). This will result in shorter, less frequent, and more localized precipitation events, along with an increase in the duration of dry periods. In particular, compared to the 1986–2005 average, decreases in precipitation are likely to be more pronounced in subtropical regions (especially southern North America, the Mediterranean basin, and southern Africa), with greater warming (IPCC, 2013). There is also some evidence that uncertainties in multi-model averaging can lead to an underestimation of drying in many regions (IPCC, 2013).

Numerous projections indicate negative impacts resulting from climate change in various agricultural domains (Fig. 4.1). Agricultural

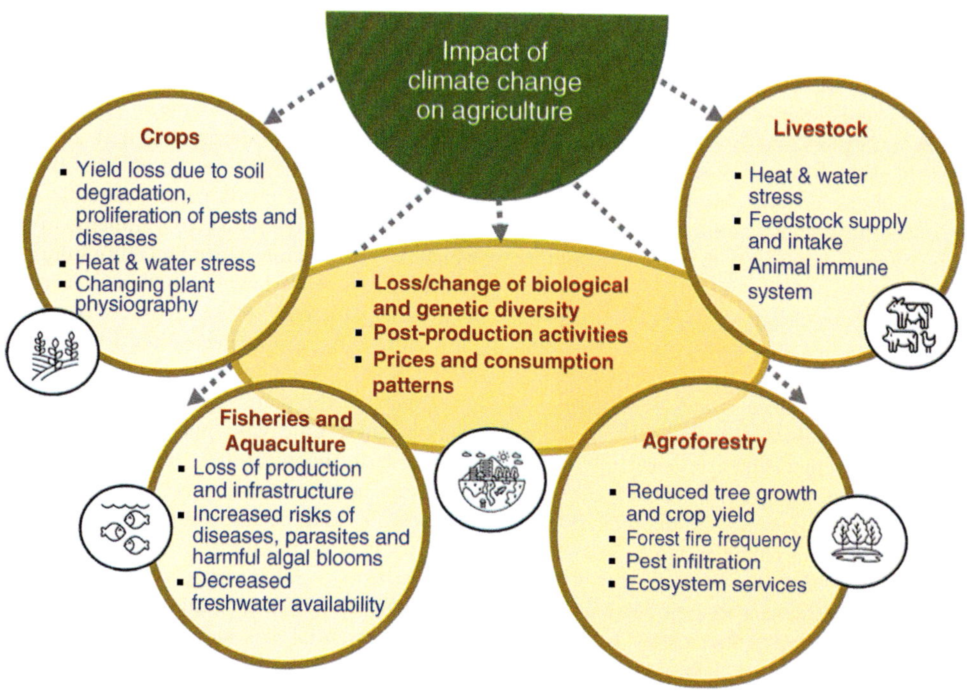


**Fig. 4.1.** Impact of climate change on agriculture.

production is highly vulnerable to changes in the weather, which ultimately affects global food production. The impact of rising temperatures, variations in precipitation, and increased carbon dioxide ($CO_2$) levels on agriculture varies depending on the specific crop, location, and magnitude of the changes in these parameters. Research suggests that temperature increases generally lead to reduced crop yields, while increased precipitation may partially offset or mitigate the effects of rising temperatures (Malhi *et al.*, 2021). An annual increase in temperature makes cold-weather crops more vulnerable (Wolfe *et al.*, 2008). An increase in temperature and $CO_2$ can increase yields in specific crops in some places. But to realize these benefits, nutrient levels, soil moisture, water availability, and other conditions must also be met (Hatfield *et al.*, 2014).

Climate change's effects on crop yields also depend on factors such as the region, irrigation practices, changes in farming practices, and technology. Expanding irrigated areas can increase crop yields but may have adverse environmental consequences (Malhi *et al.*, 2021). Rising temperatures are expected to reduce the duration and yield of many crops; for instance, the aggregate production of wheat, rice, and maize is predicted to decrease in a 2°C warming (Challinor *et al.*, 2014).

Changes in the frequency and severity of droughts and floods could pose serious challenges for farmers and threaten food safety. The consequences of the rise in temperature include melting of glaciers and changes in rainfall patterns, causing flooding and soil erosion, shifts in crop-growing seasons, which affect food security, and changes in the distribution of disease vectors. Temperature influences most plant processes, including photosynthesis, transpiration, respiration, germination, and flowering. As temperature increases (up to a point), photosynthesis, transpiration and respiration increase. When combined with day length, temperature also affects the change from vegetative (leafy) to reproductive (flowering) growth (Malhi *et al.*, 2021).

The impact of climate change on forests is generally expected to be more severe in hot, dry regions, where trees are at their adaptive limits, and in confined areas of moist forest surrounded by drier land (Nair *et al.*, 2009). Potential problems associated with climate change include more frequent fires, breakdowns in the synchrony between tree flowering periods and the presence of pollinators, alien species invasions, and more severe pest outbreaks (Nair *et al.*, 2009). The vital contributions that microorganisms and insect genetic resources make to agriculture and food production (creation and maintenance of soils, pollination, biological control of pests, etc.) were earlier often overlooked, but with the creation of ICAR-National Bureau of Agriculturally Important Microorganisms (NBIAM) and ICAR-National Bureau of Agricultural Insect Resources (NBAIR), their importance in agriculture has been realized. These organisms also play key roles in the carbon cycle and are therefore vitally important to climate change mitigation efforts. Changes to temperature and moisture regimes and atmospheric carbon dioxide levels affect these organisms and their capacities to provide ecosystem services. However, little is known about precisely how they will be affected by climate change.

The climate also affects many aspects of aquatic environments, including the temperature, oxygenation, acidity, salinity, and turbidity of seas, lakes, and rivers, the depth and flow of inland waters, the circulation of ocean currents, and the prevalence of aquatic diseases, parasites, and toxic algal blooms. Climate change poses significant challenges both to capture fisheries and to aquaculture. Problems may arise because of direct physiological effects on aquatic organisms themselves or because the habitats upon which they depend are disrupted. Aquatic organisms living in restricted environments from which they cannot migrate to find more suitable conditions (e.g. those living in shallow rivers and lakes or kept in aquaculture cages) are particularly vulnerable. Aquatic ecosystems and their biota account for the largest carbon and nitrogen fluxes on the planet and act as its largest carbon sinks. Calcifying microorganisms fall constantly to the ocean floor, and the calcium carbonate in the skeletal structures of marine invertebrates and the carbonates precipitated in the intestines of marine fish also make huge contributions to global carbon storage. Disturbances caused by climate change could adversely affect this essential ecosystem service.

With respect to India, yields of the major crops are predicted to be reduced by 9–18% due to climate change (Pathak, 2023). The Indo-Gangetic Plains (IGP) face significant risks in crop productivity, particularly for rice and wheat, while the mid-IGP and Southern Plateau are also at risk, especially for maize. Considering that agriculture contributes around 17% to the country's GDP, the impact of climate change on the crop sector alone could result in a 1% reduction in GDP (Pathak, 2023).

## Climate Change in Drylands

With climate change, it is expected that the area under dryland farming is likely to increase significantly. Observational data from 1955 to 2000 suggest that global drylands have increased in area by 4% (Feng and Fu, 2013). Further, a precipitation decrease of more than 15% per 1°C increase in annual temperature and more than 20% drying per 1°C during the dry season in certain land regions between 60°N and 60°S has been predicted (Knutti *et al.*, 2010). Dryland agriculture (including livestock rearing), thus, would be particularly vulnerable to the impacts of climate change due to their inherent climatic and environmental characteristics such as prolonged droughts, desertification, and increased temperatures, raising the risk of heat stress for both plants and animals. *De facto*, drylands are fragile ecosystems and climate change disrupts the delicate balance, leading to biodiversity loss, changes in plant and animal distribution patterns, and increased vulnerability to invasive species. The anticipated future climatic conditions have the potential to adversely impact the structures and functions of dryland ecosystems, specifically in terms of nutrient cycling, soil fertility, pollination, plant productivity, microbial communities, and, ultimately, gross primary productivity (Nath *et al.*, 2021). The predominantly rural populations in dryland areas often depend on rainfed agriculture, livestock rearing, and natural resources for their livelihoods. Climate change impacts such as reduced crop yields, livestock productivity, and water availability directly affect the socio-economic well-being of these communities in terms of food security, poverty, and migration (Huang *et al.*, 2016).

## Plant Genetic Resources and Climate Change

### Significance of PGR

Plant Genetic Resources (PGR) are the basic raw materials in food and agriculture and an essential component of agrobiodiversity. They include any plant-derived genetic material, which includes reproductive and vegetative propagating parts containing functional hereditary units (such as seeds, tissues, and other components) that possess present or future significance for humans (FAO, 2009). The broad diversity of plant species, including their wild relatives, landraces, and cultivated varieties, which have evolved through natural selection and human intervention over time, are all included under PGR (Fig. 4.2). The genetic diversity found in plant species arises from heritable variations present within and among populations, arising through genetic variation in the DNA nucleotide sequence, chromosome mutations, and recombination during sexual reproduction. This vast array of genetic diversity within plant species, which provides a wide range of traits, such as resistance to diseases, pests, and environmental stresses, can be utilized for improving yield potential, nutritional quality, adaptation to diverse climatic conditions, combating pest and disease resistance, and reducing the input requirements (e.g. water, fertilizers, pesticides) of agricultural systems (Paroda, 2022). A wide genetic base is essential for crop improvement, which includes landraces, elite cultivars, local selections, genetic stocks, advanced breeding materials, wild relatives of crop plants, and wild species. If genetic diversity is lost, the particular combination of genes in a well-adapted landrace/variety may be difficult to rebuild. Further, PGR have evolved and adapted to different ecological conditions, including diverse soil types, temperature ranges, rainfall patterns, and pest pressures. Their genetic diversity is a valuable resource for breeding programs, enabling the development of improved crop varieties with enhanced tolerance to abiotic stresses (e.g. drought, heat, salinity) and biotic stresses (e.g. pests, diseases) (Paroda and Agrawal, 2017).

PGRs also serve as a source of nutritionally rich and culturally important crops, promoting

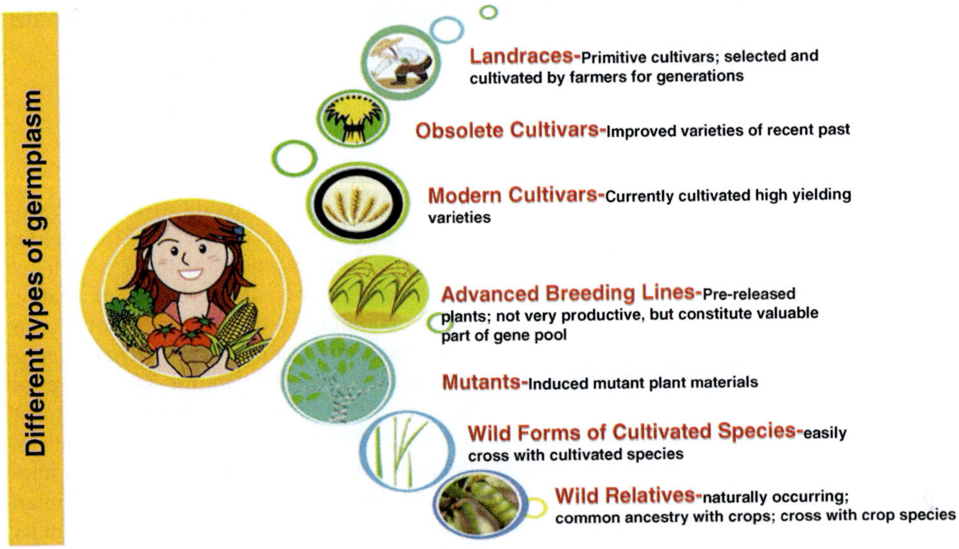


**Fig. 4.2.** Categories of plant genetic resources.

dietary diversity and food sovereignty. Biotechnological tools, such as next-generation sequencing, molecular markers, *in vitro* culture technology, cryopreservation, and gene banks aid in accurately characterizing and conserving PGR. The use of genomic tools enables the identification of quantitative trait loci (QTLs) and novel genes in plants, which can be transferred through marker-assisted selection and marker-assisted backcrossing breeding or gene-editing methods to sustain genetic diversity and support the creation of improved crop varieties. Thus, PGR provides the foundation for crop improvement programs, enabling breeders to select and cross plants with desired traits to develop new cultivars. This process enhances crop productivity, nutritional quality, and yield potential, contributing towards food security, agricultural sustainability, environmental protection, climate change adaptation, and sustainable development (Paroda and Agrawal, 2017).

## Role of PGR in accelerated climate change scenario

Since the beginning of crop domestication (> 12,000 years ago), diverse genetic resources have enabled farmers to adapt to gradual climatic changes and other shifting demands and pressures (Paroda *et al.*, 2018). Traditional crop varieties are well-adapted to current conditions in their local production environments. Diverse species, varieties, and cultivation practices allow crops to be grown across a wide range of environments. The challenge for the future is to maintain a good match between crops and production environments as the effects of climate change increase. Anthropogenic accelerated climate change calls for a much broader vision of risk management, especially given its potentially catastrophic effects on food production (FAO, 2015a). While some of the problems associated with climate change may emerge relatively gradually, action is urgently needed in order to allow enough time to build resilience into crop varieties. For this, diverse genetic traits and characteristics present within PGR are expected to play a crucial role in enhancing climate resilience in agricultural production systems (FAO, 2015b).

In the context of dryland agriculture, traits that would be needed would include short life cycles, deep root systems, waxy leaves, water storage capabilities in roots and trunks, dormancy during droughts, and tolerance to high temperatures. Similarly, livestock species and breeds would require further adaption to the limited vegetation and water resources in drylands, exhibiting traits such as reduced water loss, the ability to traverse long distances over rugged terrain, and resilience to harsh conditions.

Individual genotypes that are not useful now will become useful in future due to changes in climate or the outbreaks of diseases (Campbell *et al.*, 2008). Genetic resources from drylands hold a unique advantage as they have evolved over centuries by natural and human selection under drought, high temperature, or saline conditions (Yadav, 2019). They are better adapted to the local conditions and could be of immense importance, especially as sources of native genes with resistance to various biotic and abiotic stresses, and also make unique study material to understand the mechanism of adaptation to abiotic stresses. They could also serve as excellent genomic resources for the isolation of candidate genes responsible for tolerance to climatic and edaphic stresses for accelerating further genetic improvement (Yadav, 2019).

Crop wild relatives (CWR) will be a key resource as their genes can promote resistance to many of the environmental stresses associated with climate change. Through traditional breeding techniques and modern biotechnological approaches, traits such as improved water-use efficiency, enhanced nutrient uptake, tolerance to extreme temperatures, and resistance to pests and diseases can be incorporated into mainstream varieties of crops. By incorporating these traits, climate-resilient crop varieties can be developed which can better cope with the adverse impacts of climate change. By utilizing the diverse germplasm, farmers can adopt climate-smart practices, including crop diversification, intercropping, and agroforestry to enhance ecosystem services, conserve natural resources, and improve food security.

Crop production in all countries relies on genetic resources sourced from all over the world (Paroda and Agrawal, 2017). This interdependency is likely to increase as a result of climate change. Novel climatic conditions would mean that landraces and varieties will no longer be well-adapted to environmental conditions in the places where they have traditionally been grown. Better-adapted crops will need to be brought in from elsewhere. Climatic projections by Burke *et al.* (2009) indicate that by 2050, many countries will experience novel climates that are not currently found within their borders. Accordingly, national and international breeding programs for several crops are already seeking to develop new varieties that will be well-adapted to future climates. This is likely to increase demand for a range of PGR, including those of CWR. While this demand will be global, the natural distribution of CWR is restricted to the centers of origin of the respective crops, often specific sub-regions within continents. Hence, the greater international movement of germplasm will be essential in adapting agriculture to novel climates.

Farmers usually address gradual changes in climate either by drawing on adapted material from among the genetic resources already present locally or by seeking material from neighboring areas. However, as the climate becomes more variable and extreme events become more extreme, new strategies will be needed including greater intra-varietal diversity. Traits that contribute to phenotypic plasticity (the capacity to cope with a wide range of environmental conditions) will become increasingly important. Climatic conditions, particularly temperature and photoperiod, determine the distribution of a plant species by acting directly through physiological constraints (growth and reproduction) or indirectly through ecological factors (competition for resources). Of the three main components of climate change, temperature is known to have large effects, both in terms of lethality at high and low temperatures, and in affecting both phenology and distribution of crops (Osnato *et al.*, 2022). Recent emerging data are indicating that elevated $CO_2$ influences the outcome of plant–pathogen interactions, but no unifying framework exists to understand and predict its effects and consequences (Singh *et al.*, 2023).

## PGR Diversity in Drylands: Indian Perspective

### Status of agriculture in India

India plays a crucial role in driving inclusive economic growth, as it serves as a primary source of livelihood for millions of smallholder farmers and rural residents. Its economy relies heavily on agriculture, with over 54% of the country's land classified as arable and 54.6% of the entire workforce involved in agricultural and related activities (DAFW, 2022). The agricultural sector contributed 18.8% to gross value added (GVA), and over 24% to its gross domestic product during 2021–22 (DAFW, 2022).

Over the years, India has achieved significant progress in food production through initiatives like the Green, White, Yellow, and Blue Revolutions (Paroda, 2022). Nevertheless, the degradation of natural resources (such as soil, water, and air) combined with the detrimental effects of climate change present significant challenges to the long-term sustainability of agricultural production and the livelihoods of farmers. The challenges are particularly severe in the Green Revolution corridors of India, where issues such as hydrological imbalance, soil degradation, and water pollution have become daunting problems (Paroda, 2019b). Under these circumstances, India needs to particularly strategize about climate change due to high dependence of its population on agriculture, forestry, and fisheries for their livelihoods.

In India, the majority of agricultural practices rely on rainwater for irrigation, encompassing both dry farming and dryland agriculture. Out of the total 153.88 million hectares (mha) of cultivated land in the country, approximately 48% is rainfed and spread over 177 districts (ISRO, 2016; DAFW, 2022). Rainfed agriculture plays a crucial role in food production, contributing 44% of the country's foodgrains.

## Dryland agriculture in India

The drylands in India occupy about 80 mha and are spread over arid, semi-arid, and sub-humid climatic zones (Yadav, 2019). Dryland agriculture has a significant component in the Indian agrarian economy, as nearly 56% of the total net cultivable area falls under dryland, wherein 48% is under food crops and 68% under non-food crops (DAFW, 2022). Further, 50% of the total rural workforce and 60% of livestock are concentrated in the dry regions of India (Paroda, 2019b). The Indian arid zone, covering an area of 0.32 million km$^2$ falls in the states of Rajasthan, Gujarat, Haryana, Andhra Pradesh, and Karnataka, of which 62% lies in Rajasthan alone (Bhatt *et al.*, 2019).

The economy of the region is livestock-centric, which mainly depends upon the native rangelands, pasture grasses, cultivated fodder, and roughages. Thus, drylands always have played and will continue to play a critical role in India's food security. In terms of crops cultivated in dryland farms in India, prominent ones include coarse cereals such as pearl millet (*Pennisetum glaucum* (L.) R.Br.), barley (*Hordeum vulgare* L.), sorghum (*Sorghum bicolor* (L.) Moench, finger millet (*Eleusine coracana* (L.) Gaertn.), along with legumes including chickpea (*Cicer arietinum* L.), mungbean (*Vigna radiata* (L.) R.Wilczek), mothbean (*Vigna aconitifolia* (Jacq.) Maréchal), and clusterbean (*Cyamopsis tetragonoloba* (L.) Taub.). Oilseeds like Indian mustard (*Brassica juncea* (L.) Czern. & Coss.), groundnut (*Arachis hypogaea* L.), and soybean (*Glycine max* (L.) Merr.) are also commonly grown in these regions.

Furthermore, dryland agriculture encompasses the cultivation of numerous horticultural crops, grasses, fodder crops, bushes, medicinal plants, multi-purpose trees, underutilized potential crops, and seed spices (Yadav, 2019). The forage genetic resources (e.g. lucerne, cowpea, range grasses and legumes, trees, and shrubs) occurring in this region assumes great significance in view of its adaptability to harsh environmental conditions forming a storehouse of genes of stress tolerance (Bhatt *et al.*, 2019).

## PGR diversity in drylands

Among the 36 recognized global biodiversity hotspots, nine are in the drylands and about 0.5% of the plant species are endemic to the region (Mittermeier *et al.*, 2005; Agrawal *et al.*, 2019). In terms of agriculture, plant species endemic to the drylands make up 30% of the crops under cultivation today, including many local landraces and CWRs. However, the exact status of species in the drylands remains unknown, as no comprehensive assessment has been collated. It is also interesting that globally drylands have been the centers of origin of the most important food crops and a large number of breeds of cattle, buffaloes, goats, sheep, equines, and camels (Paroda, 2019b). Specific examples of crops originating in dryland regions are maize in Mexico, beans in Central and South America (Peru especially), tomatoes in Peru/Ecuador, potatoes in Peruvian and Bolivian Andes, millets in Africa and India, and wheat in the Fertile Crescent

(Iran-Iraq) under West Asia and North Africa (WANA), especially regions of Syria (Breda and Boider) and Turkey (Paroda, 2019b).

Rich crop diversity occurs in the Indian arid zone in terms of a number of species, genotypes, and ecotypes, encompassing a diverse range of genetic materials, including landraces, improved elite varieties, traditional cultivars, genetic stocks, and CWR (Singh *et al.*, 2019). These genetic resources hold immense potential as they possess specific phenotypic traits, genes for resistance to diseases and insect pests, as well as tolerance to various abiotic stresses prevalent in dryland environments. Dryland soils tend to be vulnerable to wind and water erosion, subject to intensive mineral weathering, and have low fertility (due to the low content of organic matter in the topsoil). Thus, the growth and development of plants are deterred by a prolonged period without rain that threatens plant life with desiccation and complete destruction. Nonetheless, India's Thar Desert and adjoining arid zone (regions of Rajasthan, parts of Gujarat, Haryana, and Punjab), are an important center of origin and diversity for many crops and plant species that can withstand its harsh environment and are listed in Table 4.1.

The Thar Desert in India is rather unique among the deserts of the world as it not only sustains the highest human and livestock population per km$^2$ but has also unique and rich flora and fauna, generally not found elsewhere. The Indian drylands are also home to numerous landraces and wild relatives of important crop plants that are still found within the prevailing traditional farming systems, which can be used as potential sources of useful genes for plant breeding. Unfortunately, many of the remaining hotspots of dryland biodiversity which have the potential to contribute to climate change solutions are under rapid erosion and facing extinction, on account of the combined effects of over-exploitation, destruction of natural habitats, modernization of traditional farming systems, lack of attention, and climate change (Paroda, 2019b). Hence, an important aspect of food security within the context of climate change will be to take measures to secure the genetic resources of agricultural drylands. Agrobiodiversity contributes to resilience through a number of, often combined, strategies: the protection and restoration of ecosystems, the sustainable use of soil and water resources, agroforestry, diversification of farming systems, various adjustments in cultivation practices, the use of stress-tolerant crops, and crop improvement.

## PGR Management for Climate-resilient Agriculture

### Germplasm exploration and collecting in drylands

Systematic collecting and conserving PGR diversity from drylands is essential for future adaptation strategies. During a satellite symposium on 'Dryland Agrobiodiversity for Climate Change Adaptation' held at Jodhpur, India, it was understood that available PGR variability in arid regions is relatively advanced. However, much work is needed particularly in the context of changing climatic conditions (Agrawal *et al.*, 2019). Hence, efforts on research, documentation, and conservation, including their protection under the existing laws, are required to be intensified as a priority (Paroda, 2019b).

India possesses a unique and highly adapted genetic diversity of crops, grasses, shrubs, and trees that can play a crucial role in mitigating the impacts of climate change. To effectively conserve the existing biodiversity in arid regions, it is essential to prioritize the species requiring immediate attention or those that are on the brink of extinction. This can be achieved by developing a 'Red Data Book' specifically dedicated to arid region agrobiodiversity, which would serve as a valuable tool for prioritization and enhancing conservation efforts. To ensure the success of this initiative, it is important to raise public awareness and advocate for supportive policies. Additionally, there is a need to identify and prioritize climate-resilient germplasm for utilization in crop improvement programs. Incentives and rewards should be provided to farmers who actively conserve the agrobiodiversity, and the valuable environmental services they provide as smallholder farmers. To accomplish this, networking and partnerships among institutes and countries in the region to transform areas that lack vegetation into green spaces is most essential.

**Table 4.1.** Some important crop and plant species of the Indian arid zone. (Adapted from Agrawal *et al.*, 2019 and Gowthami *et al.*, 2021)

| Botanical name | Common/ *vernacular* name | Botanical name | Common/ *vernacular* name |
|---|---|---|---|
| **Cereal and millets** | | **Wild edible species** | |
| *Hordeum vulgare* L. | Barley, *jau* | *Cassia tora* L.* | Sickle Senna, *chakvad* |
| *Eleusine coracana* (L.) Gaertn.** | Finger millet, *ragi* | *Caralluma edulis* Edgew.* | Edible caralluma, *mithi dudhi* |
| *Panicum sumatrense* Roth** | Little millet, *sama* | *Ceropegia bulbosa* Roxb.* | Bulbous ceropegia, *mudakathan* |
| *Paspalum scrobiculatum* L. | Kodo millet, *kodra* | *Cordia gharaf* Roxb.* | *gharaf* |
| *Pennisetum glaucum* (L.) R.Br.** | Pearl millet, *bajra* | *Grewia tenax* Forssk.* | White crossberry, *gangeran* |
| *Setaria italica* (L.) P. Beauv.** | Foxtail millet, *kangni* | *Leptadenia reticulata* Wight & Arn.* | *dori* |
| *Sorghum bicolor* (L.) Moench** | Sorghum, *jowar* | *Rhus mysorensis* Heyne ex Wight & Arn.* | Mysore Sumac, Mysore *anacardi* |
| **Legumes** | | **Tree /shrub/multi-purpose species** | |
| *Cyamopsis tetragonoloba* (L.) Taub.** | Cluster bean, *guar* | *Acacia nilotica* (L.) Delile** | *babool* |
| *Macrotyloma uniflorum* (Lam.) Verdc.** | Horsegram, *kulthi* | *Acacia senegal* (L.) Delile | Gum arabic, *khair* |
| *Vigna aconitifolia* (Jacq.) Maréchal* | Moth bean, *matki* | *Acacia persica* (L.) Willd.** | Persica tree, *desi khejri* |
| **Oilseeds** | | *Aerva persica* (Benth.) Meikle** | *desi bui, khareti, buinda* |
| *Brassica juncea* (L.) Czern. & Coss.** | Mustard, *sarson/rai* | *Azadirachta indica* L.* | *neem* |
| *Ricinus communis* L.** | Castor, *arandi* | *Calotropis procera* Aiton** | *Madar, aak* |
| **Seed spices** | | *Crotalaria burhia* Buch.-Ham. ex Roxb.** | *Sanai* |
| *Cuminum cyminum* L.* | Cumin, *jeera* | *Leptadenia pyrotechnica* (Forssk.) Decne.** | *Dagar phool* |
| *Trigonella foenum-graecum* L.* | Fenugreek, *methi* | *Prosopis cineraria* (L.) Druce* | *Khejri* |
| **Fruits and vegetables** | | *Salvadora* spp.* | *Pilu* |
| *Aegle marmelos* Correa* | Woodapple, *bael* | *Tecomella undulata* (Sm.) Seem.** | *Rohida* |
| *Balanites roxburghii* Planch.* | Desert date, *hingot* | *Vachellia nilotica* (L.) P.J.H. Hurter & Mabb** | *Babool* |
| *Capparis decidua* (Forssk.) Edgew.* | *ker* | *Ziziphus nummularia* (Burm.f.) Wight & Arn.* | *Ber* |
| *Coccinia indica* Wight & Arn.* | Ivy gourd, *kundru, tindora* | **Threatened species** | |
| *Cordia dichotoma* Forssk.* | *Gonda, lasora* | *Barleria prionitis* L.* | *Bajradanti* |
| *Ziziphus mauritiana* Lam.** | Indian jujube, *ber* | *Cenchrus prieurii* Fernald | *Khursi* |

**Table 4.1.** Continued.

| Botanical name | Common/ *vernacular* name | Botanical name | Common/ *vernacular* name |
|---|---|---|---|
| **Medicinal plants** | | *Chlorophytum borivilianum* (Santapau & Fernandes) Thomps. | *Safed musli* |
| *Plantago ovata* Forssk.* | Psyllium, *isabgol* | *Cleome gynandra* L. | *Hulki* |
| *Solanum surattense* Burm.f.* | Surattense Nightshade, *kantakari* | *Cochlospermum religiosum* (L.) Alston | Golden silk teak |
| *Withania somnifera* (L.) Dunal* | *Ashwagandha* | *Commiphora wightii* (Arn.) Bhandari* | *Guggul* |
| **Forage and fodder plants** | | *Farsetia macrantha* Decne. | *Bangariya* |
| *Calligonum polygonoides* L.* | *phog* | *Haloxylon salicornicum* (Moq.) Bunge | *Morari* |
| *Cenchrus ciliaris* L.* | Buffel grass, *dhaman* | *Peganum harmala* L. | *Isband* |
| *Cenchrus setigerus* Vahl.** | Bristle bur grass, *pulha* | *Tephrosia falciformis* (L.) Pers. | *Sarpgandh* |
| *Dichanthium annulatum* (Forssk.) Stapf. | Blady grass, *dhaman* | *Tribulus rajasthanensis* Ahuja & Gajer* | *Gokhuru* |
| *Lasiurus scindicus* Henrard* | Rajasthan Wild Rye, *sewan* grass | *Withania coagulans* (Stocks) Dunal | *Dhak* |
| *Panicum antidotale* Retz.** | *Mundi* | *Ziziphus truncata* R.Br.** | *Dhabb* |

*primary center of origin and diversity in Indian gene center
**secondary center of diversity in Indian gene center

For germplasm-collecting missions, it is important to collect unique germplasm that is well adapted to specific environments (Amri *et al.*, 2019). Further, the areas may not be disturbed, thus allowing for the natural evolution of the species. This approach would not only contribute to crop improvement but would also help identify potential sites for *in situ* conservation.

## PGR conservation strategies under climate change

The overall goal of PGR conservation is to mitigate the loss of genetic, species, and ecosystem diversity, so as to ensure food, nutrition, and health security for posterity. Humans have adopted, broadly, two approaches to save the diversity of life on Earth, namely: (i) protectionism (*in situ*), which focuses on the integral protection of nature, without human interference; and (ii) conservationism (*ex situ*), which targets saving life and ecosystems through sustainable and harmonious use of nature. Conservation, whether *in situ* or *ex situ*, involves maintaining minimum viable populations of the target taxa (Agrawal *et al.*, 2022). With respect to plants in general, and PGR in particular, a multiplicity of conservation strategies has been developed in the last two centuries and are depicted in Fig. 4.3. Success of any conservation program depends on its ability to adequately represent the species during storage and to preserve the utility of stored germplasm in future recovery efforts.

Traditionally, the Indian Genebank conserves germplasm *ex situ* (Paroda and Agrawal, 2017). Given the unprecedented challenges posed by climatic disasters, social conflicts, and political uncertainties, the integration of *in situ* and *ex situ* strategies is increasingly being recognized as a necessity for the effective conservation of PGR, especially CWR. Towards this, the concept of '*trans situ* conservation' has gained momentum, which aims to protect the genetic diversity and

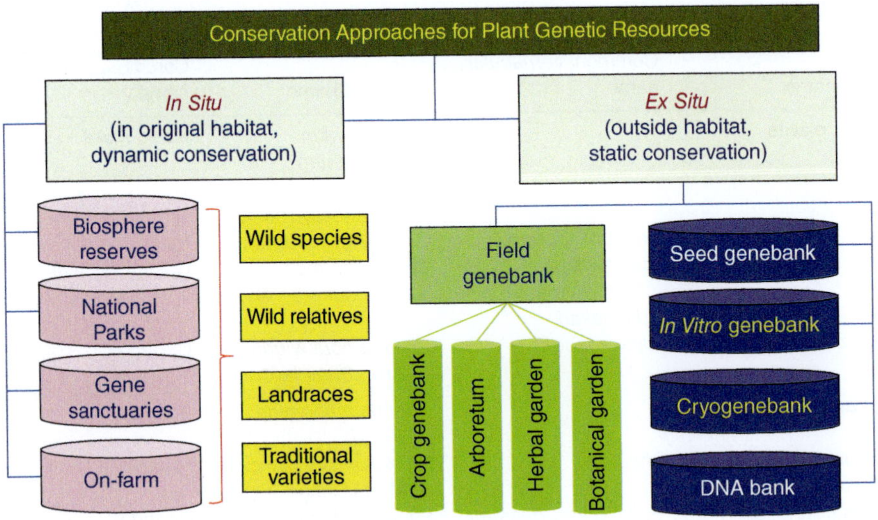


**Fig. 4.3.** Approaches for conservation of plant genetic resources.

accessibility of PGR for crop improvement (Riordan and Nabhan, 2019). *Trans situ* conservation builds upon the existing approach of combining *in situ* protection with *ex situ* back-up in gene banks but takes it a step further by dynamically integrating multiple measures spanning local to global scales, including conservation, research, and education. This 'one conservation' or 'integrated approach' develops multi-disciplinary conservation strategies based on the ecology, biology, diversity, and utility of the targeted plants.

A *trans situ* approach offers two important benefits: firstly, integrating *in situ* and *ex situ* studies of PGR genetic diversity, adaptation, and ecological interactions, advancements in crop improvement, and *in situ* management. This holistic approach allows for a comprehensive understanding of the interplay between PGR and their changing environments, leading to more effective conservation and utilization. Secondly, the complementarity, redundancy, and synergy achieved through *trans situ* conservation serve as a buffer against the uncertainties arising from climatic, economic, political, and institutional instabilities (Riordan and Nabhan, 2019).

Fig. 4.4 illustrates the key components of *trans situ* conservation: *in situ* protection on public lands/on-farm; living plant collections in botanical gardens, arboreta, nurseries, or field genebanks; and germplasm accessions in the national/international gene banks comprising seeds, *in vitro*, and/or cryobank collection. These complementary and mutually reinforcing approaches are adopted based on three criteria: (i) determination of the taxa of conservation concern (species, sub-species, variety or race); (ii) determination of the threats to the species; and (iii) consideration of the different approaches and resources that may be applied for an integrated conservation program.

## Traditional knowledge and community involvement

On-farm conservation of PGR will become increasingly important in the context of climate change. Conserving genetic resources in complex, diverse, risk-prone environments builds on natural and farmer selection and helps in mitigating climates change issues. Fortunately, farmers in the drylands of India continue to uphold their tradition of conserving agrobiodiversity. For instance, in the north-western states of Rajasthan, Haryana, and Punjab, the Bishnoi community follows a philosophy that strictly prohibits harm to both trees and fauna (Paroda, 2019b). They believe that the souls of their ancestors reside in plants and animals, with the khejri tree (*Prosopis cineraria*) and the black buck being classical examples. The khejri tree serves multiple purposes, providing food, fodder, fuel,

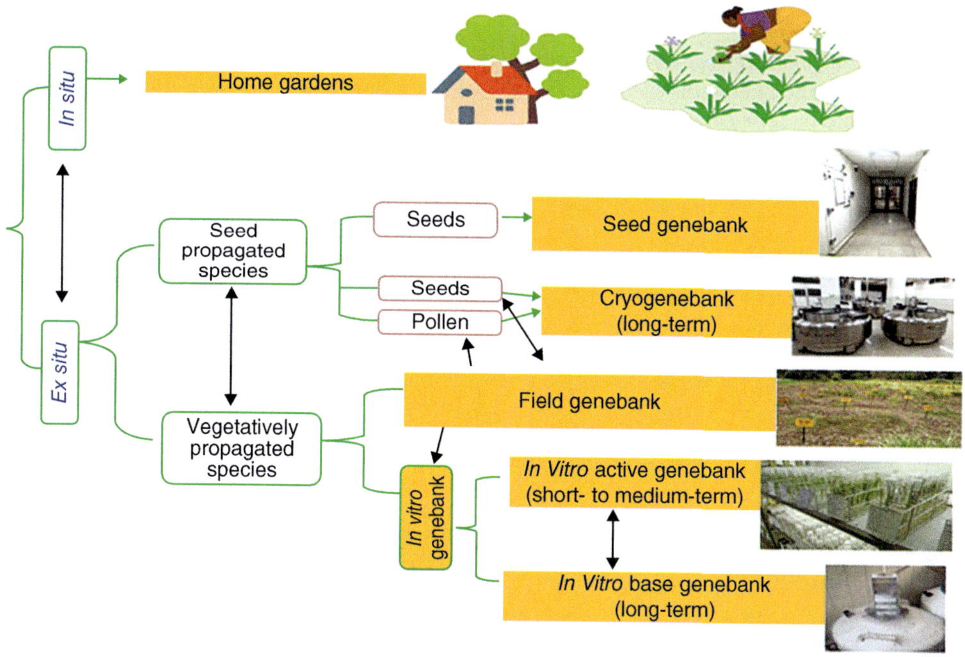


**Fig. 4.4.** Model for *trans situ* (integrated) conservation strategy of PGR.

and building materials. It also enriches the soil with nitrogen and allows crops beneath it to thrive without interference. The farmers consciously conserve valuable trees and shrubs, such as the khejri, within their crop fields, ensuring sustained cropping without any fallow periods throughout the growing season. During drought years when annual crops fail, farmers heavily rely on the products of trees, shrubs, and animals for their survival. This eco-consciousness within communities contributes to the *in situ* conservation of agrobiodiversity and fosters an eco-friendly and healthy environment for the present generation as well as future generations to come (Paroda *et al.*, 2004; Paroda, 2019b).

In the context of adaptation to climate change, seed systems play a crucial role, particularly at the local level. They provide farmers with the opportunity to exchange landraces that possess a wide range of characteristics and have been subjected to novel selection pressures associated with climate change. These diverse seeds enable farmers to adapt their agricultural practices to changing climatic conditions. However, there are limitations to the adaptive capacity of local seed systems. As the pace of climate change accelerates, the geographic reach of seed systems needs to expand. It becomes necessary to develop policies that support seed systems and facilitate long-distance seed exchange, potentially through initiatives like seed fairs and other means of connectivity. In addition to informal mechanisms, local seed systems can be enhanced through the establishment of more formal community-based seed enterprises. These enterprises can help smallholder farmers gain access to improved and adapted cultivars, as well as other inputs necessary for climate-change adaptation. By combining traditional knowledge with modern breeding techniques, these community-based enterprises contribute to building resilience and supporting farmers in their efforts to adapt to climate change.

## Utilization of PGR

### Breeding for climate resilience

In the history of plant breeding over the last century, primary efforts were concentrated on the use of PGR in major food crops such as wheat, rice, barley, tomato, soybean, and mustard. The use of PGRs lags far behind its potential due to certain hindrances such as cross-incompatibilities, infertility in the F1 and subsequent progeny, non-availability or poor conservation of CGRs, underutilized and underexplored wild germplasm, lack of reliable evaluation data, and the expression of undesirable traits in cultivated genetic background. The lack of easy accessibility of PGR to researchers and breeders is another important impediment to crop improvement. There are many examples of successful gene discovery and transfer of superior alleles from specific PGRs to domesticated crop varieties, despite the numerous impediments encountered (Goodman *et al.*, 1987). A few examples of potential climate change mitigation traits from PGR to crop varieties are given in Table 4.2.

## Use of biotechnology and molecular breeding

High-throughput next-generation sequencing technology has led to increased discovery of single nucleotide polymorphisms (SNPs) and their association with important traits. Combining the phenotypic data with genotypic data, researchers have been able to validate and fine-map previously identified genes and to discover novel genomic regions underlying valuable agronomic traits in CWRs by association mapping (Li *et al.*, 2014; Qi *et al.*, 2014; Zhou *et al.*, 2015). Thus, it is evident that the availability of genomic data along with efficient phenotypic evaluation is the key factor in the discovery of genes controlling superior traits in CWRs.

These omics approaches are particularly suitable for the dissection of complex traits such as drought tolerance and pest resistance by characterizing CWRs under diverse treatments. For example, dehydrin genes in both wild barley (*H. spontaneum*) and wild tomato (*S. chilense* and *S. peruvianum*) as well as ABA/water stress/ripening induced Asr gene family members (Asr2 and Asr4 from wild *Solanum* species) are known to be involved in drought tolerance (Fischer *et al.*, 2013). The genome-assisted breeding coupled with next-generation sequencing approaches has opened an important corridor for evaluating alternative crop genomes for different stress-tolerance mechanisms.

The discovery of the model plant, *Arabidopsis thaliana*, helped in the evaluation of alternative crop genomes for serving new gene resources, for example; diploid wild grass *Brachypodium distachyon* with fully sequenced small genome size (International Brachypodium Initiative, 2010) is a valuable model plant for wheat and barley for drought tolerance owing to high phylogenetic similarity reported by Mochida *et al.* (2013). Moreover, there was no report on the negative effects of drought stress on the growth and development of *Brachypodium distachyon* (Verelst *et al.*, 2013). *Haynaldia villosa* is another example that possesses many beneficial genes for improving resistance to powdery mildew, leaf and stem rusts, and wheat streak mosaic virus (Chen *et al.*, 2002). *Elytrigia elongata* is a potential source of the salt-tolerance gene from the tribe Triticeae. Similarly, sheepgrass (*Leymus chinensis*), and desiccation model plants *Sporobolus stapfianus* and *Agropyron cristatum* are beneficial reservoirs of stress-tolerance genes for introduction into cultivated crops such as wheat and barley.

## Identification of desirable traits and reviewing mitigation strategies

The Second Global Plan of Action for Plant Genetic Resources for Food and Agriculture (FAO, 2011), negotiated by the Commission on Genetic Resources for Food and Agriculture and adopted by the FAO Council in 2011, contains a range of priority activities related to PGR management, building sustainable institutional and human capacities that address challenges such as climate change and food insecurity, as well as novel opportunities such as the use of information and communication technologies and molecular methodologies. It highlights the following strategic elements needed to safeguard PGR and use them optimally to help cope with climate change:

- Enabling gene banks to respond to novel and increased demands for germplasm that can be used in climate change adaptation by including different characteristics in screening processes and ensuring that collections are comprehensive and include crops that are so far considered to be 'minor'.

**Table 4.2.** Examples of potential climate change mitigation traits from PGR to crops. (Adapted from Maxted *et al*., 2012)

| Crop | CWR | Uses | References |
|---|---|---|---|
| Chickpea (*Cicer arietinum* L.) | *Cicer reticulatum* Zohary & Heller | Drought and temperature tolerance | Hajjar and Hodgkin (2007) |
| Sunflower (*Helianthus annuus* L.) | *Helianthus paradoxus* Heiser | Salt tolerance | Lexer *et al.* (2004) |
| Barley (*Hordeum vulgare* L.) | *Hordeum spontaneum* C. Koch | Drought and temperature tolerance | Hadjichristodoulou (1993); Eglinton *et al*. (2001); Baum *et al*. (2003); Talamè *et al.* (2004); Chen *et al*. (2008) |
| Tomato (*Solanum lycopersicum* L.) | *Solanum cheesmanii* J.M. Edmonds, *Solanum pennellii* Correl, *Solanum peruvianum* L. | Various forms of resistance, drought and salinity tolerance, soluble solids, insect resistance | Stevens and Rick (1986); Rick and Chetelat (1995); Hajjar and Hodgkin (2007) |
| Cassava (*Manihot esculenta* Crantz) | *Manihot angustiloba* Kuntze, *Manihot davisiae* Pax & K. Hoffm., *Manihot rubricaulis* Hook | Drought tolerance, adaptation to high altitudes and cool temperatures | Jennings (1995) |
| Banana and plantain (*Musa* spp.) | *Musa balbisiana* Colla, *Musa itinerans* Cheesman, *Musa nagensium* Kuntze | Drought resistance, water-logging resistance | INIBAP/IPGRI (2006) |
| Rice (*Oryza sativa*) | *Oryza longistaminata* A. Chev. & Roehr, *Oryza rufipogon* Griff., *Oryza glaberrima* Steud., *Oryza minuta* J. Presl & C. Presl, *Oryza officinalis* Wall. ex Watt | Drought resistance and yield increase, early flowering leaf elongation and photosynthetic rates | Brar *et al*. (2005) |
| | *O. rufipogon* | Drought resistance | Zhang *et al*. (2006) |
| | *Oryza officinalis* | Mitigate high temperature-induced spikelet sterility at anthesis | Ishimaru *et al*. (2010) |
| | *Oryza meridionalis* Ng. | Heat tolerance, | Scafaro *et al*. (2010) |
| Wheat (*Triticum aestivum)* | *Aegilops* spp., *Thinopyrum* spp., *Elytrigia elongata* Nevski, *Hordeum* spp. | salt tolerance, stay-green trait, thylakoid membrane damage, chlorophyll content and yield | Farooq *et al*. (1992, 1995); Colmer *et al*. (2006) |
| | *Thinopyrum* spp. | Drought tolerance | King *et al*. (1997a, b) |
| | *Aegilops* spp. | High-temperature tolerance | Pradhan *et al*. (2012) |

*Continued*

**Table 4.2.** Continued.

| Crop | CWR | Uses | References |
|---|---|---|---|
| *Triticum durum* Desf. | *Aegilops speltoides* Tausch ex Gren. & Godr. | Salt tolerance | Noori (2005) |
| *Triticum turgidum* L. | *Aegilops tauschii* Coss. | Water-logging tolerance | Villareal *et al.* (2001) |
| Potato (*Solanum tuberosum* L.) | *Solanum gandarillasii* Cardenas | Drought resistance | Coleman (2008) |

- Reviewing breeding strategies and priorities crop-by-crop and region-by-region to ensure that the products of crop improvement programs will be relevant to the challenges the world will be facing when the products are ready for release at the end of the crop improvement cycle (five to ten years into the future).
- Greater emphasis on *in situ* conservation of genetically diverse populations, especially CWR, should be given to allow evolution to continue and thus permit the continued generation of adaptive traits.
- Significantly expanding programs on *ex situ* conservation, especially of CWR, to ensure the maintenance of diversity of species, populations and varieties, including those adapted to extreme conditions and those from areas expected to be highly affected by climate change.
- Increased research and improved availability of information on the characteristics of material held *ex situ* that will become useful under new climate conditions.
- Reviewing and strengthening policies to promote dynamic seed systems, implementing measures that enable the longer distance exchange of seed between farmers, and reviewing priorities and procedures for post-disaster seed relief.
- Targeted and increased involvement of farmers and farming communities in national and local crop improvement activities, including support for participatory research and plant breeding.
- Consolidating collections of wild species, including CWR, because of the increased likelihood that narrowly adapted and endemic species will become extinct.
- Ensuring that collections give sufficient emphasis to stress-adapted genetic material that can contribute to adaptation to climate change.
- Ensuring that policies facilitate the international exchange of the genetic resources needed to adapt farming systems to shifts in the distribution of climatic zones. Increased support for access to and movement of PGR to meet the greater interdependence of countries resulting from the new environmental conditions.
- More support for building capacity in plant breeding and seed systems management that make effective and sustainable use of PGR.

## Strategies for Climate-smart Agriculture

Agricultural adaptation to climate change is expected to occur in various ways during the 21st century, mostly as a result of two different, but complementary, processes: short-term coping strategies and longer-term planned system changes (Howden *et al.*, 2007). There are several mitigation and adaptation practices that can be effectively put to use to overcome the effects of climate change with desirable results. These methods fall into the broad categories of crop/cropping system-based technologies, resource conservation-based technologies and socio-economic and policy interventions. Climate-smart agriculture (CSA) has been defined as a transformative approach aimed at achieving a triple win of development, adaptation, and mitigation in the face of a changing climate (FAO, 2010). It encompasses sustainable agricultural practices that not only increase productivity and yields but also helps to capture carbon and reduce GHG emissions, thus enhancing adaptation and resilience to climate change (Pathak *et al.*, 2021). Climate-smart strategies include strengthening the capacity of farmers and relevant stakeholders; introducing climate-smart agricultural technologies; promoting improved climate-resilient crop varieties and alternative adaptation options; raising awareness; building partnerships to address climate change in agriculture-dependent communities; documenting and disseminating best practices; and engaging in policy advocacy. There are various agricultural practices, which are discussed hereunder (Campbell *et al.*, 2011).

### Conservation agriculture

Conservation agriculture (CA) is a novel paradigm to achieve food production targets while addressing issues of natural resource degradation. CA harnesses natural ecological processes to enhance crop productivity and sustainability. It revolves around three fundamental principles: minimizing soil disturbance, maintaining permanent soil cover, and adopting diverse crop rotations to enhance soil health, water retention, and overall ecosystem resilience (Pathak *et al.*, 2021). The practice of minimizing soil disturbance helps maintain an optimal balance of respiration gases around the plant roots, reduces the oxidation of soil organic matter, preserves soil porosity for efficient water movement, and prevents weed seeds from being exposed to sunlight, thus inhibiting germination. Additionally, it enables beneficial microbes to create stable soil aggregates, promoting effective water infiltration (biological tillage).

In dryland ecosystems, implementing CA has led to improvements in soil organic carbon, soil biodiversity, and reduced soil erosion, achieved through practices such as crop residue cover, reduced tillage, and incorporating legumes into crop rotations. By increasing soil organic carbon content, water retention is enhanced, enabling crops to better withstand mid-season droughts (Pathak *et al.*, 2021). Thus, climate-change mitigation can be realized by integrating CA as an effective and sustainable livelihood strategy to combat ecosystem degradation.

### Crop diversification

Prior to the Green Revolution, Indian agriculture exhibited greater diversity and sustainability in terms of crop and livestock choices. Practices such as silvi-pastoral systems and agroforestry were employed alongside animal-based farming systems (Paroda, 2019a). However, with scientific advancements and the availability of improved crop varieties, the focus shifted towards a limited number of high-yielding crops that could generate higher income. This approach resulted in a heavy reliance on a few crops such as wheat, rice, maize, and sugarcane.

Crop diversification involves introducing new crops into existing farming systems, either at the individual farm level or on a larger scale (Paroda, 2022). It entails cultivating multiple varieties of crops, whether from the same species or different species, within a given area using practices such as crop rotations and intercropping. Reintroducing higher levels of crop diversification holds the potential to align with the core principles of climate-smart agriculture (Paroda, 2022). Firstly, it can contribute to improving productivity and livelihood outcomes by offering a wider range of crops to meet various market demands and enhance income opportunities. Secondly, crop diversification enhances the resilience and sustainability of farming systems, enabling them to withstand climate-related challenges. Lastly, it aids in reducing carbon dioxide emissions, making a positive impact on climate change through increased carbon sequestration (Paroda, 2022). This approach is ecologically viable, cost-effective, and logical, particularly for smallholder farmers, as it helps reduce uncertainties in agriculture.

There is a need for further diversification around local food systems, as has been emphasized by the UN Food Systems Summit held in September 2021 and the second International Agrobiodiversity Congress (IAC) organized in October 2021 (Paroda, 2022). For future diversification, pseudocereals (such as grain amaranth, buckwheat, and quinoa), legumes (including rice bean, faba bean, adzuki bean, and moth bean), and small millets (such as finger millet, foxtail millet, proso millet, little millet, barnyard millet, and kodo millet) show promise due to their high nutritional value and resilience to drought and heat. There is also the need to promote the cultivation of specific crops in new niches to harness their maximum genetic potential (Paroda, 2018).

The term underutilized and neglected species refers to a category of wild and cultivated plant species characterized by a low level of investment in research and development, and a recognized but untapped livelihood potential (Dempewolf *et al.*, 2013). Many such species have a potential for improving agriculture in diversified agro-ecological niches and have great potential for exploitation because of the value of their economic products. Widespread genetic erosion of these species has been recorded worldwide despite initiatives in recent years aimed at their conservation, these wild relatives are still at risk of erosion or extinction. They have been almost entirely neglected in *ex situ* gene banks (Castañeda-Álvarez *et al.*, 2016), although some efforts to change this have begun (Dempewolf *et al.*, 2013). In addition, current *in situ* reserves often do not meet the required management standards to maintain the wild relatives and their genetic diversity for long-term use. Therefore, the impact of climate change can be overcome through the cultivation of crop varieties which are adaptable to various environmental stresses, especially in semi-arid unutilized or underutilized marginal lands.

### Other CSA approaches

Besides CA and crop diversification, CSA includes many other approaches for reshaping and evolving production systems to cope with adverse weather conditions by developing locally adapted portfolios of crops and cropping systems (Aryal *et al.*, 2020; Pathak *et al.*, 2021; Pervez Bharucha *et al.*, 2021). These include:

1. **Climate-resilient crop varieties.** Developing and promoting climate-resilient crop varieties that are better adapted to changing climatic conditions, including drought-tolerant and pest-resistant crops.
2. **Integrated pest management (IPM).** Implementing sustainable pest control practices to reduce chemical pesticide use and promote natural pest predators.
3. **Livestock management.** Adopting climate-smart livestock practices, including improved feed efficiency, breed selection, and animal health management.
4. **Renewable energy adoption.** Encouraging the use of renewable energy sources, such as solar-powered irrigation, to reduce greenhouse gas emissions from fossil fuel consumption.
5. **Efficient water management.** Implementing water-saving techniques like drip-irrigation and rainwater harvesting to cope with changing rainfall patterns and water scarcity.
6. **Climate information services.** Disseminating timely and accurate climate information to farmers to enable better decision-making and planning.
7. **Capacity building and awareness.** Enhancing the knowledge and skills of farmers, extension workers, and policymakers about climate-smart practices through training and awareness programs.

## Role of Global Forums and Treaties

Over the last three decades, land degradation, loss of agrobiodiversity and climate change at local, national, and international levels have been identified as major interrelated challenges to sustainable food security (IPBES, 2019). This recognition led to the signing of three binding international agreements, namely, United Nations Framework Convention on Climate Change (UNFCCC), United Nations Convention to Combat Desertification (UNCCD), and Convention on Biological Diversity (CBD). Amidst the growing effects of climate change, a crucial challenge being faced by global leaders and policymakers is assisting rural communities in the marginal and dryland regions to sustain their livelihoods and ensure food security, despite facing degraded land, limited water resources, and increasingly erratic rainfall and temperature patterns. Combining crop diversification and sustainable intensification in rainfed (non-irrigated) dryland agriculture is now recognized as a global priority in order to fulfil the objectives outlined in the Sustainable Development Goals (SDGs) (Pervez Bharucha *et al.*, 2021). This priority is especially crucial for addressing poverty alleviation (SDG 1), ensuring food and nutritional security (SDGs 2 and 3), promoting livelihoods and economic growth (SDG 8), taking action on climate change (SDG 13), and halting land degradation, conserving biodiversity, and restoring land productivity (SDG 15). The goal pertains to enhancing productivity and agricultural yields in rainfed dryland areas, while also ensuring environmental sustainability and promoting social and economic well-being by contributing towards improved access to nutritious food, economic growth, and mitigating climate-change impacts.

The Commission on Genetic Resources for Food and Agriculture (CGRFA) under the FAO provides an intergovernmental forum for the discussion and development of knowledge and policies related to managing genetic diversity under a multilateral system, governed under an International Treaty on Plant Genetic Resources for Food and Agriculture (ITPGRFA). It currently covers 35 of around 150 food crops traded in the world market. Thousands of other crops or plant species, though consumed and traded locally, have not entered the global trading system. Also, their genetic diversity is not properly conserved in the gene banks.

To effectively address the climate change scenario, the exchange of useful species between countries will be mutually rewarding. Hence, following the principle of fair access and sharing of benefits will be desirable so that genetic resources can be used by those who need them the most. Many global online portals serve as a source of CWR inventories such as the databases Crop Wild Relative Global Portal (http://www.cropwildrelatives.org/), Crop Wild Relatives and Climate Change (http://www. cwrdiversity.org/), Gateway to Genetic Resources (http://www.genesys-pgr.org/), Crop Trust (https://www.croptrust.org/) etc., which can be taken advantage of.

## Conclusions

Climate change will cause shifts in the distribution of land areas suitable for the cultivation of a wide range of crops and agricultural ecosystems, including drylands. Understanding climate-change impact is important to design adaptation options for dryland agricultural systems by technology-based interventions such as water conservation, supplemental irrigation, foliar sprays, integrated nutrient management, resilient crops-based cropping systems, use of artificial intelligence, and precision agriculture (modeling and remote sensing). Responding to a changing climate will require changes in PGR management to address both immediate and slow-onset changes. Genetic resource conservation and use should extend beyond crops cultivated on farmlands, to encompass other components of the dryland ecosystem, such as trees, shrubs, and grasses. To cope with problems in agriculture arising due to climate change in dryland ecosystems, it is desirable to take action on the following priorities:

**1.** There is a need to understand the importance of genetic resources for food and agriculture in the ecosystem context to provide informed knowledge on climate change mitigation and adaptation strategy.

**2.** Concerted efforts are needed on germplasm collection from dryland areas, their characterization, evaluation, and utilization, so as to develop better varieties for changing climate regimes. Priority should be given to the characterization and evaluation of crop diversity in the arid region that already demonstrate resilience to abiotic and biotic stresses.

**3.** The development of a 'Red Data Book', specifically for drylands, would be an important tool to determine conservation priorities and minimize genetic erosion.

**4.** Research efforts need to be intensified using new science, techniques, and tools such as phenomics, genomics, genome editing, artificial intelligence, drones, robotics etc., for the identification of valuable genes that are better adapted to the changing climate in drylands.

**5.** To cope with drastic climate change events, we need to utilize the genetic diversity available within crop wild relatives (CWR), an aspect which has invariably been ignored. Also, to widen the genetic base, intensive pre-breeding efforts are critical. In the past, plant breeders had invariably been facing the problem of linkage drag while using CWRs. However, DNA sequencing technology, in particular the combination of *de novo* and resequencing, is now effectively employed to investigate relevant genetic variation in CWRs. Despite compatibility difficulties, improved techniques, such as genomics-assisted breeding, are enabling the introgression of desirable traits from wild to cultivated species.

**6.** Traits-specific breeding be targeted to (i) increase water-use efficiency (WUE) in areas of frequent drought; (ii) increase heat tolerance in areas of high temperature; (iii) increase yield under conditions of high atmospheric $CO_2$; and (iv) select and breed new species and cultivars that are perennial and stress-hardy.

**7.** Promoting mixed/inter-cropping, agroforestry, and silvi-pastoral systems help in reducing the risk and provide stable income to resource-poor farmers under the adverse conditions caused by climate change. Hence, genetic resources suitable for such practices need to be identified and used.

**8.** There is a need for the critical assessment of the risk and vulnerability of agricultural systems to different climate change scenarios at the local, regional, and national level. Accordingly, location-specific eco-regional planning for biodiversity conservation through use can be the best option.

**9.** Establishing primary agro-processing units in rural and peri-urban areas is crucial to minimize post-harvest losses and provide income options to the farmers. It is therefore necessary to create entrepreneurship-related opportunities for youth and women while conserving dryland genetic resources.

**10.** Appropriate models to incentivize farmers for their ecological services using sustainable agrobiodiversity conservation approaches have now to be promoted. For this, indexing environmental services for *in situ* and on-farm conservation of agrobiodiversity in dryland ecosystems would be helpful. Towards this, mainstreaming crops and varieties adapted well to the local environments shall help in promoting climate-resilient agriculture.

**11.** Identification of adaptation measures, taking into account the diversity of agricultural systems, indigenous knowledge and their differences in scale, as well as possible co-benefits and sharing experiences both around research and development would help in protecting available rich genetic resources.

**12.** Early-warning systems and contingency plans are needed to effectively address the ill effects of extreme weather events such as drought, floods, landslides, storm surges, soil erosion, and saline water intrusion.

**13.** Needed policy support for increased funding and the creation of *sui generis* system in each country to protect, manage, and exchange genetic resources of dryland crops will ensure greater sustainability and much-needed resilience in dryland agriculture.

**14.** Building effective partnerships in a consortium mode will help in accelerating the use of genetic resources. This should include knowledge and germplasm sharing, capacity building, and sustainable resource enhancement. Collaboration among institutions and countries would also facilitate a better exchange of expertise and resources for PGR management in dryland areas.

## Acknowledgement

The authors are grateful to Dr. Bhag Mal, Secretary, Trust for Advancement of Agricultural Sciences (TAAS), New Delhi, and Dr. Umesh Srivastava, Consultant, TAAS, for their critical inputs in editing the manuscript. AA thanks Dr. R. Gowthami, Scientist, ICAR-NBPGR, New Delhi, for designing Fig. 4.4.

## References

Agrawal, A., Rajkumar, S., Rana, J.C., Tyagi, R.K., Yadav, O.P. *et al.* (2019) *Proceedings and Recommendations of the Satellite Symposium on Dryland Agrobiodiversity for Adaptation to Climate Change*. Indian Society of Plant Genetic Resources, New Delhi.

Agrawal, A., Gowthami, R., Chander, S. and Srivastava, V. (2022) Sustainability of in vitro genebanks and cryogenebank. *Indian Journal of Plant Genetic Resources* 35(3), 180–184. DOI: 10.5958/0976-1926.2022.00065.1

Amri, A., Kehel, Z., Kumar, S. and Bassi, F.M. (2019) Efficient conservation and use of genetic resources of cereals and legumes. In: Agrawal, A., Rajkumar, S., Rana, J.C., Tyagi, R.K., Yadav, O.P., Singh, K. and

Paroda, R.S. (eds) *Proceedings and Recommendations of the Satellite Symposium on Dryland Agrobiodiversity for Adaptation to Climate Change*. Indian Society of Plant Genetic Resources, New Delhi, pp. 31–35.

Aryal, J.P., Sapkota, T.B., Rahut, D.B. and Jat, M.L. (2020) Agricultural sustainability under emerging climatic variability: the role of climate-smart agriculture and relevant policies in India. *International Journal of Innovation and Sustainable Development* 14(2), 219–245. DOI: 10.1504/IJISD.2019.10020869

Baum, M., Grando, S., Backes, G., Jahoor, A., Sabbagh, A. *et al.* (2003) QTLs for agronomic traits in the Mediterranean environment identified in recombinant inbred lines of the cross 'Arta' × *H. spontaneum* 41-1. *Theoretical and Applied Genetics* 107, 1215–1225. DOI: 10.1007/s00122-003-1357-2

Bhatt, R.K., Rajora, M.P., Patidar, M., Singh, J.P. and Kalia, R.K. (2019) Forage resources of drylands of India. In: Agrawal, A., Rajkumar, S., Rana, J.C., Tyagi, R.K., Yadav, O.P., Singh, K. and Paroda, R.S. (eds) *Proceedings and Recommendations of the Satellite Symposium on Dryland Agrobiodiversity for Adaptation to Climate Change*. Indian Society of Plant Genetic Resources, New Delhi, pp. 61–65.

Brar, D.S., Toriyama, K., Heong, K.L. and Hardy, B. (2005) Broadening the gene pool of rice through introgression from wild species. In: Toriyama, K., Heong, K.L. and Hardy, B. (eds) *Rice is Life: Scientific Perspectives for the 21st Century*. International Rice Research Institute, Metro Manila, Philippines, pp. 157–160.

Burke, M.B., Lobell, D.B. and Guarino, L. (2009) Shifts in African crop climates by 2050, and the implications for crop improvement and genetic resources conservation. *Global Environmental Change* 19(3), 317–325.

Campbell, A., Chenery, A., Coad, L., Kapos, V., Kershaw, F. *et al.* (2008) *Linkages between Biodiversity and Climate Change Mitigation: A Review of the Recent Scientific Literature*. UNEP World Conservation Monitoring Centre, Cambridge, UK. Available at: https://wedocs.unep.org/20.500.11822/8566 (accessed 4 July 2024).

Campbell, B., Mann, W., Meléndez-Ortiz, R., Streck, C. and Tennigkeit, T. (2011) *Agriculture and Climate Change: A Scoping Report*. Meridian Institute, Washington, DC. Available at: https://climatefocus.com/publications/agriculture-and-climate-change-scoping-report/ (accessed 4 July 2024).

Castañeda-Álvarez, N.P., Khoury, C., Achicanoy, H., Bernau, V., Dempewolf, H. *et al.* (2016) Global conservation priorities for crop wild relatives. *Nature Plants* 2(4), 1–6. DOI: 10.1038/nplants.2016.22

Challinor, A.J., Watson, J., Lobell, D.B., Howden, S.M., Smith, D.R. *et al.* (2014) A meta-analysis of crop yield under climate change and adaptation. *Nature Climate Change* 4, 287–291. DOI: 10.1038/nclimate2153

Chen, G., Li, C., Shi, Y. and Nevo, E. (2008) Wild barley, *Hordeum spontaneum*, a genetic resource for crop improvement in cold and arid regions. *Sciences in Cold and Arid Regions* 1, 0115–0124.

Chen, Q., Conner, R.L., Li, H., Laroche, A., Graf, R.J. *et al.* (2002) Expression of resistance to stripe rust, powdery mildew and the wheat curl mite in *Triticum aestivum* × *Haynaldia villosalines*. *Canadian Journal of Plant Science* 82(2), 451–456. DOI: 10.4141/P01-062

Coleman, W.K. (2008) Evaluation of wild Solanum species for drought resistance: 1. *Solanum gandarillasii* Cardenas. *Environmental and Experimental Botany* 62(3), 221–230. DOI: 10.1016/j.envexpbot.2007.08.007

Colmer, T.D., Flowers, T.J. and Munns, R. (2006) Use of wild relatives to improve salt tolerance in wheat. *Journal of Experimental Botany* 57(5), 1059–1078. DOI: 10.1093/jxb/erj124

DAFW (2022) *Annual Report 2021–22, Department of Agriculture and Farmers' Welfare*. Ministry of Agriculture and Farmers' Welfare, Government of India. Available at: https://agriwelfare.gov.in/Documents/annual-report-2021-22.pdf (accessed 4 July 2024).

Dempewolf, H., Eastwood, R.J., Guarino, L., Khoury, C.K., Müller, J.V. *et al.* (2013) Adapting agriculture to climate change: a global initiative to collect, conserve, and use crop wild relatives. *Agroecology for Sustainable Food* 38, 369–377. DOI: 10.1080/21683565.2013.870629

Eglinton, J.K., Evans, D.E., Brown, A.H.D., Langridge, P., McDonald, G. *et al.* (2001) The use of wild barley (*Hordeum vulgare* ssp. *spontaneum*) in breeding for quality and adaptation. Proceedings of the 10th Barley Technical Symposium. Canberra, ACT, Australia, 16-20 September 2001. Available at: http://www.regional.org.au/au/abts/1999/eglinton.htm (accessed 25 September 2024).

FAO (2009) *International Treaty on Plant Genetic Resources for Food and Agriculture*. Food and Agriculture Organization of the United Nations, Rome. Available at: https://www.fao.org/3/i0510e/i0510e.pdf (accessed 4 July 2024).

FAO (2010) *The State of Food Insecurity in the World*. Food and Agriculture Organization of the United Nations, Rome. Available at: https://www.fao.org/3/i1683e/i1683e.pdf (accessed June 2023).

FAO (2011) *Draft Updated Global Plan of Action for the Conservation and Sustainable Utilization of Plant Genetic Resources for Food and Agriculture*. Food and Agriculture Organization of the United Nations, Rome. Available at: https://www.fao.org/3/be310e/be310e.pdf (accessed 4 July 2024).

FAO (2015a) *Climate Change and Food Security: Risks and Responses*. Food and Agriculture Organization of the United Nations, Rome. Available at: https://www.fao.org/3/i5188e/I5188E.pdf (accessed July 2023).

FAO (2015b) *Coping with Climate Change: The Roles of Genetic Resources for Food and Agriculture*. FAO, Rome. Available at: https://www.fao.org/3/i3866e/i3866e.pdf (accessed 4 July 2024).

Farooq, S., Iqbal, N., Asghar, M. and Shah, T.M. (1992) Intergeneric hybridization for wheat improvement. VI. Production of salt tolerant germplasm through crossing wheat (*Triticum aestivum* L.) with *Aegilops cylindrica* and its significance in practical agriculture. *Journal of Genetics and Breeding* 46, 125–132.

Farooq, S., Asghar, M., Iqbal, N., Askari, E., Arif, M. and Shah, T.M. (1995) Production of salt-tolerant wheat germplasm through crossing cultivated wheat with *Aegilops cylindrica*. II. Field evaluation of salt-tolerant germplasm. *Cereal Research Communications* 23(3), 275–282.

Feng, S. and Fu, Q. (2013) Expansion of global drylands under a warming climate. *Atmospheric Chemistry and Physics* 13(19), 10081–10094. DOI: 10.5194/acp-13-10081-2013

Fischer, I., Steige, K.A., Stephan, W. and Mboup, M. (2013) Sequence evolution and expression regulation of stress-responsive genes in natural populations of wild tomato. *PLOS ONE* 8(10), e78182. DOI: 10.1371/journal.pone.0078182

Goodman, R.M., Hauptli, H., Crossway, A. and Knauf, V.C. (1987) Gene transfer in crop improvement. *Science* 236(4797), 48–54. DOI: 10.1126/science.236.4797.48

Gowthami, R., Sharma, N., Pandey, R. and Agrawal, A. (2021) Status and consolidated list of threatened medicinal plants of India. *Genetic Resources and Crop Evolution* 68, 2235–2263. DOI: 10.1007/s10722-021-01199-0

Hadjichristodoulou, A. (1993) The use of wild barley in crosses for grain production under dryland conditions. *Euphytica* 69, 211–218. DOI: 10.1007/BF00022367

Hajjar, R. and Hodgkin, T. (2007) The use of wild relatives in crop improvement: a survey of developments over the last 20 years. *Euphytica* 156, 1–13. DOI: 10.1007/s10681-007-9363-0

Hansen, J., Ruedy, R., Sato, M. and Lo, K. (2010) Global surface temperature change. *Reviews of Geophysics* 48(4). DOI: 10.1029/2010RG000345

Hatfield, J.G., Takle, R., Grotjahn, P., Holden, R.C., Izaurralde, T. *et al.* (2014) Agriculture. In: Melillo, J.M., Richmond, T.T.C. and Yohe, G.W. (eds) *Climate Change Impacts in the United States: The Third National Climate Assessment*. U.S. Global Change Research Program, Washington DC, pp. 150–174. DOI: 10.7930/J02Z13FR

Howden, S.M., Soussana, J.F., Tubiello, F.N., Chhetri, N., Dunlop, M. *et al.* (2007) Adapting agriculture to climate change. *Proceedings of the National Academy of Sciences of the United States of America* 104(50), 19691–19696. DOI: 10.1073/pnas.0701890104

Huang, J., Yu, H., Guan, X., Wang, G. and Guo, R. (2016) Accelerated dryland expansion under climate change. *Nature Climate Change* 6(2), 166–171. DOI: 10.1038/nclimate2837

INIBAP/IPGRI (2006) *Global Conservation Strategy for Musa (Banana and Plantain)*. INIBAP/IPGRI, Montpellier/Rome.

International Brachypodium Initiative (2010) Genome sequencing and analysis of the model grass *Brachypodium distachyon*. *Nature* 463, 763–768. DOI: 10.1038/nature08747

IPBES (2019) *Global Assessment Report on Biodiversity and Ecosystem Services of the Intergovernmental Science-Policy Platform on Biodiversity and Ecosystem Services*. In: Brondizio, E.S., Settele, J., Díaz, S. and Ngo, H.T. (eds) IPBES Secretariat, Bonn, Germany. DOI: 10.5281/zenodo.3831673

IPCC (2001) *Climate Change 2001: The Scientific Basis. Contribution of Working Group I to the Third Assessment Report of the Intergovernmental Panel on Climate Change*. In: Houghton, J.T., Ding, Y., Griggs, D.J., Noguer, M., van der Linden, P.J. *et al.* (eds) Cambridge University Press, Cambridge, UK and New York.

IPCC (2013) *Climate Change 2013: The Physical Science Basis. Contribution of Working Group I to the Fifth Assessment Report of the Intergovernmental Panel on Climate Change*. In: Stocker, T.F., Qin, D., Plattner, G.K., Tignor, M., Allen, S.K. *et al.* (eds) Cambridge University Press, UK and New York.

IPCC (2014) *Climate Change 2014: Synthesis Report. Contribution of Working Groups I, II and III to the Fifth Assessment Report of the Intergovernmental Panel on Climate Change*. In: Core Writing Team, Pachauri, R.K. and Meyer, L.A. (eds). IPCC, Geneva, Switzerland.

IPCC (2021) Summary for policymakers. In: Masson-Delmotte, V., Zhai, P., Pirani, A., Connors, S.L., Péan, C. *et al.* (eds) *Climate Change 2021: The Physical Science Basis. Contribution of Working Group I to the Sixth Assessment Report of the Intergovernmental Panel on Climate Change*. Cambridge University Press, UK and New York, pp. 3–32. DOI: 10.1017/9781009157896.001

Ishimaru, T., Hirabayashi, H., Ida, M., Takai, T., San-Oh, Y.A. *et al.* (2010) A genetic resource for early-morning flowering trait of wild rice *Oryza officinalis* to mitigate high temperature-induced spikelet sterility at anthesis. *Annals of Botany* 106(3), 515–520. DOI: 10.1093/aob/mcq124

ISRO (2016) *Desertification and Land Degradation Atlas of India*. Indian Space Research Organization, Ahemdabad. Available at: www.sac.gov.in/SACSITE/Desertification_Atlas_2016_SAC_ISRO.pdf (accessed 4 July 2024).

Jennings, D.L. (1995) Cassava, *Manihot esculenta* (Euphorbiaceae). In: Smartt, J. and Simmonds, N.W. (eds) *Evolution of Crop Plants*. Longman Group, Harlow, UK, pp. 128–132.

King, I.P., Forster, B.P., Law, C.C., Cant, K.A., Orford, S.E. *et al.* (1997a) Introgression of salt-tolerance genes from *Thinopyrum bessarabicum* into wheat. *New Phytologist* 137(1), 75–81. DOI: 10.1046/j.1469-8137.1997.00828.x

King, I.P., Law, C.N., Cant, K.A., Orford, S.E., Reader, S.M. *et al.* (1997b) Tritipyrum, a potential new salt-tolerant cereal. *Plant Breeding* 116(2), 127–132. DOI: 10.1111/j.1439-0523.1997.tb02166.x

Knutti, R., Furrer, R., Tebaldi, C., Cermak, J. and Meehl, G.A. (2010) Challenges in combining projections from multiple climate models. *Journal of Climate* 23(10), 2739–2758. DOI: 10.1175/2009JCLI3361.1

Koutroulis, A.G. (2019) Dryland changes under different levels of global warming. *Science of the Total Environment* 655, 482–511. DOI: 10.1016/j.scitotenv.2018.11.215

Lexer, C., Lai, Z. and Rieseberg, L.H. (2004) Candidate gene polymorphisms associated with salt tolerance in wild sunflower hybrids: implications for the origin of *Helianthus paradoxus*, a diploid hybrid species. *New Phytologist* 161(1), 225–233. DOI: 10.1046%2Fj.1469-8137.2003.00925.x

Li, Y.H., Zhou, G., Ma, J., Jiang, W., Jin, L.G. *et al.* (2014) De novo assembly of soybean wild relatives for pan-genome analysis of diversity and agronomic traits. *Nature Biotechnology* 32(10), 1045–1052.

Malhi, G.S., Kaur, M. and Kaushik, P. (2021) Impact of climate change on agriculture and its mitigation strategies: a review. *Sustainability* 13, 1318. DOI: 10.3390/su13031318

Maxted, N., Kell, S., Ford-Lloyd, B., Dulloo, E. and Toledo, Á. (2012) Toward the systematic conservation of global crop wild relative diversity. *Crop Science* 52(2), 774–785. DOI: 10.2135/cropsci2011.08.0415

Mittermeier, R.A., Robles Gil, P. and Hoffman, M. (2005) *Hotspots Revisited: Earth's Biologically Richest and Most Endangered Terrestrial Ecoregions*. University of Chicago Press, Chicago, Illinois.

Mochida, K., Uehara-Yamaguchi, Y., Takahashi, F., Yoshida, T., Sakurai, T. *et al.* (2013) Large-scale collection and analysis of full-length cDNAs from *Brachypodium distachyon* and integration with *Poaideae* sequence resources. *PLOS ONE* 8(10), e75265. DOI: 10.1371/journal.pone.0075265

MoEFCC (2021) *India: Third Biennial Update Report to the United Nations Framework Convention on Climate Change*. Ministry of Environment, Forest and Climate Change, Government of India, New Delhi.

Morice, C.P., Kennedy, J.J., Rayner, N.A. and Jones, P.D. (2012) Quantifying uncertainties in global and regional temperature change using an ensemble of observational estimates: the HadCRUT4 data set. *Journal of Geophysical Research: Atmospheres* 117(D8), 1–22. DOI: 10.1029/2011JD017187

Nair, P.K.R., Kumar, B.M. and Nair, V.D. (2009) Agroforestry as a strategy for carbon sequestration. *Journal of Plant Nutrition and Soil Science* 172, 10–23. DOI: 10.1002/jpln.200800030

Nath, S., Kumar, R. and Yash Nath, S. (2021) Influence of anthropocene climate change on biodiversity loss in different ecosystems. In: Singh, S., Singh, P., Rangabhashiyam, S. and Srivastava, K.K. (Eds.) *Global Climate Change*, pp. 63–78. Elsevier, Amsterdam, the Netherlands. DOI: 10.1016/B978-0-12-822928-6.00001-0

Noori, S.A.S. (2005) Assessment for salinity tolerance through intergeneric hybridisation: *Triticum durum* × *Aegilops speltoides*. *Euphytica* 146, 149–155. DOI: 10.1007/s10681-005-8001-y

Osnato, M., Cota I., Nebhnani, P., Cereijo, U. and Pelaz, S. (2022) Photoperiod control of plant growth: flowering time genes beyond flowering. *Frontiers of Plant Science* 12, 805635. DOI: 10.3389/fpls.2021.805635

Paroda, R.S. (2018) *Reorienting Indian Agriculture: Challenges and Opportunities*. CAB International, Wallingford, UK. DOI: 10.1079/9781786395177.0000

Paroda, R.S. (2019a) *Report on Policies and Action Plans for a Secure and Sustainable Agriculture*. Committee Report submitted to the Principal Scientific Adviser to the Government of India, New Delhi.

Paroda, R.S. (2019b) Current threats to dryland agrobiodiversity and strategies for adaptation to climate change. In: Agrawal, A., Rajkumar, S., Rana, J.C., Tyagi, R.K., Yadav, O.P., Singh K. and Paroda, R.S. (eds) *Proceedings and Recommendations of the Satellite Symposium on Dryland Agrobiodi-*

*versity for Adaptation to Climate Change*. Indian Society of Plant Genetic Resources, New Delhi, pp. 23–27.

Paroda, R.S. (2022) Crop diversification for sustainable agriculture. *Ecology, Economy and Society INSEE Journal* 5(1), 15–21. DOI: 10.37773/ees.v5i1.611

Paroda, R.S. and Agrawal, A. (2017) Changing paradigms in managing agrobiodiversity through use: an appraisal. *Indian Journal of Plant Genetic Resources* 30(1), 5–12. DOI: 10.5958/0976-1926.2017.00014.6

Paroda, R.S., Suleimenov, M., Yusupov, H., Kireyev, A., Medeubayev, R. *et al.* (2004) Crop diversification for dryland agriculture in Central Asia. *Challenges and Strategies of Dryland Agriculture* 32, 139–150. DOI: 10.2135/cssaspecpub32.c9

Paroda, R.S., Jat, M.L., Gautam, M., Clare, S. and Bhag Mal (2018) *Agricultural Policies and Investment Priorities for Managing Natural Resources, Climate Change and Air Pollution. Policy Brief*. Trust for Advancement of Agricultural Sciences (TAAS), International Maize and Wheat Improvement Center (CIMMYT), Indian Council of Agricultural Research (ICAR), the CGIAR Research Program on Climate Change, Agriculture and Food Security (CCAFS), and the World Bank Group.

Pathak, H. (2023) Impact, adaptation, and mitigation of climate change in Indian agriculture. *Environmental Monitoring and Assessment* 195, art. 52. DOI: 10.1007/s10661-022-10537-3

Pathak, H., Srinivasrao C. and Jat, M.L. (2021) Conservation agriculture for climate change adaptation and mitigation in India. *Journal of Agricultural Physics* 21, 182–196.

Pervez Bharucha, Z., Attwood, S., Badiger, S., Balamatti, A., Bawden, R. *et al.* (2021) The Top 100 questions for the sustainable intensification of agriculture in India's rainfed drylands. *International Journal of Agricultural Sustainability* 19(2), 106–127. DOI: 10.1080/14735903.2020.1830530

Pradhan, G.P., Prasad, P.V.V., Fritz, A.K., Kirkham, M.B. and Gill, B.S. (2012) High temperature tolerance in Aegilops species and its potential transfer to wheat. *Crop Science* 52(1), 292–304. DOI: 10.2135/cropsci2011.04.0186

Qi, X., Li, M.W., Xie, M., Liu, X., Ni, M. *et al.* (2014) Identification of a novel salt tolerance gene in wild soybean by whole-genome sequencing. *Nature Communications* 5(1), 4340. DOI: 10.1038/ncomms5340

Rick, C.M. and Chetelat, R.T. (1995) Utilization of related wild species for tomato improvement. *Acta Horticulturae* 412, 21–38. DOI: 10.17660/ActaHortic.1995.412.1

Riordan, E.C. and Nabhan, G.P. (2019) Trans situ conservation of crop wild relatives. *Crop Science* 59(6), 2387–2403. DOI: 10.2135/cropsci2019.06.0356

Scafaro, A.P., Haynes, P.A. and Atwell, B.J. (2010) Physiological and molecular changes in *Oryza meridionalis* Ng., a heat-tolerant species of wild rice. *Journal of Experimental Botany* 61(1), 191–202. DOI: 10.1093/jxb/erp294

Singh, B.K., Delgado-Baquerizo, M., Egidi, E., Guirado, E., Leach, J.E. *et al.* (2023) Climate change impacts on plant pathogens, food security and paths forward. *Nature Reviews Microbiology* 21, 1–17. DOI: 10.1038/s41579-023-00900-7

Singh, O.V., Shekhawat, N. and Singh, K. (2019) Plant genetic resources of Indian hot arid zone. In: Agrawal, A., Rajkumar, S., Rana, J.C., Tyagi, R.K., Yadav, O.P., Singh, K. and Paroda, R.S. (eds) *Proceedings and Recommendations of the Satellite Symposium on Dryland Agrobiodiversity for Adaptation to Climate Change*. Indian Society of Plant Genetic Resources, New Delhi, pp. 69–71.

Stevens, M.A. and Rick, C.M. (1986) Genetics and breeding. In: Atherton, J.G. and Rudich, J. (eds) *The Tomato Crop*. Springer, Dordrecht, Th Netherlands, pp. 35–109. DOI: 10.1007/978-94-009-3137-4_2

Talamè, V., Sanguineti, M.C., Chiapparino, E., Bahri, H., Ben Salem, M. *et al.* (2004) Identification of *Hordeum spontaneum* QTL alleles improving field performance of barley grown under rainfed conditions. *Annals of Applied Biology* 144(3), 309–319. DOI: 10.1111/j.1744-7348.2004.tb00346.x

Verelst, W., Bertolini, E., De Bodt, S., Vandepoele, K., Demeulenaere, M. *et al.* (2013) Molecular and physiological analysis of growth-limiting drought stress in *Brachypodium distachyon* leaves. *Molecular Plant* 6(2), 311–322. DOI: 10.1093/mp/sss098

Villareal, R.L., Sayre, K., Banuelos, O. and Mujeeb-Kazi, A. (2001) Registration of four synthetic hexaploid wheat (*Triticum turgidum/Aegilops tauschii*) germplasm lines tolerant to waterlogging. *Crop Science* 41(1), 274–274. DOI: 10.2135/cropsci2001.411274x

Wolfe, D.W., Ziska, L., Petzoldt, C., Seaman, A., Chase, L. *et al.* (2008) Projected change in climate thresholds in the Northeastern US: implications for crops, pests, livestock, and farmers. *Mitigation and Adaptation Strategies for Global Change* 13, 555–575. DOI: 10.1007/s11027-007-9125-2

Yadav, O.P. (2019) Managing agrobiodiversity of Indian drylands for climate adaptation. In: Agrawal, A., Rajkumar, S., Rana, J.C., Tyagi, R.K., Yadav, O.P., Singh, K. and Paroda, R.S. (eds) *Proceedings and*

*Recommendations of the Satellite Symposium on Dryland Agrobiodiversity for Adaptation to Climate Change*. Indian Society of Plant Genetic Resources, New Delhi, pp. 28–30.

Zhang, X., Zhou, S., Fu, Y., Su, Z., Wang, X. *et al.* (2006) Identification of a drought tolerant introgression line derived from Dongxiang common wild rice (*O. rufipogon* Griff.). *Plant Molecular Biology* 62, 247–259. DOI: 10.1007/s11103-006-9018-x

Zhou, Z., Jiang, Y., Wang, Z., Gou, Z., Lyu, J. *et al.* (2015) Resequencing 302 wild and cultivated accessions identifies genes related to domestication and improvement in soybean. *Nature Biotechnology* 33(4), 408–414. DOI: 10.1038/nbt.3096

# 5 Agro-management Techniques and Precision Agriculture/Agriculture 4.0 in Arid Lands

**Abdel-Ghany M. El-Gindy[1], Abdel-Aziz S. Sheta[2*], and Theib Y. Oweis[3]**
[1]*Faculty of Desert Agriculture, King Salman International University, Egypt;*
[2]*Formerly Faculty of Agriculture, Ain Shams University, Egypt;*
[3]*Independent expert and specialist in water harvesting and management for agriculture, especially under conditions of water scarcity, Jordan*

**Abstract**
Soil and water resources are critical factors for the production and application of smart precision agriculture systems in arid zone regions. This technology provides comprehensive management and control of soil, air, and water by computers or smartphones. Many recently established smart systems have the ability to automatically program and control all aspects of crop production. They offer a wide range of supportive water and soil management solutions that will help farmers raise the efficiencies of water, soil, energy use, and crop yield, maintaining the environment and supporting the concept of sustainable agriculture. Dryland rainfed cropping depends on green water for economical produce without irrigation. With variations in rainfall and frequent droughts, rainfed crops often suffer from moisture stress and, consequently, low and fluctuating production. The key to improving dryland rainfed productivity is directly related to good agro-management techniques for soil, water, and plants, and adopting climate-smart digital agriculture.

## Soil-management Techniques and Precision Agriculture in Irrigated Arid Lands

Arid and semi-arid areas account for about one third of the Earth's surface. According to Perez-Aguilar *et al.* (2021), arid and semi-arid regions are geographic units that cover approximately 43% of the Earth's surface; also, conditions of extreme drought and reduced vegetation cover predominate in these regions. Naturally, irrigated agriculture and different management practices in such areas are important and critical for better cropping. Climate change (CC) leads to extensive drought and hence land degradation and extensive drought in other parts of the world. Environmental degradation and arable land are serious problems caused by pollution and global warming. The misuse or inadequate use of renewable natural resources and soil erosion leads to excessive soil depletion. However, the main cause of this phenomenon is population explosion. Harsh environmental factors, such as succession of droughts, have aggravated the problem. Desertification is affecting total areas of agricultural lands and pastures necessary to provide food, water, and air. Blackmore (1994) described precision agriculture as a

*Corresponding author: sheta11us@gmail.com


DOI: 10.1079/9781800624870.0005

term used to describe the objective of increased productivity in the administration of agri-business. This implies that precision agriculture is a developing technology that modifies existing techniques and consolidates new ones to produce a new set of tools for the manager to utilize.

Stombaugh *et al.* (2001) defined precision agriculture as an environmentally helpful system solution that optimizes production quality and quantity while lowering the cost of inputs. According to McBratney *et al.* (2005), precision agriculture, in general, is defined as 'agriculture that increases the number of (correct) decisions per unit of area of land per unit time with associated net benefits'.

Ruß (2013) stated that precision agriculture is a compelling and highly active field of research typically based on large data collections and data-based decision-making for agricultural operations. Since data collection is growing rapidly, it is essential to put research efforts into methods that deal with those datasets. Recently, precision agriculture (PA), or site-specific crop management (SSCM), was defined as a philosophy or a farming management concept aimed at increasing farm production based on observing, measuring, and responding to field variability in crops while minimizing unwanted impacts on the environment. Smart agriculture is an approach to agriculture that uses data sensors, connected devices, remote controls, and other advanced technologies to give farmers more control over the field. This technology is linked to agricultural management systems with modern technology that contributes to improving and sustaining the use of natural resources, especially soil and water, increasing the quantity of production, and improving its quality.

On the other hand, smart agriculture means dealing optimally and accurately with all agricultural production inputs from soil, water, energy, and chemicals in the appropriate places within agricultural production units, allowing economic handling of agricultural inputs. Increasing concern for the sustainability of natural resources has led to the development of a more complex concept of soil health or soil quality (Figs 5.1 and 5.2).

Karlen *et al.* (1997) proposed the following as vital soil functions: (i) sustaining biological activity, diversity, and productivity; (ii) regulating and partitioning water and solute flow; (iii) filtering, buffering, degrading, immobilizing, and detoxifying organic and inorganic materials, including agricultural, industrial, and municipal byproducts and atmospheric deposition; (iv) storing and cycling nutrients and other elements within the Earth's biosphere; and (v) providing support of socio-economic structures and protection for archaeological treasures associated

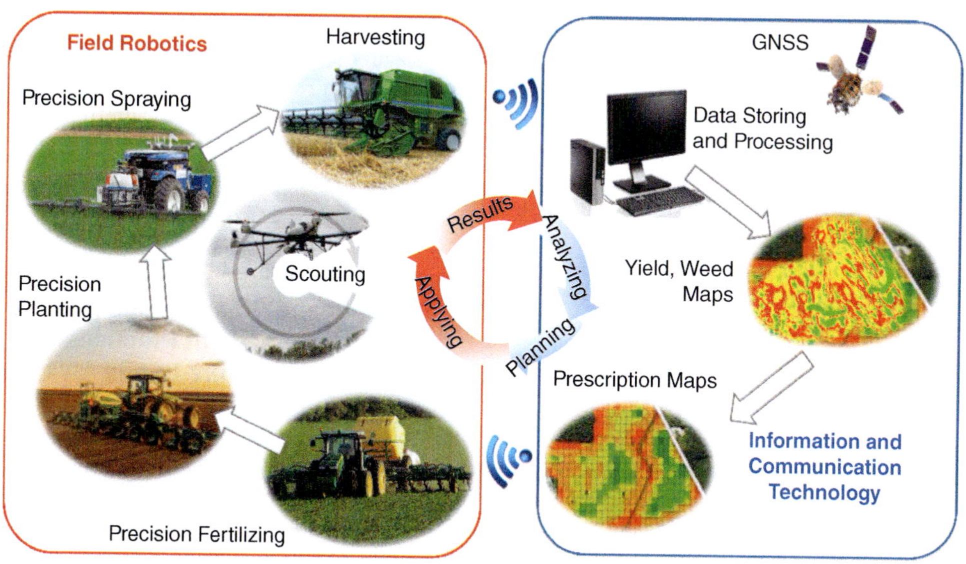


**Fig. 5.1.** Devices and tools for applying precision and digital agriculture. (From Gonzalez-de-Santos *et al.*, 2020; Open access, C.C. by 4.0)

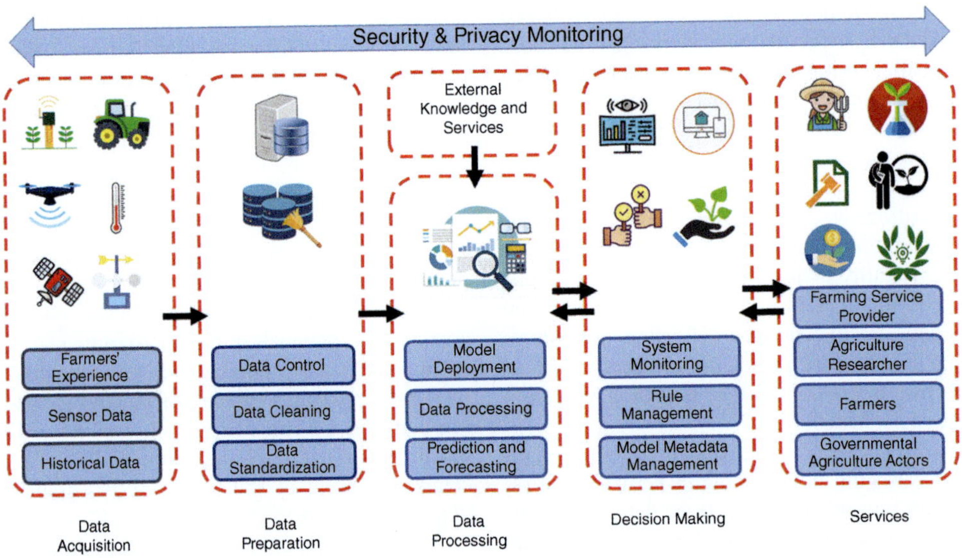


**Fig. 5.2.** Objectives, elements, and components of precision and digital agriculture. (Adapted from Amiri-Zarandi *et al.*, 2022; Open access, C.C. by 4.0)

with human habitation. Evaluating and mapping dynamic soil quality or soil health can be carried out by evaluating different indicators related to physical, chemical, and biological soil parameters. Also, data on irrigation water quality can be processed into management maps for decision makers (Simsek and Gunduz, 2007) and stakeholders in handling issues related to the soil environment and irrigation water management (Abbasnia *et al.*, 2019). Geospatial methods, including remotely-sensed imagery and geographic information system (GIS) techniques, are accurate tools, which have been widely utilized for mapping crops (Elsharkawy, 2016; Farg *et al.*, 2020), assessing land use/cover (Arafat *et al.*, 2014; Elsharkawy *et al.*, 2022a; Nabil *et al.*, 2022), and soil and water quality (Saleh *et al.*, 2021; Abdel Rahman *et al.*, 2022a, b; Elsharkawy *et al.*, 2022b). Satellite images provide an ideal characterization of high spatial features quickly and accurately at low cost, significantly over large regions, compared with classic techniques (Boori *et al.*, 2021).

The great development in remote-sensing technology, GIS, data management, and the Internet of Things (IoT) has changed farm management systems by accurately measuring differences in the field, allowing the right decision to be made at the right time and maximizing the return from available natural resources. Accordingly, farmers can improve soil quality, increase the efficiency of irrigation water addition, and selectively use fertilizers and pesticides according to spatial differences in the soil. It can be noted that the base for the application of precision agriculture in many countries is available, especially in terms of managing the quality of soil, water, and farms, and requires increasing the application of modern technologies in precision agriculture to the application of sustainable land-management systems, which results in sustainable soil use (Fig. 5.3). It shows that monitoring trends in soil quality over time can result in aggrading, sustaining, or degrading soil conditions.

## Applications of Precision Agriculture Systems and Smart Agriculture for Agro-management Practices in Arid Agriculture Lands

Precision agriculture is one of the most prominent applications of IoT, which is the approach used in farm management and crop control through information and communication technology, remote sensors, remote control systems, and autonomous machines to obtain accurate data and invest these data in directing agriculture

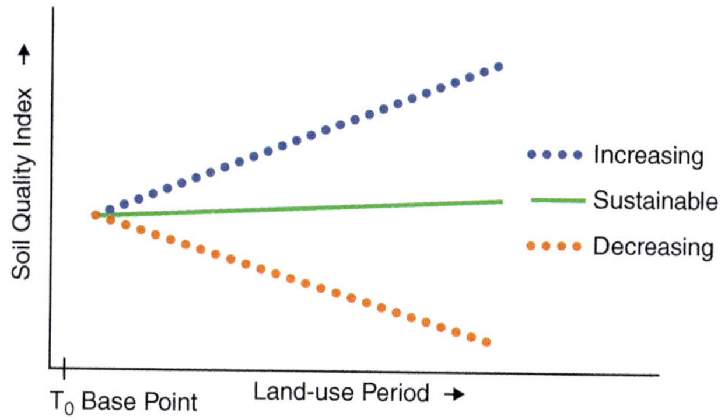


**Fig. 5.3.** Dynamic soil quality changes over time.

towards greater production at a lower cost, and the production of high-quality crops. Remote sensors are essential to obtain detailed maps of both terrain and resources in the region, as well as to measure variables such as ground humidity, acidity, alkalinity, soil temperature, salinity, and nutrients. They can also predict weather patterns for days and weeks to come. In addition, the data collected and analyzed play a major role in monitoring agricultural pests, determining the number of pesticides required accurately, as well as data collection and analysis in the rational use of irrigation water. Other important applications of IoT in smart agriculture include the use of drones to monitor and evaluate crops, photograph and map agricultural lands, measure air components, spray crops with pesticides quickly and safely, and send data instantly to software that analyzes them and instructs farmers to implement the best procedures. Therefore, PA is a management method that aims to reach the highest production using the lowest inputs while maintaining the integrity of the surrounding environment and managing fields based on soil types and units in different areas. This includes studying the impact of precision farming techniques on crop productivity and focusing on the application of GPS techniques to adjust fertilizer additives, given the soil content of available nitrogen, phosphorus, potassium, and soil micro-elements (Fig. 5.4). Thus, integrated land management in precision agriculture applications can be achieved using remote-sensing techniques and GIS by producing soil topography maps such as soil physical properties, salinity, pH, alkalinity, $CaCO_3$, available phosphorus, potassium, and trace elements such as iron, zinc, and manganese associated with crop productivity at study sites (Fig. 5.5).

Applications of GIS and RS in precision agriculture rely on the collection of climatic data and information on the type and units of different soils and vegetation cover. Samples of irrigation water and soil samples representing different soil units will be collected using GPS devices and remote-sensing satellite data, such as the field spectra radiometer. The necessary processing of satellite visuals will be carried out, such as radiometric corrections and removal of the effect of the atmosphere on the digital values of the satellite visuals used. The conversion of visuals and the process of classifying satellite visuals captured during the different stages of growth is through the use of inexpensive US satellite data (Land sat), which is characterized by the availability of thermal ranges that can be used to calculate the daily water consumption of crops with an accuracy of 30 m. In addition, satellite Land sat data can be received every few days, which allows monitoring the growth of the crop by calculating botanical evidence such as Normalized Difference Vegetation Index (NDVI) and Soil-Adjusted Vegetation Index (SAVI), which gives indications of crop productivity.

The results showed that the use of a database that includes digital soil maps and storage using GIS and processing them with predictive statistical models enables the prediction of the productivity of crops before planting with high accuracy. The results obtained by Elsharkawy *et al.* (2022b) also showed that the sampling method using GIS and the Global Coordinate Finder effectively contributed to increasing the efficiency of the systematic sampling method in

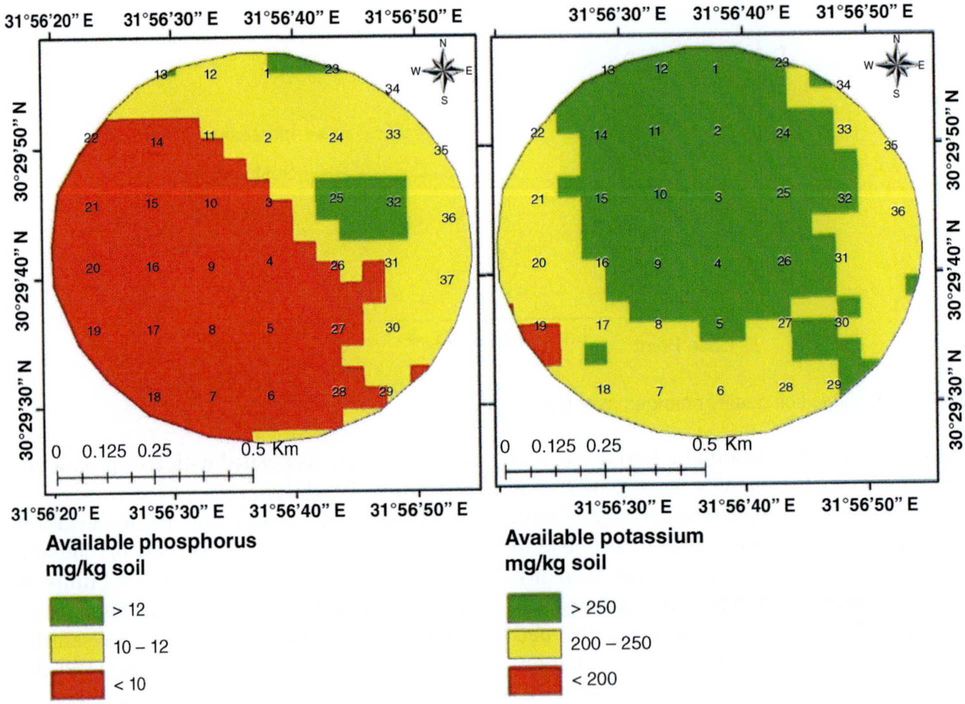


**Fig. 5.4.** Mapping available P and K in Center Pivot soil prepared for cultivation. (From Elsharkawy *et al.*, 2022b; Open access, C.C. by 4.0)

the study areas. Also, the results of the chemical analysis of irrigation water showed that the total dissolved salts EC are in the low range at a rate of 0.85 dS/m (Elsharkawy *et al.*, 2022b).

## Integration of Precision Agriculture Techniques and Smart Farming Technology to Achieve Agricultural Sustainability

Agriculture faces a major challenge at present to provide for food needs without depleting available resources of soil and water, and there is no doubt that technology and techniques of the Fourth Industrial Revolution such as artificial intelligence (AI), IoT, sensors, robotics, and other recent innovations can make current agricultural holdings more productive and efficient and their production systems more sustainable. In this regard, the development of sensors capable of examining land conditions with extreme accuracy and providing data to farmers can be very important, such as soil content sensors of elements such as nitrogen, phosphorus, and potassium; salinity and alkalinity sensors; and analog soil-moisture sensors resistant to corrosion and soil temperature. In addition, there are smart greenhouse control rooms to control temperature, humidity, lighting, and brightness. Precision agriculture allows farmers to increase productivity using minimal resources such as water, fertilizers, and seeds. By deploying sensors and mapping, farmers can begin to understand their crops on a small scale, conserve resources, and minimize impacts on the environment. Once farmers can accurately map their crop fields, they can monitor and apply fertilizer additives at accurate rates and spray pesticides only in areas that require it. Crop productivity can also be monitored, and recommendations can be made on soil alkalinity and disposal. Since more variables can be measured and introduced into crop models, more precise recommendations can be made for fertilizer use, irrigation, and even peak harvesting of yield (Fig. 5.6). Below are some of the sensors used in precision farming.

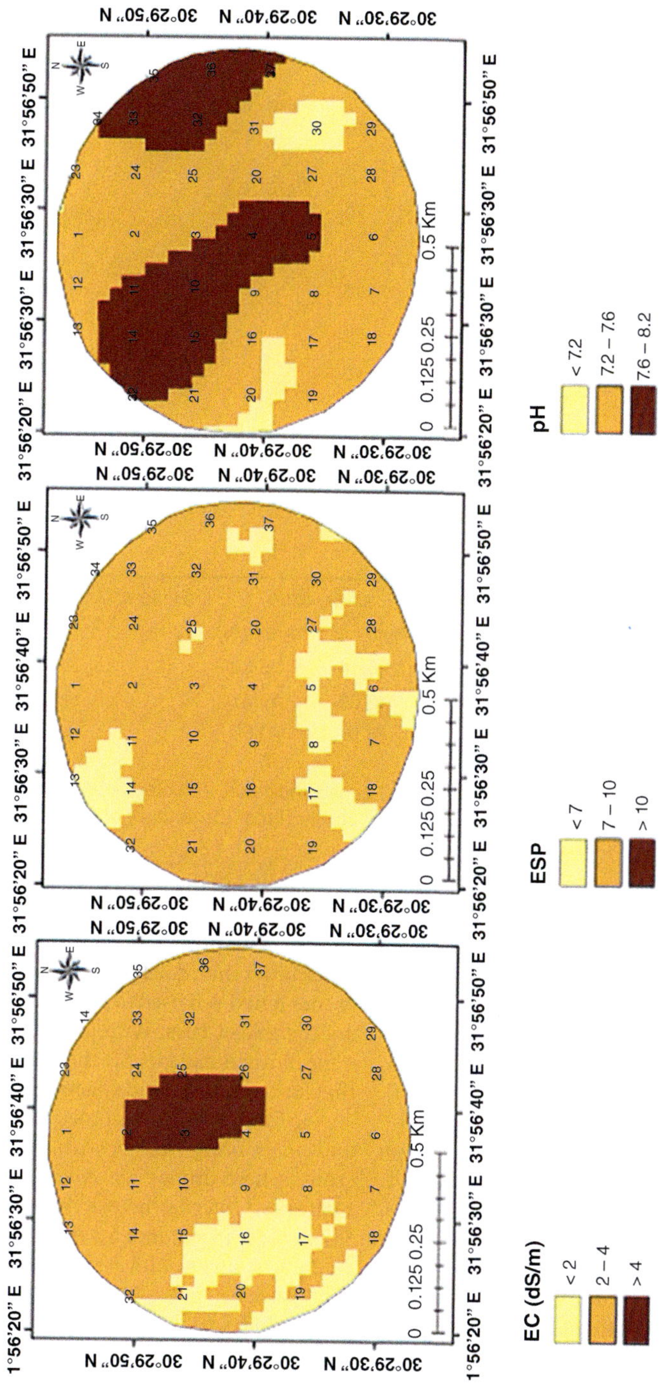


**Fig. 5.5.** Mapping EC, ESP (Exchangeable Sodium Percentage) and pH values in Center Pivot soil prepared for cultivation. (From Elsharkawy *et al.*, 2022b; Open access, C.C. by 4.0)

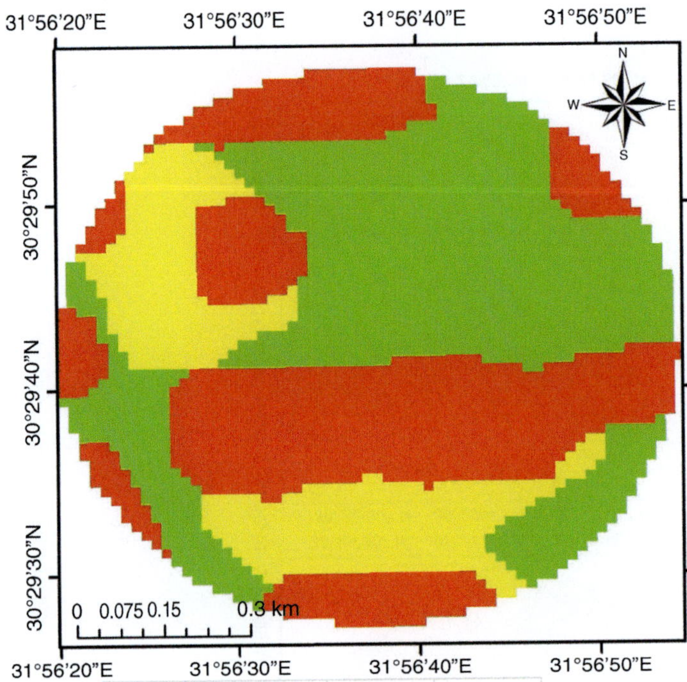


| Classes | Area ha$^{-1}$ | Area (%) |
|---|---|---|
| Management zone 1 | 26.14 | 38.88 |
| Management zone 2 | 13.09 | 19.47 |
| Management zone 3 | 28.01 | 41.65 |

**Fig. 5.6.** Different management zones based on soil-quality indicators and crop needs. Management zones 1, 2 and 3 shown in green, yellow and red respectively. (From Elsharkawy *et al.*, 2022b; Open access, C.C. by 4.0)

## Sensors used in precision agriculture

### *Sensors capable of aerial photography such as cameras carried on drones*

Many biological and physical processes occur in nature that human eyes cannot see, and thanks to multi-spectral sensors, farmers can be helped to assess the condition of their fields. These sensors are able to detect nuances in crops. These systems can be attached to drones capable of taking aerial photos. They consists of two elements: a high-resolution camera – for example, 16:40 megapixels – and a sunlight sensor; and GPS (Fig. 5.7). Both are multi-spectral and record everything from the electromagnetic spectrum to ultraviolet and infrared radiation, which is invisible to the human eye. These sensors allow collecting concise accurate information in real time about the field and compare the activity of the sun with the state of the crop. Drones also contribute to crop-irrigation and fertilizer-spraying operations, spraying them at a rate 40–60 times faster than doing it manually, and recently drones have been used to pollinate flowers, playing the role of bees. With all these functions in mind, farmers can identify the driest areas, those that lack nutrients, and systems of deficiency.

IoT also helps in the process of following up and monitoring the growth of agricultural crops, where drones are connected to today's high-speed internet networks, such as 4G and 5G, in order to conduct a comprehensive survey of the vast areas cultivated, such as pivots and fruit gardens, providing a complete database that includes the status of the crop, irrigation rates, spraying with fertilizers, and the need to spray chemical pesticides. The process of spraying the affected areas to eliminate weeds and agricultural pests is speeded up.

**Fig. 5.7.** Drones capable of carrying sensors and aerial cameras.

### *Soil-quality sensors and the condition of the elements in the plant (Elsharkawy et al., 2022b)*

Optical sensors use light to measure soil properties where sensors measure different frequencies of light reflection in visible light spectra close to infrared, medium, and ultraviolet. These sensors can be placed on drones or even satellites. The reflection of the soil and plant color data are only two variables of optical sensors that can be assembled and processed to determine the content of clay, organic matter, moisture, and elements in the soil, such as soil-content sensors of nutrients such as nitrogen, phosphorus, potassium, and other necessary elements, and salinity and alkalinity sensors, which detect the nuances of the electromagnetic spectrum in the thermal, infrared, and ultraviolet radiation ranges. IoT also helps in monitoring the condition of the soil and changes in order to follow them and solve problems immediately. Airflow sensors also measure soil-air permeability. Measurements can be made in individual locations or dynamically during movement. The desired output is the pressure required to push a predetermined amount of air into the ground at a specified depth. It produces different types of soil properties, including pressure, structure, soil type, and moisture level.

### *Electrochemical sensors*

Electrochemical sensors provide the basic information required in precision agriculture: pH and soil nutrient levels. Sensing electrodes work by detecting specific ions in the soil. Currently, sensors mounted on specially designed 'skis' help in the collection, processing, and mapping of chemical data from the soil. Mechanical sensors also measure the mechanical resistance of the soil as the sensors use a probe that penetrates the soil and records resistance forces using load cells or stress gauges. A similar form of this technique is used in large tractors to predict the requirements for towing ground interlock equipment. Tensiometers also detect the force that the roots use to absorb water and are very useful for irrigation interventions. There are also plant sensors (Figs. 5.8 and 5.9) that can detect the temperature, humidity, and growth of crops and record data used to help farmers manage the farm smartly, which helps reduce fertilizer consumption and rationalize water consumption. Predictive programs have also been developed for the level of elements in plants using sensors capable of detecting nuances in the infrared and red band, in addition to the visible spectrum band, and help monitor crop status, changes, and solve problems in real time.

### *Smart temperature sensors*

Temperature sensors help monitor the conditions surrounding the plant. High-precision temperature and humidity sensors and accurate temperature predictions are essential for the crops export industry. Temperature sensors are also necessary in the fields of precision and smart agriculture as they help food companies ensure the safety of their products by using smart thermostats that ensure the arrival of healthy foods free of contaminants, as high temperature leads to food spoilage and disease transmission. Sensors can be used in many daily applications, such as measuring the percentage of soil moisture in the field and lack of water in the roots of plants in home gardens or fields, and this application helps in rationalizing water consumption as well as maintaining the healthy state of plants in times of drought.

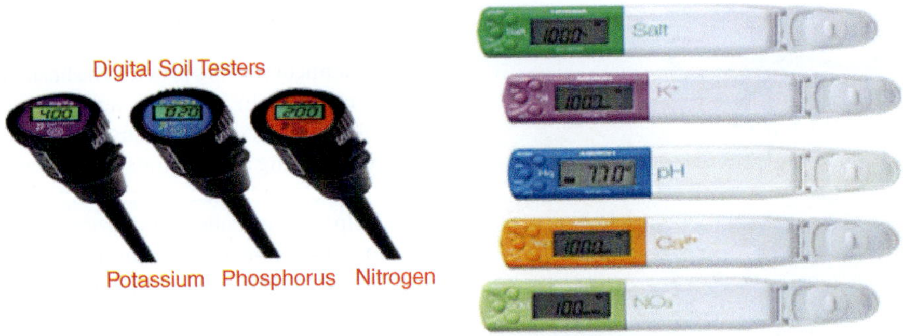


**Fig. 5.8.** Digital soil quality nutrients sensors.

**Fig. 5.9.** Plant and water sensors that detect crop and water characteristics.

## Revitalizing Dryland Rainfed Systems

### Dryland rainfed systems and productivity

Rainfed systems are established where annual rainfall is sufficient for permanent and economic cropping. The minimum amount of rainfall needed to ensure sustainable rainfed systems varies with climatic conditions, often lower in temperate with a cool winter season than in tropical with a hot summer season. Cereals, legumes, and trees are the major crops cultivated, with yields varying from country to country but generally low (0.5–2 t/ha) in dry environments and less than half of the global average and a third of its potential. Over the last few decades, the annual yield improvement in the Arab region was about 1.5% compared to 5.4%, 5.6%, and 6.1% for developing, world, and developed countries, respectively. Figure 5.10 reflects the low cereal yields and improvement rates in different regions and the world for an extended period of three decades (ESCWA *et al.*, 2017).

## Impact of Climate Change

Rainfall is expected to decrease due to climate change at different rates across the drylands. The reduction of annual rainfall may reach, in some areas, 40% by the end of the century (ESCWA *et al.*, 2017). This means that rainfed crops will suffer even higher moisture stress causing a further drop in productivity. Furthermore, some areas that are currently at the margins of the rainfed zones will lose, as a result, their traditional production potential causing a shrinkage in the cultivated rainfed areas. This is based on

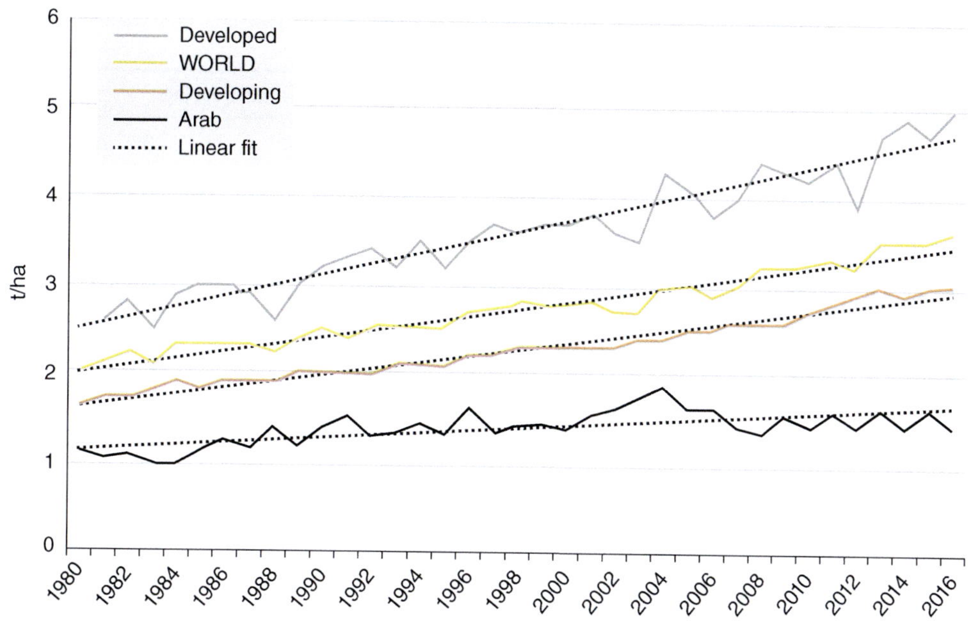


**Fig. 5.10.** Regional trends in cereal yields. (From FAO-GIEWS, 2016; based on FAO, Country Cereal Balance Sheet data; http://www.fao.org/giews/data-tools/en/ accessed 12 September 2024)

different cropping systems requiring different minimum threshold levels of annual rainfall to produce any rainfed crop. Assuming a moderate 20% precipitation reduction, a preliminary analysis to estimate the potential loss in rainfed areas using satellite data is conducted using the minimum annual rainfall cropping requirements of 200, 250, 300, and 350 mm (ESCWA, 2022). Table 5.1 presents the results of the analyses in some countries of the MENA region.

The area loss may range from zero to over 31% and 41% for the zones of > 300 and > 350 mm annually. This does not mean a total loss of land, rather those lands may be converted into rangelands or more drought-tolerant crops with lower production potential. Knowing those shifts in the rainfall zones and the impacts on the existing cropping systems would help strategize and plan for adaptation to the impacts of climate change.

## Poor Performance but Great Potential

Contrary to the limited potential for horizontal expansion in dryland rainfed areas, there is considerable potential for vertical productivity increases in most of the crops. Rainfed grain yields average around 1.0 t/ha compared with about 3.0 t/ha in irrigated agriculture. In temperate commercial agriculture, rainfed yields of significant grain crops often exceed 5 t/ha (Rockstrom *et al.*, 2010).

Actual and attainable yields of rainfed wheat in several regions of the world are analyzed by Sadras and Angus (2006) indicating a large gap between rainfed actual and achievable yields in several regions, suggesting an enormous untapped potential for yield increases.

Wheat yield gap analyses conducted in Morocco, Syria, and Turkey, (Fig. 5.11) show that farmers' yield is about half that of the neighboring research station and about one third of the simulated potential yield. This means there is a large untapped potential in rainfed agriculture in the drylands of a factor 2–4 lower than attainable yields.

Evidence indicates that yield levels in smallholder rainfed agriculture can be doubled or tripled, and, in many cases, even quadrupled (Anderson *et al.*, 2016). Yield gaps are not only associated with land productivity but also with

**Table 5.1.** Potential loss in rainfed areas in some MENA countries due to climate change is estimated with a 20% reduction in annual precipitation. (Adapted from ESCWA, 2022)

| | Zones of mean annual precipitation required for rainfed cropping | | | | | | | |
|---|---|---|---|---|---|---|---|---|
| | 200–250 mm | | 250–300 mm | | 300–350 mm | | > 350 mm | |
| Country | Potential area loss (ha) | % loss | Potential area loss (ha) | % loss | Potential area loss (ha) | % loss | Potential area loss (ha) | % loss |
| Jordan | 14,318 | 4.2 | 32,643 | 9.6 | 55,566 | 16.3 | 89,714 | 26.4 |
| Libya | 2302 | 0.26 | 180,641 | 20.4 | 277,322 | 31.3 | 120,836 | 13.6 |
| Morocco | 11,493 | 0.11 | 159,979 | 1.6 | 970,218 | 9.6 | 1,881,006 | 18.7 |
| Syria | 157,760 | 4.4 | 224,784 | 6.2 | 606,711 | 16.8 | 686,985 | 18.9 |
| Yemen | – | – | 72,405 | 10.0 | 217,620 | 30.4 | 299,840 | 41.8 |

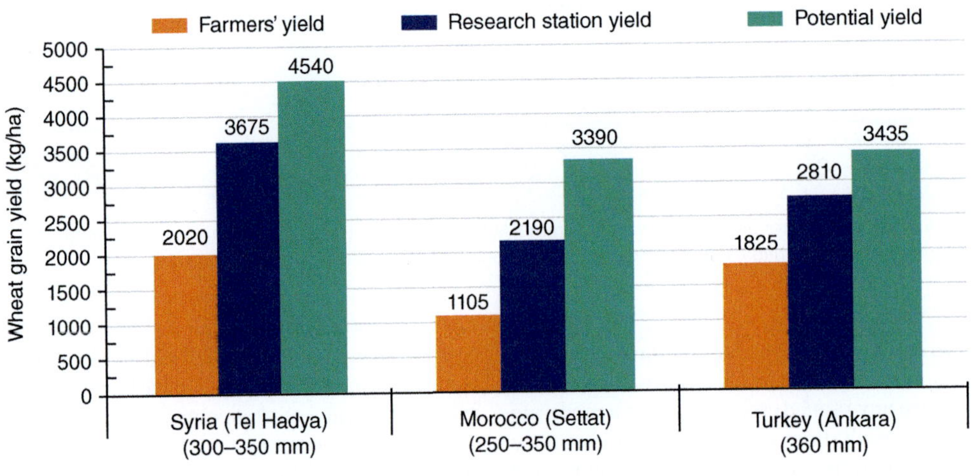


**Fig. 5.11.** Average wheat yields at farmers' research stations, and potential yields of wheat in Morocco and Syria. (From Oweis and Hachum, 2012)

rainwater productivity. In dryland rainfed environments in Australia, the wheat rainwater productivity has reached 18–20 kg grain/ha-mm of rainwater (Passioura, 2006), compared to less than 5 kg/ha-mm in the MENA region. As rainwater is the most limiting to rainfed agriculture, using rainwater productivity as a performance indicator besides yield is essential to fill this gap (ESCWA, 2022).

## Constraints to Improved Dryland Rainfed Agriculture

There are three major constraints to improved rainfed agriculture in the drylands:

**a. Climatic constraints.** The temporal and spatial variability of rainfall is the most critical constraint to rainfed agriculture improvement in the drylands. This factor is particularly important in future rainfed cropping as climate change is reducing rainfall amounts and increasing variability (Deryng, 2015). Figure 5.12 shows a typical soil-water status for rainfed wheatfields in the Mediterranean environment. Always rainfed crops in this environment face water stress, especially at the end of the season but the extent of the stress varies depending on the rainfall amount and distribution and soil-water storage capacity.

**b. Soil constraints.** The most crucial soil constraints to crop productivity are poor soil-water infiltration and storage capacity, poor soil fertility,

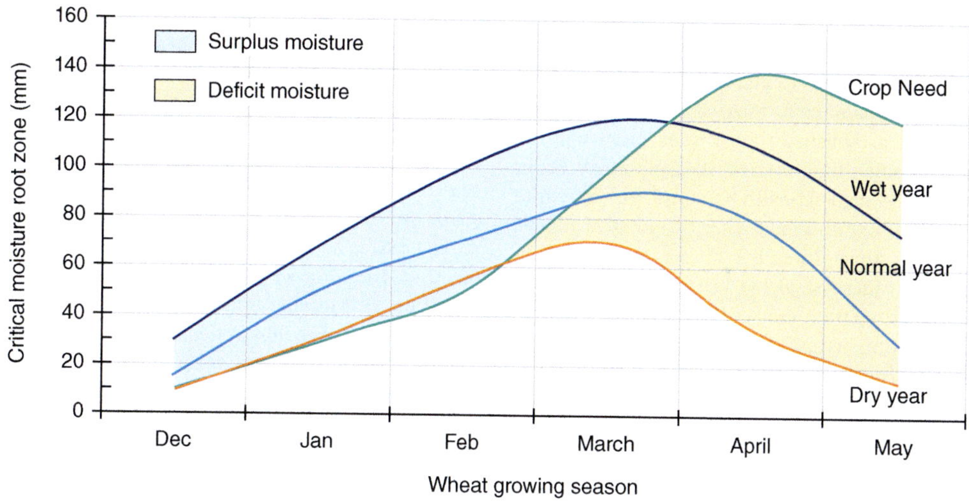


**Fig. 5.12.** Typical soil-water availability for wheat crops in a dry, average, and wet, cool winter-growing season in a Mediterranean environment. (From Oweis and Hachum, 2012)

and nutrient deficiency. Higher storage capacity can keep more water in the soil to alleviate the impacts of drought spells, and in wet years, soil can maintain more moisture to avoid terminal drought (Karrou and Oweis, 2014). In many mountainous areas, soils are shallow and degrade as a result of erosion with both water storage capacity and nutrient content limiting.

**c. Adoption constraints.** Farmers' poor adoption rates are a major constraint to improved rainfed productivity in the drylands. This can be attributed to a lack of knowledge and low financial capacity. The inability of farmers to invest in inputs, mainly fertilizers, improved varieties, pesticides, and disease control agents, is a common problem (Shideed *et al.*, 2014). The World Bank indicated that this is due to the lack of an enabling policy environment (World Bank, 2010).

## Change Strategies to Unlock the Potential

### Enhancing water availability

In dry environments, there is always some degree of water stress during one or more stages of crop growth, which negatively affects productivity. To improve rainfed agriculture, it is necessary to find ways to increase water availability through improved water and crop management. *Ex situ* water management aims at supplying water to crops from external sources and requires the reallocation of blue water resources and redirecting runoff water through practices such as supplemental irrigation (SI) and rainwater harvesting (RWH) (Oweis and Hachum, 2012; Oweis *et al.*, 2012). Using groundwater resources as a buffer against drought and climate change impacts, and supplementing rainfed systems, is a sound strategy. Breaking the governance divide between blue and green water resources will be an essential strategic step toward raising the institutional priority regarding investments in rainfed agriculture (Rockstrom *et al.*, 2010).

*In situ* water management can increase the beneficial portion of evapotranspiration by suppressing excessive evaporation and converting it into transpiration. This may be done through genetic improvement, high transpiration efficiency, improved crop geometry, high harvest index, and resistance to pests and diseases. Agronomic practices that enhance crop growth and water uptake include nutrient management, optimal crop rotation, intercropping, organic matter management, controlling pests and diseases, and conservation agriculture (Haddad *et al.*, 2011).

### Increasing crop-water uptake

This strategy aims to increase the portion of the soil-water balance to maximize its productivity. It is well established that a weak plant will not uptake soil water even if it is plentiful. Crops need to be in good health to transform water into transpiration and food when water is available. In many areas, there is enough rainfall to substantially increase yields, but it may be available at the wrong time, causing dry spells, or is lost in evaporation, runoff, or deep percolation. Improving rainfed agriculture requires investments in soil, crop, and farm management to maximize the ability of the crop to utilize the available water and nutrients for enhanced production. This may be done through integrated plant-soil water management to optimize transpiration (Wani *et al.*, 2009).

### Improving rainwater productivity (RWP)

Rainwater productivity is the benefits obtained from every millimeter of rainfall consumed as green water. The conventional objective is to increase land productivity as yield per unit of land. In rainfed systems, especially when blue water resources are used to supplement rainfall, a focus on water productivity is more relevant as it is the more limiting resource for enhanced productivity. Water productivity benefits include direct biophysical returns such as grain, biomass, milk, and meat; and economic returns such as income. Indirect benefits include nutrition such as protein and carbohydrates, energy such as calories, environmental benefits such as carbon sequestration, and social benefits such as employment (Molden *et al.*, 2010). Especially in rural areas where farmers practice subsistence farming, the indirect benefits are important to maintain sustainable and resilient agriculture. RWP framework better reflects the integration and value chain components, including markets and virtual water trade. Figure 5.13 shows the positive impacts of deficit irrigation in increasing water productivity, which also allows the use of saved water in expanding supplemental irrigation and obtaining higher overall farm production and income (Zhang and Oweis, 1999).

## Adopting Climate-smart Rainfed Agriculture

To cope with CC impacts, strategic changes must convert traditional systems to climate-smart rainfed production systems. This is associated with the three main pillars of climate-smart agriculture (CSA), namely mitigation, adaptation, and productivity enhancement. Investments in CSA can increase agricultural productivity while

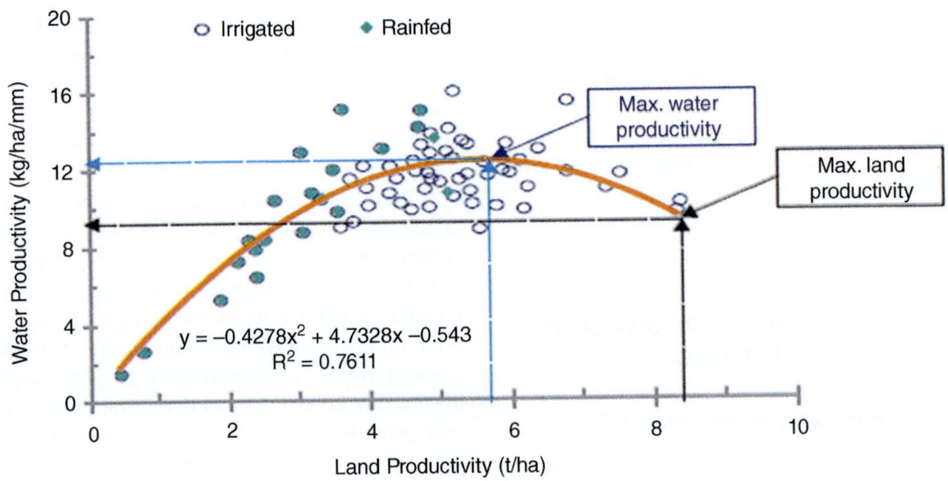


**Fig. 5.13.** Relationship between land and water productivities of wheat in northern Syria. (Adapted from Oweis and Hachum, 2012)

providing climate-related benefits, supporting adaptation, building resilience, and reducing emissions. CSA focuses on agriculture, but it is multi-sectoral and includes commitments to enhance livelihoods while ensuring food security. The challenge here is designing rainfed cropping systems from a multifunctional perspective, requiring trade-offs in changing crops or moving them to more favorable environments (Nangia and Oweis, 2016). Integration is key to climate-smart rainfed systems and to achieving sustainable development in this environment. Change of policies and empowered institutions should form the basis for creating enabling environments for investments and adoption at scale.

## Key practices and approaches

### *Supplemental irrigation*

Supplemental irrigation (SI) is the application of a limited amount of water to rainfed crops during drought spells to alleviate soil-water stress and enhance crop productivity. SI is a critical coping strategy to alleviate the negative impacts of low and variable rainfall. It reduces risks and incentivizes more investment in other production factors. SI achieves high yields and water productivity for the applied amounts and makes rainfall more productive; otherwise, crop failure may occur due to drought. Using SI would not only stop the decline in rainfed production due to climate change but can benefit from increased $CO_2$ levels and energy to gain even higher yields (Nangia and Oweis, 2016). It can alter the crop calendar to avoid unfavorable climatic conditions and match crop growing periods with good moisture availability. A well-designed SI can surpass the levels of water productivity attainable for rainfall and full irrigation systems (Fig. 5.14).

Maximizing water productivity in SI systems can be obtained by applying a deficit irrigation strategy. There are two significant constraints to the wide adoption of SI: the availability of water sources and the cost of irrigation systems. Local springs, shallow groundwater, and conventional water resources may be used during the rainy season. In addition to the possible reallocation of blue water resources, RWH and marginal-quality water can also be used. Conventional surface irrigation and low-cost mobile sprinkler systems are low-cost options for applying SI (Oweis and Hachum, 2012).

### *Rainwater harvesting*

Rainwater harvesting is another key practice for enhancing water availability for plants in rainfed agriculture. Micro-catchments increase water storage in the soil profile, refill the soil profile with water depleted during the growing season, reduce the impacts of drought spells, and ensure better soil storage for the dry period. It can recharge groundwater aquifers to be pumped later for SI. Macro-catchments provide blue water in

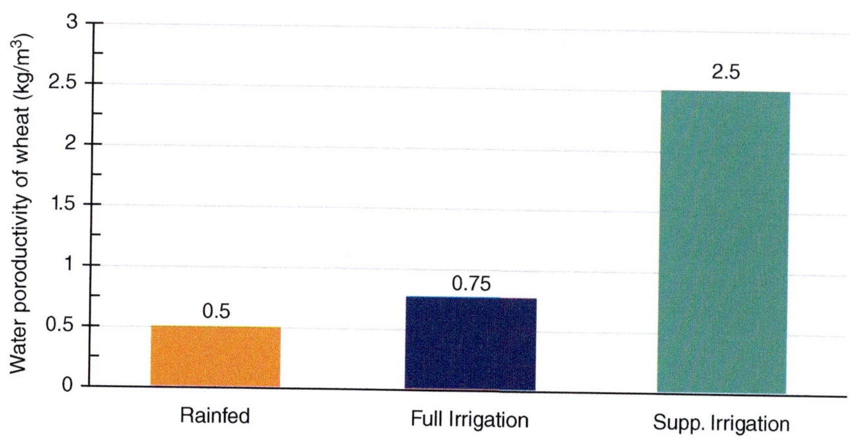


**Fig. 5.14.** Water productivity of supplemental irrigation water compared to that of rainfall and full irrigation water in wheat in northern Syria. (Oweis and Hachum, 2012)

surface reservoirs or groundwater aquifers, which provide water for SI. Downstream water rights may need to be considered when planning these reservoirs. RWH is key in combating desertification and restoring degraded rainfed agro-ecosystems. In addition to increasing productivity, indirect benefits include stabilizing rural communities and reducing the migration of rural inhabitants to cities.

The main constraint to the adoption of RWH at scale is the cost of implementing the structures. To overcome this constraint, ICARDA, in collaboration with national programs in Jordan and Syria, developed a fully mechanized laser-guided system based on the Vallerani dolphin plow (Fig. 5.15). The system can construct contour ridges and pits on over 20 ha/day with no need to allocate the contours manually. The construction cost/ha was reduced tenfold (Oweis, 2017).

## Agronomic and conservation practices

Cultural practices play an important role in increasing rainfed crop productivity; most important are nutrient management, conservation agriculture, and the use of improved varieties. Nutrient deficiency is a primary cause of low yields and declining response to enhanced practices. Other agronomic practices include optimal crop geometry, seeding rates, planting dates, integrated pest management, and integrated crop livestock management (Ryan, 2011; ESCWA, 2022).

Conservation agriculture systems are essential strategies to enhance land productivity and moisture conservation. The practice is applied as an integrated package, including minimum tillage, permanent soil cover, and intercropping. It offers multiple benefits for rainfed farmers in the region, including increased and more stabilized yields, soil erosion control, drought tolerance, and improved soil nutrients. Mulching makes soil moisture more available to crops by reducing evaporation. Adoption challenges in the region include using crop residue as a feed supplement for animals instead of keeping it as mulch; and managing weeds in the absence of tillage, especially when farmers cannot afford herbicides and the cost of machinery for seeding (Anderson *et al.*, 2016).

Genetic improvement has focused on varieties of strategic crops through conventional

**Fig. 5.15.** (a) The laser beam transmitter is mounted on a tripod on the uphill side; (b) Laser receiver mounted on a telescopic mast on the tractor; (c) Display panel mounted in front of the tractor operator; (d) Contour water-harvesting micro-catchments constructed by the Vallerani mechanized system in Jordan. (From Gammoh and Oweis, 2011)

breeding and biotechnology tools. Programs mainly use crop survival mechanisms to enhance tolerance to drought and resistance to pests and diseases while increasing yields (Tadesse *et al.*, 2016). Targeting transpiration efficiency combined with early vigor to reduce soil evaporation losses and a deeper rooting system to expand soil-water storage capacity would enhance soil-water availability. Researchers are currently working on genetically modified plants to improve transpiration efficiency and eventually save water. However, many are skeptical that those techniques come without hidden trade-offs (Jacobsen *et al.*, 2013).

### Adoption of digital rainfed agriculture

Over the last decade, digital agriculture approaches have gained huge momentum as a means of revolutionizing farming. Digital agriculture includes using precision agriculture and smart farming. Networking of the farm and the use of web-based data platforms together with Big Data are some features of digital agriculture. Rainfed agriculture can hugely benefit from the advancements of those tools by facilitating proper and timely farm decision making. Technology tools such as sensors, drones, and computer imaging, coupled with satellite imagery, machine learning, predictive analytics software, and data storage in clouds, give farmers access to timely valuable insights so that they can adopt best practices and manage farms more efficiently, thus reducing losses and maximizing profits.

Especially in rainfed systems, dealing with climatic variability is very challenging. Using digital approaches, better prediction of rainfall and droughts, and having timely quantified data can help reduce the risks of the rainfed system by avoiding failure and responding in time. Further, soil-water sensors connected to analytic software can help schedule supplemental irrigation and application of other farm inputs (Fig. 5.16). Drones equipped with sensors can reflect the farm conditions and variations in soil moisture, nutrients, pests, and diseases, giving the farmer the opportunity to target specific locations instead of applying blanket solutions.

The challenges, however, for farmers to adopt digital agriculture are many, especially in developing countries. Rainfed farmers in developing countries are the poorest and have the least capacity to digest and use those advancements. Building collective bodies to provide the outputs of modern tools to farmers is the responsibility of governments and technology organizations.

It is clear that dryland rainfed agriculture has excellent production potential and can

**Fig. 5.16.** A technician using canopy-water stress sensors to determine supplemental irrigation scheduling at an ICARDA station in northern Syria. (From T. Oweis, personal album)

reduce rural poverty. Development should aim at integrated digital farming systems to enhance productivity and minimize risks associated with climatic variabilities and change. Poor investment in rainfed agriculture is attributed to risks associated with climatic variabilities and change, but this risk should not be the reason to reduce it. Investment in zero-till seeders, supplemental irrigation, and rainwater harvesting can have a significant pay-off. Changing cropping patterns to maximize economic water productivity increases prospects for intensification and sustainable development. Enabling the environment to adopt modern solutions and attract adequate investment in rainfed agriculture is needed as farmers cannot alone drive the change. Governments should consider retargeting agricultural investment towards rainfed agriculture. Subsidies can increase investment in using advanced technologies and sustainable practices. In research, the focus should be directed at overcoming adoption constraints. Risk management, payment for environmental services, and *ex ante* and *ex post* impact assessments are other deficit research areas. Most national data on rainfed agriculture are aggregated with those on irrigated agriculture, which is a constraint to properly analyze rainfed systems and draw proper conclusions. The long-term priority must be increased investment in science and technology to drive sustainable digital rainfed agricultural development (ESCWA, 2022).

## Precision Agriculture and Smart Irrigation Systems

Water resources are the critical factor for all production processes and for sustainable development in many countries around the world, especially in the arid zone. Agriculture production of both food and fibers is the main activity related to water resources in a country's economy and it is the major consumer of water. For example, Egypt lies in the arid zone where nearly all agriculture depends on irrigation water (85% of Egypt's annual water use). Today, nearly 700 million people worldwide suffer from water scarcity. And that number is expected to grow to 1.8 billion in just ten years. Some African countries reached the water poverty line (below 1000 $m^3$/yr per capita) (Egypt 535 $m^3$/yr per capita). By 2025, approximately 29 countries in the world are expected to experience water scarcity, particularly due to the expected increase in population. Water use has grown more than twice the rate of population increase in the last century. Currently, there are 18 countries in the Middle East and northern Africa classified as water-scarce. Desalination as a non-conventional water resource is the world's only sustainable source of new freshwater supplies but cannot cover the gap and needs huge energy resources. Many countries are confronted by various water problems due to both climatic conditions and socio-economic factors. From an eco-climatic point of view, most desert regions extend across semi-arid and arid zones, wherever they have been particularly affected by cycles of drought, salinity, and desertification. Under these conditions, irrigation is considered a key issue for agricultural development. In addition, concerning the socio-economic factors, the regions are characterized by fast-increasing populations, as well as misuse of water in general and irrigation water in particular, which has resulted in a sharp decline of the quota per capita availability. However, due to increasing water scarcity and reduced quota per capita availability, many countries have recognized the urgent need to secure, rationalize, and utilize new irrigation technologies and techniques such as smart irrigation management to sustain a minimum resources base, as well as modify the traditional irrigation systems.

One of the main keys to future agriculture is closely related to agricultural service operations using modern technology that contributes to increasing the quantity of production and improving its quality. On-farm irrigation systems and their efficient management are considered one of the most important factors in agricultural sectors. Efficient and improved irrigation management plays a critical role in improving water unit productivity as well as maximizing the yield as a result of increasing demands on limited water resources. Irrigation scheduling is a method of applying water for irrigation of crops based on calculated crop-water needs. Modeling, expert systems, and simulation of irrigation requirements to ensure effective water management have been employed in many regions, and a number of irrigation schedules have been developed.

## Precision (smart) agriculture

Precision digital agriculture including smart irrigation is one of the alternatives and solutions to meet the CC challenges to optimize water unit productivity. Precision agriculture means dealing with optimal, accurate, and efficient measuring, and estimating all agricultural production inputs such as water, energy, chemicals, seeds, agricultural mechanization processes, and post-harvest services within the agricultural production units, which allows for economical handling of agricultural inputs. The digital agriculture is defined as that agriculture that uses ICT and ecosystem data to support and develop information and services to farmers in a timely manner. It is sometimes also known as smart agriculture or e-agriculture due to the use of digital technologies in agricultural production, both plant and animal.

## Smart irrigation

The concept of smart irrigation is defined as the digital and technical application of a package of databases and information that allows easy management of the water unit through an electronic system that accurately controls the irrigation process for different crops aimed at maximizing crop production with limited quantities of water. Smart irrigation offers a variety of solutions, supporting water management, which would help farmers raise the efficiency of water and energy use, especially in farms that depend on pressure irrigation or developed surface irrigation. Here are some smart-irrigation solutions that help in optimal irrigation scheduling:

- automatic control of the addition of irrigation water by relying on weather stations;
- automatic control of the addition of irrigation water by relying on the measurement of the moisture content of the soil; and
- using vegetation spectral reflection sensors to control irrigation and linking them to soil moisture and vegetation temperature sensors.

AI applications in agriculture are gaining traction owing to their ability to resolve issues such as water shortage, energy saving, and climate change. The purpose of a smart-irrigation system with controllers is the ability to automatically program and control all aspects of field irrigation, regardless of their location (Brajović *et al.*, 2015). When combined with weather-sensing instrumentation, the smart-irrigation control system can determine the daily ET rate and adjust irrigation schedules appropriately by downloading new irrigation programs to the field controllers every day (García *et al.*, 2020). Water savings can therefore be achieved by precisely computing the amount of water that plants require (Kamienski *et al.*, 2019). The developed smartphones include high-speed broadband and several built-in sensors (e.g. location, motion, cameras), that help users in completing different activities (Khelifa *et al.*, 2015). The smartphone and agriculture sensor application provides substantial benefits to precision agriculture with a network of sensors, cameras, and gadgets that all work together to assist a farmer in doing his job intelligently (Gayatri *et al.*, 2015).

Smart-irrigation technology provides comprehensive management and control of irrigation water by computers or even smartphones. Smart irrigation offers a wide range of supportive water management solutions that will help farmers increase the efficiency of their water and energy use. Although water savings are the most readily apparent benefit, a smart-irrigation control system is an ideal environmental management tool that provides several additional direct and indirect cost savings. In addition to water, labor, and financial savings, there are enormous potential savings in irrigation agricultural costs alone. These technologies will help to efficiently re-apply water to match the crops' ET usage. They will also maintain a desirable moisture balance within the root zone of the crop areas, doing much to prevent the leaching of excess fertilizers and pesticides. In addition, smart-control irrigation systems can deliver the following benefits:

- calculate the vapor-transpiration loss that has occurred since the last irrigation cycle;
- control each water distributor (sprinkler head, dripper, and bubbler) relative to the quantity of water applied and the duration of the application to replace the lost water;
- apply water at a rate that will prevent runoff; and
- regulate pressure at each head to get maximum uniformity of water distribution.

The use of ICT for digital agriculture and smart irrigation such as remote sensing, GIS, computer applications, and AI and its applications in agricultural fields will be a strong support in the expansion of the application and spread of precision and digital agriculture as well as smart irrigation. On the other hand, the introduction of simulation, modeling, and expert systems for precision and digital agriculture will facilitate the success of agricultural extension and the provision of excellent service to farmers.

IoT also helps in the process of following up and monitoring the growth of agricultural crops, as the drones are connected to today's high-speed internet networks such as 4G and 5G. These conduct a comprehensive survey of large cultivated areas such as center pivots, fruit and citrus gardens, and provide a complete database that includes crop status, irrigation, and fertilizer-spraying rates.

## Devices and Tools for Operating Smart On-farm Irrigation Systems

Smart-irrigation systems can be applied to on-farm irrigation methods and systems in which the movement of water is through closed water conductors or pipes and under pressure. It is operated and controlled by opening and closing valves as in pressurized irrigation systems (sprinkler and localized) and developed surface irrigation systems (Figs 5.17 and 5.18). It can also be applied in both protected agriculture farms/hydroponics and integrated management projects for fish farms and agricultural production (Fig. 5.19).

The devices required for the automatic operation of smart-irrigation systems are divided into the following areas:

**1. Meteorological (weather) station.** These measure weather data to estimate water consumption rates and irrigation requirements. Using evapotranspiration devices, these stations operate by solar energy and are connected to an irrigation system for wireless operation (Fig. 5.20).

**2. Sap flow sensors**. This technology is the latest technology to measure the flow of sap and monitor the movement of the internal fluid of plants to calculate the rate of water consumption for the plant through sensors that are placed inside the plants. The energy-balance sensors determine the balance of energy passing through the area of the plant section by measuring the amount of heat carried by the sap, which is converted to a real-time leachate flow in grams or kilograms per hour. These sensors are available from 2 mm to 125 mm in length. Sap flow requires no calibration as leachate flow is directly determined by energy balance and convection rates. References are available for most major crops and many tree species. This technique can be used to measure the water consumption rate of plants in greenhouses, nurseries, or natural environments.

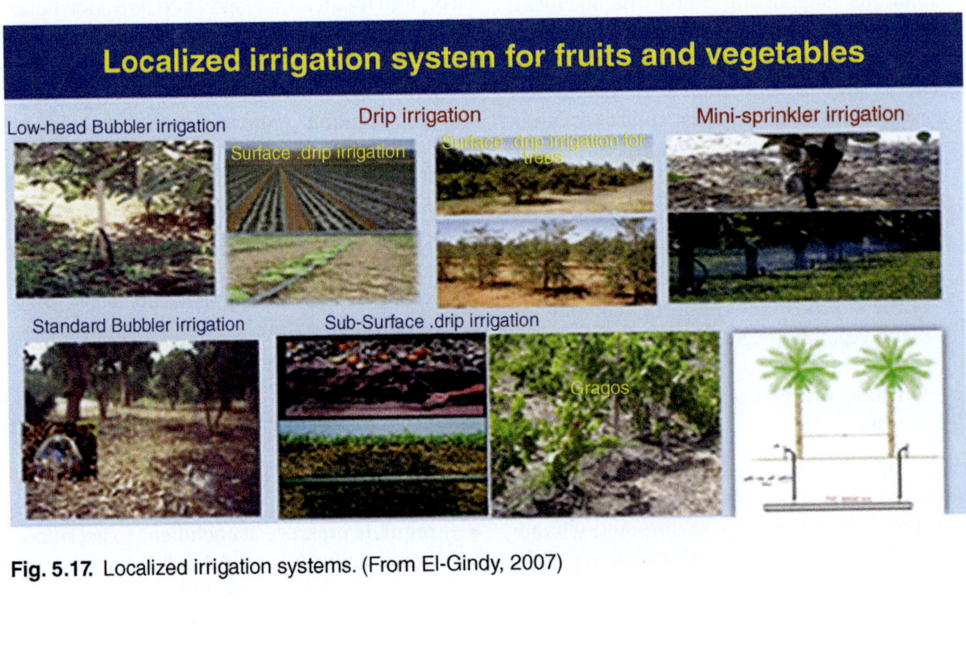


**Fig. 5.17.** Localized irrigation systems. (From El-Gindy, 2007)

**Fig. 5.18.** Sprinkler irrigation systems. (From El-Gindy, 2007)

**3. Soil-moisture sensors.** These are used to set the soil-moisture content at two levels to start and stop irrigation wirelessly (Fig. 5.20).

**4. Measuring and control devices.** These devices are used for controlling and measuring water flow through the piping network, and they consist of hydraulic or electrical valves (solenoid valves) that take signals from a programmed control panel for operating the irrigation systems and can be operated from a computer or a smartphone from any location. The electric flow meter and flow sensor are also used to control and measure the water flow through the irrigation network. All previous devices work after preparing a computer program. All information about the agricultural project is input, and the output is software for smart operation.

## Energy in Precision (Smart) Agriculture

Renewable energy (solar, wind, and green hydrogen) is utilized in SA. Solar energy is considered one of the clean energies that saves a lot of operating costs. It is a common application to operate water pumps for different water resources, controllers or control panels. Biofuel is also considered one of the clean energies. It is used in the operation of engines as a clean economical energy source that expands the cultivation of jojoba and the extraction of its oil.

## Agricultural Mechanization Processes and Post-harvest Services

Agricultural engineering plays a major contribution in precision agriculture. The availability of agricultural mechanization leads to:

**a.** opportunities to intensify production in a sustainable way;
**b.** food systems development and increased added value;
**c.** improvement in crop productivity and quality while ensuring the appropriate timing of agricultural operations and improving the efficiency of resource use; and
**d.** increased opportunities to improve local economies and livelihoods.

Agricultural mechanization processes include drones for spraying agricultural chemicals and seeds; photography; smart-harvesting machines for different crops; and robotic milking machines.

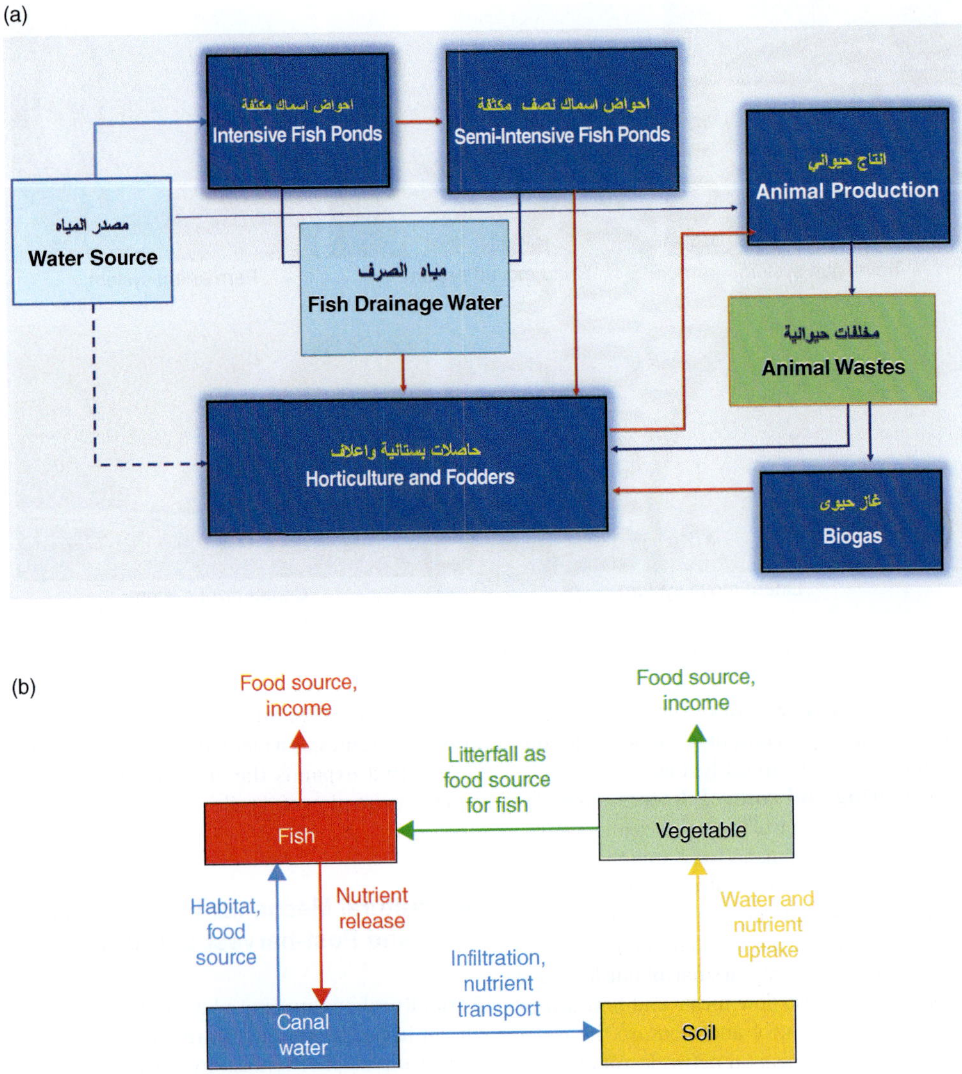


**Fig. 5.19.** (a) Integrated management projects for fish farms and agricultural production. (From Agricultural Research and Development Council, 2009). (b) Integrated management projects for fish farms and agricultural production. (Adapted from Thi Da *et al.*, 2020; Open access, C.C. 4.0)

Agricultural power and machinery has revolutionized the agricultural sector due to their success in relieving and reducing hard labor, and compensating for shortage of labor in the case of millions of agricultural families and workers, as it has improved crop productivity and quality while ensuring the appropriate timing of agricultural operations and improving the efficiency of resource use.

The introduction of aerial photography, such as cameras mounted on drones, in modern agriculture will contribute to the exploration of many fields. This application is possible in small farms by use of smartphones.

Agricultural mechanization provides opportunities to intensify production in a sustainable manner, add value and improve local economies and livelihoods. Driverless tractors used in ploughing and other agricultural tasks could improve operations on the farm and provide a safer and much less dangerous work environment. Autonomous tractors and robots are

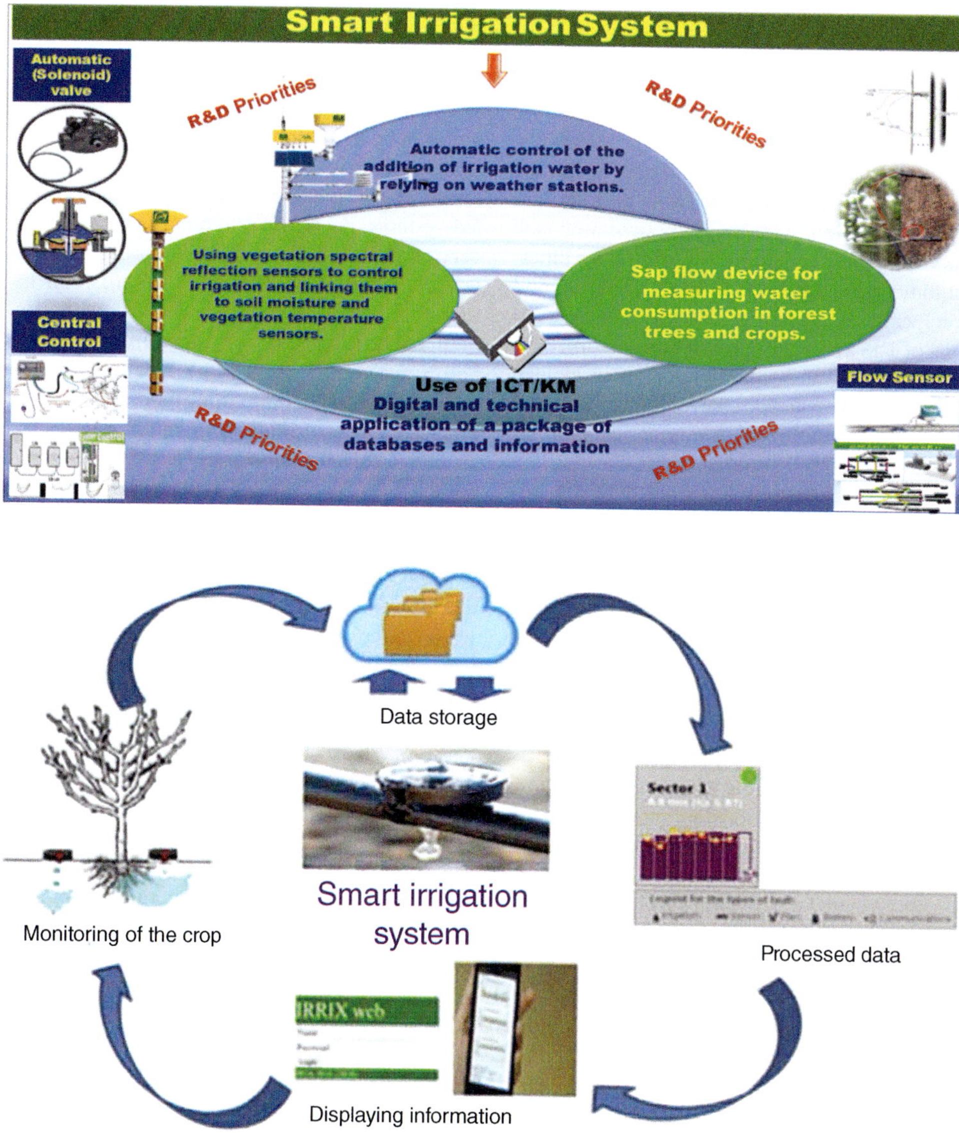


**Fig. 5.20.** Devices and tools for operating smart-irrigation systems. (From El-Gindy, personal album)

becoming more popular for certain tasks. Some are attached to a human-powered tractor, while others are highly customizable with sensors and accessories that perform specific tasks. These robots are often directed through precise GPS tracking, so they can easily navigate the narrow space between rows of crops.

However, the challenge of building exceedingly small robots is the need for a smaller controller, which was achieved in 2020.

## Conclusion

The digitization of the sector has many benefits if IoT, Big Data, and AI are combined. This can be done through many elements, e.g. building a database for the agricultural system, entering climatic data for agricultural areas to predict weather conditions, establishing an electronic agricultural marketing system, and developing a special application for farmers that enables them to follow all the details of their crops. Integrated land management in precision agriculture applications can be achieved using remote-sensing techniques and GIS by producing soil topography maps and various soil maps to map physical properties – salinity, pH, alkalinity, $CaCO_3$, phosphorus, potassium, iron, zinc, and manganese. Evaluating and mapping dynamic soil quality can be carried out by evaluating different indicators related to physical, chemical, and biological soil parameters. Therefore, applications of GIS and RS in precision agriculture include the use of recommended methodology relying on the collection of climatic data and available information on the type and units of different soils and vegetation cover. The development of sensors capable of examining land conditions with extreme accuracy and providing data to farmers can be very important. Drones equipped with sensors can reflect the farm conditions and variations in soil moisture, nutrients, pests, and diseases giving the farmer the opportunity to target specific locations instead of applying blanket solutions.

Water resources are the critical factor for all life-production processes and sustainable development in the arid zone. Many countries have recognized the urgent need for smart-irrigation management to sustain a minimum water resource base and modify traditional irrigation systems. Rainfed agriculture can hugely benefit from the advancements of those tools. Technology tools such as sensors, drones, and computer imaging coupled with satellite imagery, machine learning, predictive analytics software, and data storage in clouds give farmers access to timely valuable insights so that they can adopt best practices, manage farms more efficiently, reduce losses, and maximize profits. Especially in rainfed systems, dealing with climatic variability is very challenging. Using digital approaches, better prediction of rainfall and droughts can be made, and having timely quantified data can help reduce the risks of the rainfed system. Soil-water sensors connected to analytic software can help schedule supplemental irrigation and application of other farm inputs.

On-farm irrigation systems and their management are one of the most important agricultural sectors in applying the concept of precision and digital sustainable agriculture. It offers a wide range of supportive water management solutions that will help farmers increase the efficiency of their water and energy use. Smart-irrigation control systems allow a centrally located computer the ability to automatically program and control all aspects of on-farm irrigation controllers, regardless of their location. When combined with weather-sensing instrumentation, they can determine the daily ET rate, and adjust irrigation schedules appropriately by downloading new irrigation programs to the field controllers every day. Water savings can therefore be achieved by precisely computing the amount of water that plants require.

## References


Abbasnia, A., Yousefi, N., Mahvi, A.H., Nabizadeh, R., Radfard, M. *et al.* (2019) Evaluation of groundwater quality using water quality index and its suitability for assessing water for drinking and irrigation purposes: case study of Sistan and Baluchistan province (Iran). *Human and Ecological Risk Assessment: An International Journal* 25, 988–1005. DOI: 10.1080/10807039.2018.1458596

Abdel Rahman, M.A.E., Farg, E., Saleh, A.M., Sayed, M., Abutaleb, K. *et al.* (2022a) Mapping of soils and land-related environmental attributes in modern agriculture systems using geomatics. *Sustainable Water Resources Management* 8, 116. DOI: 10.1007/s40899-022-00704-2

Abdel Rahman, M.A.E., Saleh, A.M., Elsharkawy, M.M., Farg, E., Moustafa, M. *et al.* (2022b) Geomatics-based mapping of hydraulic soil properties for agricultural management. *Annals of Agricultural Science Moshtohor* 60, 225–238. DOI: 10.21608/assjm.2022.227887

Agricultural Research and Development Council (2009) *Sustainable Agriculture Development Strategy, SADS Towards 2030.* In: El-Beltagy A. *et al.* (eds) Ministry of Agriculture and Land Reclamation, Egypt. ISSN: 20003-2009, ISBN: 978-977-302-250-6.

Amiri-Zarandi, M., Fard, M.H., Yousefinaghani, S., Kaviani, M. and Dara, R. (2022) Platform approach to smart farm information processing. *Agriculture* 12(6), 838. DOI: 10.3390/agriculture12060838

Anderson, W., Johansen, C. and Siddique, K. (2016) Addressing the yield gap in rainfed crops: a review. *Agronomy for Sustainable Development* 36, 18. DOI: 10.1007/s13593-015-0341-y

Arafat, S.M., Afify, A.A., Nagwan, M.A. and Elsharkawy, M.M. (2014) Change detection of land use/land cover categories in the Nile Delta region using remote sensing and GIS. *SPIE Remote Sensing*. Proceedings of the SPIE Remote Sensing and SPIE Security, Amsterdam, The Netherlands, vol. 9239.

Blackmore, B. (1994) Precision farming; an introduction. *Outlook on Agriculture* 23(4), 275–280. DOI: 10.1177/003072709402300407

Boori, M.S., Choudhary, K., Paringer, R. and Kupriyanov, A. (2021) Spatiotemporal ecological vulnerability analysis with statistical correlation based on satellite remote sensing in Samara, Russia. *Journal of Environmental Management* 285, 112138. DOI: 10.1016/J.JENVMAN.2021.112138

Brajović, M., Vujović, S. and Dukanović, S. (2015) An overview of smart irrigation software. 4th Mediterranean Conference on Embedded Computing (MECO), Budva, Montenegro, pp. 353–356. DOI: 10.1109/MECO.2015.7181942

Deryng, D. (2015) Climate change impacts on crop productivity in global semi-arid areas and selected semi-arid economies. Small Grant Programme Working Paper. Overseas Development Institute (ODI), London. Available at: http://prise.odi.org/research/small-grants-programme-climatechange-impacts-on-crop-productivity-in-global-semi-arid-areas-andselected-semi-arid-economies (accessed 7 July 2024).

El-Gindy, A.M. (2007) *Sustainable Agriculture Strategy 2030. On-Farm Irrigation Systems and Technologies*. Self-published. ISBN: 1-318-237-977.

Elsharkawy, M.M. (2016) Precision agriculture using advanced remote sensing techniques for peanut crop in Arid Land. *Journal of Hyperspectral Remote Sensing* 6, 72–81. DOI: 10.5935/2237-2202.20160007

Elsharkawy, M.M., Sheta, A.S., Abd El-Wahed, M.S., Arafat, S.M. and El Behiery, O.M. (2016a) Precision agriculture using remote sensing and GIS for peanut crop production in arid land. *International Journal of Plant & Soil Science* 10, 1–9. DOI: 10.9734/IJPSS/2016/20539

Elsharkawy, M.M., Sheta, A.S., Arafat, S.M. and Abd El-Wahed, M.S. (2016b) Precition agriculture using advanced remote sensing and GIS for peanut crop production in arid lands. *Journal of Hyperspectral Remote Sensing* 6(1), 72–81 DOI: 10.9734/IJPSS/2016/20539

Elsharkawy, M.M., Nabil, M., Farg, E. and Arafat, S.M. (2022a) Impacts of land-use changes and landholding fragmentation on crop water demand and drought in Wadi El-Farigh, New Delta project, Egypt. *Egyptian Journal of Remote Sensing and Space Science* 25, 873–885. DOI: 10.1016/j.ejrs.2022.08.002

Elsharkawy, M.M., Sheta, A.S., D'Antonio, P., Abdelwahed, M.S. and Scoa, A. (2022b) Tool for the establishment of agro-management zones using GIS techniques for precision farming in Egypt. *Sustainability* 14(9). DOI: 10.3390/su14095437

ESCWA (2022) Unlocking the potential of rainfed agriculture in the Arab region. ESCWA, United Nations House, Beirut, Lebanon.

ESCWA, ACSAD, FAO, GIZ, League of Arab States *et al.* (2017) *Arab Climate Change Assessment Report – Main Report*. ESCWA, United Nations House, Beirut, Lebanon.

FAO-GIEWS (2016) *Global Information and Early Warning System on Food and Agriculture*. Country briefs. Food and Agriculture Organization of the UN, Rome.

Farg, E., Abutaleb, K., Arafat, S., Elsharkawy, M.M. and Nabil, M. (2020) Assessment of Sentinel-2 data capabilities for vegetation physiological parameters retrieving in the Nile Delta. *Bioscience Research* 17, 467–478.

Gammoh, I.A. and Oweis, T. (2011) Contour laser guiding for the mechanized 'Vallerani' micro-catchment water harvesting systems. *Journal of Environmental Science and Engineering* 5(9). Available at: https://hdl.handle.net/20.500.11766/8049 (accessed 7 July 2024).

García, L., Parra, L., Jimenez, J.M., Lloret, J. and Lorenz, P. (2020) IoT-based smart irrigation systems: an overview on the recent trends on sensors and IoT systems for irrigation in precision agriculture. *Sensors* 20, 1042. DOI: 10.3390/S20041042

Gayatri, M.K., Jayasakthi, J. and Mala, G.S.A. (2015) Providing smart agricultural solutions to farmers for better yielding using IoT. EEE Technological Innovation in ICT for Agriculture and Rural Development (TIAR), Chennai, India, pp. 40–43. DOI: 10.1109/TIAR.2015.7358528

Gonzalez-de-Santos, P., Fernández, R., Sepúlveda, D., Navas, E., Emmi, L. *et al.* (2020) Field robots for intelligent farms: inhering features from industry. *Agronomy* 10(11), 1638. DOI: 10.3390/agronomy10111638

Haddad, N., Duwayri, M., Oweis, T., Bishaw, Z., Rischkowsky, B. *et al.* (2011) The potential of small-scale rainfed agriculture to strengthen food security in Arab countries. *Food Security* 3 (Suppl 1), S163–S173. DOI: 10.1007/s12571-010-0099-7

Jacobsen, S.E., Sørensen, M., Pedersen, S.M. and Weiner, J. (2013) Feeding the world: genetically modified crops versus agricultural biodiversity. *Agronomy for Sustainable Development* 33(4), 651–662. DOI: 10.1007/s13593-013-0138-9

Kamienski, C., Soininen, J.-P., Taumberger, M., Dantas, R., Toscano, A. *et al.* (2019) Smart water management platform: IoT-based precision irrigation for agriculture. *Sensors* 19(2), 276. DOI: 10.3390/S19020276

Karlen, D.L., Mausbach, M.J., Doran, J.W., Cline, R.G., Harris, R.F. *et al.* (1997) Soil quality: a concept, definition, and framework for evaluation (a guest editorial). *Soil Science Society of America Journal* 61, 4–10. DOI: 10.2136/sssaj1997.03615995006100010001x

Karrou, M. and Oweis, T. (2014) Assessment of the severity and impact of drought spells on rainfed cereals in Morocco. *African Journal of Agricultural Research* 9(49) 3519–3530. Available at: https://hdl.handle.net/10568/68186 (accessed 7 July 2024).

Khelifa, B., Amel, D., Amel, B., Mohamed, C. and Tarek, B. (2015) Smart irrigation using Internet of Things. *2015 Fourth International Conference on Future Generation Communication Technology (FGCT), Luton, UK*, pp. 1–6. DOI: 10.1109/FGCT.2015.7300252

McBratney, A., Whelan, T.A. and Bouma, J. (2005) Future directions of precision agriculture. *Precision Agriculture* 6, 7–23. DOI: 10.1007/s11119-005-0681-8

Molden, D., Oweis, T., Steduto, P., Bindraban, P., Hanjra, M. *et al.* (2010) Improving agricultural water productivity: between optimism and caution. *Agricultural Water Management* 97(4), 528–535. DOI: 10.1016/j.agwat.2009.03.023

Nabil, M., Farg, E., Arafat, S.M., Aboelghar, M., Afify, N.M. *et al.* (2022) Tree-fruits crop type mapping from Sentinel-1 and Sentinel-2 data integration in Egypt's New Delta project. *Remote Sensing Applications: Society and Environment* 27, 100776. DOI: 10.1016/j.rsase.2022.100776

Nangia, V. and Oweis, T. (2016) Supplemental irrigation: a promising climate – resilience practice for sustainable dryland agriculture. In: Farooq, M. and Siddique, K. (eds) *Innovations in Dryland Agriculture*. Springer, pp. 549–564.

Oweis, T. (2017) Rainwater harvesting for restoring degraded dry agro-pastoral ecosystems: a conceptual review of opportunities and constraints in a changing climate. *Environmental Reviews* 25(2), 135–149. DOI: 10.1139/er-2016-0069

Oweis, T. and Hachum, A. (2012) *Supplemental Irrigation: A Highly Efficient Water-Use Practice*, 2nd edn. International Center for Agricultural Research in the Dry Areas (ICARDA), Beirut, Lebanon.

Oweis, T., Prinz, D. and Hashem, A. (2012) *Rainwater Harvesting for Agriculture in the Dry Areas*. CRC Press/Balkema, Taylor & Francis Group, London. DOI: 10.1201/b12351

Passioura, J. (2006) Increasing crop productivity when water is scarce – from breeding to field management. *Agricultural Water Management* 80(1), 176–196. DOI: 10.1016/j.agwat.2005.07.012

Perez-Aguilar, L.Y., Plata-Rocha, W., Monjardm-Amenta, S.A., Franco-Ochoa, C. and Zambrano-Medina, Y.G. (2021) The identification and classification of arid zones through multicriteria evaluation and GIS – case study: Arid regions of Northwest Mexico. *ISPRS International Journal of Geo-Information* 10(11), 720. DOI: 10.3390/ijgi10110720

Rockstrom, J., Karlberg, I., Wani, S.P., Barron, J., Hatibu, N. *et al.* (2010) Managing water in rainfed agriculture – the need for a paradigm shift. *Agricultural Water Management* 97(4), 543. DOI: 10.1016/j.agwat.2009.09.009

Ruß, G. (2013) From spatial data mining in precision agriculture to environmental data mining. In: Moewes, C. and Nürnberger, A. (eds) *Computational Intelligence in Intelligent Data Analysis. Studies in Computational Intelligence*, Vol. 445. Springer, Berlin, Heidelberg. DOI: 10.1007/978-3-642-32378-2_18

Ryan, J. (2011) Rainfed farming systems in the West Asia–North Africa (WANA) region. In: Tow, P., Cooper, I., Partridge, I. and Birch, C. (eds) *Rainfed Farming Systems*. Springer, Berlin, Germany. DOI: 10.1007/978-1-4020-9132-2_15.

Sadras, V. and Angus, J. (2006) Benchmarking water-use efficiency of rainfed wheat in dry environments. *Australian Journal of Agricultural Research* 57(8), 847–856. DOI: 10.1071/AR05359

Saleh, A.M., Elsharkawy, M.M., AbdelRahman, M.A.E. and Arafat, S.M. (2021) Evaluation of soil quality in arid western fringes of the Nile Delta for sustainable agriculture. *Applied and Environmental Soil Science* 2021, 1434692. DOI: 10.1155/2021/1434692

Shideed, K., Mazid, A., Oweis, T. and van Ginkel, M. (2014) Potential of rainfed agriculture and smallholder farmers in food self-sufficiency. *Arab Environment: Food Security*. Annual Report of the Arab Forum for Environment and Development. Technical Publications, Beirut, Lebanon.

Simsek, C. and Gunduz, O. (2007) IWQ index: a GIS-integrated technique to assess irrigation water quality. *Environmental Monitoring and Assessment* 128, 277–300. DOI: 10.1007/s10661-006-9312-8

Smith, D.M. and Allen, S.J. (1996) Measurement of sap flow in plant stems. *Journal of Experimental Botany* 47(12), 1833–1844. DOI: 10.1093/jxb/47.12.1833

Stombaugh, T.S., Mueller, T.G., Shearer, S.A., Dillon, C.R. and Henson, G.T. (2001) *Guideline for Adopting Precision Agricultural Practices*. System, College Station, Texas. Available at: http://www2.ca.uky.edu/agcomm/pubs/pa/pa2/pa2.pdf (accessed 7 July 2024).

Tadesse, W., Nachit, M., Abdalla, O. and Rajaram, S. (2016) Wheat breeding at ICARDA: achievements and prospects in the CWANA region. In: Bonjean, A.P., Angus, W.J. and van Ginkel, M. (eds) *The World Wheat Book: A History of Wheat Breeding*. Lavoisier (Hermes Science Publications), Paris. Available at: https://hdl.handle.net/20.500.11766/6069 (accessed 7 July 2024).

Thi Da, C., Anh Tu, P., Livsey, J., Tang, V.T., Berg, H. *et al.* (2020) Improving productivity in integrated fish-vegetable farming systems with recycled fish pond sediments. *Agronomy* 10(7), 1025. DOI: 10.3390/agronomy10071025

Wani, S., Rockstrom, J. and Oweis, T. (eds) (2009) *Rainfed Agriculture: Unlocking the Potential*. Comprehensive Assessment of Water Management in Agriculture Series 7. CAB International, Wallingford, UK. Available at: https://hdl.handle.net/10568/36475 (accessed 7 July 2024).

World Bank (2010) *Improving Water Management in Rainfed Agriculture: Issues and Options in Water-Constrained Production Systems*. World Bank Group, Washington, DC. Available at: http://hdl.handle.net/10986/13028 (accessed 7 July 2024).

Zhang, H. and Oweis, T. (1999) Water-yield relations and optimal irrigation scheduling of wheat in the Mediterranean region. *Agricultural Water Management* 38, 195–211. DOI: 10.1016/S0378-3774(98)00069-9

# 6 Impact of Climate Change on Inland and Ocean Fisheries and Aquaculture

**Meryl Williams***
*Faculty of Arts and Social Sciences, University of Technology Sydney, Australia*

**Abstract**
Fisheries and aquaculture produce highly nutritious foods that are generally of high value and are often targeted at export markets. Initially slow to recognize the threats from climate change, compared to other sectors, research by sectoral experts and greater attention to aquatic systems (freshwater and marine) by the IPCC rapidly improved understanding of likely productivity changes from climate change. Diverse approaches to adaptation are revealed by reviews of today's adaptation practices, the most prominent pathways promised by technological fixes and emerging more comprehensive socio-technical systems approaches. The major challenges brought by climate change are going to exacerbate the ongoing dilemmas of how adaptation will affect small-scale fisheries and aquaculture and the large-scale enterprises, and which people will be advantaged or displaced in the adaptation processes.

## Introduction

Fish and other aquatic products such as seaweeds, molluscs, crustaceans, and echinoderms are recognized as important and growing components of the world food system, vital traded commodities in the economies of many countries, and providers of livelihoods and businesses for millions of people (Table 6.1).

Since World War II, the global growth in fisheries and aquaculture production has been dramatic, increasing ninefold while human population tripled. These relative growth figures explain the large increase in per capita fish consumption (Table 6.1). Over the same period, fish also maintained its value, according to the FAO trade-based Fish Price Index (Williams and Syddall, 2022). In the mid-1980s, the production of fisheries and aquaculture in developing countries (countries of the global south) began to outstrip that of countries in the global north. Northern countries shifted to stronger management of their depleting fisheries resources and satisfied their needs for fish by importing more from countries of the south. Governments supported increasing fish production, especially through aquaculture, to garner trade benefits such as foreign exchange income. The high economic value of fish and other aquatic foods may create tensions between the drive to export, mediated through well-off firms, and the role of aquatic foods in nutrition and food security for the poor and less-well-off people. Reconciling the tensions requires economic policies that also

*Address for correspondence: meryljwilliams@gmail.com

 *Climate Change and Sustainable Agro-ecology in Global Drylands* (A.S. El-Beltagy, R. Lal and K. Malik eds.)
DOI: 10.1079/9781800624870.0006

**Table 6.1.** Global importance of fish and other aquatic products, and fisheries and aquaculture sectors.

*Production of fish and other aquatic organisms, 2020 (FAO, 2022, pp. 3, 26)*

- Capture fisheries: 90.3 million tonnes
- Aquaculture (animal species excluding aquatic mammals, crocodiles, alligators and caimans): 97.5 million tonnes
- Algae (aquaculture and fisheries): 35.1 million tonnes

*Fish as aquatic food*

- Global per capita consumption: 20.5 kg.capita$^{-1}$.year$^{-1}$; up from 9.0 in 1961 (FAO, 2022, p. 1)
- Aquatic food as share of total food protein: 7% of protein, 17% of animal protein
- Other nutrients: essential fatty acids and micronutrients, e.g. iron, zinc, calcium, iodine and vitamins A, B12 and D (FAO, 2022, pp. 85–88)

*Labor force participation and people who depend on aquatic food resources*

- Capture fisheries: 120 million people in the value chains for small- and large-scale capture fisheries (FAO *et al.*, 2023, p. 105); 94% of the workers are engaged in small-scale fisheries value chains; 492 million people engaged in subsistence fishing and small-scale fishing, or live in households depending on these
- Aquaculture: 21 million people in fish-farming node of value chain (FAO, 2022, p. 68). No estimates for whole of value chains and dependent people

*International trade*

- 59.8 million tonnes, US$ 151 billion (FAO, 2022, pp. 91, 93) (excluding algae)
- Aquatic animal products represented 49% of global value of all internationally traded animal meat products (bovine meat 19%, pig meat 18%, chicken meat 11%, other meat 3% (FAO, 2022, p. 92))

address social justice and sustainability (Gephart *et al.*, 2020). Increased production of higher-value aquatic animal products tends to be directed towards international trade. The poor may benefit if they work in the sector and/or government revenue from the sector is redistributed fairly. They are unlikely to directly consume the additional production. Indeed, based on a rigorous meta-analysis of peer-reviewed literature, the links between poverty alleviation and fisheries and aquaculture were found to be complex, although often explained by simple cause-and-effect narratives (Béné *et al.*, 2016, p. 187).

As a source of animal meat, animal-based aquaculture has had the fastest growth rate of all aquatic and terrestrial forms of animal meat. However, production of edible terrestrial meat (beef, pork, chicken, and other meats) is three times that of the combined total production of aquatic animal meat (i.e. minus shells and bones and excluding seaweeds and other plants) (Edwards *et al.*, 2019). The higher value of many aquatic animal products is reflected in strong international trade, in which aquatic products account for nearly half the value of all animal products traded (Table 6.1).

Whereas most countries produce at least some aquatic products, the bulk of fisheries and aquaculture production is highly concentrated. In 2018, the top 25 countries produced 83% of the world's production (excluding aquatic plants) (Table 6.2). Seventeen of the top 25 countries are from the global south. Looking at the top producing countries from a dryland perspective, a notable number are also countries often categorized broadly as dryland countries, e.g. China, India, Peru, Chile, Mexico, and Egypt. In the case of major capture-fisheries producers, fishing vessels from these countries may fish widely across the globe, e.g. China, or have access within their own exclusive economic zones to biologically productive upwellings that support many of the largest fisheries, e.g. the Peruvian anchoveta (*Engraulis ringens*) fishery. Where aquaculture contributes a large share of aquatic production, freshwater and land-based production is often overlooked in current estimates and in future projections that focus on the potential of marine aquaculture (mariculture) (Zhang *et al.*, 2022). In countries such as Egypt, for example, fish farming, predominantly of tilapia, including Nile tilapia (*Oreochromis niloticus*), is carried out in ponds in the highly productive Nile river delta (Nasr-Allah *et al.*, 2021).

Compared to other sectors such as agriculture, fisheries and aquaculture were slow to recognize

**Table 6.2.** Capture and aquaculture fisheries production of top 25 countries (2018). OECD member countries are considered to be in the global north. (From Global capture production 1950–2018 (FishstatJ))

| Country | 2018 production (MT) | 1 = global south<br>2 = global north |
|---|---|---|
| China | 62,206,893 | 1 |
| Indonesia | 12,642,158 | 1 |
| India | 12,386,253 | 1 |
| Peru | 7,273,403 | 1 |
| Russian Federation | 5,308,359 | 1 |
| United States of America | 5,212,603 | 2 |
| Bangladesh | 4,276,641 | 1 |
| Norway | 3,843,920 | 2 |
| Japan | 3,773,779 | 2 |
| Chile | 3,388,485 | 2 |
| Myanmar | 3,163,460 | 1 |
| Philippines | 2,875,632 | 1 |
| Thailand | 2,598,000 | 1 |
| Mexico | 1,939,237 | 2 |
| Egypt | 1,934,742 | 1 |
| Korea, Republic of | 1,904,636 | 2 |
| Malaysia | 1,675,515 | 1 |
| Morocco | 1,372,853 | 1 |
| Brazil | 1,319,292 | 1 |
| Iceland | 1,278,478 | 2 |
| Spain | 1,273,347 | 2 |
| Nigeria | 1,169,478 | 1 |
| Ecuador | 1,138,557 | 1 |
| Taiwan Province of China | 1,098,112 | 1 |
| Canada | 1,019,050 | 2 |
| United Kingdom | 897,845 | 2 |
| Denmark | 825,725 | 2 |
| South Africa | 565,917 | 1 |
| Production of top 25 countries as % 2018 world total | 83 | |
| **Grand Total** | 178,528,817 | |

> 10 m MT
> 5 m and < 10 m MT
> 2 m and < 5 m MT
> 1 m and < 2 m MT
< 1 m MT

the challenges of climate change (Brugère and De Young, 2015), but now climate change is reckoned to be one of the most important sources of uncertainty in fish production over the next decade (OECD and FAO, 2023, p. 222). Part of the slow realization of the importance to the sectors was due to the lag in including marine and freshwater data series by the Intergovernmental Panel on Climate Change (IPCC) Assessment Reports (AR). In the fourth AR (2007), for example, only 85 marine and freshwater time series were used in examining impacts, compared to 28,586 time series from terrestrial system (IPCC, 2007, p. 4, Fig. SPM.2). Nevertheless, AR4 did engender a greater interest in climate change among fisheries and aquaculture experts (Cochrane *et al.*, 2009). Since 2016, the aquatic sectors literature has grown exponentially (FAO *et al.*, 2023, p. 74). In the 2023 IPCC Assessment Report (AR7), projections for marine fisheries yields were highlighted in the main synthesis report (IPCC, 2023, p. 16), and oceans and dependent sectors were given detailed treatment in the report of Working Group II – *Impacts, Adaptation and Vulnerability* (IPCC, 2022).

From around the time of the IPCC AR4, small-scale fisheries and aquaculture experts were already considering pathways through which small-scale fishers and fish farmers would experience vulnerabilities to climate change (Allison *et al.*, 2009), being particularly vulnerable because of where they live and their poverty levels (Barange *et al.*, 2018, p. 612).

The present chapter provides an overview of the impacts of climate change on marine and inland fisheries and aquaculture, and of the aquatic production sectors on climate change, drawing on available syntheses, and highlights the fundamental factors of climate change impacts, especially from biophysical perspectives as socio-economic studies are less developed. A brief review is given of current knowledge on greenhouse gas (GHG) emissions from fisheries and aquaculture. The chapter will focus on adaptation pathways and deal little with mitigation, although the production of certain aquatic foods and the technological directions of the global aquatic food system contributes to GHG emissions, certain species more than others (Gephart *et al.*, 2021). Adaptation pathways are grouped into three categories: adaptation of today's practices, pathways relying on major technology developments, and pathways that take a socio-technical systems (STS) approach to change. In each category, selected examples are presented, illustrating the range and complexities of situations.

## Impacts of Climate Change on Inland and Marine Fisheries and Aquaculture

For more than 150 years, fisheries scientists have struggled to understand the distribution, abundance, and fluctuations in stocks of harvested aquatic species (Smith, 1994). For many fish stocks, the scientists' endeavors have brought success, depending upon the biology and ecosystem complexities of the harvested species and ecosystems, and the data available for assessments. Where the advice from the assessments has been followed by sound fisheries management, fish stocks are improving and are sustainably managed. For marine fisheries, however, this situation applies only to about half of the catch (Hilborn *et al.*, 2020), leaving many fish stocks poorly understood and poorly managed. In the case of inland fisheries, answers – typically for the decline of catches – were sought from understanding the decline in water quality from competing uses (Welcomme, 2016).

Climate has long been recognized as a factor in fish stock fluctuations. Often, climate variability must be distinguished from the impacts of fishing on the stocks. Whether in marine, brackish, or fresh water, climate change, rather than climate variation over time around a stable long-term pattern, raises further challenges. The major El Niño event of 1997–1998 was one of the first alerts to scientists such as Guerrero (1999), who estimated substantial production losses in the Philippines of nearly 300,000 tonnes, and some minor gains, from an annual expected production of about 2.3 million tonnes. Clearly, the impacts of such severe events indicated the scale of possible impacts on aquatic production of the changing climate should the frequency and intensity of such events change.

As the IPCC periodic assessments became more inclusive, fisheries researchers began collaborating with climate modelers to better understand the challenges of integrating fisheries impact studies into climate projections (Stock *et al.*, 2011). Studies on local, national, and regional changes in fisheries resources due to climate were undertaken, e.g. southeast Australia (Last *et al.*, 2011), West Africa (Lam *et al.*, 2012), and tropical Pacific fisheries and aquaculture (Bell *et al.*, 2011). Among other actions, FAO summarized the relevant parts of IPCC AR5 for fisheries and aquaculture (Seggel and De Young, 2016).

The upswing in climate change and fisheries assessments for a variety of geographies enabled globally coordinated assessments, notably through FAO whose mandate necessitated a focus on the impacts on people engaged in fisheries and aquaculture and consumers reliant on aquatic foods. Following preparatory desk and field studies that focused on vulnerability assessment approaches (Brugère and De Young, 2015), in 2017, FAO embarked on a full expert consultation process leading to a global synthesis (Barange *et al.*, 2018). The synthesis by Barange *et al.* (2018) concluded that climate change would cause considerable changes in harvestable aquatic resources, affecting availability and trade, and having economic impacts, especially in countries most dependent on the resources. The synthesis used experts' studies of major marine regions, inland systems, and aquaculture, complemented, in the case of marine fisheries, by model projections of potential impacts, based on two IPCC scenarios (Cheung *et al.*, 2018).

Barange *et al.* (2018) summarized the overall impacts of climate change on aquatic products from marine ecosystems, inland waters, and aquaculture as follows. Climate change will impact aquaculture and fisheries in ways that are short- and long-term, occurring at multiple scales. Key biophysical factors are changes in temperature, precipitation, ocean pH – typically acidification that is dangerous for organisms including plankton with calcareous skeletons – increasing hypoxic areas, and sea-level changes. Changes may mean food safety is affected, by increased growth of pathogenic bacteria, and aquatic organisms will suffer physiological stresses from temperature regimes. Vulnerabilities, resilience, and adaptation options for people, communities, and economies engaged in or consuming the products from aquatic systems vary with specific local and geographic conditions.

For marine regions, by 2050 the synthesis concluded that the catch potential in the world's exclusive economic zones (EEZ) would decline by between 2.8% and 5.3% under GHG emission scenario RCP2.6, and between 7.0% and 12.1% under GHG emission scenario RCP8.5. Regional variation is high and therefore many projected regional impacts are much greater than these global projections imply. The biggest decreases are projected for the tropics, especially in the South Pacific, whereas harvestable stocks may increase in some high-latitude countries or show less decrease.

For inland fisheries systems, the largest threats are expected to come from competition for scarce water by other powerful sectors such as those responsible for municipal water use, energy production, and agriculture. Pakistan, Iraq, Morocco, and Spain are facing high inland water stresses, and these are projected to become more severe under climate change. Some areas will receive higher precipitation, but to capture the benefits of this from an inland fisheries perspective will require additional investment, technologies, and management developments advantageous to fisheries within inland water systems.

In the short term, production may decrease, and infrastructure suffer damages from extreme temperatures and events such as floods and droughts. Diseases, parasites, and harmful algal blooms may increase in frequency under changing climate conditions. Long-term impacts may arise from reduced water availability and/or more competition for available water, and, for some cultured species, less wild seed.

In freshwater aquaculture in Asia, China, Bangladesh, Lao People's Democratic Republic, and Vietnam were projected to be the most vulnerable countries. In the Americas, the most vulnerable are Belize, Costa Rica, Ecuador, and Honduras. In Africa, Egypt, Nigeria, and Uganda are particularly vulnerable.

In brackish water aquaculture production, Egypt, Thailand, and Vietnam were projected to be highly vulnerable. For marine aquaculture, Chile and Norway were the most vulnerable, due to their high production that focuses on a limited selection of species, with the more diversified countries of China, Madagascar, the Philippines, and Vietnam also highly vulnerable.

In reviewing the evidence for projecting the likely impacts of climate change on capture fisheries, the dominant focus has been on biophysical changes, whereas in aquaculture, relatively greater emphasis has been given to economic and social impacts.

In biophysical terms, the fundamental factors that underlie the impacts of CC are many, complex, and frequently related. The four key factors are: (i) water, its availability and chemistry; (ii) temperature, which has a particularly direct effect on utilized aquatic animals, which are all poikilotherms that do not regulate body temperature; (iii) the occurrence of extreme events such as storms, floods, and droughts; and (iv) the conditions of supporting ecosystems that may be subject to effects such as sea-level changes, and changes in their extent, from inundation and other water-depth changes, pollution, and deliberate destruction, such as for land reclamation. How these come together in specific regions, fisheries, and aquaculture settings has been explored in detailed expert studies and reviews (see, e.g., Barange *et al.*, 2018; Phillips and Pérez-Ramírez, 2018).

Phillips and Pérez-Ramírez (2018, pp. 959–961) concluded that the productivity of harvestable aquatic organisms may be changed by biodiversity of the target and supporting species, e.g. through shifts to higher latitudes and deeper, cooler waters. Larvae and juveniles may be affected by ocean current changes, as demonstrated by the major impacts of periodic ocean changes such as the El Niño southern oscillation, the Southern Oscillation Index, the Pacific Decadal Oscillation, and the Indian Ocean Dipole.

Acidification resulting from the uptake of GHG in the oceans is affecting calcification rates in marine organisms such as corals, molluscs, and crustaceans. The distribution and abundance of small (e.g. herrings, anchovies, sardines) and large (e.g. tuna and mackerel) pelagic fish species are affected by changing currents and temperatures and access to nutrients for food and reproduction. Spatial shifts are already being recorded. Each of the major oceans is exhibiting different types, complexities, and magnitudes of CC, e.g. the Pacific and Atlantic oceans show different types and magnitudes of climate change.

In the intersection of marine and freshwater ecologies, estuaries are often critical feeding and breeding spaces for many species, including offshore species, but their biological productivity is disrupted when freshwater flow patterns change (Gillanders *et al.*, 2011), and estuaries warm and acidify (Scanes *et al.*, 2020).

In aquaculture, Galappaththi *et al.* (2020) surveyed the literature on CC adaptation. In 44 studies that met their selection criteria for cases, the authors found a focus on extreme events such as floods, droughts, and cyclones that affect the aquaculture enterprises, and multiple and general impacts. More than half the studies used integrated approaches. Reflecting the less-ecological focus of aquaculture research, the cases reviewed tended to look at impacts that were simultaneous, such as heatwaves and extreme weather events; or interrelated, such as disease outbreaks and economic impacts to supply chains; or geographically-specific, such as particular storms (Galappaththi *et al.*, 2020).

## Greenhouse Emissions from Fisheries and Aquaculture

Aquaculture and fisheries are not only impacted by climate change but their operations throughout the value chains also contribute to GHG emissions and environmental changes. These industries have boomed partly because many of the enterprises have been mechanized, scaled up, intensified, digitized, and the products have been traded over long distances, all of which has progressively increased energy needs and therefore carbon footprint. This increasing trend is recognized by experts in the fish sectors, but often addressed in a defensive way by comparing the carbon footprints of different aquatic production with those from terrestrial animal products, (e.g. Parker and Tyedmers, 2015, Fig. 2, showing median expected GHG emissions from fuel use in different fisheries, aquaculture, pork and chicken).

In marine fisheries, trawl fisheries have often been the target of conservation campaigns because of their operations. In these fisheries, fishing trawlers pull ropes or chains across the seafloor, herding fish and other mobile organisms into large nets that are then hauled up onto the vessel. These fisheries land 26% of global marine catches. Hilborn *et al.* (2023) conducted a comprehensive study of the environmental impacts of trawl fisheries, comparing them to other fishing methods with respect to impacts on sustainability of target species, impacts on the seafloor, bycatch and discards, and carbon footprint. Carbon footprint was standardized on kilograms of $CO_2$ produced per kg of processed product from life cycle analyses for foods ranging from corn (lowest at 0.10) to beef (19.2) (Hilborn *et al.*, 2023, Table 3). The bottom-trawl fisheries average was 4.65 compared to 5.5 for Atlantic salmon. The trawl fisheries average carbon footprint is higher than those for chicken (2.28), pork (2.92), and a plant-based meat alternative called Impossible Burger (3.50).

In aquaculture, MacLeod *et al.* (2020) estimated the global total GHG emissions for aquaculture, based on the production node rather than a life-cycle analysis, and compared emissions across aquatic (fisheries, aquaculture) and terrestrial animal protein types. For 2018, global aquaculture emissions (aquatic animals only) were estimated at approximately 0.49% of anthropogenic emissions. The authors noted that aquaculture occurs across a wide range of intensities of energy use. Compared to terrestrial livestock, particularly cattle, sheep and goats, aquaculture emissions are low because the cultured organisms do not emit enteric methane, have high fertility rates, and, as poikilotherms supported in water, have low feed-conversion ratios.

## Diverse Adaptation Pathways

The types of adaptation pathways that aquaculture and fisheries have may still be in their early stages of development. This section uses examples to illustrate possibilities, some taken from work directed to climate change adaptation, and others from adaptations to major environmental change from natural and anthropogenic causes. The examples are organized into three categories. The first comprises those pathways that are adaptations of current practices. As such, these pathways incrementally push the envelope available for today's practices, stretching out the time for which people may still conduct their enterprises. The second category is for bolder technological fixes, trying to establish fundamentally different options for production. The third category takes a more inclusive and nuanced approach by proposing socio-technical changes that aim to create solutions that are more sensitive to people.

## Adaptation of today's practices

Many of the articles on adaptation are written from current perspectives that confront short-term realities and options. Transformative changes are major and therefore not achievable in the short term. Vulnerable people must survive the near term without recourse to radical options, other than outmigration. In the aquaculture literature, Galappaththi *et al.* (2020) identified three adaptation pathways: coping mechanisms, adaptive strategies, and management approaches. The adaptation measures found in the literature review ranged from near term 'no regrets' coping strategies to far-reaching and much more complex adaptative measures.

Studies on coping mechanisms were mainly focused on marine and coastal aquaculture, including shrimp farming. Other studies were on inland aquaculture. The mechanisms, depending on the aquaculture operations of interest, relied on knowledge of adaptation strategies that is mediated by access to extension, consultants, workshops and conferences, farmers' access to early-warning information, and access to credit facilities. Adaptive strategies were all derived from studies carried out on aquaculture in the global south, many initiated by researchers from the global north. In the national natural resource management plans of Fiji, Solomon Islands, Timor Leste, and Vietnam, aquaculture is named as a marine strategy for adapting to climate change (Galappaththi *et al.*, 2020, p. 2166). The third adaptation pathway was management plans, which varied from expert-led climate-adaptation planning, to bottom-up community-based adaptation, iterative planning, through to top-down government-provided support.

Based on the findings of this literature review, four main areas of focus for aquaculture emerge (Table 6.3). The dominant area for action is managing water and farm infrastructure at all scales – pond, farm, ecosystem, catchment, and national. The second area is the need for climate and weather information and warning systems at local levels. The third area for adaptation action concerns better, more affordable farm inputs (seed, breeds, disease management) to meet the climate challenges. Finally, cost-cutting and income-securing measures for the farmers are needed for enterprises to survive, including insurance and financial support.

Embedded within all types of adaptation suggestions are some that appear logical and straightforward yet, due to historical issues and matters of the balance of power, are proving immensely difficult. The adaptive strategy 'integrated water resource management recognizing aquaculture stake', for example, throws up a challenge to existing water users in countries such as Egypt where, under law, aquaculture is only permitted in brackish and marine waters, and infertile lands not suitable for agriculture. Irrigation water (freshwater) cannot be used for aquaculture (Cai *et al.*, 2017, p. 39).

Geographic and farmer surveys of aquaculture vulnerability may help guide decision makers on climate change adaptation. Islam *et al.* (2019) applied 21 climatic, environmental, and socio-economic indicators in all 64 districts of Bangladesh, finding that aquaculture was highly vulnerable in eight inland and four coastal districts. In India, Adhikari *et al.* (2018) surveyed farmers in five major freshwater aquaculture states. The farmers reported on the impacts of extreme flooding, cyclone, and hot- and cold-weather events leading to loss of fish or fingerlings, ponds, and increases in pollution and fish diseases. Measures to cope with the risks echo many mentioned in Table 6.3.

In fisheries, recent expert assessment of best practice in climate-adaptative management,

**Table 6.3.** Adaptive strategies mentioned in the aquaculture literature, paraphrased and reorganized from the measures presented in the literature review by Galappaththi *et al.* (2020, Table 3).

| | |
|---|---|
| 1. COPING STRATEGIES | |
| *No regrets* | Supplementary pond aeration |
| | Harvest fish early to reduce pending losses |
| | Monitoring water conditions and fish behaviour frequently |
| | Actively adjust pond water levels during rains |
| | Reducing disease risk by better pond and feed management |
| | Share knowledge in fish-farming groups and networks |
| *Low regret* | Shift stocking dates and adjust stocking density to conditions |
| | Buy fingerlings nearby to reduce heat stress during transport |
| | Use groundwater to refill ponds |
| | Adjust stocking tanks to regulate pond water supply |
| | Shade hatchery tanks/cages; grow aquatic weeds in ponds for shelter |
| | Strengthen cages to lessen risks of damage in extreme events |
| | Stock and harvest multiple fish species to reduce risk or switch species |
| | Dip ice bags in pond/hatchery to reduce water temperature |
| | Exchange pond water frequently |
| | Plant fruit trees on pond dikes and vegetation on pond slopes |
| | Seek compensation assistance following disaster-related losses |
| | Enter contract farming arrangements, e.g., leases |
| 2. ADAPTIVE STRATEGIES | |
| *Future benefit strategies* | Early-warning systems informing about floods, droughts and heatwaves |
| | Research and development to improve climate risk information systems and accessibility |
| | New farm-level technology to improve water productivity |
| | Rainwater harvesting tanks to use rainwater for fish culture and pond-dike cropping |
| | Protect and restore ecosystems for flood protection, water storage, and water-quality services |
| | Community-based watershed management |
| | Zone production so that aquaculture has sufficient water (volume/quality) |
| | Integrated water resource management recognizing aquaculture stake |
| | Reuse waste and integrate resources into the farm to reduce input costs and dependencies on input suppliers |
| | Broad range of higher thermal tolerance breeds |
| | Diversify into other business/income sources to subsidize risk-reduction investments |
| | On-farm value-added processing |
| | Increase savings to buffer household from losses and still make risk-reduction investments |
| *Up-front strategies* | Water-treatment equipment in storage ponds with recirculating technology |
| | Protective flood dike around aquaculture ponds |
| | Large-scale water storage and infrastructure development to consider aquaculture uses of water |
| | Standards to improve climate- and water-related risk management |
| | Avoid risk-prone areas and shift production site to a lower-risk location |
| | Seek opportunities for floodplain aquaculture |
| | Develop new export markets and strengthen existing markets for farmed fish products, to create higher farm prices |
| | Establish mutual or weather-indexed insurance for aquaculture |
| 3. MANAGEMENT APPROACHES | |
| | Climate adaptation planning for aquaculture based on expert knowledge |
| | Community-based adaptation using bottom-up approach |
| | Adaptive management using iterative approach |
| | Government support involving active governmental provisioning of resources, such as technical trainings or financial subsidies for individuals or communities |

such as that led by FAO (Bahri *et al.*, 2021) has reinforced mainstream fisheries management practices applied to addressing the CC impacts. The FAO assessment drew from the evidence of 13 case studies of fisheries management under CC conditions and supported 15 measures. The case studies are drawn from all continents and cover a wide range of fished species. The measures address one or more of the significant CC impacts on fished species, namely: distributional change, productivity change, and species composition change. Each measure was required to meet three essential criteria for good practice: explicitly addresses CC impacts; sufficient evidence to ensure/infer its effectiveness; return a win-win (costs and benefits) or lose-win outcome. In addition, two beneficial criteria were also considered: flexible/responsive, and socially acceptable. A literature review by Galappaththi *et al.* (2022) found many of these measures also, and several additional suggestions, with particular relevance to the conduct of fishing operations. The two sets are merged in Table 6.4.

Some of the adaptative management measures would be more appropriate for fisheries with advanced fisheries management capacity, e.g. tradable quotas. Small-scale fisheries (SSF) often present additional challenges because they are often data-poor and have limited financial and management resources. More work is needed to develop short-term no-regrets or low-regrets policies suitable for SSF (Gutierrez *et al.*, 2023, p. 79).

Regional studies have addressed CC adaptation for specific water bodies and regions. For the central and western Pacific, Bell *et al.* (2011) examined the CC outlook for industrial and small-scale coastal fisheries and aquaculture noting that the projected changes in the distribution and abundance of tuna required industry and country flexibility. Some more eastern countries such as Kiribati, Nauru, Tuvalu, and Tokelau may benefit from more tuna. The more westerly countries with major tuna fisheries were also large and more diverse economies and considered more capable of handling changes. Aquaculture of freshwater fish may benefit from greater rainfall predicted in some countries that typically suffer water scarcity. Nearshore coastal fisheries including reef lagoons where women, men, and children spend considerable time gleaning for subsistence, will suffer from coastal erosion, rising sea levels, and heating reefs. Those people will need to consider making more use of nearshore pelagic fish, including skipjack and yellowfin tuna, through establishing local anchored fish-aggregating devices. Despite efforts in some countries, such as Solomon Islands, to engage women in fishing from boats around these anchored devices, their involvement has been minimal, tends to deepen gender inequalities within the household, and can lead to conflicts (Labuinao, 2019).

For Africa, Kolding *et al.* (2016) reviewed the long-term climate-linked patterns of water availability and fisheries production – 'fish come with the rains' – in rivers and lakes in African dryland regions. The results of quantitative studies from such water bodies as Lakes Chad, Malawi, Malombe, Victoria, and Turkana, and River Niger showed how irregular rainfall was the most critical driver of fisheries in the drylands through its impacts on surface water availability. The dryland water bodies are unstable ecosystems fed by pulses of unpredictable rainfall. Their fish biota tends to be productive, comprising small-sized, opportunistic, resilient fish species that have 'boom and bust' abundance patterns. People living in the arid and semi-arid areas and using fish from these water bodies will be subject to increasing variability in these already erratic climate patterns. However, with greater recognition of the benefits of fish when available, Kolding *et al.* (2016) stressed that such fish may still provide some greater productivity, but the people using them would need to adopt diversified livelihood strategies to match the irregular production.

Lubembe *et al.* (2022) reviewed the African literature for fisheries and aquaculture, concluding the confluence of overfishing and climate stresses in the case of capture fisheries, and climate factors such as rising sea levels, rising water temperatures, rising water salinity, ocean acidification, changes in precipitation patterns, disease, and algal blooms. Fish and fishing communities have both adapted to the changes, but often to the detriment of sustainability, such as fishing more intensively on already depleted or shifting fish stocks. Some mitigation and adaptation are possible such as through mangrove replanting and coral reef restoration, methods for which are being tested in over 50 countries. Despite the challenges, many adaptative measures

**Table 6.4.** Climate-adaptative fisheries management good-practice measures recommended in Bahri *et al.* (2021, p. 25, Table 5) and those by Galappaththi *et al.* (2022, p. 11, Table 2) from a literature review of coping responses and common adaptive strategies. The measures have been ordered into six categories from monitoring and early warning through to market and financial measures. Some have been paraphrased.

**1. Monitoring and early warning**
- Enhance monitoring programs through community-based approaches
- Update with weather information before fishing trips
- Establish early-warning systems for extreme events
- Adjust spatial scale of monitoring to be responsive to shifting stocks

**2. Fisheries assessments**
- Incorporate environmental variables and risk into fisheries assessment and management advice
- Use multiple knowledge systems, e.g. indigenous, local and western science
- Use new technology for collaboration, e.g. location-aware mobile devices, cloud-based computing, and visualization and querying of geographical data over the web to capture, visualize, and share logbook data

**3. Practical fisheries management and fishing measures**
- Flexible and adaptable fishing seasons
- Tradable fishing rights/allocations to allow flexibility in response to stocks shifting across international borders
- Close fishery during climate-driven events to support resilience and recovery
- In-season management systems that are responsive to rapid climate-driven stock changes
- Conserve keystone species complexes to avoid ecological tipping points and related changes in target species abundance
- Develop new fishery opportunities to capitalize on distribution shifts or enhance productivity (including for 'new' species)
- Increase fishing-gear diversity or use a different technology, including revival of traditional fishing techniques
- Improving vessel fuel-efficiency
- Capacity building to inform/train fishers (e.g. workshops, education, research)

**4. Manipulating the resources**
- Relocate fishery species to more productive areas to compensate for changes in productivity
- Plant trees to reduce river temperature

**5. Household and community measures**
- Collaborating with fellow fishers and sharing food
- Building community resilience by increasing social capital (social cohesion, local fundraising, maintaining cultural identities); strengthening local institutions and establishing new institutions
- Using family members as labor
- Taking out small loans or pawning gold jewelry
- Temporarily migrating (low regrets); permanently migrating (up-front strategy)
- Livelihood diversification inside and outside fisheries, including quitting fisheries altogether

**6. Market, economic support, and infrastructure measures**
- Relocate landing and processing practices
- Source more diverse supplies of seafood for processing facilities
- Develop new products and markets to maximize fishery value as catches decline
- Make the sale price of fish more responsive to fuel-price fluctuations
- Develop insurance schemes that protect fishers against loss and damage after climate events or due to 'forced' practice changes or exit from the industry

rest with the local fishers and their collective understanding of CC measures and adaptive management options. Unfortunately, support to local adaptative management is through a patchwork of ministerial arrangements such as wildlife, water, and fisheries agencies and rarely becomes part of national adaptation plans (Lubembe *et al.*, 2022, p. 67).

## The technological fix pathways

As the foregoing section outlines, experts including fishers, farmers, researchers, and policymakers have begun to incorporate CC impacts into their work considerations. In parallel, the stresses on current operations, and the emerging and pending stresses of CC have pushed better-endowed fishing and farming enterprises, research and development institutes, and entrepreneurs to more cutting-edge technological fixes. Some of these were hinted at in Tables 6.3 and 6.4. In aquaculture, the technological fixes either address current bottlenecks, such as the supply and cost of feeds in aquaculture, or the desire for much greater control over the aquaculture process. In fact, most of the technology pathways are not being designed as fixes for CC but as growth pathways for production. They are, however, happening in an era of greater sensitivity to climate and environmental challenges, and frequently promote their superiority in these regards (Ahmed and Turchini, 2021). In fisheries, the fixes are more likely to be concerned with manipulating the ecosystem or improving knowledge for fishers, firms, and managers. In marine fisheries, fisheries management and fish-stock assessment technologies are paramount. In inland fisheries, technology pathways are more likely to be concerned with how fish and fishers can cope with major waterway changes, such as from dams and competition with urban and irrigation water use.

## Aquaculture

Aquaculture is particularly focused on technology options and performance. As a major growth area, aquaculture is subject to the entry of many new adoptees on steep learning curves, as well as established actors expanding their operations and restlessly seeking new technology options to give them a competitive edge. With respect to technologies, each of the items in Table 6.5 merits deeper treatment that is beyond the scope of the present chapter, but three items are singled out for mention, because each is receiving public attention beyond fisheries and aquaculture. These areas are recirculating aquaculture systems (RAS), seaweed farming, and the application of genetic technologies to aquaculture. They illustrate the challenges and opportunities indicative of those facing aquaculture more generally.

RAS and other contained systems are receiving the greatest commercial attention; offshore cage installations are also in the news for some of the same reasons. These are driven by onshore constraints for suitable farming space and water and the need for greater water and nutrient recycling (circular) (Verdegem *et al.*, 2023). RAS is attractive for its high productivity but it is also complex, highly capital-intensive, requires low-emissions energy solutions, and is rarely considered a suitable option for community-based adoption, especially in developing countries (Ahmed and Turchini, 2021). Despite not arousing some of the same animosity shown to inshore aquaculture, RAS nevertheless competes for onshore space and resources, natural and human-constructed. Proponents of RAS facilities have to build local legitimacy, credibility, and trust, arising from concerns over its local sources of energy, competition for space, integration with local practices and culture, and potential animal health and welfare issues (Fudge *et al.*, 2023).

Seaweed farming is receiving attention as creating large investment opportunities as a multi-purpose solution to environment, climate, and development. A typical statement is that recently given in a World Bank document: 'With its ability to sink carbon, sustain marine biodiversity, employ women, and unlock value chains, seaweed farming demonstrates how development, climate, and nature work together to generate value and uplift communities' (World Bank, 2023, p. xi).

The World Bank market report canvassed a range of market sectors and ecosystem services that seaweed production could serve (Table 6.6). It noted the current concentration of seaweed production in Asia, especially brown algae or kelps in the genera *Saccharina* and *Laminaria*, tropical red seaweeds comprising eucheumatoid seaweeds of the genus *Eucheuma* and *Kappaphycus*, and *Gracilaria* species, and the major growth in seaweed production of the last decade.

To the present, seaweed has been serving three major markets – direct human food (85%), aquaculture animal feeds, and the hydrocolloid markets (carrageenan, agar, alginates) (World Bank, 2023, p. 4). The hydrocolloids markets are dominated by about ten major firms, located

**Table 6.5.** Innovative aquaculture technology systems and core strategic technologies under development and implementation, with a focus on those with the potential to cope with or reduce CC impact. Descriptions are taken, where possible, from the FAO Aquaculture Glossary (https://www.fao.org/faoterm/collection/aquaculture/en/) and supplemented with information from Yuan (2022).

| Aquaculture system | Description |
|---|---|
| Recirculating aquaculture systems (RAS) | A closed or partially closed system employed in aquaculture production where the effluent water from the system is treated to enable its reuse. Highly capital-intensive. |
| Shipping container-based fish culture | The fish are intensively raised in steel tanks modified from shipping containers. RAS models have been designed for use in shipping containers. |
| Pond-based recirculation systems | Reservoir for water storage, pond(s) for water supply, and culture ponds. Tend to use large areas for water treatment, often using biological methods (bivalves, plants, and filter-feeding fish). |
| In-pond raceway aquaculture | Based on water structure, usually above-ground, with a long, linear configuration; high water turnover rate; highly controlled environment; often terraced with water reuse. In-pond systems integrate traditional pond culture and intensive raceway culture. |
| Aquaculture and photo-voltaic integration (API) | In China, water beneath photo-voltaic panels is stocked with fish that are raised in intensive ways, such as using raceways. |
| Offshore cage culture | Large-scale cage, and even ship-based culture, are being tested, mainly in China. Highly capital- and technology-intensive, often combined with tourism for multiple income streams and targeted at higher-value species. |
| Polyculture | Rearing of two or more non-competitive species in the same culture unit, e.g. shrimp-fish, shrimp-plants. Rice-fish and rice-shrimp are another form of polyculture in which land use may rotate between rice and aquaculture or may carry out the two simultaneously. |
| Integrated multi-trophic aquaculture | Species from different trophic levels are raised in proximity to one another and the co-products (organic and inorganic wastes) of one cultured species are recycled to serve as nutritional inputs for others (Knowler *et al.*, 2020). Attractive environmentally, but knowledge-intensive and not widely practiced. |
| Aquaponics | Biointegrated system that links recirculating aquaculture with hydroponic vegetable, flower, and/or herb production. |
| Seaweed culture | Seaweed is often touted as a major solution to carbon sequestration. |
| **Strategic technologies** | **Description** |
| Genetics | Whole genome sequences done for 52 aquaculture animal species, large-scale SNP genotyping including constructing genetic linkage maps, QTL mapping, genome-wide association study and genomic selection for, e.g., growth and body shape, stress, salinity tolerance, disease resistance (You *et al.*, 2020); genetically modified fish including using CRISPR (Yuan, 2022; Okoli *et al.*, 2022). |
| Feeds | Novel feed ingredients replacement for fish meal and fish oil, functional feed additives, improved on-farm feed management, dietary fortification for improving nutrient value of aquatic products, prebiotics, and probiotics to improve aquatic animal nutrition. |
| Information, digital, and artificial (AI) technologies | Supports automation, process controls, trading, and communications. |
| Management technologies | Farm management technologies, often involving combinations of other strategic technologies are critical for aquaculture adaptation and mitigation measures. |

**Table 6.6.** Extent of genetic improvement of the world's top 15 cultured aquatic species or species groups (plants and animals). The information in this table is summarized from Table 3 of Sonesson *et al*., 2023, Table 3).

| Genetic improvement status | Species or species groups |
|---|---|
| Very high | Atlantic salmon – *Salmo salar* |
| | Whiteleg shrimp – *Penaeus vannamei* |
| High | Common carp – *Cyprinus carpio* |
| | Gracilaria seaweeds – *Gracilaria* spp. |
| | Goldfish and crucian carps – *Carassius* spp. |
| Moderately high | Cupped (Pacific) oysters – *Magallana gigas* |
| Moderate | Nile tilapia – *Oreochromis niloticus* |
| | Rohu carp – *Labeo rohita* |
| Low | Japanese kelp – *Laminaria japonica* |
| | Silver carp – *Hypophthalmichthys molitrix* |
| | Manila clam/Japanese carpet-shell – *Ruditapes philippinarum* |
| | Bighead carp – *Hypophthalmichthys nobilis* |
| Very low | Eucheuma seaweeds – *Kappaphycus alvarezii, Eucheuma cottonii, Eucheuma denticulatum, Eucheuma* spp. |
| | Grass carp – *Ctenopharyngodon idellus* |
| | Catla – *Catla catla* |

in China, Denmark, France, Philippines, and USA (Porse and Rudolph, 2017).

As the multiple benefits of seaweed production are promoted, e.g. *Seaweed Revolution: A Manifesto for a Sustainable Future* (Doumeizel and Aass, 2020), the distribution of benefits along the value chains are rarely analyzed. In 2019, first sale prices of various species of seaweed varied from US$ 0.21 to 0.89 $kg^{-1}$ (wet weight) (Cai *et al*., 2021, p. 8). Most tropical seaweeds are sold dry, with the primary producers providing the labor for drying (Ramirez *et al*., 2020). Much of the value in seaweed production, therefore, is captured after the first point of sale, raising questions about the extent to which the expansion of seaweed production advantages human development.

Forward-looking market reports project that existing complex, high-value-added markets, and other emerging ones will become more important. The World Bank (2023) examined ten emerging and potential market sectors: biostimulants, animal feed additives, pet food, methane-reducing feed supplements, nutraceuticals, alternative proteins, fabrics, bioplastics, pharmaceuticals, and construction materials. Predicting market size and time to maturity, nutraceuticals are judged the most promising market and construction the longest-term bet (World Bank, 2023, Fig. 32). The market and non-market potential and challenges for realizing ecosystem services include blue carbon, seaweed bioremediation, methane reduction, seaweed for biodiversity, and other ecosystem services.

Other experts express much more cautious views on the promises and consequences of simultaneously growing seaweed to feed the world, provide animal feeds promising lower livestock emissions, and mitigation through mass-scale automated farming of seaweed for ocean carbon sequestration. Considerable uncertainty surrounds the timescales of carbon cycling of seaweeds, the economics, ecological impacts, and ethical considerations (Troell *et al*., 2023).

Genetic improvement of farmed crops and terrestrial livestock species have been at the core of agricultural progress for millennia. Widespread application of genetic technologies is more recent in aquatic animals, although fish culturists altered the genetic composition of farmed fish to suit their needs since the start of fish culture thousands of years ago (Dunham, 2023, Ch. 1). Despite significant progress in genetic improvement, the production of many commonly farmed aquatic animal and plant species is based on species and species groups that are little improved over wild types (Table 6.6). Many seaweeds are little genetically improved. An exception is the agar-producing *Gracilaria* species, which are favored in the hydrocolloid industries (Porse and Rudolph, 2017).

From species selection through to genetically improved farmed varieties, aquaculture production faces the challenge of where to invest safely in genetic advances. A large number of aquatic species are farmed. Metian *et al.* (2020) noted 462 identified species and 145 groups not elsewhere included (nei) but often reported at the genus level, making for more than 600 farmed species. Aquatic animal production (by volume), however, comes mainly from about 20 species accounting for 70% of the global total. Of these, whole genomes are available for 52 species of aquaculture animals (You *et al.*, 2020). The aquaculture industry has embraced high-throughput sequencing technologies, producing many genetic markers especially to find and apply single-nucleotide polymorphisms (SNPs) to create genetic linkage maps, map quantitative trait loci (QTL), study genome-wide associations (GWAS), and perform genomic selection (GS).

QTLs identified in aquaculture species were mainly focused on growth and related factors, but climate-linked factors such as osmoregulation, temperature tolerance, and several types of disease resistance received significant attention (You *et al.*, 2020, Table 7). GWAS focused on growth, but salinity, temperature, oxygen, and disease factors also received some attention (You *et al.*, 2020, Table 8). GS studies are focusing almost equally on growth and disease-management factors, depending on the species used (You *et al.*, 2020, Table 9). Current applications of genetic technologies to aquaculture species hold considerable promise that knowledge gained on a species may be transferrable to other species (You *et al.*, 2020). The overall dominance of growth and associated factors, including disease management, does not create a strong pipeline of targeted work to food security species, i.e. those of lower market value, and climate adaptation characteristics. However, the rapid progress with genetic technologies does offer the possibilities that, with strategic planning and the right incentives, a pivot towards genetically improved aquatic species that could better adapt to CC would be possible. Concomitant targeting of genetic solutions to food security needs is economically and policy-wise more complex.

Finally, in light of an expected upswing in applying genetic technologies, stakeholders of access- and benefit-sharing (ABS) arrangements and patents must comprehend the options available and how these apply to aquatic genetic resource use, if they are to operate legally (Sonesson *et al.*, 2023). As indicated by the relatively low level of genetically improved stocks used in aquatic farming, relative to crop and livestock farming, aquatic activities still rely considerably on native germplasm and biodiversity and ABS arrangements are less mature or well understood in aquaculture.

## Fisheries

In considering adaptation in capture fisheries, the major focus has been on the technologies of fisheries management, i.e. the strategies, rules, and regulations operating on the fishers, their operations, the fish value chains, and fish stock assessment. Empirical studies verify that managed fisheries are more sustainable, yet many fisheries are not under the most effective management (Hilborn *et al.*, 2020). Fish stock assessment is central to fisheries management, although the social and economic context of management is also germane to success (FAO, 2022, pp. 126–136). Fish stock assessment gauges the state and projected sustainable productivity of aquatic resources. If suitable data are available, stock assessment typically combines increasingly sophisticated mathematical-modeling approaches, predicting the impacts of fishing on the stocks and disentangling the effects of weather, climate, and environmental variables including pollution and supporting habitat states. Stock assessment good practice has been codified by experts, e.g. Punt (2023). Ideally, the results of fish stock assessments are heeded and used in making fisheries-management decisions.

Fisheries management is considered critical to achieving key targets in SDG 14 (Life below water) and a Blue Transformation (see FAO, 2022, pp. 126–136), which is defined as a 'set of actions, policies and strategies aimed at sustainably expanding and enhancing aquatic food systems and increasing their contribution to affordable and accessible healthy diets, while fostering equitable growth' (FAO, 2022, p. 225). Fisheries-management strategies for climate adaptation are also critical to the sector, as illustrated in the case studies reviewed by Bahri *et al.* (2021), and the resulting adaptation strategies included in Table 6.4, especially in parts 1–4 of the table.

For many fisheries in the global south and the global north, however, data for stock assessment and management are limited. In many countries, the capacity for stock assessment and fisheries management is also limited. Data- and management-capacity-limited fisheries are characterized by poor data collection; and management arrangements that are not fit for local social and economic contexts, e.g. ignoring local knowledge and poorly applied through having weak and ineffective monitoring, surveillance, and enforcement. Fortunately, long-term capacity-building support, often provided through regional fisheries management and technical agencies, can provide cost-effective solutions (FAO, 2022, pp. 135–136). With respect to data for management, my own experience covers nearly 50 years of working in and keeping in touch with fisheries data arrangements at levels from small-scale local fisheries to international industrial-scale (tuna) fisheries. It has shown that modern information technology – computing hardware, software, and connectivity – has brought into reach real solutions. This was well illustrated in the recent newsletter of the Philippines National Fisheries Research and Development Institute that covered articles on capacity building in data enumeration, field fish taxonomy, stock assessment, and harvest strategies for management (FISEARCH, 2023).

Due to progress in decadal and seasonal climate predictions for the oceans, some of these are now at a scale usable to regional fish stock assessments for fisheries management, according to Tommasi *et al.* (2017) who also recommend practices for assessing forecast needs, developing the forecast, and delivering the results. The authors concluded that although a start has been made on integrating fisheries and climate forecasts, further institution building is needed to achieve a mutual understanding of needs between climate and fisheries scientists, develop shared understanding of prediction uncertainties, and means for coping, *post hoc*, with failures in predictions. Practitioners will also have to cope with what is realistic in terms of management reform, but noting that climate-adaptive fisheries management, according to modeling by Free *et al.* (2020), is more profitable than business as usual.

Practices for integrating fisheries and climate assessment for many small-scale fisheries face the challenges of having schemes that are relevant to their geographic scales, much as terrestrial farmers are linking with the most local of weather, climate, and CC forecasts available.

Mainstream fish stock assessment is mainly the domain of marine fisheries scientists, whereas adaptation technology pathways for inland fisheries are more likely to be concerned with the major disruptions to water bodies caused by other uses of water as well as by climate change. To cope with these disruptions, and the creation of new water bodies, technologies such as stock enhancement (stocking reservoirs, dams, rivers, and lakes with fry or more advanced specimens of desirable species), fish migration systems to overcome barriers in waterways, and other water engineering structure are being utilized. The case of the Aral Sea in Central Asia presents a case in which catastrophic degradation from human intervention was ameliorated in one part of the ecosystem by remedial technology pathways.

The plight of the drying Aral Sea has captured world attention for its loss of fisheries and fish livelihoods in the decades since 1960 (Aladin *et al.*, 2018; Micklin *et al.*, 2020). The Aral Sea, in Kazakhstan and Uzbekistan, is a major brackish water lake in the Central Asian desert, formed in a closed basin fed by the rivers Amu Darya (feeding the southern part of the sea) and Syr Darya (feeding the northern part). Excessive draining of the rivers for agricultural irrigation during the Soviet Union era caused, from the 1960s, progressive desiccation and salinization of the sea and severe toxic salt and dust storms. When the Aral separated into a small northern (Small Aral) and larger southern (Large Aral) part in 1987–89, an earthen dike was constructed in 1992 to prevent backflow of water from the Small Aral. After the temporary dike was breached in 1999, a more permanent structure was constructed in the early 2000s, deepening the Small Aral and greatly improving its ecology and fisheries. The Large Aral has been desiccated and salinized to the extent that even the remaining water cannot support fish but only some introduced brine shrimp that are harvested for their eggs, whereas the engineered hydrological solution has helped the Small Aral achieve a 1960s-like ecosystem and fisheries (Aladin *et al.*, 2018). Furthermore, Aladin *et al.* (2018) believe that with further engineering structure, a partial recovery of the Large

Aral may be possible. An international coalition of multi-disciplinary, multi-sectoral actors including researchers, artists, policymakers, and industry representatives have formed to promote the possibilities for finding solutions and raising awareness of the issues and options, as the declaration from the 2019 conference on the Aral Sea problems reports (https://www.zin.ru/conferences/Aral2019/index.html). Among the headline statements coming from the conference were that '*reports of the death of the Aral Sea are premature*' and that 'scientists, artists and cultural experts all have important roles in the preservation and rehabilitation of the Aral Sea and its region'. In specific terms, cooperative management of the flow of the Amu Darya and Syr Darya rivers was essential, but (heroic) engineering solutions, such as redirecting major rivers (from outside the catchment), is not feasible for solving the problems. The role of cross-cultural communication and its emotional content have been recognized (Dumetz and Aladin, 2021).

## Socio-technical systems approaches

The technology-fix pathways discussed in the previous section were mainly based on the idea that a single technology would be critical to adaptation. Many of these examples, however, soon reached implementation challenges that caused them to invoke the need for new institutions, networks, legal instruments, and other contextual elements to succeed. Early in their development, climate mitigation and adaptation solutions have increasingly been acknowledged as requiring development, testing, and dispersion in highly complex systems, named socio-technical systems (STS).

Geels (2002, 2004) developed methodology to encompass interactions among actors, levels, and institutions in STS, concerned across the spectrum, from creation, production, and distribution to use of technologies. Of particular interest in this work was understanding the longer-term forces acting, as technologies and society jointly changed, often with one STS changing into a new one.

Geels recognized not only the material technology but also the multi-layered systems within which STS operate (Geels, 2004, 2019). The layers are (i) the niches in which innovations are tested and developed; (ii) the legal, management, and social regimes that form the current systems; and (iii) the higher-level landscape that constructs the regimes and over which the lower layers have little influence. In a political economy sense, this is the set of organizing institutions of the political economy.

A major field of interest for social scientists working with STS is the issue of how they transition from one regime to another. In this regard, the study of transitions in CC-mitigating technologies is a major theme among STS studies (Hess and Sovacool, 2020). STS approaches have been rarely used to date in fisheries and aquaculture, but they are nevertheless drawing attention from an emerging rank of interdisciplinary social scientists and hold promise for providing insights and pointers to action in more inclusive and positively directed STS transitions (Williams and Syddall, 2022). Rather than full STS approaches, other researchers are modeling complex fisheries scenarios under CC using such methodologies as game theory (strategic interactions approaches, e.g. Vogel *et al.*, 2023) in examining options for managing fisheries conflicts arising from CC impacts on transboundary fish stocks. Multi-disciplinary and transdisciplinary capacity in research into complex systems is recognized as fundamental, although research tends to focus on socio-ecological systems (Cooke *et al.*, 2023), rather than socio-technical systems.

Our exploration of the possibilities of applying STS transitions approaches to improve the inclusion of women's interests and gender equality in fisheries/aquaculture technology transitions indicated that *post hoc* and anticipatory socio-technical systems transition research could both be useful (Williams and Syddall, 2022). The same applies in the case of social inclusion more generally under technology transitions in fisheries/aquaculture affected by CC. *Post hoc* studies illuminate the current situation with respect to who benefits, i.e. Are the benefits reaching or bypassing the poor, and why? Anticipatory research may guide researchers and policymakers on how to achieve a more inclusive political economy outcome.

Many of the technological-fix pathways outlined in the previous section do not take an expansive systems approach, but will, if successful, diffuse and transform industries and enterprises

**Table 6.7.** Technology-fix pathways, selected from examples in the preceding section, which, if successfully developed and widely dispersed, will lead to socio-technical systems transitions, some of which may have unintended consequences, creating winners and losers.

|  |
|---|
| Recirculating aquaculture systems (RAS) |
| Seaweed culture as blue carbon solution |
| Genetic aquaculture technologies |
| New aquaculture feeds |
| Information, digital and artificial (AI) technologies in aquaculture |
| New aquaculture management technologies |
| Improved marine fish-stock assessment and management of data-limited and management capacity-limited fisheries |
| Improved fisheries and climate oceanography projections to support fisheries management |
| Engineering and biological enhancement solutions in inland fisheries |

(Table 6.7). These are the types of technological fixes that need research and monitoring so that their transition impacts can be understood and adjustments made to help achieve justice in access to the benefits and in mitigating any negative impacts.

## Conclusions

This chapter has attempted to provide a snapshot of fisheries and aquaculture research and strategy development in the face of climate change. Relevant research has mainly been conducted in the last decade, coinciding with tangible evidence that resource changes and challenges due to CC are already strongly affecting fisheries and aquaculture, plus IPCC's increasing efforts to include the oceans, freshwater, and aquatic production systems in their assessment reports. The impacts of CC are specific to each geographic area and the aquatic systems in which fish are produced naturally or in culture. Projections of future impacts are available for a few aquatic systems, but more modeling of aquatic resources and climate changes are needed, especially at locally meaningful scales.

Reviews of how fisheries and aquaculture are adapting to CC show a wide set of strategies being applied by key actors (Tables 6.3 and 6.4). The fish sectors are choosing technology-fix pathways as the main adaptation approach. Fisheries, which have already faced three decades of total production limits, have been more accommodating of the need to factor in CC in stock assessment and management than has aquaculture. The aquaculture sector tends to be still concerned with maintaining its rapid growth rate but also recognizes CC as a rising challenge. Both sectors will benefit from emerging collaborations with climate oceanography. Many of the technology fixes being sought are leading to appreciation of the complexities of major technology transitions, whether driven by sectoral growth or climate change. Little work is yet being done using socio-technical transitions approaches.

The major challenges brought by CC are going to exacerbate the ongoing dilemmas of whether adaptation will affect small-scale fisheries and aquaculture more than the large-scale enterprises, and which people will be displaced in the adaptation processes. People in dry areas will certainly struggle more than those in areas with greater water availability, but disasters such as extreme weather (storms, floods, droughts, fires) appear to be affecting those in even the most well-endowed areas. Climate adaptation options, technical-fix pathways and the pathways of transitions occurring in systems that are complex in socio-technical and ecological senses will all be called on for fisheries and aquaculture to continue to help feed the world.

## References

Adhikari, S., Keshav, C.A., Barlaya, G., Rathod, R., Mandal, R. *et al.* (2018) Adaptation and mitigation strategies of climate change impact in freshwater aquaculture in some states of India. *Journal of Fisheries Sciences* 12(1), 16–21. DOI: 10.21767/1307-234X.1000142

Ahmed, N. and Turchini, G.M. (2021) Recirculating aquaculture systems (RAS): environmental solution and climate change adaptation. *Journal of Cleaner Production* 297, 1–14. DOI: 10.1016/j.jclepro.2021.126604

Aladin, N.V., Chida, T., Chuikov, Y.S., Ermakhanov, Z.K., Kawabata, Y. *et al.* (2018) The history and future of the biological resources of the Caspian and the Aral Seas. *Journal of Oceanography and Limnology* 36, 2061 –2084. DOI: 10.1007/s00343-018-8189-z

Allison, E.H., Beveridge, M.C. and Van Brakel, M. (2009) Climate change, small-scale fisheries and small-holder aquaculture. In: Wramner, P., Ackefors, H. and Cullberg, M. (eds) *Fisheries, Sustainability and Development*. Royal Swedish Academy of Agriculture and Forestry, pp. 109–122.

Bahri, T., Vasconcellos, M., Welch, D., Johnson, J., Perry, R.I. *et al.* (eds) (2021) Adaptive management of fisheries in response to climate change. *FAO Fisheries and Aquaculture Technical Paper No. 667*. FAO, Rome. DOI: 10.4060/cb3095en

Barange, M., Bahri, T., Beveridge, M., Cochrane, K., Funge-Smith, S. *et al.* (2018) Impacts of climate change on fisheries and aquaculture. *FAO Fisheries and Aquaculture Technical Paper 627*. FAO, Rome. Available at: https://www.fao.org/3/i9705en/I9705EN.pdf (accessed 8 July 2024).

Bell, J.D., Johnson, J.E. and Hobday, A.J. (2011) *Vulnerability of Tropical Pacific Fisheries and Aquaculture to Climate Change: Summary for Pacific Island Countries and Territories*. Secretariat of the Pacific Community, Noumea, New Caledonia.

Béné, C., Arthur, R., Norbury, H., Allison, E.H., Beveridge, M. *et al.* (2016) Contribution of fisheries and aquaculture to food security and poverty reduction: assessing the current evidence. *World Development* 79, 177–196. DOI: 10.1016/j.worlddev.2015.11.007

Brugère, C. and De Young, C. (2015) Assessing climate change vulnerability in fisheries and aquaculture: available methodologies and their relevance for the sector. *FAO Fisheries and Technical Paper No. 597*. FAO, Rome.

Cai, J., Quagrainie, K. and Hishamunda, N. (2017) Social and economic performance of tilapia farming in Africa. *FAO Fisheries and Aquaculture Circular C1130*. FAO, Rome.

Cai, J., Lovatelli, A., Aguilar-Manjarrez, J., Cornish, L., Dabbadie, L. *et al.* (2021) Seaweeds and microalgae: an overview for unlocking their potential in global aquaculture development. *FAO Fisheries and Aquaculture Circular No. 1229*. FAO, Rome. DOI: 10.4060/cb5670en

Cheung, W.W., Bruggeman, J. and Butenschön, M. (2018) Projected changes in global and national potential marine fisheries catch under climate change scenarios in the twenty-first century. In: Barange, M., Bahri, T., Beveridge, M.C.M., Cochrane, K.L., Funge-Smith, S. and Poulain, F. (eds) *Impacts of Climate Change on Fisheries and Aquaculture: Synthesis of Current Knowledge, Adaptation and Mitigation Options*. FAO, Rome, pp. 63–85.

Cochrane, K., De Young, C., Soto, D. and Bahri, T. (2009) Climate change implications for fisheries and aquaculture. *FAO Fisheries and Aquaculture Technical Paper No. 530*. FAO, Rome.

Cooke, S.J., Fulton, E.A., Sauer, W.H., Lynch, A.J., Link, J.S. *et al.* (2023) Towards vibrant fish populations and sustainable fisheries that benefit all: learning from the last 30 years to inform the next 30 years. *Reviews in Fish Biology and Fisheries* 33(2), 317–347. DOI: 10.1007/s11160-023-09765-8

Doumeizel, V. and Aass, K. (eds) (2020) *Seaweed Revolution: A Manifesto for a Sustainable Future*. Lloyd's Register Foundation. Available at: https://unglobalcompact.org/library/5743 (accessed 8 July 2024).

Dumetz, J. and Aladin, N. (2021) Saving the Aral Sea: a cross-cultural communication success. *Academia Letters* 360. DOI: 10.20935/AL360

Dunham, R. (2023) *Aquaculture and Fisheries Biotechnology: Genetic Approaches*. CAB International, Wallingford, UK.

Edwards, P., Zhang, W., Belton, B. and Little, D.C. (2019) Misunderstandings, myths and mantras in aquaculture: its contribution to world food supplies has been systematically over reported. *Marine Policy* 106, 103547. DOI: 10.1016/j.marpol.2019.103547

FAO (2022) *The State of World Fisheries and Aquaculture 2022: Towards Blue Transformation*. Food and Agriculture Organization, Rome. DOI: 10.4060/cc0461en

FAO, Duke University and WorldFish (2023) *Illuminating Hidden Harvests – The Contributions of Small-Scale Fisheries to Sustainable Development*. Food and Agriculture Organization, Rome. DOI: 10.4060/cc4576en

FISEARCH (2023) *FISEARCH: Quarterly Publication of the DA – National Fisheries Research and Development Institute* 5(2). Available at: https://drive.google.com/file/d/1yRsQO4LnHsvKB44WjzGlgZwgvNGb3cvl/view?fbclid=IwAR3ApnLIbkDIHFOUpjCiOqLDWR7j0Xnx2g967DkDEWpA6GU7y6oxW-JPqa0 (accessed 8 July 2024).

Free, C.M., Mangin, T., Molinos, J.G., Ojea, E., Burden, M. *et al.* (2020) Realistic fisheries management reforms could mitigate the impacts of climate change in most countries. *PLOS ONE* 15(3), e0224347. DOI: 10.1371/journal.pone.0224347

Fudge, M., Higgins, V., Vince, J. and Rajaguru, R. (2023) Social acceptability and the development of commercial RAS aquaculture. *Aquaculture* 739295. DOI: 10.1016/j.aquaculture.2023.739295

Galappaththi, E.K., Ichien, S.T., Hyman, A.A., Aubrac, C.J. and Ford, J.D. (2020) Climate change adaptation in aquaculture. *Reviews in Aquaculture* 12(4), 2160–2176. DOI: 10.1111/raq.12427

Galappaththi, E.K., Susarla, V.B., Loutet, S.J., Ichien, S.T., Hyman, A.A. *et al.* (2022) Climate change adaptation in fisheries. *Fish and Fisheries* 23(1), 4–21. DOI: 10.1111/faf.12595

Geels, F.W. (2002) Technological transitions as evolutionary reconfiguration processes: a multi-level perspective and a case-study. *Research Policy* 31(8–9), 1257–1274. DOI: 10.1016/S0048-7333(02)00062-8

Geels, F.W. (2004) From sectoral systems of innovation to socio-technical systems: insights about dynamics and change from sociology and institutional theory. *Research Policy* 33(6–7), 897–920. DOI: 10.1016/j.respol.2004.01.015

Geels, F.W. (2019) Socio-technical transitions to sustainability: a review of criticisms and elaborations of the Multi-Level Perspective. *Current Opinion in Environmental Sustainability* 39, 187–201. DOI: 10.1016/j.cosust.2019.06.009

Gephart, J.A., Golden, C.D., Asche, F., Belton, B., Brugere, C. *et al.* (2020) Scenarios for global aquaculture and its role in human nutrition. *Reviews in Fisheries Science & Aquaculture* 29(1), 122–138. DOI: 10.1080/23308249.2020.1782342

Gephart, J.A., Henriksson, P.J., Parker, R.W., Shepon, A., Gorospe, K.D. *et al.* (2021) Environmental performance of blue foods. *Nature* 597(7876), 360–365. DOI: 10.1038/s41586-021-03889-2

Gillanders, B.M., Elsdon, T.S., Halliday, I.A., Jenkins, G.P., Robins, J.B. *et al.* (2011) Potential effects of climate change on Australian estuaries and fish utilising estuaries: a review. *Marine and Freshwater Research* 62(9), 1115–1131. DOI: 10.1071/MF11047

Global capture production 1950–2018 (FishstatJ). FAO Fisheries Division, Rome [online]. Available at: www.fao.org/fishery/statistics/software/fishstatj/en (accessed 9 July 2024).

Guerrero, R.D.I. (1999) The impacts of El Niño on Philippine fisheries. *Naga, The ICLARM Quarterly* 22(3), 14–15. Available at: https://hdl.handle.net/20.500.12348/2436 (accessed 8 July 2024).

Gutierrez, N.L., Funge-Smith, S., Giulia, G., Mancha-Cisneros, M., Defeo, O. *et al.* (2023) Production and environmental interactions of small-scale fisheries. In: FAO *et al.* (ed.) *Illuminating Hidden Harvests: The Contributions of Small-Scale Fisheries to Sustainable Development*. FAO, Duke University and WorldFish. DOI: 10.4060/cc4576en

Hess, D.J. and Sovacool, B.K. (2020) Sociotechnical matters: reviewing and integrating science and technology studies with energy social science. *Energy Research & Social Science* 65, 101462. DOI: 10.1016/j.erss.2020.101462

Hilborn, R., Amoroso, R.O., Anderson, C.M., Baum, J.K., Branch, T.A. *et al.* (2020) Effective fisheries management instrumental in improving fish stock status. *Proceedings of the National Academy of Sciences of the United States of America* 117(4), 2218–2224. DOI: 10.1073/pnas.1909726116

Hilborn, R., Amoroso, R., Collie, J., Hiddink, J., Kaiser, M. *et al.* (2023) Evaluating the sustainability and environmental impacts of trawling compared to other food production systems. *ICES Journal of Marine Science* 1–13. DOI: 10.1093/icesjms/fsad115

IPCC (2007) *Climate Change 2007: Synthesis Report*. Contribution of Working Groups I, II and III to the Fourth Assessment Report of the Intergovernmental Panel on Climate Change (Pachauri, R.K and Reisinger, A. (eds)). IPCC, Geneva, Switzerland.

IPCC (2022) *Climate Change 2022: Impacts, Adaptation and Vulnerability*. Contribution of Working Group II to the Sixth Assessment Report of the Intergovernmental Panel on Climate Change (Pörtner, H.-O., Roberts, D.C., Tignor, M., Poloczanska, E.S., Mintenbeck, K. *et al.* (eds)). Cambridge University Press. DOI: 10.1017/9781009325844.

IPCC (2023) *Climate Change 2023: Synthesis Report*. Contribution of Working Groups I, II and III to the Sixth Assessment Report of the Intergovernmental Panel on Climate Change (Lee, H. and Romero, J. (eds)). IPCC, Geneva, Switzerland. DOI: 10.59327/IPCC/AR6-9789291691647

Islam, M.M., Barman, A., Kundu, G.K., Kabir, M.A. and Paul, B. (2019) Vulnerability of inland and coastal aquaculture to climate change: evidence from a developing country. *Aquaculture and Fisheries* 4(5), 183–189. DOI: 10.1016/j.aaf.2019.02.007

Knowler, D., Chopin, T., Martínez-Espiñeira, R., Neori, A., Nobre, A. *et al.* (2020) The economics of integrated multi-trophic aquaculture: Where are we now and where do we need to go? *Reviews in Aquaculture* 12(3), 1579–1594. DOI: 10.1111/raq.12399

Kolding, J., van Zwieten, P.A., Marttin, F. and Poulain, F. (2016) Fisheries in the drylands of sub-Saharan Africa: 'Fish come with the rains': building resilience for fisheries-dependent livelihoods to enhance food security and nutrition in the drylands. *FAO Fisheries and Aquaculture Circular 1118*. Food and Agriculture Organization of the United Nations, Rome.

Labuinao, E.S. (2019) Kastom, gender and economic development: the case of the fish aggregating device in To'baita, Solomon Islands. *Women in Fisheries Information Bulletin* 32, 13–18.
Lam, V.W., Cheung, W.W., Swartz, W. and Sumaila, U.R. (2012) Climate change impacts on fisheries in West Africa: implications for economic, food and nutritional security. *African Journal of Marine Science* 34(1), 103–117. DOI: 10.2989/1814232X.2012.673294
Last, P.R., White, W.T., Gledhill, D.C., Hobday, A.J., Brown, R. *et al.* (2011) Long-term shifts in abundance and distribution of a temperate fish fauna: a response to climate change and fishing practices. *Global Ecology and Biogeography* 20(1), 58–72. DOI: 10.1111/j.1466-8238.2010.00575.x
Lubembe, S.I., Turyasingura, B. and Chavula, P. (2022) Reflection on impacts of climate change on fisheries and aquaculture: sub-Sahara Africa. *International Journal of Academic Pedagogical Research* 6(10), 62–72.
MacLeod, M.J., Hasan, M.R., Robb, D.H. and Mamun-Ur-Rashid, M. (2020) Quantifying greenhouse gas emissions from global aquaculture. *Scientific Reports* 10(1), 11679. DOI: 10.1038/s41598-020-68231-8
Metian, M., Troell, M., Christensen, V., Steenbeek, J. and Pouil, S. (2020) Mapping diversity of species in global aquaculture. *Reviews in Aquaculture* 12(2), 1090–1100. DOI: 10.1111/raq.12374
Micklin, P., Aladin, N.V., Chida, T., Boroffka, N., Plotnikov, I.S. *et al.* (2020) The Aral Sea: a story of devastation and partial recovery of a large Lake. In: Mischke, S. (ed.) *Large Asian Lakes in a Changing World: Natural State and Human Impact*, pp. 109–141. DOI: 10.1007/978-3-030-42254-7_4
Nasr-Allah, A., Dickson, M., Al-Kenawy, D., Ahmed Ibrahim, N., Ali, S. *et al.* (2021) Better management practices for tilapia culture in Egypt. In: CGIAR Research Program on Fish Agri-Food Systems. Manual: FISH-2021-03. Available at: https://hdl.handle.net/20.500.12348/4698 (accessed 8 July 2024).
OECD and FAO (2023) *OECD-FAO Agricultural Outlook 2023–2032*. OECD Publishing, Paris.
Okoli, A.S., Blix, T., Myhr, A.I., Xu, W. and Xu, X. (2022) Sustainable use of CRISPR/Cas in fish aquaculture: the biosafety perspective. *Transgenic Research* 31(1), 1–21. DOI: 10.1007/s11248-021-00274-7
Parker, R.W. and Tyedmers, P.H. (2015) Fuel consumption of global fishing fleets: current understanding and knowledge gaps. *Fish and Fisheries* 16(4), 684–696. DOI: 10.1111/faf.12087
Phillips, B.F. and Pérez-Ramírez, M. (2018) *Climate Change Impacts on Fisheries and Aquaculture: A Global Analysis*, Vol. 2. John Wiley & Sons, Chichester, UK.
Porse, H. and Rudolph, B. (2017) The seaweed hydrocolloid industry: 2016 updates, requirements, and outlook. *Journal of Applied Phycology* 29(5), 2187–2200. DOI: 10.1007/s10811-017-1144-0
Punt, A.E. (2023) Those who fail to learn from history are condemned to repeat it: a perspective on current stock assessment good practices and the consequences of not following them. *Fisheries Research* 261, 106642. DOI: 10.1016/j.fishres.2023.106642
Ramirez, P.J., Narvaez, T.A. and Santos-Ramirez, E.J. (2020) Gender-inclusive value chains: the case of seaweed farming in Zamboanga Peninsula, Philippines. *Gender, Technology and Development* 24(1), 110–130. DOI: 10.1080/09718524.2020.1728810
Scanes, E., Scanes, P.R. and Ross, P.M. (2020) Climate change rapidly warms and acidifies Australian estuaries. *Nature Communications* 11, 1803(1), 1–11. DOI: 10.1038/s41467-020-15550-z
Seggel, A. and De Young, C. (2016) Climate change implications for fisheries and aquaculture: summary of the findings of the Intergovernmental Panel on Climate Change Fifth Assessment Report. *FAO Fisheries and 54 Aquaculture Circular 1122*. FAO, Rome.
Smith, T.D. (1994) Scaling fisheries: the science of measuring the effects of fishing, 1855–1955. Cambridge University Press, Cambridge, UK. DOI: 10.1017/CBO9780511470868
Sonesson, A.K., Hallerman, E., Humphries, F., Hilsdorf, A., Leskien, D. *et al.* (2023) Sustainable management and improvement of genetic resources for aquaculture. *Journal of the World Aquaculture Society* 54(2), 364–396. DOI: 10.1111/jwas.12968
Stock, C.A., Alexander, M.A., Bond, N.A., Brander, K.M., Cheung, W.W. *et al.* (2011) On the use of IPCC-class models to assess the impact of climate on living marine resources. *Progress in Oceanography* 88(1–4), 1–27. DOI: 10.1016/j.pocean.2010.09.001
Tommasi, D., Stock, C.A., Hobday, A.J., Methot, R., Kaplan, I.C. *et al.* (2017) Managing living marine resources in a dynamic environment: the role of seasonal to decadal climate forecasts. *Progress in Oceanography* 152, 15–49. DOI: 10.1016/j.pocean.2016.12.011
Troell, M., Henriksson, P.J., Buschmann, A., Chopin, T. and Quahe, S. (2023) Farming the Ocean: seaweeds as a quick fix for the climate? Vol. 31. Taylor & Francis, pp. 285–295. DOI: 10.1080/23308249.2022.2048792
Verdegem, M., Buschmann, A.H., Latt, U.W., Dalsgaard, A.J. and Lovatelli, A. (2023) The contribution of aquaculture systems to global aquaculture production. *Journal of the World Aquaculture Society* 54(2), 206–250. DOI: 10.1111/jwas.12963

Vogel, J.M., Longo, C., Spijkers, J., Palacios-Abrantes, J., Mason, J. *et al.* (2023) Drivers of conflict and resilience in shifting transboundary fisheries. *Marine Policy* 155, 105740. DOI: 10.1016/j.marpol.2023.105740

Welcomme, R. (2016) Inland fisheries: past, present, and future. In: Taylor, W.W., Bartley, D.M., Goddard, C.I., Leonard, N.J. and Welcomme, R. (eds) *Freshwater, Fish and the Future: Proceedings of the Global Cross-Sectoral Conference*. FAO and Michigan State University, pp. 7–14. DOI: 10.47886/9789251092637

Williams, M.J. and Syddall, V. (2022) Women, fisheries technology and development: toward new research approaches. *Gender, Technology and Development* 1–28. DOI: 10.1080/09718524.2022.2125456

World Bank (2023) *Global Seaweed: New and Emerging Markets Report, 2023*. World Bank, Washington, DC. Available at: http://hdl.handle.net/10986/40187 (accessed 8 July 2024).

You, X., Shan, X. and Shi, Q. (2020) Research advances in the genomics and applications for molecular breeding of aquaculture animals. *Aquaculture* 526. DOI: 10.1016/j.aquaculture.2020.735357

Yuan, D. (2022) In: *6th International Conference for Fisheries and Aquatic Sciences*. Iloilo, Philippines.

Zhang, W., Belton, B., Edwards, P., Henriksson, P.J., Little, D.C. *et al.* (2022) Aquaculture will continue to depend more on land than sea. *Nature* 603(7900), E2–E4.

# 7 Impact of Climate Change on Livestock in Drylands

**Adel M. Aboul Naga[1]* and Mohamed K. Yousef[2]**
[1]*Animal Production Research Institute, ARC, Egypt;* [2]*University of Nevada, Las Vegas, USA*

**Abstract**

Climate change affects livestock production in drylands in multiple ways, both directly and indirectly. Increasing temperatures and reducing precipitation in drylands have a negative impact on both animal and crop yields. Animals are a major contributor to greenhouse gas (GHG) emissions. Alternatives have been offered by various research groups to reduce enteric GHG, and to reduce GHG from livestock waste.

Livestock animals are known to have remarkable capacity to survive and produce in various stressful climatic regions of the world. Environmental stress (heat and drought) and climate change are the major problems affecting animal production in drylands, where ambient temperature exceeds the upper critical temperature of livestock and negatively affects its production and increases its losses. Mitigation measures to help animals cope with, and/or minimize the adverse effects of, heat stress (HS) have been developed, together with a number of measures to cope with the interrelated processes of climate change and livestock production.

## Introduction

Livestock is a rapidly growing economic sector, with around 40% of the global agricultural gross domestic product (GDP). It is a key contributor to food security in different regions of the world (FAO, 2012). It produces 13% of the globally consumed kcal and 25% of the consumed protein. In drylands, livestock is the only means of turning sparse and erratic biomass resources into edible products. Livestock consumes crop residues and its byproducts to produce meat, milk, eggs, and manure, and to provide transport and draught power. Livestock is a major asset for rural communities, providing a range of essential services including saving, credit, and buffer against crises, including climatic shocks. Income from livestock contributes directly to livelihood and human health. More than one in two citizens in arid sub-Saharan Africa (SSA) are livestock keepers and one in three citizens is considered a poor livestock keeper (FAO, 2012). Livestock, especially small ruminants, and chickens are also key to women's empowerment and gender equity. Food security and better nutrition in drylands are dependent on livestock farming.

Climate change (CC) is no longer a serious problem only for the future; now, most people are already experiencing CC in many different forms such as global warming (GW), severe heatwaves, extreme drought, unpredictable rainstorms and rainfall patterns, and wildfires. If not combated,

*Corresponding author: adelmaboulnaga@gmail.com

 *Climate Change and Sustainable Agro-ecology in Global Drylands* (A.S. El-Beltagy, R. Lal and K. Malik eds.)
DOI: 10.1079/9781800624870.0007

these symptoms will result in substantial risk to food production, specifically in drylands, resulting in serious economic losses. The interrelationships between animal agriculture and CC are best described as a 'double-edged sword' where both impact the other adversely. Aboul-Naga *et al.* (2011) summarized the effect of environmental stress and CC on livestock in four main paths: (i) impact on feed-grain availability; (ii) impacts on ranges and pastures; (iii) direct effects of weather and extreme events on animal production and health; and (iv) changes in the distribution and incidence of livestock diseases. A major contribution to CC is the emissions of greenhouse gases (GHGs). The GHGs trap some of the energy that each emits back to space, thus causing GW. Primary GHGs emitted by animal agriculture are carbon dioxide ($CO_2$), methane ($CH_4$), and nitrous oxide ($N_2O$). Each of these gases differs in its capacity to absorb and remit radiation leaving the Earth. The global warming potential (GWP) of each gas is equated to the GWP of $CO_2$, which equals one, where $CH_4$ (=23) and $N_2O$ (=295) (IPCC, 2007). The present chapter is not intended to present all current knowledge related to the impact of CC on livestock farming but focuses on some consequences and corrective measures that have a realistic chance of maintaining adequate food security in the future, especially in drylands.

## How Does Climate Change Affect Livestock Farming?

Climate change affects livestock production in drylands in multiple ways, both directly and indirectly (Table 7.1). Crescio *et al.* (2010) reported that high environmental temperatures and humidity led to 60% cattle mortality in sub-Saharan Africa, and 20–60% of losses in all animals that were recorded during serious drought events in the past two or three decades. In South Africa, Niang *et al.* (2014) reported a decrease of 10–25% in dairy yield under some CC scenarios. Furthermore, impacts of CC on animal health are well documented, especially for vector-borne diseases. Rising air temperatures increase winter survival of vectors and pathogens; outbreaks of Rift Valley fever in East Africa were associated with increased rainfall and floods in El Niño-Southern Oscillation events (Rosenthal, 2009; Porter *et al.*, 2014). Despite uncertainties resulting from complex interactions between climatic factors, mainly air temperatures and $CO_2$ concentrations, impacts of CC on field crops and forages have also been quantified (Table 7.1).

Increasing temperatures and reducing precipitation in drylands have negative impacts on both animal and crop yields. CC also affects quality of fodder through shifts from C3 to C4 plants and shrub cover. Increases in fungi and mold infestation in feed resources further impact feed production and food safety. Under current climate constraints, there are no assessments of livestock production to support policies that would improve resilience in the sector (IPCC, 2014). Several challenges will need to be overcome to quantify the many impacts on livestock production systems (Thornton *et al.*, 2015). In extensive grazing and pastoral systems, impacts on rangeland productivity and carrying capacity remain unknown. Similarly, animal diseases are affected by CC, but future patterns should be modeled to understand their impacts on rangeland productivity and carrying capacity.

Impact of groundwater availability is another area where assessments are needed. Livestock's vulnerability depends first on its exposure to climate stress with emphasis on its duration, frequency, and severity. Rising temperatures increase survival of vectors and pathogens where diseases such as West Nile virus and Schistosomiasis are projected to expand, and new ones such as Bluetongue or Lyme are on the rise and known to be climate-sensitive.

Climate change is a key driver in the erosion of genetic resources for food and agriculture, the raw materials that local communities rely upon to improve quality and output of their food production. The risks posed by CC on genetic resources for food and agriculture remain an open question for future research. Animal breeds that will be required for the changing climate conditions will have to come from the existing pool of genetic resources (Aboul-Naga *et al.*, 2021). Increased tolerance to biotic and abiotic stresses will be needed. Already new breeds and populations adapted to higher temperatures and decreased rainfall have been developed around the world. However, CC is threatening the strategic

**Table 7.1.** Direct and indirect pathways of impacts of climate change on livestock. (Modified from Thornton *et al*., 2009; IUCN, 2010; Niang *et al*., 2014)

| CC symptoms | Animals | Forages and feed crops | Labor force and capital |
|---|---|---|---|
| Severe variability in rainfall | Shortages in drinking and servicing water<br>Increased pathogens, parasites, and vectors<br>Changed distribution and transmission of diseases | Decreased crop yields<br>Changes in forage and pasture composition<br>Changes in production system (mixed crop-livestock to rangelands) | Altered human health and livestock resources<br>Decreased productivity<br>Migrations<br>Conflicts |
| Extreme air temperatures | Decreased feed intake and livestock yields<br>Decreased conception<br>Altered metabolism and increased mortality<br>Transmission of pathogens and vectors<br>Decreased resistance<br>Domestic biodiversity losses | Decreased yields<br>Decreased forage quality<br>Change in pasture composition | |
| Increased $CO_2$ in the atmosphere | | Partial stomata closure<br>Reduced transpiration<br>Change in pasture composition | |

reservoir of the livestock genetic resources from which new breeds could be adapted for production systems to meet future challenges. As conditions change, indigenous breeds and populations may be lost forever. Extreme weather events such as floods and droughts, which are expected to become more frequent with CC, can pose an immediate threat to the survival of breeds and populations that are raised in specific geographical areas.

Heat stress is one of the primary CC challenges and strain on livestock farming in both intensive and extensive production systems. Livestock animals are known to have remarkable capacity to live in various stressful climatic regions of the world. In hot climates, these animals adjust and minimize the adverse effects using various physiological, biochemical, and behavioral mechanisms. Much work has been done to review and analyze the adaptive responses to heat stress (HS) by many scientists worldwide (Leonard, 1986; Collier and Collier, 2012; Sejian *et al*., 2012, 2021). In general, as HS exceeds the adaptive limits of livestock animals, significant losses of production (growth rate, milk, meat, and egg production) and reproduction (processes that span from follicular development to parturition and lactation) occur (Fig. 7.1). Furthermore, HS increases susceptibility to diseases.

In severe HS situations, death of animals occurs. The magnitude of the economic losses resulting from HS and effects of extreme weather events such as heatwaves await further research. An important contribution to evaluate the economic loss of HS has been the development of a software program that can be downloaded from the Florida Dairy Extension website (http://dairy.ifas.ufl.edu/tools [accessed 9/7/24]). This program is continuously updated, user-friendly, and allows easy comparison of two scenarios side by side (DeVries, 2006).

Full understanding of the effect of CC on livestock production requires a good and accurate measure of environmental warmth, performance, and well-being of animals and corrective measures to cope with CC.

## Environmental Stresses in the Drylands

### How can environmental warmth be measured?

Climate change contributes significantly to the prevailing heat-stress level in drylands as well as

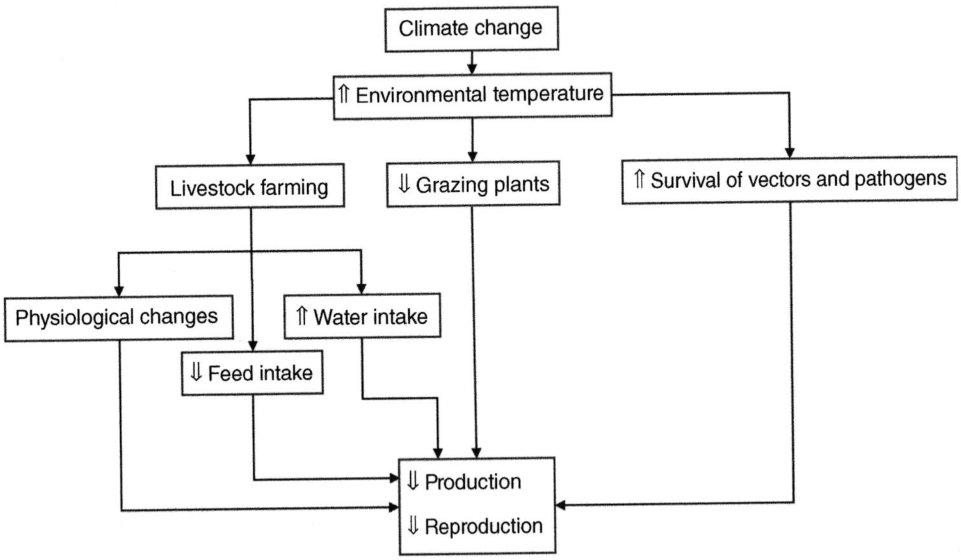


**Fig. 7.1.** Direct and indirect effects of climate change on animal production.

other regions of the world. The higher the level of HS, the greater the risk on livestock productivity depending on its intensity and duration of exposure. The need for an index to accurately describe and measure HS has been investigated for several decades (Yousef, 1990). Elements of the thermal environment, i.e. air temperature; relative humidity (RH) or wet bulb temperature (WBT); air movement; and radiant heat are considered the key parameters for development of any HS index. Several indices that combine two or more of these environmental elements were developed. Each of these indices has its advantages and limitations.

Livestock farming systems vary from one location to another and there is no one specific HS index that can fit to all regions or locations. One of the first indices developed was the temperature-humidity index (THI). The THI is used by the US National Weather Service for advisories to animal and plant agriculture farming (Hahn *et al.*, 2009). It should be noted here that THI may not be suitable for regions that encounter high wind speeds and/or where solar radiation is intense. Recently, Gaughan *et al.* (2006) introduced a new index named 'Heat Load Index' (HLI) in which wind speed and solar radiation were added to the THI equation. The development of the HLI for different regions of drylands is a more suitable index for weather service organization and can be used to advise farmers of expected extreme weather changes and alert them to potential HS conditions.

## How can HS be measured in livestock animals?

Several physiological measures have been developed to measure HS in farm animals (Hahn *et al.*, 2009; Khalifa *et al.*, 2014). Many of these indices are partly useful. The need for an index to help define the threshold levels for decline in productivity and thus provide guidelines for tactical management of livestock in drylands is badly needed. Eigenberg *et al.* (2004) reported that measurements of respiratory rate provide a non-invasive and practical assessment of stress in feeding-lot cattle held under hot environmental conditions. Respiration rate increases in a non-linear fashion in response to increasing air temperature, with a breakpoint or a threshold of between 20°C and 25°C (Hahn *et al.*, 1997). Along with respiration rate, Young and Hall (1993) listed behaviors exhibited by cattle experiencing heat stress including: onset of open mouth breathing; labored panting; and excessive salivation (drooling). Using this information, Mader and Davis (2002) developed the 'panting score' index. The panting score index ranges from 0 (indicating a non-stressed animal) to 4 (describing an animal suffering from heat stress, as shown in Table 7.2.

## Corrective measures for coping with HS: mitigation scheme

Climate and farm systems vary from one location to another, which means that any specific corrective measures must be specially developed for that location/region. There is no one solution that can fit everywhere. Mitigation measures to help animals cope with and/or minimize the adverse effects of HS have been developed. The selection of an appropriate countermeasure must be economically and environmentally sound for productive performance and well-being of the animals. Available countermeasures for reducing the effects of thermal stress on livestock have been extensively reviewed by Moberg (1985). These mitigation schemes fall into four categories:

**1.** *Physical modification of the environment.* A broad alternative to physically reduce intensity of heat stress include providing shade for animals outdoors and/or to totally house the animals. Adequate shade reduced effects of solar radiation coupled with high air temperature and humidity. Production loss of shaded cows was reduced to 50% of those with no shade (Hahn, 1985). The most practical and effective shades are trees and hay or straw shades. Solid shade provided by sheet metal painted white on top is next in effectiveness. Shades should be 3.6–4.2 m in height in areas with clear sunny afternoons; however, in areas with cloudy afternoons, shades of 2.1–2.7 m in height are effective, and 1.8–2.5m$^2$ per head for cattle.

**Table 7.2.** Description of panting scores in feed-lot heifers.

| Score | Description |
|---|---|
| 0 | Normal respiration rate, ~60 or less/min. |
| 1 | Slightly elevated respiration rate, ~60–90/min. |
| 2 | Moderate panting, ~90–120/min. |
| 3 | Heavy open-mouthed panting, ~120–150/min. |
| 4 | Severe open-mouthed panting, protruding tongue, excessive salivation |

**2.** *Alteration to the microclimate of animals.* Augmenting natural air flow with fans; increasing air vapor pressure via evaporation water (e.g. sprinklers); and increasing heat loss by conductivity of surfaces that animals might contact (e.g. wetness of the pen surfaces) are all measures that have been used with varied success (Davis *et al.*, 2003).
**3.** *Improvement of nutritional and water management techniques.* Changing time of feeding and/or limiting the feed intake of feed-lot cattle enhanced animal comfort and reduced the adverse effects of thermal stress (Mader *et al.*, 2002; Davis *et al.*, 2003; Das, 2018). Providing an adequate water source that provides at least 15 liters/100 kg body weight daily is essential to prevent dehydration.
**4.** *Genetic development of less sensitive animals to heat.* Selection and breeding of livestock tolerant to heat has been a long-sought-after solution. Success in this effort has been limited (Johnson, 1987).

Differences in thermal tolerance exist between livestock species/breeds, which may provide markers to select heat-tolerant animals using modern genetic tools. Identification of the genetic markers associated with high production traits have become possible with the availability of animal genome sequences.

## How Does Livestock Farming Contribute to CC?

Livestock animals are not a wholly benign factor in CC. These animals are a major contributor to GHG emissions (Table 7.3) and in turn contribute to GW. It is said that if cattle were a nation, they would be the world's third largest emitter of GHG after the USA and China. FAO (2006) reported that livestock animals contribute 18% of GHG emissions. However, a few years later they revised this number down to 14.5% (FAO, 2012). Regardless of the correct estimate, no one would be able to ignore the adverse impacts of GW without consideration of emissions of GHG from livestock animals.

The direct and indirect adverse effects of livestock farming on CC are illustrated in Fig. 7.2. The percentage contribution of major livestock sources to GHG emissions are summarized in Table 7.4.

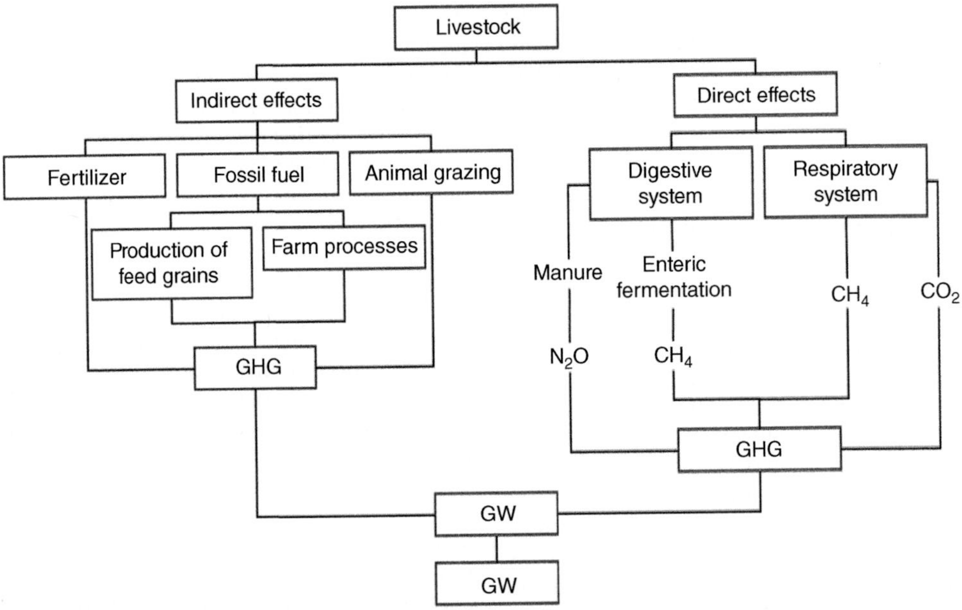


**Fig. 7.2.** An illustration of the impact of livestock farming on climate change.

**Table 7.3.** Estimates of farm animal methane emissions/million metric tons (Tg).

| Animal | $CH_4$/Tg | Reference |
|---|---|---|
| Dairy cattle | 18.9 | McMichael *et al.* (2007) |
| Beef cattle | 55.9 | McMichael *et al.* (2007) |
| Sheep and goats | 9.5 | Johnson and Ward (1996) |
| Buffalo | 6.2–8.1 | Johnson and Ward (1996) |
| Camel | 0.9–1.1 | Johnson and Ward (1996) |
| Pigs and horses | 0.9–1.7 | Johnson and Ward (1996) |

**Table 7.4.** Contribution of major sources of livestock to greenhouse gas emissions. (From IPCC, 2007)

| Source | % |
|---|---|
| Enteric (fermentation) | 25.3 |
| Manure | 30.5 |
| Deforestation | 34.0 |
| Fertilizer | 6.2 |
| Other | 4.1 |

Emissions of $CH_4$ from various farm animals have been estimated (Table 7.3). Can livestock industry reduce the GHG burden? The answer is yes, only if producers, consumers, and policymakers can come together to support innovative and appropriate technical solutions. In the absence of corrective measures, GW will worsen dramatically. The following suggestive corrective measures address reduction of GHG per unit of animal production.

## Reducing enteric greenhouse gas emission

Enteric fermentation is a process by which ruminants digest fibrous feed and release $CH_4$. Most strategies developed to reduce or prevent emission of $CH_4$ resulting from rumen fermentation are focused on technical solutions to reduce excessive fermentation and regulate metabolic activity of rumen bacteria (Koneswaran and Nierenberg, 2008).

Several alternatives have been offered by various research groups:

**1.** *Promotion of high-quality forages in ruminant feeding and grazing systems.* Feeding high-quality pasture plants reduces $CH_4$ emissions by 50% (Boadi *et al.*, 2000). Furthermore, scientists from Gramina, a joint biotechnology venture by Australia's Molecular Plant Breeding Cooperative Research

Center and New Zealand rural services group Plant Genomic Wrightson Genomics developed a burpless grass that reduces enteric $CH_4$ emission and grows well in hotter climates.

**2.** *Improved animal efficiency, i.e. decreasing $CH_4$ emission per unit product using better animal management techniques and innovative formulation of rations.* Improved feed formulations can alter fermentation in the rumen (Ellis *et al.*, 2007; Eugene *et al.*, 2008). High-grain rations for cattle decrease $CH_4$ emissions as compared to low-quality forage diet. Furthermore, addition of fats or oils to grain diets decreases $CH_4$ emission by variable amounts (McGinn *et al.*, 2004). Fats act through inhibition or reduction of the cellulolytic bacteria and/or as a binder of hydrogen in the rumen, thus resulting in decreased $CH_4$ production. Other feed additives that are known to reduce and/or inhibit $CH_4$ production by decreasing activity of the enzyme methyl transferase and/or decrease rumen pH have been under investigation. Examples of such additives are: ionophores (monensin, lasolocid); bromoethanosulfonate 'BBS' and bovine growth hormone, which were reported to decrease $CH_4$ production by 9–28% (Guan *et al.*, 2006). More recently, the use of essential oils, tannin-rich plants, and biodiesel co-products for reducing enteric $CH_4$ production have been reported with positive and promising results (Chesson, 2006; Benchaar *et al.*, 2008; Sallam and Abdelgaliel, 2010; Abdalla *et al.*, 2012).

**3.** *Breeding farm animals for low $CH_4$ production.* Future genetic selection of ruminants using modern bioengineering technologies for low $CH_4$ enteric fermentation must be considered seriously. Growth in ruminant populations, particularly in large, industrialized operations, would be more environmentally friendly if individual animals were bred to produce lower levels of $CH_4$.

**4.** *Limitations on livestock populations.* Recently, the IPCC's 4th report concluded that emissions of livestock GHG in western Europe are decreasing and are predicted to continue to fall up to 2020. This decrease has occurred due to a reduction in the size of the animal production industry and to reducing meat consumption. Reduction of 10% of livestock numbers in Belgium reduced annual GHG emissions by 0.242 Tg $CO_2$ equivalent. In another study researchers calculated that an overall cut of 10% in global meat consumption to 90 g per person/day would merely prevent an increase in GHG from livestock (McMichael *et al.*, 2007). It should be noted that a target of 90 g of animal protein is equal to a reduction of meat consumption between 55% and 64% in the industrialized nations. In developing countries, where the average consumption is only 10–20% of that in developed nations (FAO, 2006), the stated target of 90 g/person/day would allow continued growth in consumption. In fact, the proposed target would not lower GHG emissions that come from animal production in developing countries. Reduction in livestock populations may be considered a simple, quick, and effective measure to GHG emissions to the extent that it is necessary to prevent future GW. Such a measure may be an oversimplification of the problem, specifically in the developing nations where we have the world's fastest-growing populations of affluent middle- to high-level income class, which is no longer confined to developed countries, and increased demand of animal protein in the diet.

## Reducing GHG from livestock waste

Animal manure emits gases such as $CH_4$, $N_2O$, ammonia, and carbon dioxide, depending on the way it is produced (solid or liquid) and managed (collection, storage, or spreading). Emissions of $CH_4$ from anaerobic decomposition of manure are estimated to be about 4% of global anthropogenic $CH_4$ emissions (US-EPA, 2006). Western Europe, China, and North America have the highest emission from manure. Of farm animals, pig production contributes the largest share followed by cattle (Table 7.5). In the USA, between 1990 and 2005, $CH_4$ emissions from the manure of dairy cattle and pigs increased by 50% and 37%, respectively (US-EPA, 2007). When manure is used to fertilize crops, soil is enriched, resulting in healthy sustainable farms. However, when manure production exceeds the needs of available farmland, as is the case of industrialized animal production, i.e. landless or intensive systems, then manure becomes a hazardous waste that threatens soil, water, and air quality. Emission levels of GHG from anaerobic decomposition of manure are dependent on manure management systems, i.e. slurry pits, solid-manure piles, and on wet soil following incorporation of manure. Manure that is managed in slurry pits or lagoon systems produces the highest levels of $CH_4$ but $N_2O$ emissions can be important when ponds become encrusted or are covered (Rotz, 2004). However, solid manure piles increase $N_2O$ emissions while suppressing that of $CH_4$ (Hole *et al.*, 2005). There is no silver-bullet solution to reduce GHG from livestock manure, but corrective measures urgently need to be addressed.

**1.** *Bring waste generated into line with the capacity of accessible land to absorb that waste.* Intensive animal production farms must be located as much as possible where crop land can utilize the waste effectively without creating problems with soil nutrient loading, i.e. better integration of crop and livestock activities.

**2.** *Mandate the use of anaerobic digester biogas systems.* Biogas is produced by controlled anaerobic fermentation of organic matter in a closed vessel. Biogas is typically made up of 65% $CH_4$ and 35% $CO_2$. This gas can be burned directly for heating or light or in modified gas boilers to run internal combustion engines or generators. In recent years, available technology has allowed for building small anaerobic digesters in India, China, and Kenya. Such digesters are available commercially in different sizes and can be afforded economically by small farmers.

**3.** *Decrease the pH of slurry pits.* There is evidence that a decreased pH of slurry pits is effective in lowering both $CH_4$ and $N_2O$ emissions (Berg *et al.*, 2006).

**4.** *Intensify managed animal grazing systems where agro-ecological potential permits and where soil-nutrient balances are still negative.* Cattle raised on pasture, eating natural low-energy diet composed of grasses and other forages, produce manure with about half of the potential to generate $CH_4$ (US-EPA, 1998). Most existing grazing systems are present in marginal lands and livestock GHG emissions per unit of product is relatively high.

**5.** *Reverse deforestation.* Amazon deforestation is related to expansion of livestock production. Therefore, livestock contributes to increased GHG by decreasing the size of the natural sink capacity of these gases. Reforestation in tropical regions results in restoring carbon in soils; however, this must be done without threatening food security in the region.

## Responding to Climate Change: Future Perspectives

Climate mitigation and adaptation should not be seen as alternatives to each other, but rather as a combined set of actions within a comprehensive strategy to reduce GHG emissions and GW. These processes present new challenges and opportunities for the scientific community, decision makers, and producers at any given location or region. Animal production scientists must abandon their own 'silos' and formulate cooperative and interdisciplinary research projects with agronomists, ecologists, meteorologists, and bioengineers to develop better measures to combat the adverse effects of CC on livestock farming. Few of the major key issues that await immediate attention, especially in drylands, include the following.

**Table 7.5.** Methane emissions from livestock species in different world regions by 2004 (Tg/year).

| Region | Cattle | Buffalo | Sheep & goats | Pigs | Poultry |
|---|---|---|---|---|---|
| Sub-Saharan Africa | 0.42 | 0.00 | 0.08 | 0.03 | 0.04 |
| India | 0.54 | 0.19 | 0.04 | 0.17 | 0.01 |
| China | 0.19 | 0.05 | 0.05 | 3.43 | 0.14 |
| Asia* | 0.39 | 0.09 | 0.03 | 0.50 | 0.13 |
| Central & S. America | 0.46 | 0.00 | 0.02 | 0.74 | 0.19 |
| WANA | 0.15 | 0.01 | 0.05 | 0.00 | 0.11 |
| North America | 1.57 | 0.00 | 0.00 | 1.65 | 0.16 |
| Western Europe | 2.45 | 0.00 | 0.02 | 1.52 | 0.09 |
| Oceania & Japan | 0.19 | 0.00 | 0.03 | 0.10 | 0.03 |
| Eastern Europe | 1.11 | 0.00 | 0.01 | 0.19 | 0.06 |
| Total | 7.45 | 0.34 | 0.33 | 8.34 | 0.95 |

*Not including India and China (FAO, 2006)

## Development of HS early-warning system for drylands

There is a need to provide livestock farming with HS information from an onsite real-time weather-monitoring system. The weather station collects and predicts future weather data (Ta, RH, windspeed, and solar radiation) where these data can be used to generate a physiological-based stress index. This index in the form of a given value alerts farmers early enough to prepare and select an appropriate measure to reduce HS impacts on their animals. The development of an early-warning index for livestock producers based on 'real-time' weather conditions was reported by Eigenberg *et al.* (2004). Similarly, livestock safety monitors must be developed for various regions of drylands and extensive farming systems to combat the adverse effects of CC on drylands.

## Development of new farm animals and feed plants for drylands

Integration of new technologies into future research on development of animal- and plant-breeding programs and nutritional interventions are urgent actions to be taken to lower the harmful effects of CC on livestock farming, especially in the harsh dryland regions. According to the high genetic variability between and within breeds, there is no doubt that it is feasible to select for tolerance to HS. Future research on identification of genes associated with thermal and reduced GHG emissions must be fully supported. Using these genes as markers to develop new breeds of livestock should be considered seriously. Similarly, new feeder plants with high-quality nutritive value, tolerant to HS and producing less GHG during fermentation, are the need of the hour for sustainability of livestock farming in the drylands.

## Utilization of gene editing in livestock to reduce methane emission

Genome editing offers a new opportunity to mitigate $CH_4$ production by manipulating the genome of methanogenic archaea. It was found that it is possible to delete genomic sequences of *Methanosarcina acetivorans* to generate frameshift mutations using Cas9-mediated genome editing, and also introduce multiple deletions. The genome sequence of *Methanobrevibacter ruminantium* is available. Further studies with rumen methanogens are required to provide information about genes that play a role in $CH_4$ production in the rumen. Gene editing targets multiple candidate genes to generate mutant methanogenic Archaea that produce less $CH_4$. Karavolias *et al.* (2021) and Rumah *et al.* (2023) reviewed successful application trials that have been carried out globally for gene editing in livestock, to resist CC, including, e.g., reducing methane emission by methanotrophic bacteria, enhancing thermotolerance in cattle; raising animal disease resistance, raising tuberculosis and mastitis resistance; and enhancing livestock yield.

Gene editing applies to reducing methane production directly in the rumen and can potentially contribute indirectly to reduce cattle methane emission per unit of protein produced (Camargo *et al.*, 2022). The Archaea genome can be manipulated using the CRISPR/Cas system. One option could be to knock down key genes in the methanogenesis pathway. The transcript abundance of genes-encoding enzymes involved in the hydrogenotrophic methanogenesis pathway is lower in the rumen methanogens with low methane emissions compared with those with high emission. Based on that, one can infer that downregulation of genes involved in hydrogenotrophic methanogenesis can be a good strategy to reduce rumen methane emission from livestock, especially in drylands where large quantities of poor roughage are used.

## Conclusion

Drylands occupy over 40% of the world's terrestrial surface (IUCN, 2010). Climate change scientists predict that global dryland will expand by 10% by the end of the 21st century. It is expected that CC will present a severe threat to drylands and food security for the inhabitants. The interrelationship between livestock farming and CC is documented to be an adverse one. On one hand, CC has an adverse impact on livestock farming resulting in costly economic losses for producers, from reduced production and reproduction performance to increased mortality. On the other hand, livestock farming accounts for a large share of global GHG emission, which contributes to GW.

In this chapter, a number of possible measures to cope with the interrelated processes of CC and livestock production have been summarized. Implementation of any of these measures needs to be cost-effective and suitable for the available infrastructure of the location or region. Major reductions in the negative impact of CC and livestock farming may be accomplished by management systems and practices as CC occurs. In drylands, livestock farming is primarily extensive in nature and small ruminants (sheep and goats) represent a major component of the farming system. Such farming systems contribute less emission of GHG as compared to the intensive farming system used in temperate regions of the world. However, the fast increase in population coupled with a significant demand for animal proteins will demand an increase in the numbers of livestock animals and probably a shift to intensive-farming systems surrounding large cities. These possible changes will exacerbate CC and increase the frequency and severity of climate shocks. Therefore, the corrective measures proposed in this article are considered urgent action to be implemented sooner rather than later in drylands.

## References


Abdalla, A.L., Louvandini, H., Sallam, S.M., Bueno, I.C., Tsai, S.M. *et al.* (2012) In vitro evaluation, in vivo quantification, and microbial diversity studies of nutritional strategies for reducing enteric methane production. *Tropical Animal Health and Production* 44(5), 953–964. DOI: 10.1007/s11250-011-9992-0

Aboul-Naga, A.M., Khalifa, H.H., Elbeltagy, A.R., Khalek, T.A., Elshafie, M.H. *et al.* (2011) Tolerance to abiotic stresses in Egyptian Barki desert sheep and goats raised under hot-dry conditions: individual variations. In: El-Beltagy, A. and Saxena, M.C. (eds) Proceedings of the Tenth International Conference on Development of Drylands, Cairo. *Meeting the Challenge of Sustainable Development in Drylands under Changing Climate-Moving from Global to Local*. International Dryland Development Commission (IDDC), Cairo, pp. 445–459.

Aboul-Naga, A.M., Elshafie, M., Khalifa, H., Osman, M., Abdel Khalek, T.M. *et al.* (2021) Tolerance capability of desert sheep and goats to exercise heat stress under hot dry conditions, and its correlation with their production performance. *Small Ruminant Research* 205, 106550. DOI:10.1016/j.smallrumres.2021.106550

Benchaar, C., Calsamiglia, S., Chaves, A.V., Fraser, G.R., Colombatto, D. *et al.* (2008) A review of plant-derived essential oils in ruminant nutrition and production. *Animal Feed Science and Technology* 145, 209–228. DOI: 10.1016/j.anifeedsci.2007.04.014

Berg, W., Brunsch, E. and Pazsiczki, I. (2006) Greenhouse gas emissions from covered slurry compared with uncovered during storage. *Agriculture, Ecosystems & Environment* 112, 129–134. DOI: 10.1016/j.agee.2005.08.031

Boadi, D., Wittenberg, K.M. and McCaughey, W.P. (2000) Effect of energy supplementation on methane production in grazing steers. *Canadian Journal of Animal Science* 82(2).

Camargo, L.S.A., Saraiva, N.Z., Oliveira, C.S., Carmickle, A., Lemos, D.R. *et al.* (2022) Perspectives of gene editing for cattle farming in tropical and subtropical regions. *Animal Reproduction* 19(4), e20220108. DOI: 10.1590/1984-3143-AR2022-0108

Chesson, A. (2006) Phasing out antibiotic feed additives in the EU: worldwide relevance for animal food production. In: Barug, D.,de Jong, J., Kies, A.K. and Verstegen, M.W.A. (eds) *Antimicrobial Growth Promoters*. Wageningen Academic, Leiden, The Netherlands, pp. 69–80. DOI: 10.3920/9789086865703_005

Collier, R.J. and Collier, J.L. (eds) (2012) *Environmental Physiology of Livestock*. John Wiley & Sons, Inc., Chichester, UK. DOI: 10.1002/9781119949091

Crescio, M.I., Forastiere, F., Maurella, C., Ingravalle, F. and Ru, G. (2010) Heat-related mortality in dairy cattle: a case crossover study. *Preventive Veterinary Medicine* 97(3), 191–197. DOI: 10.1016/j.prevetmed.2010.09.004

Das, S.K. (2018) Impact of climate change (heat stress) on livestock: adaptation and mitigation strategies for sustainable production. *Agricultural Reviews* 39(2), 130–136. DOI: 10.18805/ag.R-1777

Davis, M.S., Mader, T.L., Holt, S.M. and Parkhurst, A.M. (2003) Strategies to reduce feedlot cattle heat stress: effects on tympanic temperature. *Journal of Animal Science* 81, 649–661. DOI: 10.2527/2003.813649x

DeVries, A. (2006) *The Dairy VIP Program to Evaluate the Consequences of Changes in Herd Management and Prices on Dairy Farms*. University of Florida. Available at: http://edis.ifas.ufl.edu (accessed 11 July 2024).

Eigenberg, R.A., Brown-Brandl, T.M. and Nienaber, J.A. (2004) Development of a livestock weather safety monitor for feed-lot cattle. *Applied Engineering in Agriculture* 23, 657–660. DOI: 10.13031/2013.23666

Ellis, J.L., Kebreab, E., Odongo, N., McBride, B., Okine, E. *et al.* (2007) Prediction of methane production from dairy and beef cattle. *Journal of Dairy Science* 90(7), 3456–3466. DOI: 10.3168/jds.2006-675

Eugene, M., Masse, D., Chiauette, J. and Benchaar, C. (2008) Meta-analysis on the effect of the effects of lipid supplementation on methane production in lactating dairy cows. *Canadian Journal of Animal Science* 88, 331–337. DOI: 10.4141/CJAS07112

FAO (2006) Livestock a major threat to the environment: remedies urgently needed. Available at: https://www.thepigsite.com/news/2006/12/livestock-a-major-threat-to-environment-remedies-urgently-needed-1 (accessed 11 July 2024).

FAO (2012) *Greening the Economy with Agriculture*. FAO, Rome. Available at: https://www.fao.org/3/i2745e/i2745e00.pdf (accessed 11 July 2024).

Gaughan, J.B., Mader, T.L., Holt, S.M. and Lisle, A. (2006) A new heat load index for feed lot cattle. *Journal of Animal Science* 86, 226–234. DOI: 10.2527/jas.2007-0305

Guan, H., Wittenberg, K.M., Ominski, K.H. and Krause, D.O. (2006) Efficacy of ionophores in cattle diets for mitigation of enteric methane. *Journal of Animal Science* 84, 1896–1906. DOI: 10.2527/jas.2005-652

Hahn, G.L. (1985) Management and housing of farm animals in hot environments. In: Yousef, M.K. (ed.) *Stress Physiology in Livestock. Vol.* II. *Ungulates*. CRC Press, Boca Raton, Florida, pp. 151–174.

Hahn, G.L., Parkhurst, A.M. and Gaughan, J.B. (1997) *Cattle Respiration Rate as a Function of Ambient Temperature*. Paper #MC97-121. ASAE, St. Joseph, Michigan.

Hahn, G.L., Gaughan, J.B., Mader, T.L. and Eigenberg, R.A. (2009) Thermal indices and their applications for livestock environments. In: DeShazer, J.A. (ed.) *Livestock Energetic and Thermal Environmental Management*. ASABE, St. Joseph, Michigan, pp. 113–130.

Hole, D.G., Perkins, A., Wilson, J., Alexander, I., Grice, P. *et al.* (2005) Does organic farming benefit biodiversity? *Biological Conservation* 122(1), 113–130. DOI: 10.1016/j.biocon.2004.07.018

IPCC (Intergovernmental Panel on Climate Change) (2007) *Climate Change 2007: Synthesis Report. Contribution of Working Groups I, II and III to the Fourth Assessment Report of the Intergovernmental Panel on Climate Change*. Pachauri, R.K and Reisinger, A. (eds). IPCC, Geneva, Switzerland.

IPCC (Intergovernmental Panel on Climate Change (2014) *Climate Change 2014: Impacts, Adaptation, and Vulnerability. Part A: Global and Sectoral Aspects. Contribution of Working Group II to the Fifth Assessment Report of the Intergovernmental Panel on Climate Change*. Field, C.B., Barros, V.R., Dokken, D.J., Mach, K.J., Mastrandrea, M.D. *et al.* (eds), Cambridge University Press, Cambridge, UK and New York.

IUCN (2010) *Building Climate Change Resilience for African Livestock in Sub-Saharan Africa. World Initiative for Sustainable Pastoralism (WISP): A Program of IUCN*. The International Union for Conservation of Nature, Eastern and Southern Africa Regional Office, Nairobi.

Johnson, D.E. and Ward, G.M. (1996) Estimates of methane emission. *Environmental Monitoring and Assessment* 42, 133–141. DOI: 10.1007/BF00394046

Johnson, H.D. (1987) *Progress in Animal Biometeorology*. Swets and Zeitlinger, Amsterdam.

Karavolias, N.G., Horner, W., Abugu, M.N. and Evanega, S.N. (2021) Application of gene editing for climate change in agriculture. *Frontiers in Sustainable Food Systems* 5, 685801. DOI: 10.3389/fsufs.2021.685801

Khalifa, H.H., Aboul-Naga, A.M., Abdel Khalek, T.M.M., El Shafie, M.H., El Beltagy, A.R. *et al.* (2014) Biological and mathematical analysis of desert sheep and goats' responses to natural heat stress, in Egypt. *Egyptian Journal of Animal Production, Egyptian Society of Animal Production* 52(sup), 45–51.

Koneswaran, G. and Nierenberg, D. (2008) Global farm animal production and global warming: impacting and mitigating climate change. *Environmental Health Perspectives* 116, 578–582. DOI: 10.1289/ehp.11034

Leonard, B.E. (1986) Book review. *Stress Physiology in Livestock. Vol. 1: Basic Principles*. Yousef, M.K. (ed.). CRC Press. Florida. *Stress Medicine* 2(4), 357–358. DOI: 10.1002/smi.2460020413

Mader, T.L. and Davis, M.S. (2002) Wind speed and solar radiation corrections for the temperature-humidity index. In: *Proceedings of 15th Conference on Biometeorology and Aerobiology Joint with 16th International Congress on Biometeorology*. American Meteorological Society, pp. 154–157.

Mader, T.L., Holt, S.M., Hahn, G.L., Davis, M.S. and Spiers, D.E. (2002) Feeding strategies for managing heat load in feedlot cattle. *Journal of Animal Science* 80, 2373–2382. DOI: 10.2527/2002.8092373x

McGinn, S.M., Beauchemin, K.A., Coates, T. and Colombatto, D. (2004) Methane emissions from beef cattle: effects of Monessen, sunflower oil, enzymes, yeast and fumaric acid. *Journal of Animal Science* 82(11), 3346–3356. DOI: 10.2527/2004.82113346x

McMichael, A.J., Powles, J.W., Buttler, C.O. and Uauy, R. (2007) Food, livestock production, energy, climate change, and health. *The Lancet* 370, 1253–1263. DOI: 10.1016/S0140-6736(07)61256-2

Moberg, G.P. (1985) *Animal Stress*. American Physiological Society, Bethesda, Maryland.

Niang, I., Ruppel, O.C., Abdrabo, M.A., Essel, A., Lennard, C. *et al.* (2014) Africa. In: Barros, V.R., Field, C.B., Dokken, D.J., Mastrandrea, M.D., Mach, K.J. *et al.* (eds) *Climate Change 2014: Impacts, Adaptation, and Vulnerability. Part B: Regional Aspects. Contribution of Working Group II to the Fifth Assessment Report of the Intergovernmental Panel on Climate Change*. Cambridge University Press, Cambridge, UK and New York, pp. 1199–1265.

Porter, J.R., Xie, L., Challinor, A.J., Cochrane, K., Howden, S.M. *et al.* (2014) Food security and food production systems. In: Field, C.B., Barros, V.R., Dokken, D.J., Mach, K.J., Mastrandrea, M.D. *et al.* (eds) *Climate Change 2014: Impacts, Adaptation, and Vulnerability. Part A: Global and Sectoral Aspects. Contribution of Working Group II to the Fifth Assessment Report of the Intergovernmental Panel on Climate Change*. Cambridge University Press, Cambridge, UK and New York, pp. 485–533.

Rosenthal, J. (2009) Climate change and the geographic distribution of infectious diseases. *EcoHealth* 6, 489–495. DOI: 10.1007/s10393-010-0314-1

Rotz, C.A. (2004) Management to reduce nitrogen losses in animal production. *Journal of Animal Science* 82, E119–E137. DOI: 10.2527/2004.8213_supple119x

Rumah, B.L., Stevens, B.H.C., Yeboah, J.E., Stead, C.E., Harding, E.L. *et al.* (2023) In vivo genome editing in type I and II methanotrophs using a CRISPR/Cas9 system. *ACS Synthetic Biology* 12, 544–554. DOI: 10.1021/acssynbio.2c00554

Sallam, S.M.A. and Abdelgaliel, S.A.M. (2010) Effect of different levels of citrus essential oil and its active component on rumen microbial fermentation and methane production in vitro. *Cuban Journal of Agricultural Science* 44(4), 367–372. Available at: https://cjascience.com/index.php/CJAS/article/view/200 (accessed 11 July 2024).

Sejian, V., Naqvi, S.M.K., Ezeji, T., Lakritz, J. and Lal, R. (eds) (2012) *Environmental Stress and Amelioration in Livestock Production*. Springer-Verlag GMH Publishers, Germany. DOI: 10.1007/978-3-642-29205-7

Sejian, V., Chauhan, S.S., Devaraj, C., Malik, P.K. and Bhatta, R. (eds) (2021) *Climate Change and Livestock Production: Recent Advances and Future Perspectives*. Springer Nature, Switzerland AG. DOI: 10.1007/978-981-16-9836-1

Thornton, P.K., Van de Steeg, J., Notenbaert, A. and Herrero, M. (2009) The impacts of climate change on livestock and livestock systems in developing countries: a review of what we know and what we need to know. *Agricultural Systems* 101(3), 113–127. DOI: 10.1016/j.agsy.2009.05.002

Thornton, P.K., Boone, R.B. and Ramirez-Villegas, J. (2015) *Climate Change Impacts on Livestock. CCAFS Working Paper no. 120. CGIAR Research Program on Climate Change*. Agriculture and Food Security (CCAFS), Copenhagen. Available at: https://ccafs.cgiar.org/resources/publications/climate-change-impacts-livestock (accessed 11 July 2024).

US-EPA (1998) *Inventory of U.S. Greenhouse Gas Emissions and Sinks: 1990–1996*. U.S. Environmental Protection Agency office of Policy, Planning, and Evaluation. EPA 236-R-98-006. Washington, DC. Available at: https://www.epa.gov/sites/default/files/2015-12/documents/98cr.pdf (accessed 11 July 2024).

US-EPA (2006) *Inventory of U.S. Greenhouse Gas Emissions and Sinks: 1990–2004*. U.S. Environmental Protection Agency, EPA 430-R-06002. Washington, DC. Available at: https://www.epa.gov/sites/default/files/2015-12/documents/06_complete_report.pdf (accessed 11 July 2024).

US-EPA (2007) *Inventory of U.S. Greenhouse Gas Emissions and Sinks: 1990–2005*. U.S. Environmental Protection Agency, EPA 430-R-07-002. Washington, DC. Available at: https://www.epa.gov/sites/default/files/2015-12/documents/07cr.pdf (accessed 11 July 2024).

Young, B.A. and Hall, A. (1993) Heat load in cattle in the Australian environment. In: Coombes, R. (ed.) *Australian Beef*. Morescope Publishing, Melbourne, Australia, pp. 143–148.

Yousef, M.K. (1990) Importance of field studies in stress physiology. In: Anderson, D.M. *et al.* (eds) *Proceedings of Stress and Free-Ranging Animal Symposium. Agricultural Experimental Station Research Report*, Vol. 646. New Mexico State University, Las Cruces, New Mexico, pp. 15–30.

# 8 Climate Change and Food Security in Africa's Drylands

**Hafez Ghanem[1]*, Simeon Ehui[2], Steven Were Omamo[3], and Chakib Jenane[2]**
[1]*Brookings Institution, Paris, France;* [2]*International Institute of Tropical Agriculture, Ibadan, Nigeria;* [3]*International Food Policy Research Institute, Nairobi, Kenya;* [4]*The World Bank Group, Washington, DC, USA*

**Abstract**

Intensifying climate-related events like droughts, floods, and disease outbreaks are impacting Africa's food security across dimensions: availability, access, utilization, and stability. Improved food security aligns with economic growth, poverty reduction, and system stability. Climate hazards strongly influence dryland food systems. More resilient and climate-adapted food systems are therefore central policy and investment priorities.

This chapter examines the impacts of major climate trends on each dimension of food security: availability, access, utilization, and stability. It delves into food insecurity challenges in the Sahel and Horn of Africa. Policy and investment implications for resilient, climate-adapted food systems in Africa are discussed. Action is crucial from African nations and the global community. Enhanced investment in agriculture, water, and environment sectors, coupled with local capacity development, is necessary. Given individual countries' limitations, a pan-African climate adaptation and food security initiative is urged.

## Introduction

Since 2021, every part of Africa has been affected by extreme weather events: heatwaves and wildfires in North Africa; drought in Eastern Africa; rising temperatures, droughts and destructive floods in Western and Central Africa; cyclones and catastrophic flooding in Southern Africa (Dunne, 2022; FEWSNET, 2023a). Millions of people have been displaced, with concomitant disruptions to livelihoods; hundreds of lives have been lost (OCHA, 2022). In every instance, acute food insecurity has surged, with Africa's dryland areas especially hard hit (GCA, 2021a; FEWSNET, 2023a).

Nearly a third of global drylands fall in Africa, where they cover 19.6 million km$^2$, and nearly two thirds of southern, western, eastern and northern Africa. Africa's drylands are home to over 525 million people (40% of the population) – a number that is growing by 3% per year (compared to the African average of 2.5%) (GCA, 2021a). Land use in these areas is broadly dominated by nomadic, transhumant or sedentary pastoralism, rainfed cropping and agroforestry, and localized areas of irrigated farming; the livelihoods of over 200 million people in Africa's drylands depend on cropping (GCA, 2021a). The Intergovernmental Panel on Climate Change (IPCC) projects that regional warming in African drylands may be up to twice the global average temperature. Key impacts of climate change include reduced water availability, increased occurrence of vector- and waterborne diseases,

*Corresponding author: hafez.ghanem@gmail.com

 *Climate Change and Sustainable Agro-ecology in Global Drylands* (A.S. El-Beltagy, R. Lal and K. Malik eds.)
DOI: 10.1079/9781800624870.0008

reduced crop and livestock productivity, and damage to transportation infrastructure and buildings.[1]

These impacts have direct implications for food security across the continent. In drought-sensitive African countries, the number of undernourished people increased by 40% between 2012 and 2021 (GCA, 2021a). At the beginning of 2023, following five consecutive below-average rains, 22 million people were acutely food insecure, and 5.1 million children were acutely malnourished in drought-affected areas of Kenya, Ethiopia, and Somalia (WFP, 2023). In Benin, Cote d'Ivoire, Ghana, Guinea, Guinea-Bissau, Liberia, Sierra Leone, and Togo, multiple climate shocks combined with persistent insecurity, high food prices, the economic fallout from COVID-19, and the impact of the conflict in Ukraine to increase food insecurity in the last quarter of 2022 by 20%, compared to the same period in 2021 (Africa Renewal, 2022).

The policy response is emerging. In response to these challenges, leaders from various countries and organizations convened in September 2022 during the United Nations General Assembly meetings and issued a declaration on global food security. Discussions at the World Bank and International Monetary Fund Annual Meetings in October 2022 also emphasized the importance of addressing food security, with Africa being a central focus due to its high incidence of undernourishment. An African Development Bank-led High-Level Summit of 34 African heads of state was held in Dakar, Senegal, in January 2023, to mobilize political, financial, and technical support for agricultural transformation on the continent at scale. The recommendations from these discussions highlight the need for African countries to strengthen social safety nets, invest more in research and development, and avoid trade restrictions.

But more needs to be done. Addressing the deteriorating food security situation in Africa, compounded by climate change (CC), declining agricultural productivity, and rapid population and urbanization growth, should be a top priority for the continent over the next decade. Encouragingly, initiatives such as the Comprehensive Africa Agriculture Development Program (CAADP), adopted by African Union member states, and United Nations Sustainable Development Goal 2, which aims to end hunger and malnutrition by 2030, reflect this priority. However, progress has been limited, with only 9 out of 55 African countries currently on track to significantly reduce undernutrition by 2025. This lack of progress underscores the need to intensify efforts.

Science holds immense potential for providing sustainable solutions for food security, including innovations to enhance CC adaptation, science-based management of productive resources (such as land, soil, and water), and improvements in food storage and transportation to reduce waste. For every US$ 1 of investment in agricultural research, society reaps $10 in benefit (Alston *et al.*, 2020). For every $1 invested in research by the Consultative Group for International Agricultural Research (CGIAR), at least $9 worth of additional food is produced in Africa (IFPRI, 2013). Effectively harnessing scientific advancements requires the translation of research findings into practical packages that can be disseminated and adopted by farmers on a large scale, both at the farm and landscape levels. This endeavor necessitates strong collaborations between international, regional, and national science organizations, as well as farmers and extension systems. By co-generating solutions through the collaboration of researchers and farmers, local challenges to resilience can be addressed in a demand-driven and knowledge-intensive manner.

Following an examination of major CC trends and projections, this chapter examines emerging and potential impacts on food security on the continent, considering each dimension of food security: availability, access, utilization, and food system stability. Climate-related food insecurity challenges in two especially threatened areas are considered in detail: the Sahel and Horn of Africa.

Implications for policy and investment toward resilient and climate-adapted food systems in Africa are also presented. Action is needed by African countries as well as by the international community. African countries need to invest more in three key sectors: agriculture, water and environment. They also need to develop local capacities and invest in science and technology. The ability of each of the 54 African countries to deal with CC and achieve food security on its own is limited. Therefore, we also call for a pan-African climate adaptation and food security initiative.

CC is a global phenomenon and Africa is the continent most impacted by it, although it is only responsible for 3.9% of global greenhouse gas GHG) emissions. The international community needs to help Africa finance climate action in adaptation and mitigation; and this finance needs to be additional to existing development finance. Money spent on climate action in Africa obviously helps Africans, but it also helps the rest of the world as it contributes to a global public good. We argue here for the creation of an international Green Bank that would fund private sector green projects in adaptation and mitigation, and for the adoption of a new system of debt for climate swaps that would reduce Africa's debt burden while releasing resources for much needed green investments.

## Climate Change in Africa: Trends, Projections, and Impacts

### Trends and projections

Africa accounts for just 3.9% of carbon dioxide ($CO_2$) emissions, but under all temperature scenarios over 1.5°C, it is the most vulnerable continent to CC impacts (IPCC, 2022; USDA, 2023). Between 1991 and 2021, Africa warmed by an average of 0.3°C per decade. This trend, which held across the continent (Fig. 8.1), was quicker than the warming of +0.2°C per decade between 1961 and 1990; −0.04°C per decade between 1931 and 1960; and +0.08°C per decade between 1901 and 1930; and was faster than the global average (WMO, 2022).

Africa's drylands, which cover three-fifths of the continent, are warming at double the global rate (GCA, 2021b). Although there are some differences among regions, the implications of a rise of 2°C or more are significantly worse in all regions than are those of a 1.5°C rise (Fig. 8.2) (GCA, 2021b).

The rise in temperatures across Africa has caused unprecedented changes in the climate system, increasing the intensity and frequency of weather and climate extremes such as longer and more severe droughts and floods, storms and cyclones, and heatwaves (Eckstein *et al.*, 2020). These are summarized below.

### Droughts and floods

As in other parts of the world, droughts and floods are interconnected in Africa. Rising temperatures increase the amount of moisture the atmosphere

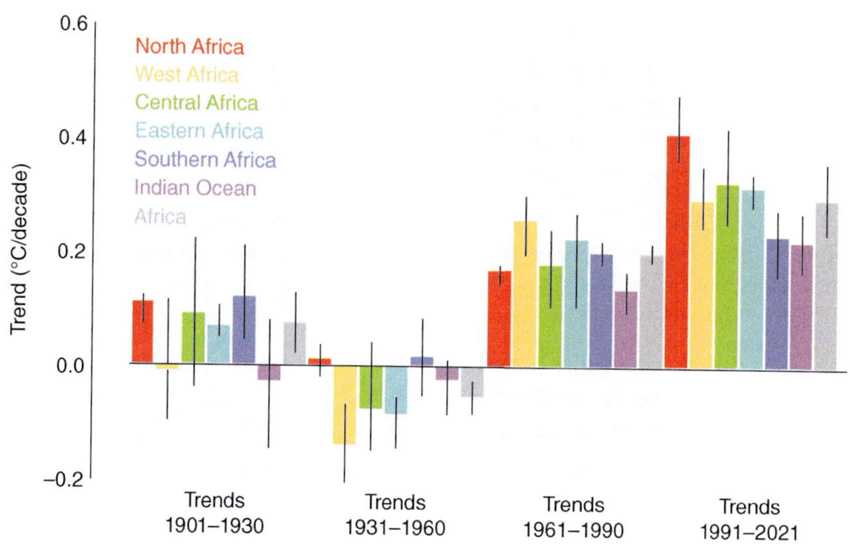


**Fig. 8.1.** Trends in the area average temperature in °C/decade for the six African sub-regions. North Africa (red), West Africa (yellow), Central Africa (green), East Africa (light blue), Southern Africa (dark blue), and Indian Ocean Island countries (purple), and the whole of Africa (grey) over four 30-year periods: 1901–1930, 1931–1960, 1961–1990, and 1991–2021. (From WMO, 2022)

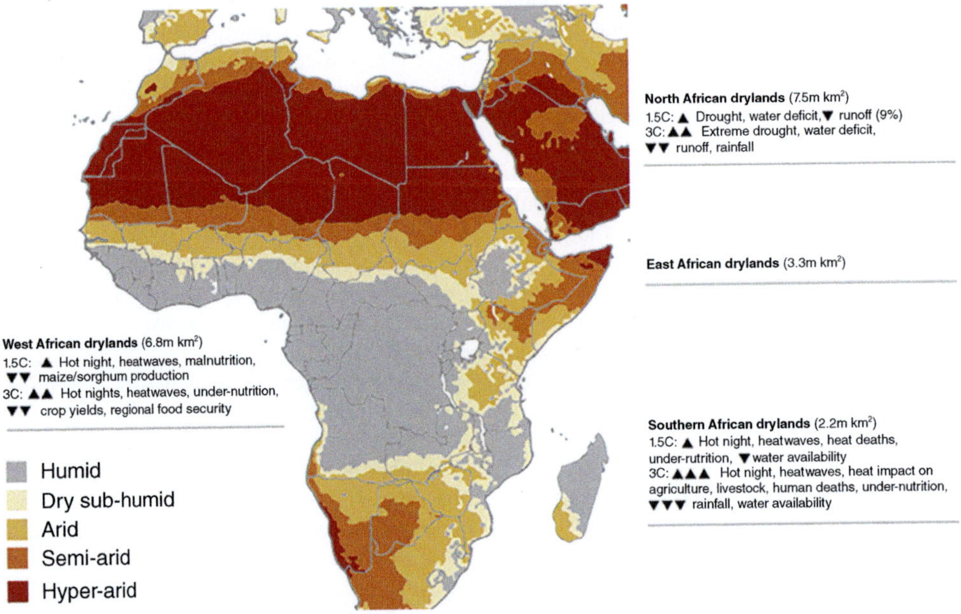


**Fig. 8.2.** Projected changes in drylands of Africa under 1.5°C and 3°C global average rises in temperature. (Based on data from IPCC, 2018 and IPCC, 2019)

can hold, resulting in storms and heavy rains, but, paradoxically, also more intense dry spells as more water evaporates from the land and global weather patterns change. These changes to the hydrological cycle can deliver stronger, more frequent, and longer droughts and floods. These dynamics are affecting Africa more than any other continent. There is an annual average of 40 flood events and seven drought events (Fig. 8.3), with devastating impacts on human health and safety, food and water security, livelihoods, and socio-economic development.

Floods and droughts are expected to become more frequent and severe as temperatures continue to rise on the continent. Between 2002 and 2021, floods (n=793) and droughts (n=137) accounted for 55% of natural hazards in Africa (n=1693), leading to 14,053 and 20,821 deaths, respectively. From the 1982–2001 period, flooding and drought events in Africa increased by 180% and 30%, respectively (CRED, 2022).

Between 2011 and 2020, severe droughts occurred across the Sahel, the Greater Horn of Africa, and Southern Africa (UNCCD, 2022). From 2020 to 2022, Ethiopia, Kenya, and Somalia in the Horn of Africa experienced five failed rainy seasons, resulting in the longest and most severe drought on record. The Shabelle and Juba river basins in Somalia saw their lowest rainfall totals since 1981 (FEWSNET, 2023b).

Severe and devastating flooding occurs across the continent, but especially in the dryland areas of the Sahel and the Horn of Africa, where floods are more frequent, widespread, and devastating (Elagib *et al*., 2021). Between 2019 and 2022, destructive flooding occurred in Burkina Faso, Cameroon, Chad, Congo Republic, Ethiopia, Ghana, Kenya, Niger, Mali, Nigeria, Niger, Senegal, Somalia, South Sudan, and Sudan, affecting millions of people and causing hundreds of deaths (Mayans, 2020; Climate Center, 2022; FEWSNET, 2023b).

## Cyclones and storms

With increasing sea-surface temperatures, the frequency and intensity of storms and cyclones in Africa will increase (IPCC, 2022). Since the 1970s, Africa has experienced twofold and fourfold increases in storms and cyclones, respectively (Fig. 8.4). On average, three cyclones form in the Indian Ocean each cyclone season.

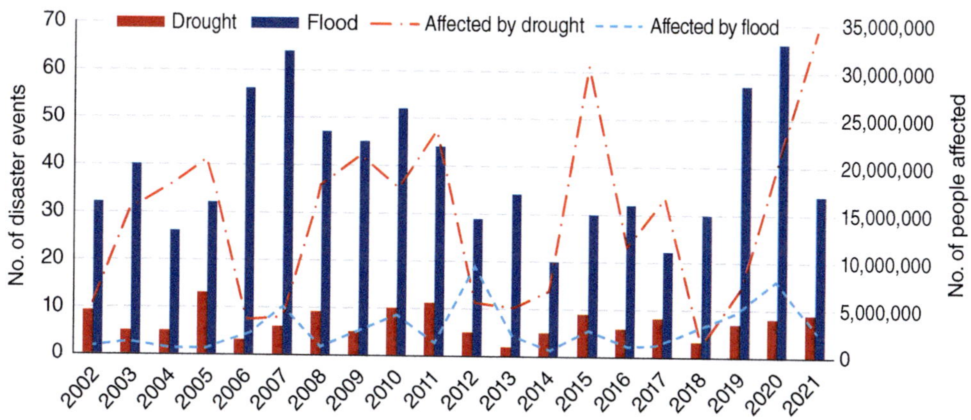


**Fig. 8.3.** The interplay of drought – flood extreme events in Africa over the last 20 years (2002–2021). (From CRED, 2022 – EM-DAT, CRED/UCLouvian, Brussels; www.emdat.be)

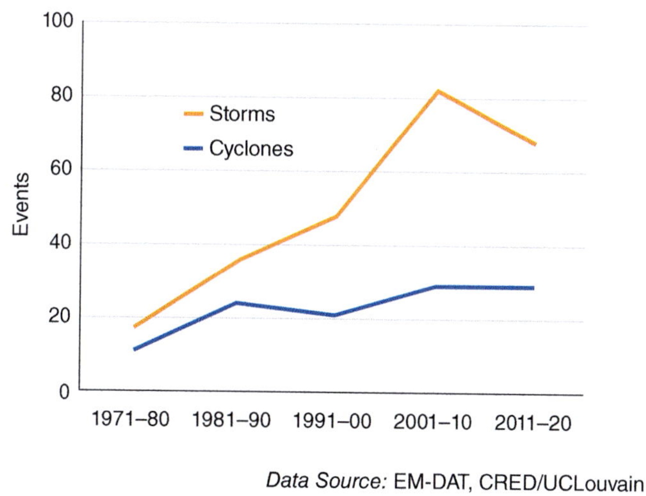


**Fig. 8.4.** Increasing storms and cyclones in Africa. (From Mbiyozo and Maunganidze, 2023).

Six were recorded in the 2021–2022 season (ACSS, 2022). The 2018–2019 season was the deadliest since 1967, with nine intense tropical cyclones in Madagascar, Mozambique, Zimbabwe, Comoros, Malawi, and Somalia. In 2019, Cyclone Idai devastated Mozambique, Malawi, and Zimbabwe, affecting almost 2 million people and killing more than 1200 (WMO, 2022). Cyclone Kenneth followed and was the strongest cyclone ever recorded on the African continent, with wind speeds of 225 kph (140 mph) and floods reaching 2.5 m above normal water levels (ACSS, 2022). In 2020, Cyclone Gati made landfall in northern Somalia as the first hurricane-equivalent storm to hit the Horn of Africa, depositing more than a year's worth of rain in 72 hours (IOM, 2021; ReliefWeb, 2021). And early in 2023, Cyclone Freddy hit southern Africa between February and March, setting several records for strength, length, and cycles of intensification (Mbiyozo and Maunganidze, 2023).

## Heatwaves

Climate change will increase the frequency, duration, and intensity of heatwave events in Africa

(IPCC, 2022). Fivefold increases are expected in a world 3°C hotter (Dunne, 2022). Across the Sahel and the Horn of Africa, heatwaves are expected to exceed human physiological and social limits. Data from 1980–2018 reveal an increasing trend of heatwaves over all regions of Africa (Engdaw *et al.*, 2022). Increases in duration and intensity have been noted, too, particularly in the southern, eastern, and northern parts of the continent. In 2018, a historic heatwave brought Africa its hottest temperature on record: 51.3°C, in Ouargla, Algeria (Masters, 2018). In July 2022, a heatwave struck North Africa, resulting in fires that severely damaged Tunisia's grain crop (Reuters, 2022). Temperatures reached 48°C (118°F) in the capital Tunis, breaking a 40-year record (NASA, 2022). Heatwaves are also becoming more common in East Africa, particularly Uganda, Ethiopia, and Kenya, and on South Africa's Atlantic coast (Dunne, 2022). In March 2022, Kenya experienced uncommonly hot daytime and nighttime temperatures, exceeding 40°C, documented in the northwestern town of Lodwar (Dunne, 2022). Heatwaves are expected to become more common in African countries located within 15 degrees of the equator. In Central Africa, countries such as Cameroon, Equatorial Guinea, Gabon, and the Republic of the Congo already experience eight to ten heatwaves every year, on average.

### Desertification

Climate-induced desertification is having adverse effects on the African continent each year the Earth continues to warm. It affects the everyday lives of Africans, from their crops, livestock, and housing to wildlife and biodiversity. The Sahara is growing by 48 km/year. The Sahel region, between the Sahara Desert and the Sudanian Savannah, is the region most vulnerable to desertification because of constant stress from frequent droughts, soil erosion, and population growth, which has increased logging, illegal farming, and land clearing for housing. Forty-six of the 55 countries in Africa are vulnerable to desertification, with some already feeling the effects (Fig. 8.5). Moderate and severe degradation in recent decades have been identified in many river basins, including the Nile (42% of its area), Niger (50%), Senegal (51%), Volta (67%), Limpopo (66%), and Lake Chad (26%).

The Horn of Africa is getting drier, exacerbating desertification. The observed decline in

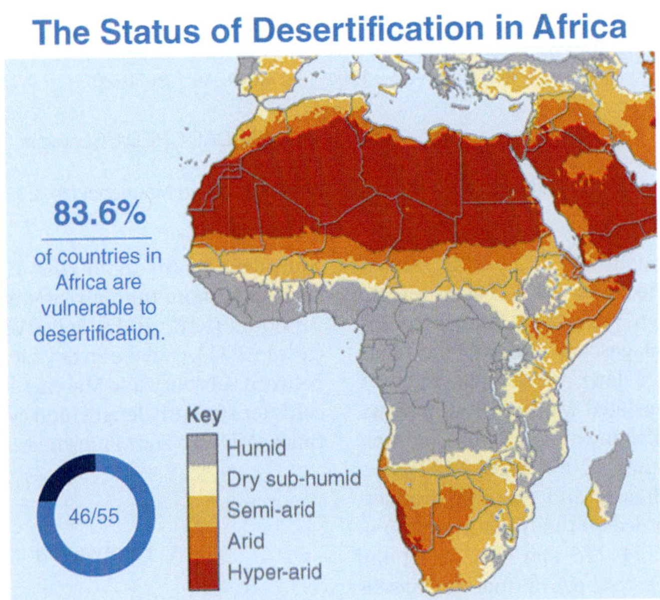


**Fig. 8.5.** Desertification in Africa. (From AGNES, 2020)

vegetation cover is diminishing ecosystem services. Fragmentation of habitats, reduction in the grazing range of livestock, and higher stocking rates are the main drivers of vegetation structure loss in the rangelands of vulnerable dryland areas. Cropland areas in the Sahel region of West Africa have doubled since 1975, with settlement area also increasing by about 150%. In Burkina Faso, from 1984 to 2013, bare soils and agricultural lands increased by 18.8% and 89.7%, respectively, while woodland, gallery forest, tree savannahs, shrub savannahs, and water bodies decreased by 18.8%, 19.4%, 4.8%, 45.2% and 31.2%, respectively (IPCC, 2018; AGNES, 2020).

## Emerging and Potential Impacts on Food Security

Food security exists when all people at all times have physical and economic access to sufficient safe and nutritious food to meet their dietary needs and preferences for an active and healthy life (FAO, 2006). Food security hinges on food availability, access, and utilization, and the stability of availability and access. The rising frequency and intensity of droughts, floods, cyclones, heatwaves, and disease outbreaks associated with CC are affecting all four dimensions (Baptista *et al.*, 2022).

### Food availability

Since 1961, rising temperatures have contributed to a 34% decline in agricultural production on the African continent. This trend is projected to continue, raising the prospect of acute food instability and malnutrition. A 1.5°C increase in global temperature is expected to result in a 9% decline in maize output in West Africa and a 20–60% fall in wheat yield in southern and northern Africa (WMO, 2022). Crop yields in Africa are expected to drop by 5–17% by 2050, particularly in commodities that dominate African diets (Baptista *et al.*, 2022). Rising temperatures and rainfall volatility are key contributors to the shrinking of growing seasons and arable land area, resulting in reduced productivity from overuse. If global warming exceeds 3°C, all current maize-, millet-, and sorghum-growing regions in Africa will become unfit for production. Sorghum and millet are staples across the continent, and maize provides more than half the calories and protein consumed in southern and eastern Africa (Carleton, 2022). Livestock are also threatened, given their sensitivity to high temperatures (FAO, 2015). The increasing number of days each year with acute heat stress is making it harder to rear livestock outside, potentially rendering large sections of the continent unfit for livestock production by the end of the century. By 2050, fish production in coastal West Africa is projected to decline by 21%, with a 50% decline in fishery-related jobs, and it is expected to decline in Lake Tanganyika by almost 30% with adverse consequences across East Africa (Burundi, Democratic Republic of Congo, Tanzania, and Zambia) (Baptista *et al.*, 2022).

Across the continent, recent crop losses due to climate shocks have been staggering. With an especially dry 2021–2022 agro-pastoral season in the Sahel, cereal output declined by 12% and 7% from the 2020 season and the five-year average, respectively. In comparison to the 2020 season, the most significant decreases were in Niger (–36%), Mauritania (–16%), Burkina Faso (–9.1%), Gambia (–8.5%), Mali (–6.5%), and Chad (–6.3%) (RPCA, 2022). Flooding in the Sahel region in 2022 resulted in massive widespread crop losses (FAO and WFP, 2022). Nigeria lost nearly 110,000 ha of farmland devoted to sorghum, maize, rice, and vegetables, with yield losses of 60–75% (Africa News, 2022). More than 465,000 ha of crops were destroyed in Chad (ReliefWeb, 2022). In 2022, as rains failed for a fifth consecutive season in the Horn of Africa, southern Somalia's Gu harvest (June/July), which generates 60% of the annual national output, was 50% lower than the 1995–2021 average. Somalia's Deyr harvest (November/December) was 58% lower than the 1995–2020 average, making it the third lowest Deyr harvest since 2010. In Kenya, harvests of maize, green grams, and cowpeas in arid and semi-arid zones were 78–95% below normal (FAO and WFP, 2022). In southern Africa, the 2022 tropical cyclones Idai and Kenneth destroyed 480,000 ha of crops in Mozambique alone (WMO, 2022).

Livestock losses have also been significant. The five consecutive failed rains in the Horn of Africa devastated livestock populations.

In Kenya, between October 2021 and July 2022, 2.43 million livestock perished, reflecting a 3.55% overall death rate. In pastoral regions, livestock milk output was almost 60% lower than usual in July 2022. Between July 2021 and April 2022, Somalia lost 3 million livestock. In Ethiopia, livestock fatalities topped 3.5 million between late 2021 and late 2022 (FEWSNET, 2023b). In the 2010 Sahelian drought, Niger alone lost 4.8 million head of cattle, roughly 25% of the herd, representing a loss of more than $700 million for the country's economy.

Rising temperatures and water levels are causing insects and weed seeds to migrate into Africa. The 2019–2020 locust infestations in East Africa devastated crops and pastureland as they swept over the region, affecting 26 million farmers in nine countries. In Ethiopia, Kenya, and Somalia, 1.25 mha of cropping and pastureland were destroyed (Stone, 2020; FAO and WFP, 2022).

## Food access

Households across Africa depend heavily on markets for food, so household income is a first-order determinant of food access (Mather *et al.*, 2011). Climate shocks affect both the supply of food and the source of income for rural households. In the agricultural households that dominate most African economies, climate variability and extreme weather events are likely to affect incomes through reduced agricultural outputs, strongly affecting food access.

Adverse weather events are central to the current unprecedented conflation of threats to food access across the continent. These weather events hurt food production, create shortages, and lead to higher market prices, lowering access to food from market sources. Ninety percent of Africans cannot afford a healthy diet (Fig. 8.6). Countries facing frequent

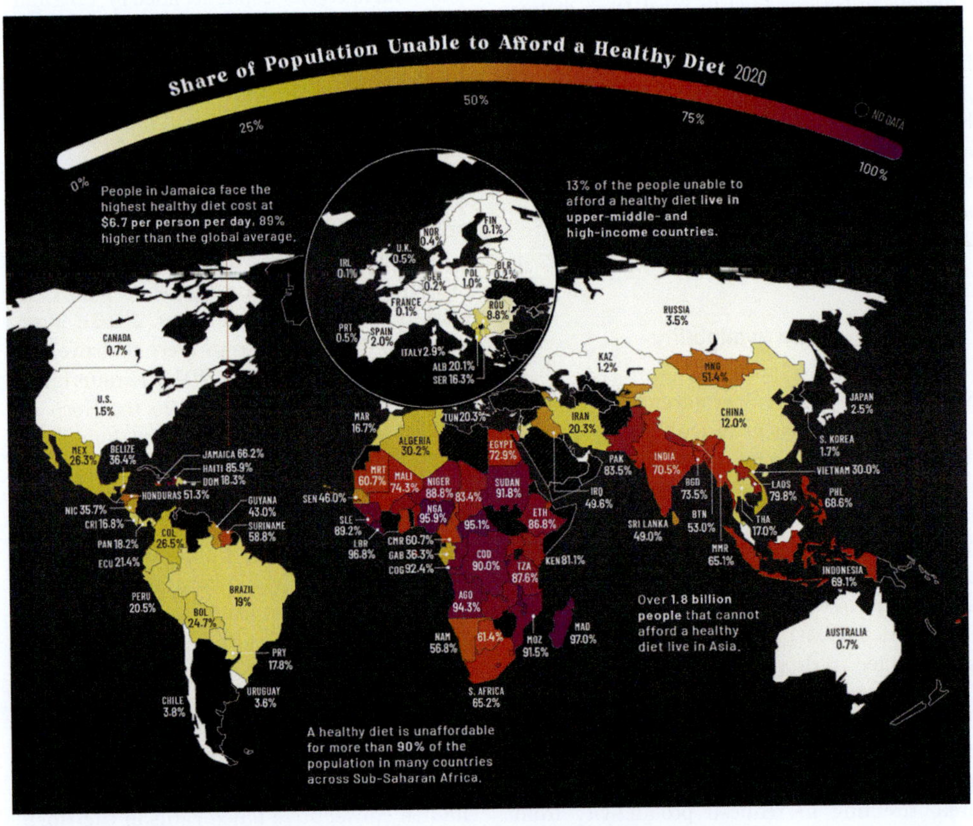


**Fig. 8.6.** Incidence of inability to afford a healthy diet. (From Bhutada, 2022)

extreme weather events such as droughts (Horn of Africa and Sahel), floods (West and Central Africa), cyclones (southern Africa) have especially high population shares unable to afford healthy diets.

### Food utilization

Child undernutrition is a powerful indicator of the nutritional quality of food in households and the bioavailability of nutrients in those foods, two vital aspects of food. Across Africa, 59 million children under age 5 are stunted and 11 million are wasted. In nine countries in the Sahel (Burkina Faso, northern Cameroon, Chad, the Gambia, Mauritania, Mali, Niger, northern Nigeria, and Senegal), health facility admissions for severe wasting in 2022 increased by 31%, equivalent to 1.9 million children, over 2021 and the previous three-year average. In the Horn of Africa, more than 1.7 million children in Ethiopia, Kenya, and Somalia require urgent treatment for severe acute malnutrition. Every second child who dies in Africa dies because of hunger. Hunger is responsible for 45% of all childhood mortality in Africa (Filipenco, 2022).

### Food system stability

The stability of Africa's food system is defined chiefly by exposure to climate and economic shocks. African food systems are highly exposed to negative economic impacts of CC due to low capacity for combating the effects of CC and high exposure to extreme weather events, notably because of the large number of Africans with livelihoods directly dependent on agriculture.

The pronounced vulnerability to CC of Africa's food production, quality, and distribution reflects a lack of resilience. Reliance on rainfed crop production prevails across Africa. Less than 1% of arable land is equipped for irrigation, leaving most farming and livestock areas prone to volatile rainfall, rising temperatures, droughts, and price volatility. Fig. 8.7 illustrates the case of Somalia in the Horn of Africa.

Drainage systems for flood protection are also lacking. Poor storage capacity of households and warehouses results in significant food loss, averaging 9% a year in Africa. This reflects both a lack of physical storage structures and access to electricity (less than half the population has access) in areas where temperature control is critical, especially for perishables. Food distribution and supplies of agricultural inputs

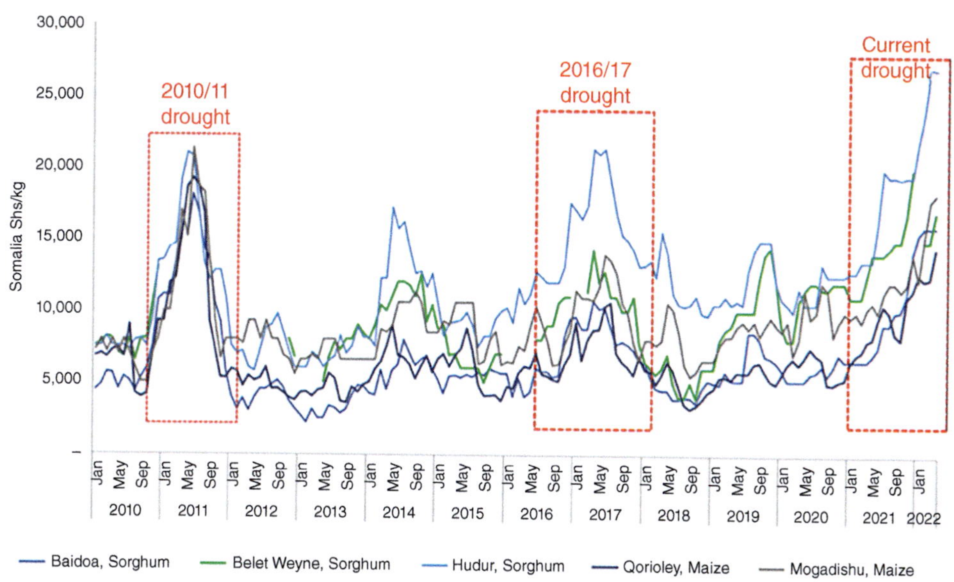


**Fig. 8.7.** Sorghum and maize prices in Somalia, 2010–2022. (From WMO, 2022)

(e.g. fertilizers, equipment) are hampered by poor quality and coverage of transport infrastructure. Roads, bridges, train tracks, and ports are easily damaged or destroyed by severe weather. This hinders not only trade between countries within SSA but also food imports from other parts of the world. Conversely, reliance on imported food is susceptible to weather events elsewhere in the world, derailing transit or damaging agricultural production (Baptista *et al.*, 2022).

In 2021 and 2022, at least 28 million people became acutely food insecure in Africa due to successive shocks that raised food prices (Fig. 8.8) and depressed incomes, especially for the most vulnerable (Baptista *et al.*, 2022). The cyclicality of food inflation reflects transitory mismatches between food market supply and demand. Negative food production shocks due to adverse climate conditions that depress output and increase input costs, and transport bottlenecks, can severely constrain supply and push food prices up (Baptista *et al.*, 2022).

## Food insecurity hotspots

In 2022, FAO and WFP identified 24 countries as hunger hotspots (FAO and WFP, 2022). Sixteen of those countries were in Africa, and all of them had significant dryland areas (Fig. 8.8). Africans account for 62% of food insecure people in hotspot nations. Food insecurity is especially prevalent in the Horn of Africa and the Sahel. Food systems in those areas depend heavily on precipitation and optimal temperatures. Agricultural productivity is reduced by CC, resulting in the loss of income, assets, and purchasing power, driving up malnutrition and leading to food insecurity. Organized violence and conflict remain the primary drivers of acute hunger worldwide, but weather extremes such as heavy rains, tropical storms, hurricanes, floods, droughts, and increased climate variability are critical drivers in Africa's drylands.

Chronic poverty and vulnerability, low and erratic rainfall, uneven access to feed in depleted

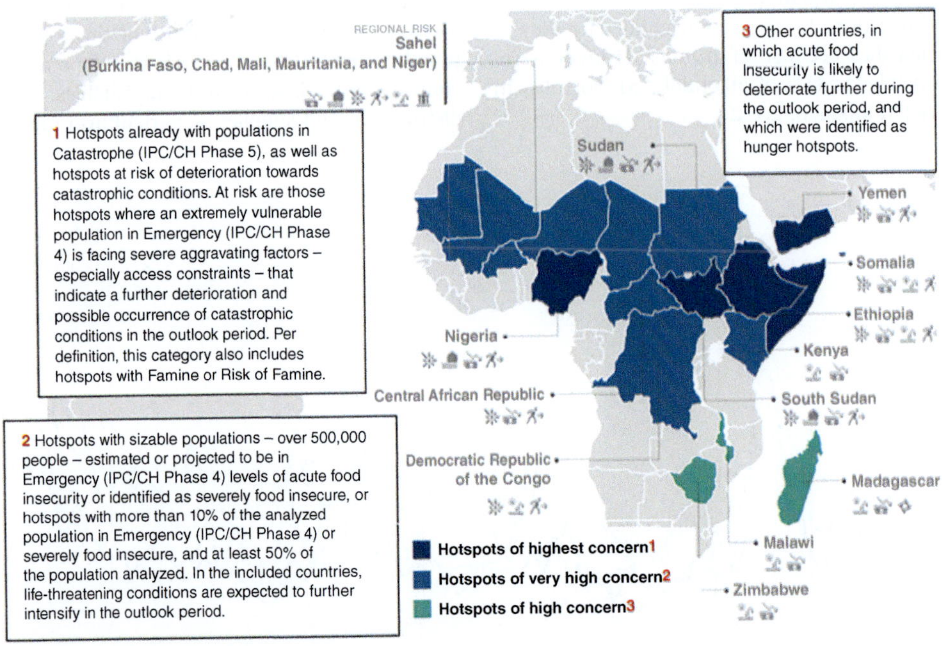


**Fig. 8.8.** Food insecurity in hotspot countries in the Sahel and Horn of Africa (WFP and FAO, 2022). Hunger Hotspots; FAO WFP early warnings on acute food insecurity; October 2022 to January 2023 Outlook (FAO and WFP, 2022). Available at: https://docs.wfp.org/api/documents/WFP-0000142656/download/?_ga=2.32491966.1228271778.1665837952-1777148390.1664459656&_gac=1.60432863.1664877592.CjwKCAjws--ZBhAXEiwAv-RNL0zmWrOLrFqMqUX1p84euaalqq7cuuz-T2QzBxv-YoXDU5Zporb2IYhoCbGAQAvD_BwE (accessed 12 July 2024).

rangelands, flooding, pest attacks, and domestic food price inflation are combining to generate severe and widespread vulnerability to food and nutrition insecurity. Conflict and displacement are sharpening and deepening these forces. As a result, increasing numbers of people are facing severe food and nutrition insecurity. Food insecurity in the Sahel region (Burkina Faso, Chad, Mali, Mauritania, and Niger) has increased by 50% since 2021, and 18 million people in the region are acutely food insecure. The Horn of Africa has been experiencing its worst drought in recent years, affecting areas of Kenya, Ethiopia, and Somalia and rendering 24 million people acutely food insecure.

## Policy Implications

Facing the challenge of climate change in Africa will require actions by Africans as well as by the international community. Clearly, Africans are responsible for their own food security and need to carry out the necessary reforms and investments to assure food security for the present as well as future generations. But climate is a global challenge and Africa has only contributed some 3.9% of world carbon emissions and yet is the continent most severely impacted by CC. The costs of adaptation and mitigation to CC in Africa are huge – between $120 and $240 billion a year or 4–8% of Africa's GDP (Songwe *et al.*, 2022). It is hard to see how African countries can meet this cost while financing other development priorities like health, education, and infrastructure. Countries of the global north have been the main emitters of GHG, and there is a general agreement, formalized by different summit declarations, that they need to contribute to financing climate action in the global south, particularly in Africa.

## Priority Actions and Investments by African Countries

In most African contexts, greater food security signals and echoes rapid growth in the macro-economy, poverty reduction through rural economic growth, and stability of the food system (AGRA, 2022). Growth and poverty reduction are also closely intertwined with effective climate action in Africa's drylands. Rapid, resilient, and inclusive growth is key to CC adaptation and mitigation in those regions (World Bank, 2022a). The same is expected to hold for food security: rapid, resilient, and inclusive growth is necessary for food security in the drylands. But such overlap and complementarity between climate action and food security action are not automatic. Policies and investments to boost food security in dryland areas are equivalent to those that promote CC adaptation and mitigation only to the extent that they address and overcome structural vulnerabilities to climate shocks and hazards in those regions. And policies and investments to enhance CC adaptation and mitigation in drylands are contiguous with those that increase food security only insofar as they advance climate adaptation and the resilience of dryland food systems.

High-performing food systems are central to enhancing and sustaining food security (von Braun *et al.*, 2023). Such systems support the core functions of food production, transformation, and consumption efficiently and predictably. They provide adequate incentives and returns to food producers, processors, and distributors. And they deliver sufficient, safe, and nutritious food to well-informed consumers with minimal delay and spoilage. The specter of millions of chronically and acutely hungry people in Africa's drylands suggests that dryland food systems are not functioning in ways that meet the food security needs of a broad cross-section of society. They are also not contributing adequately to growth and poverty reduction in these vast regions.

Climate hazards are first-order determinants of food system performance in dryland areas (World Bank, 2022b). The central policy and investment challenge for boosting food security in these regions is, thus, to make dryland food systems more resilient and better adapted to climate change. Three sectors are fundamental to climate adaptation, mitigation, and resilience in drylands: agriculture (cropping, herding, and fishing); water; and the environment (including forests, savannahs, and shrubby woodlands). Trends and conditions in these three sectors circumscribe food availability and usage, and food system stability in Africa's drylands.

## Agriculture

Rainfall is increasingly erratic and poorly distributed, with extreme droughts and floods becoming more widespread and frequent. Productivity in major cropping, livestock, and fishing systems is stagnant or falling due to lack of climate-resilient and climate-smart technologies and management practices, high pest-related production losses, and declining agrobiodiversity and loss of local varieties and breeds. Large- and small-scale irrigation potential is underexploited. Food production deficits are leading to low stocks, high demand, and high prices. Agricultural research and extension systems are poorly resourced. Trained veterinary and fishery personnel, materials, and equipment are lacking, leading to poor animal health and overexploitation of fish habitats. Rangelands and watering points are increasingly depleted, sharpening the competition between pastoralists and farmers for land and water in remote areas. Farmer organizations are weak. Women have unequal access to productive resources and technologies. Within and across borders, trade is highly informal and opaque. Market and value-chain infrastructure, organization, and information capacities are weak, leading to limited private sector investment.

The agriculture sector must contribute to reduced GHG emissions while making major sustained increases in production through expanded sustainable land management, increased availability and adoption of climate-resilient production technologies and practices in cropping, livestock, and fishing systems, expanded access to finance and financial risk management services, and effective preservation of agrobiodiversity. Given the prominence of livestock in rural livelihoods, effectively stabilized and expanded availability of and access to sustainable livestock feed and water resources and improved livestock health are vital. Overfishing and degradation of fish habitats must cease. Increased access to markets based on improved infrastructure and market information systems is critical to value-chain development for expanded livelihoods. Efficient and effective research and extension systems are critical, as are effective disaster early-warning, preparedness, and response systems.

Priority actions and investments in the sector include:

- enhancing the financial, technical, and organizational capacities of national agricultural research and extension systems to develop and deliver energy-efficient climate-resilient cropping, livestock rearing, fishing, and forestry technologies and practices, emphasizing tolerance to heat, drought, and flooding;
- strengthening regional and international agricultural research and development organizations and consortiums (such as the Consultative Group for International Agricultural Research (CGIAR)) to gather evidence and support the adoption of climate-smart agricultural policies, technologies, practices, and services, taking into account nationally determined contribution (NDC) targets and associated national adaptation plans;
- strengthening the structure and functioning of national and regional farmer and trader organizations to increase their engagement in the development and dissemination of energy-efficient, climate-resilient agricultural production, processing, and trading technologies and practices;
- boosting private sector investment in the development and dissemination of energy-efficient, climate-resilient agricultural production, processing, and trading technologies and practices, including innovative platforms and partnerships for meeting NDC targets, along with financing, technology, and policy catalyzation for prioritized long-term, low-carbon, and resilient investment projects; and
- expanding agricultural market and value-chain infrastructure and services, prioritizing climate-smart common assets for value addition (such as milking equipment; cold storage for perishable crops; cooling centers for livestock; weighbridges; paddocks; veterinary, sanitary and phytosanitary outposts; and transport services), and improved market information systems.

## Water

In rural areas, agriculture, pastoralism, and other economic activities remain reliant on highly vulnerable water-dependent ecosystems. Surface and groundwater sources are largely uncontrolled and their potential mostly unseized. Declining pastoral lands and water shortages are leading to increased migration by pastoralists to agricultural areas. In burgeoning urban areas, there is a widening gap between water demand and supply, alongside a lack of urban-planning mechanisms to meet growing water demands. Water utilities are characterized by low throughput, low efficiency, lack of investment, weak institutional capacity, and poor maintenance. Access to water-supply and sanitation services is therefore low and unequal, with strong negative impacts on women and with remote rural areas significantly deprived. Water treatment infrastructure and access to basic drinking water services are inadequate. Aquifers and rivers are increasingly polluted by uncontrolled discharge of untreated wastewater and fecal sludge, with strong impacts on urban slums. Sectoral strategies are poorly harmonized and implemented. Major water sources straddle borders, but limited attention is given to water in regional development strategies.

Per capita water availability must increase, water insecurity must drop, and sustainable water sources must be diversified and expanded. Flood prevention and erosion protection must be enhanced, including through nature-based solutions such as forest and wetland restoration, to regulate flows and reduce floods. Transboundary water resources and basins must be managed effectively. In rural areas, the potential for large- and small-scale irrigation must be fully exploited, including improved water harvesting and modernized and refurbished irrigation and drainage infrastructure. Water-use efficiency must improve. In urban areas, capacities for low-energy water treatment, water supply and sanitation, and wastewater treatment must expand, with access for women prioritized. Contamination of surface and groundwater resources and pollution of aquifers and rivers must be curtailed.

Priority actions and investments in the sector include:

- promoting the transfer and adaptation of scalable and resilient energy-efficient ground- and surface-water harvesting, storage, and management technologies and practices, emphasizing flood prevention and control and water-use efficiency in crop, livestock, and fish production;
- expanding irrigated cropland through energy-efficient, small-scale, and large-scale approaches, including sprinkler and drip-irrigation systems, solar tube wells, cemented wells, boreholes for pastoral pumping stations, watering ponds, and establishment of effective management structures;
- preserving, regenerating, and expanding rangelands and forage, water resources, watering points, and veterinary services, emphasizing drought-tolerance, water infiltration, climate-smart production techniques and technologies, replanting of native grasses (e.g. borgou) in flood-prone lands, community-based interventions to protect natural pastoral areas against wildfires (fire strips) and recovering degraded land, and strengthening the capacities of women and youth producer groups;
- expanding, rehabilitating, and maintaining infrastructure for and access to resilient energy-efficient ground- and surface-water management technologies and practices, emphasizing conservation and efficiency, low-cost and low-maintenance irrigation and drainage technologies and services, flood prevention and control, water supply and sanitation services, and wastewater management.

## Environment

There is a rapid degradation of natural capital due to a lack of capacity for the design and implementation of environmental conservation measures at landscape scale and a low understanding and limited capacity to apply nature-based solutions to the deterioration of landscapes, forests, and watersheds. Despite their pivotal role in natural resource management, women face major impediments to accessing and controlling natural capital. Financing mechanisms for adaptation, resilience, and disaster risk management are uneven across the region. Environmental and social impact assessment regulations covering climate and disaster risk and climate resilience are similarly patchy, leading to poorly functioning or inoperative regulatory systems and regulatory vacuums regarding climate and disaster risk assessment and climate resilience. Financial and technical capacities and databases in regulatory bodies are thin, especially for climate analysis, including early-warning systems for climate risks and projections at national and subnational levels.

Reduced GHG emissions must be accompanied by sustainable livelihoods based on effective exploitation of natural endowments of space, solar energy, minerals, water resources, biodiversity, and cultural heritage. The deterioration of the natural resource base and the loss and fragmentation of forests, savannah, and woodland areas must be arrested. Nature-based management and restoration of landscapes, forests, and watersheds must expand. Green technologies, green businesses, and clean energy infrastructure must increase. An effective and reliable centralized environmental information system is vital.

Priority actions and investments in the sector include:

- supporting the development and diffusion of technologies for and approaches to enhancing conservation (reducing degradation and overexploitation of landscapes, forests, and watersheds), and strengthening associated livelihoods;
- developing and implementing regulations that improve incentives for the design and application of nature-based solutions to the deterioration of landscapes, forests, and watersheds, focusing on reducing trade-offs between agricultural productivity growth and livelihood promotion, on the one hand, and protection and conservation of biodiversity, landscapes, forests, and watersheds on the other; and
- supporting sustainable development and agricultural uses of renewable energy assets such as solar, hydro, and geothermal energy sources.

### Cross-sectoral actions

The multisectoral imperative to produce resilient, climate-adapted food systems is expressed partly through existing policies and investments and partly through new ones. One cross-cutting theme is therefore the need to fill policy gaps and promote country-driven capacity development. Energy fuels and regulates all economic activity. Affordable access to safe, efficient, and sustainable energy thus lies at the core of climate-adapted food systems. A second theme is energy efficiency and access. Physical and organizational barriers to value addition and conservation in dryland food systems can be profound. A third theme is the establishment of priority infrastructure and services to expand entry and participation across food systems. Food security has impacts with strong spatial and temporal expressions, so a fourth theme is regionality and collective action across national borders, and regional specificity between rural and urban areas within national borders (see Box 8.1). Digital technologies have a crucial role to play in addressing the limitations of existing extension services and meeting the evolving needs of farmers. These technologies offer valuable solutions to enhance labor and land productivity and provide real-time monitoring of climate risks, enabling proactive measures for building resilience. Seizing opportunities offered by digital technologies thus represents a fifth theme.

### Implementation considerations

The effective implementation of these proposed policies and investments rests on pragmatic prioritization and focus based on the principles of proportionality (directing resources toward the greatest needs) and feasibility (aligning goals with available resources and capacities). The challenge is that in all countries and across all sectors, needs are immense and surging, and resources and capacities are limited and shrinking. The delivery model for climate action must recognize and overcome huge technical, organizational, institutional, and policy gaps in overburdened public sectors. These gaps must be systematically identified, assessed, and filled. Given the urgent need to fill these gaps, in some especially complex and novel areas (e.g. development of renewable energy assets or nature-based integrated landscape management), governments might deliberately delegate oversight and leadership roles to capable stakeholders, while retaining accountability through institutional and information management innovations. Enhanced collective action through a pan-African initiative would be appropriate (Box 8.1).

It is crucial that governments and the private sector have dedicated individuals with the right knowledge and skills. Currently, only 2% of Africa's students specialize in agriculture, even

**Box 8.1.** Towards a pan-African climate adaptation and food security initiative. (From Ghanem, 2022)

Recognizing the limitations of expecting each of the 54 African nations to achieve food security independently, collaborative action is appropriate. Africa, with its population of 1.4 billion people and abundant natural resources, possesses the potential to ensure food security for all. A pan-African food security initiative is proposed, built on five pillars:

1. Empowering the Secretariat of the African Continental Free Trade Agreement (AfCFTA) with a specific mandate to facilitate the smooth flow of food products across African borders. By promoting trade and fostering the exchange of agricultural goods among African nations, this measure would enhance food security.
2. Enhancing collaboration among the African Union (AU) Commission for Agriculture and institutions such as the African Development Bank (AfDB), the Food and Agriculture Organization of the United Nations (FAO), and the World Bank to identify and support regional infrastructure projects. With a particular focus on irrigation, these projects would contribute to agricultural development and the establishment of prosperous agricultural regions across Africa.
3. Enhancing collaboration between the private sector branches of the AfDB and the World Bank and African fertilizer producers, including OCP in Morocco and Dangote in Nigeria. By establishing additional fertilizer factories throughout the continent, this partnership aims to improve fertilizer availability and enhance agricultural yields.
4. Capitalizing on the expertise of the Consultative Group on International Agricultural Research (CGIAR) to establish an African Council of Agricultural Research (ACAR). This council would play a crucial role as a coordination body for research and development (R&D) endeavors across the continent. It is worth noting that the framework for such a council already exists through the Forum for Agricultural Research in Africa (FARA). By utilizing the CGIAR's research and initiatives in Africa, Africa can effectively leverage the presence of four CGIAR centers headquartered on the continent: the International Institute of Tropical Agriculture (located in Nigeria); Africa Rice (located in Cote d'Ivoire); the World Agroforestry Center; and the International Livestock Research Institute (both located in Kenya). These centers contribute valuable knowledge and resources to support agricultural research and development efforts in Africa.
5. Building on the efforts of the African Export-Import Bank (AFREXIMBANK) to secure adequate funding for food imports during crises in vulnerable countries. By expanding the scope of AFREXIMBANK's work, the availability of necessary financial resources to support countries experiencing food shortages in times of need can be assured.

By implementing these five pillars of the pan-African food security initiative, Africa can make substantial progress towards achieving food security for its entire population. Collective action and collaboration across various sectors will play a pivotal role in addressing the intricate challenges of food production, distribution, and availability on the continent.

though agriculture contributes 32% of the continent's GDP. Although there is low enrollment in agricultural programs, the demand and opportunity within the food system remain high, and this trend is expected to continue in the future. In fact, farming alone accounts for 60% of total employment in sub-Saharan Africa, and when considering jobs in the entire food system, the total number of employment opportunities is significantly higher. Furthermore, several countries (e.g. Ethiopia and Zambia) are projected to generate more jobs in the food system than in the rest of their economies between 2010 and 2025. This highlights the need for Africa to cultivate a larger pool of homegrown food producers, food processors, food system entrepreneurs, agronomists, and agricultural scientists. It is essential for countries to invest more in developing a skilled workforce that can effectively address the challenges of Africa's food system.

The establishment of the African Council of Agricultural Research (ACAR) should be facilitated through the progressive evolution of the Forum for Agricultural Research in Africa (FARA). The remarkable strides achieved by FARA lay a solid foundation for the potential establishment of ACAR.

FARA's inception can be traced back to a compelling motivation – the imperative to address the fragmentation of over 50 national research systems in Africa, each grappling with distinct limitations. Recognizing the advantages of cooperation, sub-regional entities endeavored to harness economies of scale. On a continental scale, FARA emerged as a central entity entrusted with coordination, networking, and advocacy within the domain of agricultural research. It played a crucial role in translating continental policy frameworks into actionable agricultural research programs while concurrently representing Africa in the Global Forum for Agricultural Research (GFAR).

Despite undergoing various phases, FARA's fundamental objectives and functions have remained unwavering. In its current phase (2019–2023), FARA's pivotal role as a hub for all stakeholders involved in agricultural research and innovation has gained prominence. Collaborative endeavors are concentrated on enhancing knowledge access, fostering innovation partnerships, championing proven policies, and amplifying innovation capabilities. FARA's strategic vision, mission, and goals align harmoniously with its dedication to harnessing science and technology to uplift livelihoods, fortify food security, and contribute to the aspirations of the Malabo Declaration and the Comprehensive Africa Agriculture Development Programme (CAADP).

The establishment of ACAR carries the potential to substantially invigorate agricultural development and food security throughout Africa. A robust funding model that effectively leverages both public and private support is imperative. Drawing inspiration from institutions like the Indian Agricultural Research Institute (ICAR) and the Brazilian Agricultural Research Corporation (EMBRAPA), crafting a comprehensive funding strategy that amalgamates diverse funding sources is crucial to establish a robust financial base for ACAR's initiatives and undertakings.

The success of ACAR hinges on its capacity to institute a multifaceted and sustainable funding model encompassing government allocations, international grants, private sector partnerships, donor collaborations, fee-based services, and more. This approach not only assures financial stability but also resonates with ACAR's core mission of collaboration. By ensuring transparency through regular reporting on research impact and financial transparency, ACAR can bolster the confidence of donors and stakeholders, solidifying its role as a catalyst for transformative agricultural research and development across the African continent. Several imperatives emerge and are detailed below.

- **Government funding.** An essential pillar of ACAR's funding framework rests upon contributions from African governments. By allocating a portion of their budgets to agricultural research and development through ACAR, governments could showcase their commitment to nurturing agricultural innovation. This funding stream would provide stability and serve as the bedrock of ACAR's financial structure.
- **International grants.** Pursuing international grants from institutions like the World Bank, United Nations agencies, and the African Development Bank constitutes another pivotal facet of ACAR's funding strategy. These grants could be earmarked for specific research projects or aligned with ACAR's broader goals. Additionally, collaborations with donor organizations that share a vested interest in advancing Africa's agricultural sector could infuse substantial financial support.
- **Private-sector collaboration.** Engaging the private sector through partnerships with agricultural companies, agri-businesses, and technology firms offers a dynamic avenue for funding. Private sector entities could contribute funds and expertise in exchange for access to ACA's research findings and innovative solutions. This collaborative partnership would not only bolster ACAR's financial resources but also ensure the practical application of its research outcomes.
- **Competitive research grants, contracts, and fee-based services.** ACAR could actively compete for research grants and contracts from both public and private research funding bodies. These competitive funding opportunities could provide supplementary financial support while recognizing ACAR's pivotal role in advancing agricultural knowledge. Furthermore, ACAR could offer fee-based services such as training, consultation, and technical support to agricultural enterprises, generating revenue while fulfilling a critical knowledge-sharing function.

- **Membership contributions and intellectual property ventures.** To foster a sense of ownership and collective commitment, ACAR could establish a membership structure encompassing research institutions, universities, and agricultural organizations across Africa. Membership fees could contribute to ACAR's funding pool and reflect stakeholders' dedication to its success. Furthermore, ACAR's creation of intellectual property, including novel crop varieties or innovative technologies, could generate revenue through licensing agreements.
- **Exploration of innovative funding mechanisms and public engagement.** Exploring innovative funding mechanisms like impact investing, green bonds, and social impact bonds could attract investment from those seeking both financial returns and social impact. ACAR could also engage the public through crowdfunding campaigns and individual donations from individuals passionate about supporting African agricultural development.
- **Drawing insights from Fontagro.** The Fontagro model, a mechanism that supports agricultural research and development projects in Latin America, provides valuable insights that could be adapted to the context of ACAR. Although adjustments may be needed to align with ACAR's specific goals and challenges, certain principles from the Fontagro model can undoubtedly enhance ACAR's funding strategy. Situated within the Inter-American Development Bank (IaDB), Fontagro operates as a multi-donor fund, pooling financial contributions from diverse sources to underpin agricultural projects. This diversified funding approach mitigates risks and ensures a steady stream of resources for projects. Fontagro bestows grants upon projects, allowing research institutions to implement initiatives without the burden of repayment. This approach fosters innovation and encourages research collaboration. Projects funded by Fontagro undergo a rigorous selection process that emphasizes feasibility, potential impact, and alignment with agricultural development goals. This ensures that resources are directed toward projects with the highest potential for success and impact. Fontagro places significant emphasis on monitoring and evaluating project outcomes, facilitating the assessment of project effectiveness and enabling continuous learning and improvement. The AfDB could play a pivotal role analogous to the IaDB. While directly translating the Fontagro model to an AfDB-led model might necessitate adjustments to suit the African context, the fundamental principles of diversified funding, grant-based support, project selection, and monitoring and evaluation can serve as a valuable framework. The AfDB could tailor these principles to establish a funding mechanism that addresses the unique challenges and opportunities within the African agricultural landscape, ultimately contributing to heightened food security and agricultural development.
- **Engaging with CGIAR.** The potential for collaboration with CGIAR, drawing inspiration from ICAR's cooperative endeavors, offers another avenue for enriching ACAR's research impact. The evolution of ACAR presents a promising opportunity for collaboration with CGIAR, fostering mutual enhancement of research impacts for both organizations. The experiences gleaned from the cooperative interactions and operational frameworks of the Indian ICAR and CGIAR offer invaluable insights for shaping ACAR. Similar to ICAR, ACAR could derive significant benefits from accentuating research collaboration, capacity development, knowledge exchange, policy advocacy, public-private partnerships, and inclusivity of gender and social considerations to advance agricultural progress and ensure food security across Africa. ACAR may consider forging affiliations with both local and global agricultural research institutions, drawing inspiration from CGIAR's collaborative research model. This strategic approach will empower ACAR to tap into a wider reservoir of expertise and resources, effectively addressing the distinctive challenges confronted by African agriculture.

In addition to investing in skill development, new talent must be attracted to the food system. Convincing Africa's future generations that the agriculture and food sector offers promising career opportunities is vital for sustainable growth. These challenges require ambitious action, and it is encouraging that Africa's leaders have committed themselves to adapting the continent's food system to climate change. The recent Africa Food Security Leadership Dialogue on Climate Adaptation of Africa's Agriculture, attended by 34 African countries and 42 organizations, marked a promising start to this important conversation.

There is a pressing need to increase long-term investments in research and development (R&D) within Africa. Currently, R&D expenditures per farmer in Africa are significantly lower compared to developed countries, with a difference spanning two orders of magnitude. The required increase in R&D investments is substantial. Instead of each of the 54 African countries attempting to establish individual world-class R&D centers, it may be more sensible to explore alternative approaches. While context-specific R&D is necessary due to variations in soils, water, and climatic conditions, there are also advantages to be gained from economies of scale in research. Africa can draw valuable lessons from India's agricultural R&D experience. With a population larger than Africa's and considerable heterogeneity among its 28 states, India has maintained a sustained R&D effort since the Green Revolution. This commitment has resulted in the development of multiple generations of new crop varieties, consistent yield increases, and improvements in nutritional outcomes. Notably, India has been producing 20–40 new varieties of rice annually since 1970. The Indian Council of Agricultural Research (ICAR), established in 1929 as an autonomous body responsible for coordinating agricultural education and research across the country, spearheads this work. ICAR oversees 14 national research centers. Establishing an African Council of Agricultural Research, tasked with coordinating R&D activities across the continent, could be a viable solution (see Box 8.1). The aim would be to elevate African yields to global levels and to support the agricultural sector's adaptation to CC. An initial step would involve creating five R&D centers of excellence in southern, Central, East, West, and North Africa. Financing for the council could be sourced from contributions by African Union member states and through partnerships with external funders.

The power of the digital revolution must be seized and leveraged. Digital tools enable better food availability, accessibility, and safety by effectively monitoring and mitigating food hazards. E-commerce platforms are instrumental in integrating smallholder farmers into value chains, eliminating transaction costs associated with locating demand, determining prices, and improving service-delivery efficiency. Organizations like CGIAR and private sector players are championing these technologies, offering innovative solutions that can revolutionize the agricultural sector. By harnessing the power of Big Data, Africa can make informed decisions at the farm level, optimizing water management practices and fertilizer usage. However limited internet access in rural areas poses a significant barrier to technology adoption by farmers. Governments should prioritize the expansion of internet infrastructure and connectivity in rural regions, enabling farmers to leverage digital tools, access information, and participate in online marketplaces. This would provide them with valuable resources and opportunities to enhance their agricultural practices and increase their income.

## Actions by the International Community

Africa needs the support of the international community to finance the necessary investments to deal with CC and ensure food security. The IMF estimates that sub-Saharan Africa needs about $50 billion a year for adaptation and an additional $190 billion a year for mitigation. So far, this support has been hugely inadequate. Countries of the global north are facing increasing demands on their tight budget. The current multilateral financial system can do more if appropriate reforms are put in place, but it would not be able to provide the trillions that the global south needs for climate action (IMF, 2023).

In 2020, concessional climate flows to Africa, from all sources, were less than $16 billion; nowhere near the more than $240 billion required. At the same time, total development assistance to Africa has been on a declining trend, falling from more than 6% of recipient country GDP in the early 1990s to about 2.5% currently. To make things worse, Africa is being squeezed by a high debt overhang, rising interest rates, and currency depreciations. The continent is currently paying about $70 billion of debt service every year and that figure is expected to rise. As a result, most African countries have very little fiscal space to finance necessary social sector and infrastructure investment, not to mention climate action (IMF, 2023).

The vast majority of climate expenditures – especially on mitigation – will need to be carried out by the private sector. The challenge, therefore, is to attract international and domestic investors to Africa, and especially to green sectors. This will require policy reforms to make investment in those sectors, especially green energy, attractive. It will also require support in project formulation and identification, financing, and risk mitigation.

In this paper we propose two solutions to increasing climate financing for Africa while protecting general development financing that goes mainly to health, education, and infrastructure. The first proposal is to create a new institution specialized in climate financing, an International Green Bank, that could be completely independent or could be part of the World Bank Group but with independent financing and separate governance. The aim here is to support private sector green investments. The second proposal is to agree to reduce Africa's foreign debt provided African countries use the savings for climate action, a debt-for-climate swap. The aim is to provide more resources for public climate investments, which will be mostly in adaptation.

## The Green Bank

There is a growing realization that, under the current international financial framework, the global north is unable to provide low- and middle-income countries with the $2 trillion a year they need to meet their climate finance needs. Given the way the bank is structured, the World Bank could not support the transition on its own. Firstly, reaching the needed level of climate financing would require a substantial capital increase. This would be a huge burden on World Bank shareholders, who are all fiscally stretched sovereign states. Secondly, the bank is not able to mobilize sufficient private-sector climate funding, and governments are very unlikely to be able to come up with the trillions required. Finally, the governance structure of the bank, which only includes sovereign states and does not provide sufficient voice to countries of the south, is not conducive to increasing climate financing.

That is why some voices are calling for the creation of a new institution –that conceivably could be part of the World Bank Group – whose only mission would be to combat climate change. It would be a repository of knowledge on the green and net-zero transition. It would develop, finance, and support the implementation of private sector projects to mitigate and adapt to CC. Furthermore, it would be a public-private partnership, a place for all actors to come together on an equal footing to forward climate action – from the private sector and civil societies to donor and non-donor countries around the world. Unlike current international institutions, all partners would have a strong voice in the decision-making bodies of the institution.

The Green Bank would use four instruments to achieve these goals: knowledge products that help develop effective green projects, support advocacy, and provide policy advice; direct equity contributions to private companies investing in green projects; loans to private sector projects; and guarantees against sovereign risk for green investments (Ghanem, 2023). Africa would aspire to more than $100 billion per year of green private investments facilitated by the Green Bank.

## Debt-for-climate swaps

According to UNDP, 24 of the 54 African countries are in debt distress, including the continent's two largest economies, Nigeria and Egypt (UNDP, 2022). Four countries – Chad, Ethiopia, Ghana, and Zambia – have applied for debt reduction under the G20 Common Framework. But the Common Framework has turned out to be very slow and difficult to implement, mainly due to the difficulty of coordinating Paris and non-Paris club members and private bondholders.

Of the nearly $70 billion paid by African countries in debt service in 2021, about $30 billion was due to official lenders (divided more or less evenly between multilaterals and bilaterals), and about $40 billion to private creditors. Using part or all those funds for climate action in Africa could be a win-win – a win for the planet and a win for Africa. It would be an excellent contribution to an important global public good. One option would be for the official creditors to forgive all or part of their debt, while asking the private creditors to transform part of their debt into equity in green projects.

A system of debt-for-climate swap could be led and coordinated by the International Monetary Fund (IMF) and the World Bank. It could be inspired by the Highly Indebted Poor Countries (HIPCS) initiative of the late 20th century. The idea behind HIPCS was that the savings from debt reduction were to be used for fighting poverty. Countries were asked to prepare Poverty Reduction Strategy Papers (PRSPs) to explain how the money will be used, and the World Bank and IMF supervised implementation. A debt-for-climate swap could work more or less the same way. Countries have already produced their nationally determined contributions (NDCs). Those could then be discussed with the creditors and the two Bretton Woods Institutions and agreement reached on which NDC activities will be financed through the savings on debt service. The Bretton Woods institutions would then monitor implementation exactly as in the case of HIPCS.

## Urgency of Translating Science into Action

As highlighted earlier, science holds immense potential for providing sustainable solutions to ensure food security. These solutions encompass innovative approaches that enhance adaptability to CC, scientific management of productive resources like land, soil, and water, as well as advancements in food storage and transportation to minimize wastage. However, harnessing the power of scientific advancements requires the translation of these solutions into practical packages that can be effectively disseminated and embraced by farmers on a large scale, both at the individual farm and broader landscape levels. Achieving this aspiration necessitates robust connections between international, regional, and national science organizations, farmers, and extension systems. It is crucial for these solutions to be collaboratively crafted by researchers and farmers, thereby effectively addressing local resilience challenges through a demand-driven and knowledge-intensive approach (Ehui *et al.*, 2020).

In regions where climate-smart agriculture is being practiced, farmers are reaping tangible benefits in the form of heightened food security and enhanced resilience. For instance, in Rwanda, the Land Husbandry, Water Harvesting, and Hillside Irrigation project has successfully managed erosion, amplified yields on existing land, and provided superior safeguards against droughts. This initiative yielded an impressive 2.6-fold increase in maize yields between 2009 and 2018, accompanied by even more substantial enhancements for beans, wheat, and potatoes (Ehui *et al.*, 2020).

Similarly, in Senegal, the West Africa Agricultural Productivity Program, funded by the World Bank, has introduced new high-yielding, early-maturing, drought-resistant crop varieties like sorghum, millet, groundnuts, and cowpeas. These adapted varieties are being widely disseminated to farmers, contributing to an exceptional average yield increase of 30%, even in the face of reduced and erratic rainfall. Even during challenging periods such as 2014, marked by delayed rains and half the usual precipitation, farmers cultivating improved sorghum and millet varieties experienced noteworthy yield increments (Ehui *et al.*, 2020).

Acknowledging these imperatives, the partnership between the AfDB and CGIAR has led to the creation of the Technologies for African Agricultural Transformation (TAAT) program. This initiative aims to elevate agricultural productivity and ensure food security across Africa by promoting the adoption of established agricultural technologies. Scaling up this program is critical.

The TAAT model involves identifying fitting technologies, adapting them to local contexts, and disseminating them widely through collaborative efforts with governments, research institutions, and private sector stakeholders. This model focuses particularly on pivotal crops and livestock that possess the potential to drive transformative progress in agriculture across the continent. It addresses challenges like access to inputs, mechanization, and post-harvest losses to facilitate enduring growth within the agricultural sector.

The AfDB has committed substantial financial resources to support the implementation and initiatives of the TAAT program. Furthermore, TAAT engages in partnerships with a diverse array of stakeholders, spanning governments, international organizations, research institutions, and private enterprises, amplifying funding and resources for its undertakings. Given the impactful outcomes achieved, there is a compelling case for scaling up this model (see Box 8.2).

Through the World Bank-funded project known as Accelerating Impacts of CGIAR Climate Research for Africa (AICCRA), top-tier CGIAR scientists are leading an effort to facilitate substantial capacity building endeavors across Africa. By leveraging cutting-edge scientific expertise and innovation, these endeavors strengthen the capacity of regional and national stakeholders to proliferate climate-smart agriculture (CSA) innovations and climate-related services throughout the African continent.

Central to the project's objectives is the explicit support for farmers and vital stakeholders embedded within African value chains and food systems. This support focuses on the sustainable adoption of these innovations on a significant scale and rapid pace. The project employs a strategic 'science of scaling' approach, fostering collaborative capacity-building relationships between CGIAR centers across Africa and various regional and national entities. These encompass universities, extension agencies, civil society organizations, farmer associations, and private sector entities.

Several concrete instances highlight how AICCRA translates scientific insights in climate-related information services into tangible outcomes for farmers:

**1.** AICCRA spearheaded the development of the NextGen Agricultural Drought Monitoring and Warning System in Ethiopia. This innovative system empowers policymakers at the Ministry of Agriculture, Ethiopian Meteorological Institute, and National Disaster Risk Management Commission to effectively monitor agricultural regions for signs of water stress. Upon full implementation, the system is anticipated to benefit over 18 million smallholder farmers and livestock keepers.

**2.** In Ethiopia, AICCRA collaborated on creating the Climate Risk Management in Agricultural Extension (CRMAE) curriculum. This curriculum targets the training of 15,000 agricultural

**Box 8.2.** Transformative impacts: TAAT's success stories across Africa's agriculture.

The impacts of the TAAT program span various African nations, resulting in noteworthy achievements across several domains:

- **Cassava.** In Nigeria, TAAT introduced improved cassava varieties and technologies that led to a substantial surge in cassava yields. The adoption of the Pro-Vitamin A Cassava variety, for instance, yielded yield increases of up to 2.5 times compared to traditional varieties, significantly enhancing food security and income for local farming communities.
- **Maize.** In Kenya, TAAT promoted the adoption of drought-tolerant maize varieties alongside optimal agronomic practices. This translated into heightened maize yields, even during periods of limited rainfall. Farmers gained increased resilience against climate variations, leading to improved maize production.
- **Rice.** In Sierra Leone, TAAT facilitated the adoption of advanced rice-production technologies. Through the implementation of modern rice varieties and effective water management practices, farmers achieved greater yields per ha, thus bolstering food security and livelihoods.
- **Sorghum.** In Mali, TAAT's efforts in promoting improved sorghum varieties and agronomic techniques resulted in elevated sorghum production. This not only addressed food security concerns but also contributed to local socio-economic advancement.

These instances underscore the substantial impact of the TAAT program in reshaping agricultural landscapes and fostering sustainable development across the African continent. Scaling up TAAT's initiatives will undoubtedly play a pivotal role in translating science into impactful action.

extension workers annually through Agricultural Technical Vocational Education and Training colleges. Additionally, it provides yearly in-service training for 72,000 qualified extension workers. These CRMAE curricula are being integrated into the undergraduate programs of 24 Ethiopian universities, and a parallel program is launching in Senegal.

**3.** In Mali, the RiceAdvice app has proven instrumental in aiding 17,000 farmers (including 5000 women) in managing climate-induced risks. This app has led to an average increase of 0.9 tons per ha in rice yields and boosted profits by $320/ha. This is achieved through personalized recommendations on crop scheduling, fertilizer application, and weed management. The app's implementation has also contributed to a reduction in the use of chemical fertilizers in most cases.

**4.** Zambia is host to an accelerator program under AICCRA's purview, which supports a consortium of start-ups. These start-ups work towards scaling integrated packages of climate-smart agriculture practices and climate information services. Categorized into five thematic 'bundles', each group within the consortium receives $50,000 in funding for projects aimed at reaching 75,000 beneficiaries, respectively. Notably, one of these bundles focuses on gender equity and social inclusion, emphasizing best practices in these areas.

## Concluding Remarks

Climate change is a global problem caused mostly by more developed economies. Africa's contribution to CC has been negligible, yet the continent is the biggest victim of this phenomenon. Moreover, agriculture and food security are the areas most impacted by global warming and changing weather patterns. We already see the numbers of food insecurities rising in Africa; and, perhaps more tragic, we see increasing numbers of wasted and stunted African children.

Africans need to work together to implement climate actions and ensure food security for all. More investments are needed in agriculture, water, and the environment, as well as in agricultural research and education. The Continental Free Trade Agreement is an opportunity to increase regional trade in food and agricultural inputs, which would enhance food security for all. In addition, African countries could seize the initiative and develop a homegrown climate adaptation and food security initiative. The international community should help Africa by providing adequate financing for climate action. Two options could be explored: creating a Green Bank to finance private sector green projects; and debt-for-climate swaps to provide African governments with adequate fiscal space to finance climate investments.

## Endnote

[1] For instance, mean warming in southern African drylands is projected to be 3°C when global average temperature rises by 1.5°C, and 3.2–4°C if it rises by 2°C. In all regions, the effects of a rise of 2°C or more are significantly worse than those of a 1.5°C rise (IPCC, 2018).

## References

ACSS (2022) *Cyclones and More Frequent Storms Threaten Africa*. Africa Center for Strategic Studies. Available at: https://africacenter.org/spotlight/cyclones-more-frequent-storms-threaten-africa/ (accessed 1 July 2023).

Africa News (2022) Food crisis looms in Nigeria as floods destroy crops. Available at: https://www.africanews.com/2022/10/20/food-crisis-looms-in-nigeria-as-floods-destroy-crops/ (accessed 1 July 2023).

Africa Renewal (2022) UN agencies warn again of record hunger next year in West and Central Africa. Available at: https://www.un.org/africarenewal/magazine/november-december-2020/un-agencies-warn-again-record-hunger-next-year-west-and-central (accessed 1 July 2023).

AGNES (Africa Group of Negotiation Expert Support) (2020) Desertification and climate change in Africa: based on Chapter 3 of the IPCC Special Report on Climate Change, Desertification, Land Degradation, Sustainable Land Management, Food Security and Greenhouse Gas Fluxes in Terrestrial Ecosystems, 2019, 2020. Available at: https://agnesafrica.org/download/policy-brief-no-1-desertification-and-climate-change-in-africa/ (accessed 1 July 2023).

AGRA (2022) *Africa Agriculture Status Report 2022: Accelerating African Food Systems Transformation*. Available at: https://agra.org/wp-content/uploads/2022/09/AASR-2022.pdf (accessed 1 July 2023).

Alston, J., Pardey, P. and Rao, X. (2020) The magic of CGIAR research: harvesting $10 from every $1 invested. Supporters of Agricultural Research (SoAR) Foundation. DOI: 10.22004/ag.econ.337030

Baptista, D., Farid, M., Fayad, D., Kemoe, L., Lanci, L. *et al.* (2022) *Climate Change and Chronic Food Insecurity in Sub-Saharan Africa, 2022*. International Monetary Fund. Available at: https://www.imf.org/en/Publications/Departmental-Papers-Policy-Papers/Issues/2022/09/13/Climate-Change-and-Chronic-Food-Insecurity-in-Sub-Saharan-Africa-522211 (accessed 1 July 2023).

Bhutada, G. (2022) Mapped: the 3 billion people who can't afford a healthy diet. Available at: https://www.visualcapitalist.com/mapped-the-3-billion-people-cant-afford-a-healthy-diet/ (accessed 1 July 2023).

Carleton, E. (2022) *Climate Change in Africa: What Will It Mean for Agriculture and Food Security?* International Livestock Research Institute. Available at: https://www.ilri.org/news/climate-change-africa-what-will-it-mean-agriculture-and-food-security (accessed 1 July 2023).

Climate Center (2022) COP27: climate change exacerbated heavy rainfall and massive floods this year in West Africa – Study. Available at: https://www.climatecentre.org/9568/study-climate-change-exacerbated-heavy-rainfall-and-massive-floods-this-year-in-vulnerable-communities-of-west-africa/ (accessed 1 July 2023).

CRED (Centre for Research on the Epidemiology of Disasters) (2022) *The Interplay of Drought-flood Extreme Events in Africa Over the Last Twenty Years (2002–2021), EM-DAT: The CRED/OFDA International Disaster Database*. Available at: https://cred.be/sites/default/files/CredCrunch69.pdf (accessed 1 July 2023).

Dunne, D. (2022) Analysis: Africa's unreported extreme weather in 2022 and climate change. *Prevention Web Carbon Brief*. Available at: https://www.preventionweb.net/news/analysis-africas-unreported-extreme-weather-2022-and-climate-change (accessed 1 July 2023).

Eckstein, D., Künzel, V., Schäfer, L. and Winges, M. (2020) *Global Climate Risk Index 2020, Briefing Paper*. Germanwatch e.V. Available at: https://www.germanwatch.org/sites/default/files/20-2-01e%20Global%20Climate%20Risk%20Index%202020_14.pdf (accessed 1 July 2023).

Ehui, S., Kray, H. and Mghenhi, E. (2020) Policy priorities for achieving food and nutrition security in Africa by 2030. Available at: https://blogs.worldbank.org/voices/policy-priorities-achieving-food-and-nutrition-security-africa-2030 and https://www.brookings.edu/articles/policy-priorities-for-achieving-food-and-nutrition-security-by-2030/ (accessed January 2020).

Elagib, N., Al Zayed, S., Saad, S., Mahmood, M., Basheer, M. *et al.* (2021) Debilitating floods in the Sahel are becoming frequent. *Journal of Hydrology* 599, 126362. DOI: 10.1016/j.jhydrol.2021.126362

Engdaw, M., Ballinger, A., Hegerl, G. and Steiner, A. (2022) Changes in temperature and heat waves over Africa using observational and reanalysis datasets. *International Journal of Climatology* 42(2), 1165–1180. DOI: 10.1002/joc.7295

FAO (2006) Food security. *FAO Policy Brief*, June 2006, Issue 2. UN Food and Agriculture Organization, Rome. Available at: https://www.fao.org/fileadmin/templates/faoitaly/documents/pdf/pdf_Food_Security_Cocept_Note.pdf (accessed 1 July 2023).

FAO (2015) *Climate Change and Food Security: Risks and Responses*. UN Food and Agriculture Organization, Rome. Available at: https://www.fao.org/3/i5188e/I5188E.pdf (accessed 1 July 2023).

FAO and WFP (2022) *Hunger Hotspots, FAO-WFP Early Warnings on Acute Food Insecurity: October 2022 to January 2023 Outlook*. UN World Food Programme and UN Food and Agriculture Organization, Rome. DOI: 10.4060/cb8376en

FEWSNET (2023a) Famine Early Warning Systems Network. Available at: https://fews.net (accessed 1 July 2023).

FEWSNET (2023b) Famine and Early Warning Systems Network: Horn of Africa Experiences Five Consecutive Seasons of Drought for First Time in History. Available at: https://fews.net/horn-africa-experiences-five-consecutive-seasons-drought-first-time-history and https://earthobservatory.nasa.gov/images/151208/heavy-rains-hit-drought-stricken-horn-of-africa (accessed 1 July 2023).

Filipenco, D. (2022) Children hunger in Africa – statistics and facts. Available at: https://www.developmentaid.org/news-stream/post/146514/children-hunger-in-africa-statistics-and-facts (accessed 1 July 2023).

GCA (2021a) *State and Trends in Adaptation Report 2021: Africa*. Global Center on Adaptation. Available at: https://gca.org/reports/sta21/ (accessed 01 July 2023).

GCA (2021b) *GCA State and Trends in Adaptation Report 2021: How Adaptation Can Make Africa Safer, Greener and More Prosperous in a Warming World*. Available at: https://gca.org/wp-content/uploads/2022/07/03_WTW_14855_GCA_2021_Sect2_DRYLANDS_v7.pdf (accessed 1 July 2023)

Ghanem, H. (2022) *Towards a Pan-African Approach to Food Security. Policy Brief 62/22*. Policy Center for the New South. Available at: https://www.policycenter.ma/sites/default/files/2022-11/PB_62_22_Hafez%20 Ghanem.pdf (accessed 1 July 2023).

Ghanem, H. (2023) *The World Needs a Green Bank. Policy Brief 06/23*. Policy Center for the New South. Available at: https://www.policycenter.ma/publications/world-needs-green-bank (accessed 13 July 2024).

IFPRI (2013) Highlights of recent IFPRI food policy research for DFID: reducing poverty and hunger through food policy research. Available at: https://www.ifpri.org/publication/highlights-recent-ifpri-food-policy-research-dfid (accessed 1 July 2023).

IMF (2023) *The Big Funding Squeeze*. Regional Economic Outlook, Sub-Saharan Africa. Available at: https://www.imf.org/en/Publications/REO/SSA/Issues/2023/04/14/regional-economic-outlook-for-sub-saharan-africa-april-2023 (accessed 1 July 2023).

IOM (2021) *IOM Offers Humanitarian Assistance to Thousands of People Affected by Cyclone Gati*. International Organization for Migration, Somalia. Available at: https://eastandhornofafrica.iom.int/news/somalia-iom-offers-humanitarian-assistance-thousands-people-affected-cyclone-gati (accessed 1 July 2023).

IPCC (2018) *Global Warming of 1.5°C: An IPCC Special Report on the Impacts of Global Warming of 1.5°C Above Pre-Industrial Levels and Related Global Greenhouse Gas Emission Pathways, in the Context of Strengthening the Global Response to the Threat of Climate Change, Sustainable Development, and Efforts to Eradicate Poverty*. DOI: 10.1017/9781009157940

IPCC (2019) *Climate Change and Land: An IPCC Special Report on Climate Change, Desertification, Land Degradation, Sustainable Land Management, Food Security, and Greenhouse Gas Fluxes in Terrestrial Ecosystems*. DOI: 10.1017/9781009157988.014

IPCC (2022) *Climate Change 2022: Impacts, Adaptation, and Vulnerability. Contribution of Working Group II to the Sixth Assessment Report of the Intergovernmental Panel on Climate Change*. Pörtner, H.-O., Roberts, D.C., Tignor, M., Poloczanska, E.S., Mintenbeck, K. *et al.* (eds). Cambridge University Press, Cambridge, UK and New York. DOI: 10.1017/9781009325844

Masters, J. (2018) Africa's hottest reliably measured temperature on record: 124.3°F on Thursday in Algeria, 2018. Available at: https://www.wunderground.com/cat6/Africas-Hottest-Reliably-Measured-Temperature-Record-1243F-Thursday-Algeria (accessed 1 July 2023).

Mather, D., Boughton, D. and Jayne, T. (2011) Smallholder heterogeneity and maize market participation in southern and eastern Africa: implications for investment strategies to increase marketed food staple supply, 2011. Available at: https://www.canr.msu.edu/resources/smallholder-heterogeneity-and-maize-market-participation-in-southern-and-eastern-africa-implications-for-investment-strategies-to-increase-marketed-food-staple-supply (accessed 1 July 2023).

Mayans, J. (2020) The Sahel at the heart of climate change issues. Available at: https://www.solidarites.org/en/live-from-the-field/the-sahel-in-the-midst-of-climate-change/ (accessed 1 July 2023).

Mbiyozo, A.N. and Maunganidze, O.A. (2023) *Is Cyclone Freddy's record-breaking devastation the new normal?* Institute for Security Studies. Available at: https://issafrica.org/iss-today/is-cyclone-freddys-record-breaking-devastation-the-new-normal (accessed 1 July 2023).

NASA (2022) Heatwaves and fires scorch Europe, Africa, and Asia. Available at: https://earthobservatory.nasa.gov/images/150083/heatwaves-and-fires-scorch-europe-africa-and-asia (accessed 1 July 2023).

OCHA (Office of the UN Coordinator for Humanitarian Affairs) (2022) *Global Humanitarian Overview 2022*. Available at: https://www.unocha.org/sites/unocha/files/Global percent20Humanitarian percent20 Overview percent202022.pdf (accessed 1 July 2023).

ReliefWeb (2021) They lost everything: recovering from strongest storm ever recorded in the Horn of Africa. Available at: https://reliefweb.int/report/somalia/they-lost-everything-recovering-strongest-storm-ever-recorded-horn-africa (accessed 1 July 2023).

ReliefWeb (2022) Chad floods deepen humanitarian crisis with high risk of disease outbreaks. Available at: https://reliefweb.int/report/chad/chad-floods-deepen-humanitarian-crisis-high-risk-disease-outbreaks (accessed 1 July 2023).

Reuters (2022) Heat wave and fires damaging Tunisia's grain harvest. Available at: https://www.reuters.com/business/environment/heat-wave-fires-damaging-tunisias-grain-harvest-2022-06-27/ (accessed 1 July 2023).

RPCA (2022) *37th Annual Meeting Summary of Conclusions*. Réseau de Prévention des Crises Alimentaires (PRCA). Available at: https://www.food-security.net/wp-content/uploads/2021/12/RPCA-DEC2021_RELEVE-CONCLUSIONS_EN.pdf (accessed 1 July 2023).

Songwe, V., Stern, N., Bhattacharya, A., Soubeyran, E., Macquarie, R. *et al.* (2022) Finance for climate action: scaling up investment for climate and development. Available at: https://www.lse.ac.uk/granthaminstitute/publication/finance-for-climate-action-scaling-up-investment-for-climate-and-development/ (accessed 13 July 2024).

Stone, M. (2020) A plague of locusts has descended on East Africa: climate 2 change may be to blame. *National Geographic.* Available at: https://www.nationalgeographic.com/science/article/locust-plague-climate-science-east-africa (accessed 1 July 2023).

UNCCD (United Nations Convention to Combat Desertification) (2022) Drought in numbers. Available at: https://www.unccd.int/sites/default/files/2022-05/Drought%20in%20Numbers.pdf (accessed 1 July 2023).

UNDP (UN Development Programme) (2022) Calls for debt relief now for 54 countries. Available at: https://news.un.org/en/story/2022/10/1129427 (accessed 1 July 2023).

USDA (2023) *The Global Carbon Atlas*. Available at: https://www.fs.usda.gov/ccrc/tool/global-carbon-atlas (accessed 1 July 2023).

Von Braun, J., Afsana, K., Fresco, L.O. and Hassan, M.H.A. (2023) Food systems: seven priorities to end hunger and protect the planet. In: Von Braun, J., Fresco, L.O., Afsana, K. and Hassan, M.H.A. (eds) *Science and Innovations for Food Systems Transformation.* Springer, Cham, Switzerland), pp. 3–10.

WFP (World Food Programme) (2023) Regional drought response plan for the Horn of Africa. Available at: https://www.wfp.org/publications/regional-drought-response-plan-horn-africa-2023 (accessed 1 July 2023).

WMO (2022) State of climate in Africa 2021. Available at: https://public.wmo.int/en/our-mandate/climate/wmo-statement-state-of-global-climate/Africa (accessed 1 July 2023).

World Bank (2022a) *G5 Sahel Country Climate Development Report*. World Bank, Washington DC. Available at: https://www.worldbank.org/en/news/infographic/2022/09/19/g5-sahel-region-country-climate-and-development-report (accessed 1 July 2023).

World Bank (2022b) *African Food Systems: The Importance of Climate Adaptation*. World Bank, Washington DC. Available at: https://www.worldbank.org/en/news/infographic/2022/05/10/african-food-systems-the-importance-of-climate-adaptation (accessed 1 July 2023).

# 9 Case Studies in America on Carbon Management and Sequestration in Agriculture

**João Carlos de Moraes Sá[1,2], Rajan Ghimire[3], Prakriti Bista[3], Juan Andrés Quincke[4], and Mario Allegri[3]***

[1]*The Ohio State University, Columbus, USA;* [2]*State University of Maringá, Brazil;* [3]*New Mexico State University, Las Cruces, USA;* [4]*National Agricultural Research Institute, Uruguay*

**Abstract**

Agriculture, energy, and the environment are interdependent dimensions essential for agro-ecosystem sustainability. This chapter focuses on biomass carbon (C)-based practices as the primary pathway to develop resilient, regenerative, profitable, and environmentally friendly production systems. It includes case studies on C balance with biomass-C inputs across production systems in South and North America. The case studies include no-till production systems in Brazil, diverse cropping practices in the western United States, and long-term crop rotation systems experiments in Uruguay. Adopting, expanding, and consolidating C management practices will be the most prominent challenge farmers will face in the coming years worldwide. These case studies provide the consistent message that C management and soil health-oriented practices are central to implementing transitions of farming systems to become both sustainable and carbon-neutral. Cutting-edge technologies enhancing C capture and greenhouse gas sequestration, providing ecosystem services, and maximizing soil-water storage contribute to climate change mitigation and improved food security.

## Introduction

The increasing global need for food production with a rapidly growing population and increasing climate change (CC) impacts in agriculture calls for strategies that increase crop productivity while minimizing greenhouse gas (GHG) emissions. This is more important in arid and semi-arid regions where precipitation limits biomass production and carbon inputs. The dual goal of enhancing food production and mitigating CC impacts on agriculture could be achieved by sequestering C in plants and soils. The benefits of soil C sequestration include advancing food and nutritional security, increasing water conservation and quality, improving biodiversity, and strengthening nutrient cycling (Lal, 2015). Soils in arid and semi-arid regions are often unprotected, loosely aggregated, and coarse-textured, with inherently low fertility status. Continuous cropping is challenging because of low precipitation in the areas. Besides, intensive cultivation in these soils led to severe land degradation and soil organic carbon (SOC) loss (Rasmussen *et al.*, 1998; Sherrod *et al.*, 2003). Although the magnitude of impact varies between soil types, loca-

*Corresponding author: mario.allegri@gmail.com


DOI: 10.1079/9781800624870.0009

tions, and cropping systems, up to 63% depletion in SOC has been reported from intensively cultivated fields compared to undisturbed native prairies in arid and semi-arid western United States (Bowman *et al.*, 1990; Norton *et al.*, 2012; Ghimire *et al.*, 2015; Ghimire and Khanal, 2020), leading to significant negative impacts on agricultural production and the environment. Similar depletion in SOC and soil health degradation have been reported in South America (e.g. Oldeman *et al.*, 1991). This calls for management alternatives that sequester SOC and support sustainable crop production.

Studies show that soil management practices that reduce the frequency and intensity of soil disturbance and enhance C inputs through crop intensification and diversification, soil amendments, and other innovations in cropping strategies can enhance SOC sequestration (Ghimire *et al.*, 2019; Thapa *et al.*, 2022). This chapter will discuss alternative management strategies, including reduced and no-till, cropping system intensification and diversification through cover cropping, crop rotation, planting perennial crops, and opportunities and challenges of SOC sequestration in arid and semi-arid regions across the Americas. The experimental information in this chapter, extracted from case studies in different ecoregions of South and North America, is oriented towards an integrated approach capable of coping with the progressive adverse effects of CC and variability worldwide, contributing to a sustainable agro-ecology, particularly in arid and semi-arid areas with water limitations and vulnerability. We provide examples from studies to demonstrate SOC sequestration practices and their benefits to the region.

Soil health and C sequestration-oriented practices are central to implementing transitions of farming systems to become both sustainable and C-neutral. Integrating cutting-edge technologies enhancing C capture and GHG sequestration, providing ecosystem services, and maximizing soil-water storage further mitigate CC and improve food security. Studies suggest that agricultural strategies that increase the amount and diversity of C inputs, improve crop nutrient availability, and minimize soil disturbance can sequester C in soils (Ghimire *et al.*, 2022). Therefore, the adaptation of sustainable, regenerative, and resilient production systems is promoted to maintain a wise balance of C sequestration and GHG emissions mitigation to enhance adaptive capacity for global CC.

This topic is a high priority for research in the Americas for the transformation and development of sustainable and inclusive agri-food systems contributing to face CC and food security through integrated, inter-institutional, and multidisciplinary efforts. In this sense, the Inter-American Institute for Cooperation on Agriculture (IICA) plays a strategic role by partnering with national scientific and technological institutions, such as those included in this chapter, promoting 'Living Soils of the Americas', led by Prof. Rattan Lal, IICA Goodwill Ambassador.

## Case Studies

The case studies presented here will provide examples of biomass-C agriculture as the primary pathway to developing resilient, profitable, and low-environmental-impact production systems. The first case study discusses biomass-C input interaction and C balance across production systems under no-till (NT) in Brazil. The second case study provides examples of diverse crop and soil management practices from the western USA. The third case study includes evidence from different crop rotations in a long-term experiment in Uruguay.

### Case study 1: Interacting biomass-C input and C balance through production systems in sub-tropical and tropical ecoregions in Brazil

In the coming years, the biggest challenge for the agricultural sector will be to upscale the cropping systems that can produce large amounts of biomass-C input in soil (Sá *et al.*, 2022) to improve crop productivity, mitigate CC, and restore native C. Soil is essential to sustain life and basic human needs such as food, energy, clean water and air, and biodiversity (Keesstra *et al.*, 2016). The primary driving forces for the decline of SOC by plowing are the disruption of soil aggregates and marked changes in the soil environment (temperature, moisture, and oxygen), thus affecting microbial activity and the consequent greater access of SOC to microbial processes. In addition to accelerated soil erosion (Lal, 1976), the bare soil surface under plowed soil is exposed to frequent wet-dry cycles, enhancing the aggregate turnover rate (Beare *et al.*, 1994).

In the tropical and sub-tropical regions of Brazil, the continuous use of intensive plow and long-term monocropping grain crops by replacing native vegetation (forest or grassland) has led to the depletion of SOC, a decline in soil health (SH), and a reduction in the provision of essential environmental services (Sá *et al.*, 2015, 2017; Wingeyer *et al.*, 2015). The major drivers for the decline of SOC under conventional tillage (CT) are the disruption of soil aggregates and marked changes in the soil environment (temperature, moisture, and oxygen), which affect the microbial activity and SOC processes.

The adoption of conservation agriculture (CA) throughout NT with no soil disturbance (restricted to sowing line), where inputs of plant residues (above-ground biomass and roots) and associated turnover of soil biological activity are enhanced, can increase the SOC pool and its stability by physical protection within stable aggregates (Six *et al.*, 2000; Tivet *et al.*, 2013b; De Oliveira Ferreira *et al.*, 2018a). Recognizing the role of soils as a potential sink for atmospheric $CO_2$ (Lal, 2004, 2006; Lal *et al.*, 2018), there is a growing interest in adopting NT to store C in soils for offsetting anthropogenic emissions of $CO_2$, as well as restoring the SOC pool lost by the conversion of native vegetation to improve and sustain the productivity of agro-ecosystems (Sá *et al.*, 2013, 2015).

## No-till System and SOC Sequestration

The no-till system (NTS) was adopted 50 years ago in Brazil. Paraná state, followed by Rio Grande do Sul (southern region), were the eco-regions where the NTS was initially adopted to control soil erosion. Over this time, the NTS proved to be one of the most important technologies that supported upscaling of Brazilian agriculture. The NTS contributes to the net removal of C from the atmosphere (Bayer *et al.*, 2000; Sá *et al.*, 2017, 2022), restoration of SOC (Amado *et al.*, 2006; De Oliveira Ferreira *et al.*, 2021a, b; Sá *et al.*, 2006, 2015, 2018, 2022), improved SH, and increased crop yield (Bolliger *et al.*, 2006; Sá *et al.*, 2014, 2015). Application of NTS with minimal soil disturbance, permanent soil biomass mulch cover, diversified crops, and improved plant nutrient availability is essential to the sustainability of soil productivity (Kassam *et al.*, 2009).

Brazil currently has 36.8 million ha under NTS (Sá and De Oliveira Ferreira, 2018), corresponding to around 55.8% of the total grain production area. Paraná and Rio Grande do Sul states pioneered South American NTS adoption in the early 1970s, but the system began upscaling during the mid-1990s. In 1990, the land area under NTS growth was 0.9 million ha, with an annual increase of 0.081 million ha $yr^{-1}$. By 2000, there was a significant upscaling, reaching 14 million ha, with an annual rate of 1.64 million ha $yr^{-1}$. The NTS adoption area has doubled from 30.3 million ha in 2010 (1.51 million ha $yr^{-1}$), to 36.8 million ha in 2016/2017 (Sá and De Oliveira Ferreira, 2018) (Fig. 9.1).

Of the 36.8 Mha managed under NT in the country (Sá and De Oliveira Ferreira, 2018), only 10–15% have successfully adopted NTS based on the three pillars (absence of soil disturbance, permanent soil cover, and diverse crop systems). About 85–90% of the area, which today is under NTS, has failed to use one or two pillars. Therefore, significant portions of farmers are not applying the three fundamental principles properly, resulting in the occurrence of old problems, such as soil compaction, soil erosion, and low soil C restoration common in the conventional tillage system. However, the mission of adopting the NTS since the 1970s has been to avoid the loss of 7.1–21.7 billion tons of soil (Cogo *et al.*, 1984, 2003; Baptista and Levien, 2010; Sá *et al.*, 2022), which is equivalent to protecting 2.8–8.4 Mha of land (Sá *et al.*, 2022). The main challenge under a warm and wet climate is the fast turnover of soil organic matter (SOM) and nutrient leaching, which necessitate the adoption of cropping systems with characteristics of high biomass-C inputs that could be reached with complex crop rotation that provides closed nutrient cycling combined with minimal soil disturbance by NTS adoption (Tivet *et al.*, 2013a; Sá *et al.*, 2015).

The data on SOC sequestration rates for tropical (−0.03–2.15 Mg $ha^{-1}$ $yr^{-1}$ (Bayer *et al.*, 2006; Corbeels *et al.*, 2006; Cerri *et al.*, 2007; Sá *et al.*, 2015, 2022) and sub-tropical (−0.07–1.4 Mg $ha^{-1}$ $yr^{-1}$ (Sá *et al.*, 2001, 2006, 2014, 2015; Amado *et al.*, 2006; Cerri *et al.*, 2007; Boddey *et al.*, 2010; De Oliveira Ferreira *et al.*, 2012) regions of Brazil vary widely because of differences in the amount of biomass-C input and the

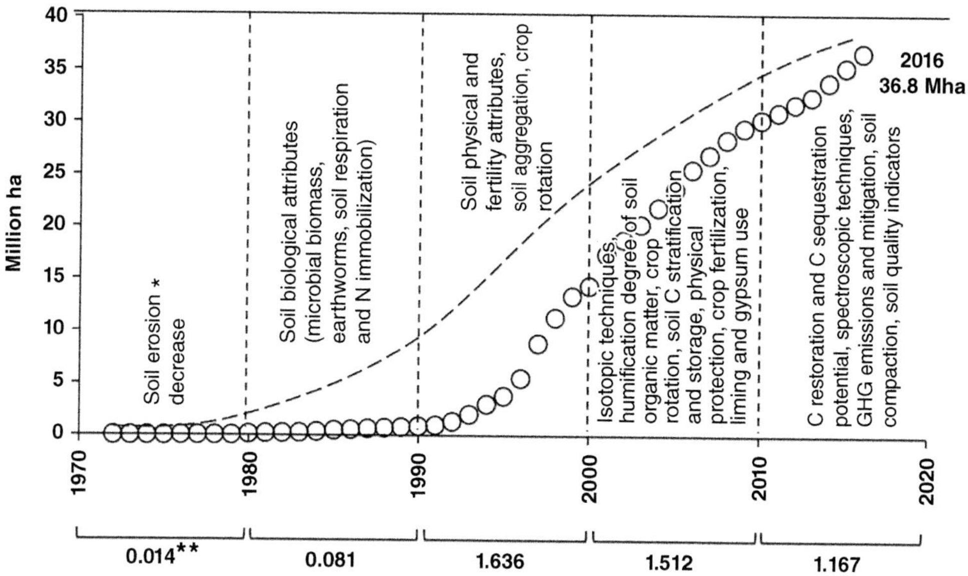


**Fig. 9.1.** Most relevant events in each decade (*) focusing soil attributes and management and no-till expansion rate by decade (**) based on no-till surface area (Mha $yr^{-1}$). (From Sá and De Oliveira Ferreira, 2018). Circles mean no-till surface area evolution and dotted lines mean the evolution of the events over the years.

agronomic potential of agro-ecosystems (Bayer *et al.*, 2006; Corbeels *et al.*, 2006; Cerri *et al.*, 2007). Batlle-Bayer *et al.* (2010) reported that changes in SOC stock, when converting from CT to NTS, ranged from 0.13 to 1.91 Mg $ha^{-1}$ $yr^{-1}$ for 0–30 cm depth and 1.98 Mg $ha^{-1}$ $yr^{-1}$ for 0–40 cm depth (Sá *et al.*, 2014). Large variability in SOC sequestration rate may be attributed to a high diversity of cropping systems, the amount and frequency of biomass-C input, and the difference in inherent soil properties (Batlle-Bayer *et al.*, 2010). For example, the rate of SOC sequestration is negligible in areas with low biomass-C input and only one crop (summer) per year (Bayer *et al.*, 2006; Jantalia *et al.*, 2007; Carvalho *et al.*, 2009), while the sequestration rate is high in ecosystems with high biomass C input such as sub-tropical and tropical ecosystems, where a high biomass-C input along with a summer crop followed by one or a combination of several winter crops are planted (Séguy *et al.*, 2006). Recently, Sá *et al.* (2022) demonstrated that when the cropping systems are intensified throughout quantity, quality, and frequency of biomass-C input, it is possible to reach a C sequestration rate of 2.15 Mg C $ha^{-1}yr^{-1}$ and to recover C lost by conversion of native vegetation to cropland in 49 years.

De Oliveira Ferreira *et al.* (2012) reported that a minimum amount of biomass-C input of 3.21 Mg $ha^{-1}$ $yr^{-1}$ (~7.13 Mg $ha^{-1}$ $yr^{-1}$ of plant biomass) is required to maintain SOC stock. Subsequently, Sá *et al.* (2015) reported that biomass-C input of 4.15 Mg $ha^{-1}$ $yr^{-1}$ (~9.23 Mg $ha^{-1}$ $yr^{-1}$ of plant biomass) under a long-term (20 years) NTS was necessary to maintain the steady state of SOC stocks for sub-tropical ecoregions. However, for a tropical climate, Sá *et al.* (2015, 2022) estimated a minimum rate of biomass-C input of 5.1–5.8 Mg $ha^{-1}$ $yr^{-1}$ to achieve a dynamic C equilibrium according to the specific soil characteristics (Sá *et al.*, 2022), which is equivalent to 11.7–13.3 Mg $ha^{-1}$ $yr^{-1}$ of crop residues (shoot + roots) input. The cropping systems should be redesigned to attain values above the minimum crop residues required to maintain the steady state of SOC and enhance soil health.

## Crop Rotation and High Carbon Input

Crop rotation and cover crops to fill up short time windows between cash crop plantings is one strategy that enhances NTS performance in tropical environments because of its positive impacts on SOC, nutrient loss control, soil physical attributes, soil temperature and moisture regulation, and soil biota and C inputs (Amado *et al.*, 2006; De Oliveira Ferreira *et al.*, 2013, 2018a, b; Sá *et al.*, 2014, 2015, 2022). These positive soil effects lead to increased crop productivity (De Oliveira Ferreira *et al.*, 2009; Sá *et al.*, 2014). However, such benefits (e.g. improved soil health and productivity, and increased SOC stock) are realized gradually over a long time period.

Making Brazilian agriculture emission negative can be achieved by maximizing the soil C sink by increasing the $CO_2$ fixed by crops during photosynthesis and its accumulation in soil profile upon land application of biomass-C. Thus, C accumulation in soil occurs only when the biomass-C input is greater than the losses of C by erosion, decomposition, and leaching (Sá *et al.*, 2015; Lal *et al.*, 2018). The challenge will be to manage the biomass-C input in a way that exceeds the amount of C to reach the potentially minimum level of C in the soil. For the tropical environment, the amount of C input is estimated at 5.1–5.8 Mg $ha^{-1}$ $yr^{-1}$ (Sá *et al.*, 2015). An example of a farming system in Lucas do Rio Verde-MT (tropical) and Piraí do Sul-PR (sub-tropical) documents that achieving a positive C balance is feasible (Table 9.1 and 9.2), which can be calculated by using the following expression:

Soil C balance (net C sequestration rate) =
(Biomass-C input × Humification rate) −
(Soil C decomposition rate × Soil C stock)

where the C is the balance represented by the net C sequestration rate considering the biomass-C input as the amount via biomass (shoots + roots), the humification rate represented by the % of C conversion from the plant biomass to soil C at the Lucas do Rio Verde site (19% or 0.19 annually, respectively); the soil C stock was calculated from the data obtained from soil analyses (soil C content and soil BD) for a specific layer.

The positive C balance in Lucas do Rio Verde-MT (Fig. 9.2), Capuaba Farm (Table 9.2) results from the combination of quantity of biomass added + quality of biomass + frequency of addition (Sá *et al.*, 2022). This means that we will comply with what is recommended in the rationale for the NTS: permanently covered soil and with diversification in crop rotation.

In brief, the key component to reach a positive C balance in farming production systems is to manage biomass-C input in association with NTS envisaging to accumulate and stabilize C throughout three main mechanisms: (i) stimulate a continuous macro-aggregation formation through soil covered year-long; (ii) control the SOC oxidation; and (iii) stabilize the C intra-aggregate in mineral organic C associate. This means that

**Table 9.1.** Carbon balance in a no-till production system under a five-year crop rotation in Lucas do Rio Verde (Capuaba Farm, 18°34'16.3" SL; 56°35'36" WL), Brazil.

| C Balance period | Year 1 | Year 2 | Year 3 | Year 4 | Year 5 |
|---|---|---|---|---|---|
| Crop rotation | Soybean/corn+ Brq Rz§ | Soybean/corn + Crtochrol | Soybean/mix of CC | Rice + radish | Soybean/corn+ Brq Rz |
| C Balance components | | | Mg $ha^{-1}$ | | |
| Annual dry matter input (shoots)* | 15.94 | 14.21 | 9.03 | 5.84 | 17.69 |
| Annual dry matter input (root) | 3.94 | 3.19 | 2.54 | 1.49 | 3.92 |
| Total annual dry matter input | 19.88 | 17.40 | 11.56 | 7.32 | 21.61 |
| Total C input by biomass | 7.02 | 6.12 | 5.01 | 3.14 | 9.59 |
| Gross C sequestration rate § | 1.33 | 1.16 | 0.95 | 0.60 | 1.82 |
| Soil C stock | 40.50 | 41.35 | 42.02 | 42.46 | 42.55 |
| Annual C decomposition rate (losses) | 0.49 | 0.50 | 0.50 | 0.51 | 0.51 |
| Net C sequestration rate = C balance | **0.84** | **0.66** | **0.45** | **0.09** | **1.31** |

§Brq Rz = *Brachairia ruziziensis;* Crtochrol = *Crotalaria ochroleuca;* Mix of CC = blend of mix of plants for cover crop

**Fig. 9.2.** Crop sequence of Capuaba Farm comprising the production system with high biomass-C input located at Lucas do Rio Verde-MT, Brazil.

we have to look forward to adding C via biomass based on quantity, quality, and frequency of input above the dynamic equilibrium level of C in the soil. The key strategy to restore C and be agriculture emission-negative is to manage the sequestration rate of the production systems through biomass-C input.

## Case study 2: Soil carbon sequestration dry environments of the western United States

Soil C storage is the primary mechanism for sequestering C in agricultural lands. Atmospheric $CO_2$ assimilated by plants through photosynthesis is ultimately sequestered as SOC. Croplands in the USA can sequester 190–610 Tg C $y^{-1}$ by adopting improved management practices (Fargione *et al.*, 2018). Several factors, such as soil temperature, moisture, particle size, soil contact, and biochemical composition of organic matter, determine the rate of decomposition and SOC sequestration. Crop growth characteristics, cropping intensity, and soil disturbance frequency and intensity also affect the rate of soil C sequestration by regulating biomass inputs. In the intensively cultivated lands of the western USA, SOC has declined to low to very low levels (i.e.,$<$1% by soil volume), which provides enormous

**Table 9.2.** Carbon balance in a crop production system under a no-till system in Pirai do Sul-PR (Paiquere Farm, 24°18'51.5" SL; 50°07'49.9" WL), Brazil (Fig. 9.3).

| C Balance period | Year 1 | Year 2 | Year 3 | Year 4 | Year 5 |
|---|---|---|---|---|---|
| Crop rotation | Wheat/ soybean | B. Oat†/corn | Wheat/ soybean | B. Oat/corn | Wheat/ soybean |
| C Balance components | | | Mg ha$^{-1}$ | | |
| Annual dry matter input (shoots)* | 7.80 | 17.75 | 7.89 | 17.32 | 8.34 |
| Annual dry matter input (root) | 1.60 | 4.05 | 2.29 | 4.49 | 1.71 |
| Total annual dry matter input | 9.40 | 21.80 | 10.18 | 21.81 | 10.06 |
| Total C input by biomass | 3.96 | 9.42 | 4.29 | 9.42 | 4.24 |
| Gross C sequestration rate § | 1.05 | 2.50 | 1.14 | 2.50 | 1.12 |
| Soil C stock | 43.00 | 43.62 | 45.68 | 46.36 | 48.39 |
| Annual C decomposition rate (losses) | 0.43 | 0.44 | 0.46 | 0.46 | 0.48 |
| Net C sequestration rate = C balance | 0.62 | 2.06 | 0.68 | 2.03 | 0.64 |
| | | | | | 1.21 |

†B. oats = Black oats (*Avena sativa*)

capacity for additional C sequestration through improved management practices (Bowman *et al.*, 1990; Norton *et al.*, 2012; Ghimire *et al.*, 2015; Ghimire and Khanal, 2020). Improved management practices such as reduced- and no-till, increased cropping intensity through cover cropping and crop rotation, perennial cropping, and improved nutrient management through smart fertility practices and the use of organic amendments (biochar, compost, etc.) have been shown to sequester C (Ghimire *et al.*, 2015; Ghimire and Khanal, 2020). The effects of these management practices are often complementary and result in many additional benefits beyond C sequestration, such as increased crop yields and economic benefits to farmers, improved soil health, and clean air for the public. For example, no-till sequesters C, saves tillage costs due to fewer field passes, and reduces soil erosion. Cover crops increase soil C input and protect the soil from erosion during the fallow period. The impacts of various conservation practices are discussed in the following sections.

Studies revealed that SOC storage in water-limited environments is strongly influenced by biomass C and N inputs and their quality (Brown and Huggins, 2012; Ghimire *et al.*, 2015; Ghimire *et al.*, 2019). Crop residue production in arid and semi-arid regions is often lower than the biomass input required to maintain SOC in the soil profile (Machado, 2011), which may have caused the considerable loss of the SOC from arid and semi-arid regions of the western United States. Here we present the evidence from studies comparing tillage systems, crop rotation, cover cropping, and perennial cropping systems in semi-arid eastern New Mexico and other studies from across the western USA.

## Reducing Soil Disturbance

Tillage is a fundamental agricultural practice in which the soils are physically manipulated to make favorable conditions for seed germination, establishment, growth, and development. Tillage type affects soil C storage and loss (i.e. oxidation of organic C); conventional tillage typically leads to SOC loss, while reduced tillage frequency and intensity maintain or enhance SOC. Conventional tillage uses a moldboard plow or rototiller, which generally loosens and disturbs the soil with an inversion of the sub-surface layer up to 20–30 cm exposing the sub-surface layer for oxidation (Abdalla *et al.*, 2013). Also, conventional tillage

**Fig. 9.3.** Crop sequence of intensive cropping system developed at Paiquere Farm, Pirai do Sul-PR, Brazil.

involves a series of disking and furrowing activities followed by leveling and smoothing of agricultural fields in which the soil is disturbed heavily. A series of disturbance and incorporating residues with conventional tillage increases dissolved organic C and N up to the tilled soil layer, which are substrate and energy sources for heterotrophic microbes. Rapid decomposition of crop residues leads to SOC loss from conventional tillage systems (Abdalla *et al.*, 2013).

Reduced- and no-till systems, also called conservation tillage systems, increase SOC sequestration because they reduce soil disturbance. Our study comparing conventional tillage, strip tillage, and no-till for four years in semi-arid dryland conditions revealed 1.21–1.45 Mg $ha^{-1}$ SOC with reduced and no-till systems (Fig. 9.4). Reduced disturbance avoids aggregate breakdown, limits contact of microorganisms with crop residues, and leaves more residues on the soil surface, providing food for microorganisms for a long time. The rate of SOC sequestration with conservation tillage varied among studies, depending on the study location, cropping practices, and other management differences (Thapa *et al.*, 2023). However, more residue on the soil surface and less fluctuation in soil temperature often led to higher SOC storage under conservation tillage than conventional tillage. It also reduces $CO_2$ fluxes and improves biological activity in the soil

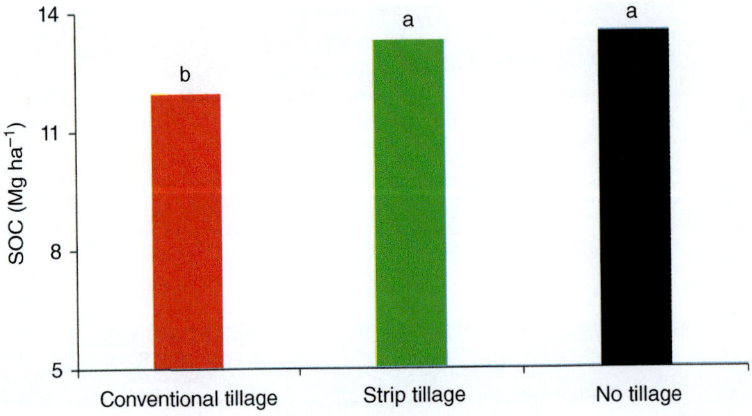


**Fig. 9.4.** SOC under different tillage systems in semi-arid eastern New Mexico. Different lower-case letters represent a significant difference between treatments ($p \leq 0.05$). (Data from Thapa *et al.*, 2022).

(Abdalla *et al.*, 2013). Conservation tillage systems (e.g. no-till, strip tillage) often maintain more than 30% residue on the soil surface, reducing soil erosion and improving soil biological activity, leading to more SOC accumulation despite dry conditions in arid and semi-arid regions.

## Intensifying and Diversifying Cropping Systems

Cropping systems in arid and semi-arid regions often involve crop–fallow systems to allow recharge of soil moisture for the succeeding crop. The practice of extended fallow periods between crops can deplete SOC because microorganisms continue to utilize native organic matter during fallow in the absence of fresh organic residues and root exudates. A decline in SOC reduces crop productivity of arid and semi-arid cropping systems, which further depletes SOC storage (West and Post, 2002). Therefore, identifying options to diversify and intensify cropping systems that curtail the negative effect of extended fallow periods, increase SOC, and improve water-use efficiency are needed. Utilizing desert-adapted crops in crop rotation or as a cover crop could provide continuous ground cover, supply of root exudates, and decomposing root, and aboveground residues could contribute to SOC accumulation. We collected metadata from dry environments of the western USA, which show 1–1.2 Mg ha$^{-1}$ more SOC accumulation with cover cropping and crop rotation than crop–fallow systems (Fig. 9.5). The input of fresh organic matter by cover cropping, crop rotation, and other forms of cropping system diversification and intensification efforts can curtail SOC depletion, provide nutrients for soil microbes, and improve soil-water storage capacity (Paye *et al.*, 2022; Thapa *et al.*, 2022). Limited available water is the main constraint for dryland cultivation, and the question is: Is cropping system intensification possible under dryland conditions in semi-arid agriculture?

The long fallow period between the cash crop is aimed at soil moisture conservation. However, precipitation storage capacity is generally low (10–35%) for the fallow (Nielsen and Vigil, 2010). In addition, fallow, under conventional practice, undergoes repeated tillage for weed control and surface residue mixing, which accelerates soil moisture and SOC loss (Shaver *et al.*, 2002; West and Post, 2002; Kahlon *et al.*, 2013). No-till management that allows growers to control weeds through herbicides instead of tillage reduces soil disturbance, can enhance water storage, and enables farmers to grow additional crops via partial or full replacement of fallow. Hence, the rising interest among semi-arid farmers in transforming crop–fallow agroecosystems to intensified/diversified cropping systems can be supported by combining alternative cropping practices such as continuous cropping, diverse crop rotation, cover cropping, etc., with no-till management.

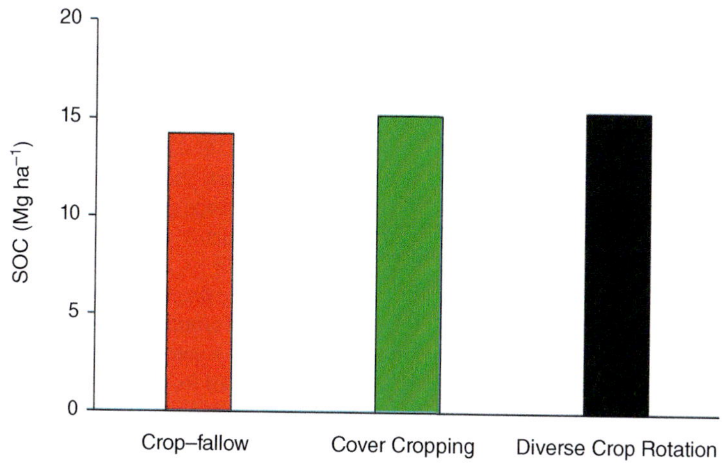


**Fig. 9.5.** SOC storage in the upper 30 cm soil depth due to cover cropping and diverse crop rotation in water-limited western USA. (Data from Thapa *et al.*, 2023).

Combining the multiple soil conservation practices such as no-till, cover cropping, and crop rotation with diverse crops like brassica and legumes in a primarily cereal-based system can promise several benefits such as improving soil tilth and soil health, enhancing fertilizer-N use efficiency (with legumes), increasing SOC accumulation, and reducing GHG balance in arid and semi-arid regions around the globe (Abdalla *et al.*, 2013; Mitchell *et al.*, 2017; Thapa *et al.*, 2019; Nicoloso and Rice, 2021; Acharya *et al.*, 2022). A study comparing 96 dryland no-till fields in semi-arid eastern Colorado and western Nebraska reported that continuous cropping increased SOC concentration by 17% and 12% in the top 0–10 cm and 10–20 cm, respectively, compared to wheat–fallow (Rosenzweig *et al.*, 2018). Similarly, a recent review by Thapa *et al.* (2023) reported that cover cropping in semi-arid western USA could increase SOC by 201 kg C ha$^{-1}$ yr$^{-1}$, which is equivalent to 7% more SOC stock with cover crops compared to crop–fallow (Fig. 9.5). The increased SOC from these crop intensification practices is associated with increased crop biomass production and additional C inputs to the soil, rhizodeposition, and increased soil aggregation.

## Perennial Cropping Systems and Grassland Restoration

The cultivation of high-yielding annual crops in conventional systems requires frequent tillage to aid in crop establishment and substantial use of herbicide to curtail weed pressure. However, these practices have several detrimental effects on soils and the environment, including nutrient outflow, soil erosion, and reduction in SOC levels. Perennial crops could be a more sustainable alternative to annual crops in arid and semi-arid regions because they have extended growing seasons and continuous ground cover, which help to sequester more C in soils and avoid nutrient leakage. Perennial crops increase SOC sequestration by twofold effects, i.e. minimizing soil disturbance and increasing biomass inputs. Perennial crops avoid disturbance because, once planted, no further cultivation is needed. They also have deeper and denser root systems than annual row crops, producing three to ten times more below-ground biomass (DuPont *et al.*, 2014). Studies also show that the roots of perennial crops decompose slowly because they have a higher C:N ratio than annual crops (Dietzel *et al.*, 2015). In arid and semi-arid regions, such as the western USA, the roots of perennial crops can utilize water and nutrients from a greater soil depth, resulting in greater total root biomass production and supporting more SOC storage in the deeper soil depths (Skinner and Comas, 2010). Therefore, breeding for perennial crops suitable to arid and semi-arid regions should be promoted to enhance SOC sequestration in these regions. Researchers are also looking for options to increase root biomass density in annual crops (Paustian *et al.*, 2019), which could also be beneficial for arid and semi-arid regions as long as they could maintain crop yield and productivity.

Grassland restoration has been promoted as an ideal method for controlling wind and water erosion, enhancing soil C sequestration, and recovering degraded ecosystems. Studies show that soil nutrient losses due to leaching, volatilization, and runoff are lower in perennial-based cropping systems than in intensive annual agricultural systems, primarily due to poor nutrient synchronization with crop requirement and reduced plant nutrient-use efficiency in annual systems (Crews, 2005; Hussain *et al.*, 2019). Converting croplands to perennial grassland significantly improved SOC sequestration under semi-arid conditions of eastern New Mexico (Ghimire *et al.*, 2019; Fig. 9.6). Further, perennial grasses increased aggregate stability, reducing susceptibility to wind erosion more effectively than annual crops (Thapa *et al.*, 2018). Such information is critical for the western USA, specifically in the hot, dry southwest, because the region faces more frequent heat and drought, high precipitation variability, and soil degradation due to erosion and SOC depletion (Ghimire *et al.*, 2019).

## Exploring Dual Opportunities of Carbon and Water Management

Since water is the main limiting factor for biomass C input in arid and semi-arid regions, identifying soil C sequestration strategies that improve soil-water storage could provide double benefits. Available evidence shows that improved farming practices such as cropping system intensification and diversification, cover cropping, crop rotation, mixed cropping, and conservation tillage can improve water-use efficiency and enhance biomass C input for enhancing SOC sequestration and soil-water conservation benefits (Ghimire *et al.*, 2022; Figs 9.6, 9.7). Integrating reduced and no-till with cropping systems intensification, diversification, and enhanced residue inputs could increase SOC in arid and semi-arid regions (Thapa *et al.*, 2019) and improve soil-water conservation (Paye *et al.*, 2022).

In recent years, frontier options such as advanced breeding for deeper and more recalcitrant rooting systems and high C soil amendments such as biochar are also explored to attain the higher SOC sequestration goal (Fig. 9.8). Studies show that biochar application in cropland soils and breeding development of perennial

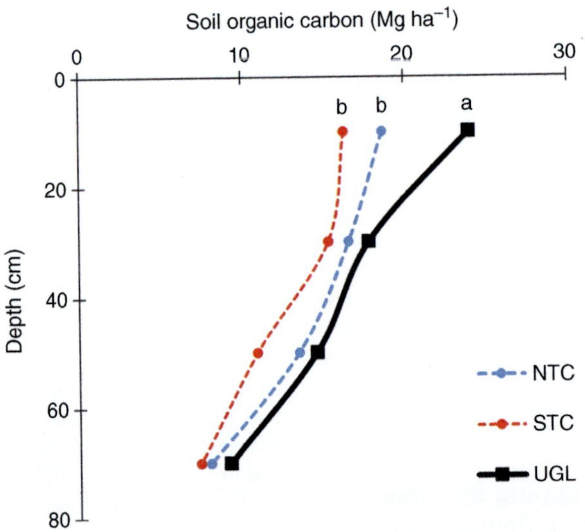


**Fig. 9.6.** Soil profile carbon under strip-tilled cropland (STC), no-till cropland (NTC), and ungrazed native grassland (UGL). Different lower-case letters on top represent a significant difference between treatments ($p \leq 0.05$) at 0–20 cm depth. (Data from Ghimire *et al.*, 2019).

**Fig. 9.7.** A conservation agriculture system field at eastern New Mexico demonstrating cover crop and no-tillage effects on grain sorghum production.

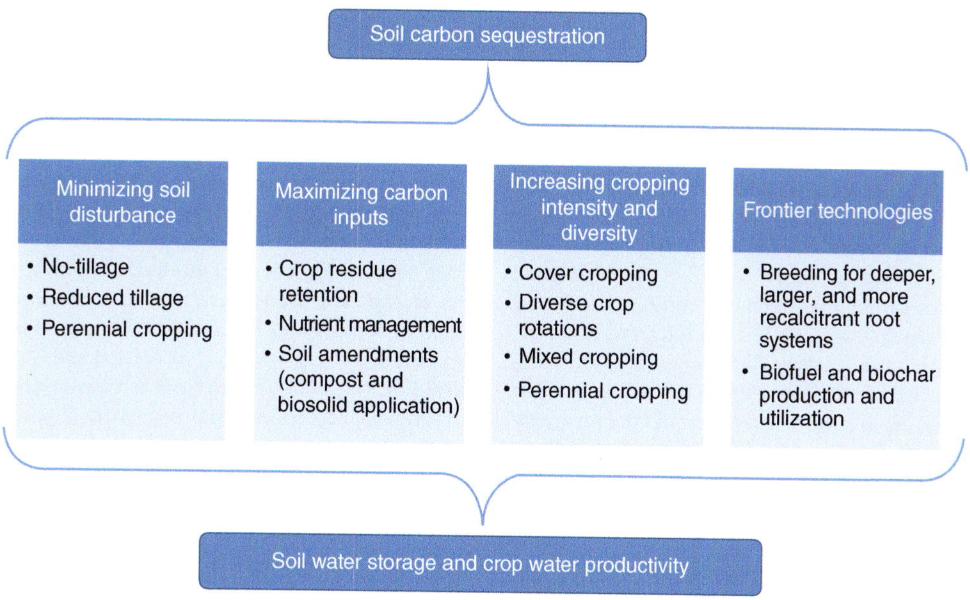


**Fig. 9.8.** Management strategies for increasing SOC sequestration and soil-water storage in water-limited environments. (Adapted from Ghimire *et al.*, 2022).

crops and annual crops that produce deeper and denser rooting systems could increase SOC quicker than business-as-usual conservation practices (Paustian *et al.*, 2019). The use of pyrolyzed crop residues, manure, and compost-biochar mixture has been more effective for SOC sequestration and stabilization (Lee *et al.*, 2010; Spokas *et al.*, 2012) and could be more valuable in dry environments. The biochemically recalcitrant and predominantly aromatic C present in pyrolyzed material can store SOC for a long time (Keith *et al.*, 2011). In addition to its recalcitrance, biochar interacts with the native SOC to stabilize it in the soil profile (Paustian *et al.*, 2019). Biochar, along with cropping system intensification, has greater potential to increase SOC in arid and semi-arid regions because of the slow decomposition of biochar C under water-limited environments. In addition, it has a high $N_2O$ emissions mitigation (9–50%) and enhanced water storage and infiltration potential compared to non-biochar amended soils (Cayuela *et al.*, 2014; Verhoeven *et al.*, 2017).

## Case study 3: SOC content under different crop–pasture rotations in a long-term experiment in Uruguay

Uruguay is located at 30–35°S and 53–58°W, limited to the south by the Rio de la Plata estuary and to the southeast by the Atlantic Ocean. Its mean annual temperature is 24°C in summer and 12°C in winter, whilst annual average precipitation is about 1150mm, although with considerable variation both between years (approximately 20–25%) and within years. The topography is gently rolled, with dominant slopes ranging between 3% and 6%. According to Paruelo *et al.* (2001), Uruguayan territory belongs to the Río de la Plata grasslands. The main soils are Mollisols and Vertisols, although other great groups may also cover significant areas. Nevertheless, Uruguay's prime cropland belongs basically to USDA class III, with no appreciable areas of class I or II soils.

In the middle of the 20th century, the dominant cropping system was based on continuous crops with conventional tillage (i.e. primary tillage with a moldboard plow and secondary tillage with a disc harrow). By the mid-1960s, water erosion and soil degradation affected approximately 30% of the country. This led to the adoption of crop–pasture rotations during the 1970s, consisting of three to four years of cash crops followed by three to four years of mixed pastures for direct grazing. Due to the pasture in the rotation, a significant erosion rate reduction was possible (García-Préchac *et al.*, 2004, 2022). No-till technology began to be adopted during the 1990s, which allowed for further improvement in soil conservation. But during the 2000s, Uruguay underwent an accelerated process of agricultural intensification, which increased continuous agriculture and drastically reduced the area under crop–pasture rotations (Ernst *et al.*, 2018).

Soils in different agro-ecosystems continuously develop changes in response to applied management practices. Among the various soil attributes that are affected, SOC requires special attention. A loss in SOC may reduce future productivity due to its direct association with the physical, chemical, and biological properties of the soil. Also, it may reduce a soil's performance regarding its environmental functions.

## The Study Approach

Different crop or crop–pasture rotations have been studied for 60 years in the oldest long-term experiment in Uruguay, initiated in 1963 and located at the INIA Experimental Station La Estanzuela (34°20′S, 57°41′W). The main purpose was to evaluate seven cropping systems with different amounts of time spent under pasture. Although modifications were made in 1974, 1983, and 2008, the main concepts in the treatment design have been maintained. A description of the crop sequences for each treatment (since 2008) is given in Table 9.1. More details can be found in Grahmann *et al.* (2019) and Rubio *et al.* (2021). Briefly, the basic crop sequence is a three-year rotation containing two cool- and three warm-season crops: maize, barley–sorghum, wheat–soybean. The sorghum and soybean are planted as double crops after barley and wheat, respectively. A second crop sequence relies heavily on soybean instead of sorghum and maize. The seven cropping systems were established in a randomized complete block design with three blocks. Plots are 25m × 200m and are situated lengthwise on a 2–4% slope. The soil is a silty clay loam Typic Arguidoll, with an estimated SOC gravimetric content of 3.17% (in the top 20 cm) in the pristine condition and 2.2% at the start of the experiment (Liu *et al.*, 2012). Soil samples are collected yearly in March or April to a depth of 15cm. Conventional tillage was used for seedbed preparation, although a gradual reduction in tillage intensity occurred from the mid-1990s onwards until full conversion to NT in 2008 (Franzluebbers *et al.*, 2014; Baethgen-*et al.*, 2021). For practical reasons, the pastures are not directly grazed. Instead, defoliation management of pastures is practiced with mowing and leaving the biomass on the soil surface to mimic an extensive, low-technology (hence, low-extraction) grazing operation (Fig. 9.10).

For the graphical display of results, SOC data from annual soil sampling were smoothed with a moving window within plots and averaged across blocks. The SOC decreased nearly constantly in continuous cropping without N and P fertilization ('ContCrop, No Fert' in Fig. 9.9). For similar cropping sequences under continuous cropping but with N and P fertilization ('ContCrop' in Fig. 9.9), SOC declined at a lesser rate and appeared to become stable almost at a SOC level of 1.5%. A crop–pasture rotation with red clover (*Trifolium pratense*; '4Crop-2Past' in Fig. 9.9) resulted in a relatively steady SOC content over the long term compared to the starting condition. Moreover, with less intensive crop–pasture rotations that had 50% crops and 50% pastures ('3Crop-3Past' in Fig. 9.9), SOC slightly increased over the long term, especially after 1983 when the cropping phase was reduced from five to three years. Finally, a crop–pasture rotation that presented only two years of crops rotating with two types of pastures (one lasting one year and the other three years) was established in 1983 ('2Crop-4Past' in Fig. 9.9). Before 1983, this treatment was a crop–pasture rotation where crops rotated with only annual forage crops. The change in the management imposed on these plots was reflected in the SOC content, which increased steadily from 1983 onwards. Furthermore, a pattern of decrease during cropping phases and increase during pastures can be recognized for several crop–pasture rotations within the rotation cycles of this long-term experiment.

In brief, this long-term experiment has allowed for observing and documenting how, under the described soil and climate conditions, SOC evolved differently in a soil that already had lost about a third of its original SOC content due to previous agricultural land use. While continuous cropping resulted in further SOC losses, certain crop–pasture rotations reverted the process and caused the SOC content to gradually increase, even under tillage conditions. Nevertheless, the original SOC level (estimated at ~3.1%) could not be achieved. Moreover, adequate

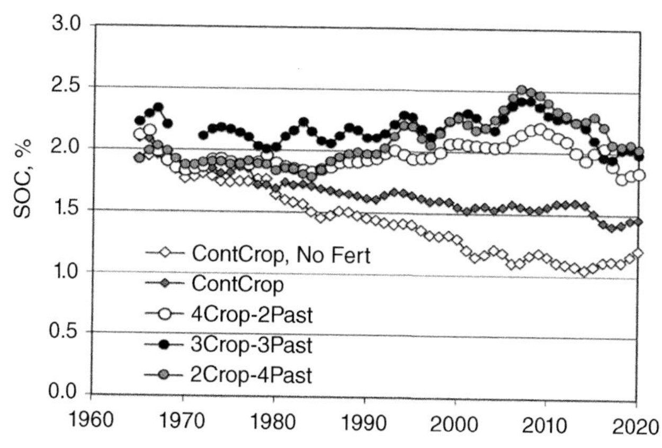


**Fig. 9.9.** SOC content as a function of time in five systems studied in a long-term experiment started in 1963 at INIA La Estanzuela, Uruguay. Sampling depth: 15 cm. ContCrop is continuous cropping; ContCrop/ NoFert is continuous cropping with no addition of fertilizers; and, for example, 2Crop-4Past is a rotation with 2 years under annual crops and 4 years under perennial pastures. (Modified after Franzluebbers *et al.*, 2014).

**Table 9.3.** Crop and pasture sequences in seven rotation systems at the mixed crop–pasture rotation experiment at INIA La Estanzuela. These sequences were in place from 2008 onwards.

| Abbreviation | Year 1 | Year 2 | Year 3 | Year 4 | Year 5 | Year 6 | Treatment concept | System number |
|---|---|---|---|---|---|---|---|---|
| ContCrop | Maize | Barley–sorghum | Wheat–soybean | Maize | Barley–sorghum | Wheat–soybean | Continuous cropping | 2 |
| ContCrop/No Fert | Maize | Barley–sorghum | Wheat–soybean | Maize | Barley–sorghum | Wheat–soybean | Same as above but with **no fertilizer addition** | 1 |
| ContCrop (2) | Oats–soybean | Oats–soybean | Wheat–soybean | Oats– soybean | Oats– soybean | Wheat–soybean | Continuous cropping, intense soybean | 6 |
| 4Crop–2Past | **Red clover** | Sorghum | Wheat–soybean | **Red clover** | Sorghum | Wheat–soybean | Rotation with a short pasture | 7 |
| 3Crop–3Past | Maize | Barley–sorghum | Wheat–soybean | **Mixed pasture** | **Mixed pasture** | **Mixed pasture** | Rotation with a mixed pasture* | 5 |
| 3Crop–3Past(L) | Maize | Barley–sorghum | Wheat–soybean | **Lotus** | **Lotus** | **Lotus** | Rotation with a pasture of *Lotus corniculatus* | 3 |
| 2Crop–4Past | Maize | Red clover | Sorghum | **Mixed pasture** | **Mixed pasture** | **Mixed pasture** | Rotation with two types of pasture | 4 |

*The mixed pasture system 5 (3Crop-3Past) is composed of fescue, white clover, and lotus. Treatment 6 has been based on soybean since 2008. In the remaining treatments, soybean replaced sunflowers in 2017.

**Fig. 9.10.** Aerial view of the 21 plots that comprise the 10.5 ha of the crop–pasture rotation experiment initiated in 1963 at INIA La Estanzuela experimental station in Uruguay. The picture depicts a harvesting operation in plot 6 (counting from the left or west), while plots 3, 14, 15, 19, and 20 are under pastures of different duration and/or species composition. (Photo: INIA).

fertilization is also a relevant long-term management practice, as shown in the comparison of continuous cropping with or without sustained fertilizer additions.

## Summary, Conclusion, and Way Forward

The restoration of soil health evaluated by natural capital should be based on the following strategies: (i) increasing the biomass-C input according to the three pillars that support conservation agriculture, adding the amount at the rate above that of the dynamic equilibrium, guaranteeing a controlled soil C oxidation rate, and promoting the continuous C and N flow to reorganize the aggregates and the new structure; (ii) stimulating macro-aggregation and enhancing C protection within intra-aggregates and stabilizing it over time and restoring soil functionality by improving the soil's chemical, physical, and biological attributes; and (iii) managing the rate of input of biomass-C as a key strategy to decrease C turnover time, leading to the restoration of natural capital and ecosystem services to society.

Available evidence from various western United States studies shows that adopting conservation and regenerative farming practices could sequester SOC in arid and semi-arid regions. The major conservation practices adopted include conservation tillage systems, cover cropping, cropping system intensification and diversification, and perennial cropping. There is increasing attention on perennial cropping, transition to grasslands, and more recalcitrant soil amendments such as biochar and other pyrolyzed materials to rapidly enhance and stabilize C in arid and semi-arid landscapes of the western USA. Since water is a limiting factor for SOC sequestration in arid and semi-arid regions, exploring the dual option of improving water-use efficiency while improving SOC sequestration could be a more sustainable option for enhancing SOC sequestration in arid areas across the world, including the western USA.

To better cope with the risk of soil-water erosion and soil degradation, Uruguayan field crop production shifted from continuous cropping under conventional tillage to systems based on important conservation practices, mainly introducing mixed pastures in the rotation and/or no-till systems. A long-term experiment on crop–pasture rotations, started in 1963 in the southwest of Uruguay, demonstrated the effects of different crop–pasture rotations on soil properties and crop performance. Later, the experiment was also shifted to no-till. This case study briefly presented and discussed the results from SOC annual data.

These case studies provide evidence from all across the Americas that no-tillage and other conservation cropping systems mitigate CC by improving SOC sequestration. In addition, the strategic role of science and technology is widely recognized across these studies for providing continuous data collection, management, and assessment of the dynamic impact of CC on ecosystems, informing policymakers in an easy and timely manner, and contributing to formulating public policies. Improving climate resilience while supporting sustainable crop production is complex and requires a holistic approach involving inter-institutional multi-disciplinary teams. Therefore, such integrated strategies should consider not only water management as the most limiting resource and the carbon cycle; the development and incorporation of digital technologies in agriculture and food systems, well-adapted genetic resources, alternative resources of renewable energy and bioeconomy, promoting innovative proposals for increasing productivity, and preserving natural resources (environmental, technological, economic, and social dimensions) will provide next-generation solutions to food security and CC mitigation.

## References


Abdalla, M., Osborne, B., Lanigan, G., Forristal, D., Williams, M. *et al.* (2013) Conservation tillage systems: a review of its consequences for greenhouse gas emissions. *Soil Use Manage* 29, 199–209. DOI: 10.1111/sum.12030

Acharya, P., Ghimire, R., Paye, W.S., Ganguli, A. and DelGrosso, S.J. (2022) Net greenhouse gas balance with cover crops in semi-arid irrigated cropping systems. *Scientific Reports* 12, 12386. DOI: 10.1038/s41598-022-16719-w

Amado, T.J.C., Bayer, C., Conceição, P.C., Spagnollo, E., De Campos, B.-H.C. *et al.* (2006) Potential of carbon accumulation in no-tillage soils with intensive use and cover crops in southern Brazil. *Journal of Environmental Quality* 35, 1599–1607. DOI: 10.2134/jeq2005.0233

Baethgen, W.E., Parton, W.J., Rubio, V., Kelly, R.H. and Lutz, S. (2021) Ecosystem dynamics of crop–pasture rotations in a fifty-year field experiment in southern South America: Century model and field results. *Soil Science Society of America Journal* 85(2), 423–437. DOI: 10.1002/saj2.20204

Baptista, J. and Levien, R. (2010) Methods of soil tillage and its influence in water erosion and initial development of *Eucalyptus saligna* on the inceptisol of the central depression of the State of Rio Grande do Sul. *Revista Arvore* 34, 567–575. DOI: 10.1590/S0100-67622010000400001

Batlle-Bayer, L., Batjes, N.H. and Bindraban, P.S. (2010) Changes in organic carbon stocks upon land use conversion in the Brazilian Cerrado: a review. *Agriculture, Ecosystems and Environment* 137(1–2), 47–58. DOI: 10.1016/j.agee.2010.02.003

Bayer, C., Mielniczuk, J., Amado, T.J.C., Martin-Neto, L. and Fernandes, S.V. (2000) Organic matter storage in a sandy clay loam Acrisol affected by tillage and cropping systems in southern Brazil. *Soil and Tillage Research* 54, 101–109. DOI: 10.1016/S0167-1987(00)00090-8

Bayer, C., Martin-Neto, L., Mielniczuk, J., Pavinato, A. and Dieckow, J. (2006) Carbon sequestration in two Brazilian Cerrado soils under no-tillage. *Soil and Tillage Research* 86, 237–245. DOI: 10.1016/j.still.2005.02.023

Bayer, C., Dieckow, J., Amado, T.J.C., Eltz, F.L.F. and Vieira, F.C.B. (2009) Cover crop effects increasing carbon storage in a subtropical no-tillage sandy Acrisol. *Communications in Soil Science and Plant Analysis* 40, 1499–1511. DOI: 10.1080/00103620902820365

Beare, M.H., Hendrix, P.F., Cabrera, M.L. and Coleman, D.C. (1994) Aggregate-protected and unprotected organic matter pools in conventional-and no-tillage soils. *Soil Science Society of America Journal* 58(3), 787–795. DOI: 10.2136/sssaj1994.03615995005800030021x

Blanchart, E., Bernoux, M., Sardi, X., Neto, M., Cerri, C. *et al.* (2007) Effect of direct seeding mulch-based system on soil carbon storage in Central Brazil. *Agriculturae Conspectus Scientificus* 72, 81–87.

Boddey, R.M., Jantalia, C.P., Conceicao, P.C., Zanatta, J.A., Bayer, C. *et al.* (2010) Carbon accumulation at depth in Ferralsols under zero-till subtropical agriculture. *Global Change Biology* 16(2), 784–795. DOI: 10.1111/j.1365-2486.2009.02020.x

Bolliger, A., Magid, J., Amado, J.C.T., Skora Neto, F., Ribeiro, M. *et al.* (2006) Taking stock of the Brazilian 'Zero Till Revolution': a review of landmark research and farmers' practice. *Advances in Agriculture* 91, 47–110. DOI: 10.1016/S0065-2113(06)91002-5

Bowman, R.A., Reeder, J.D. and Lober, R.W. (1990) Changes in soil properties in central plains rangeland soil after 3, 20, and 60 years of cultivation. *Soil Science* 150, 851–857.
Briedis, C., Sá, J.C.M., Lal, R., Tivet, F., De Oliveira Ferreira, A. *et al.* (2016) Can highly weathered soils under conservation agriculture be C saturated? *Catena* 147, 638–649. DOI: 10.1016/j.catena.2016.08.021
Briedis, C., Sá, J.C.M., Lal, R., Tivet, F., Franchini, J.C., De Oliveira Ferreira, A. *et al.* (2018) How does no-tillage deliver carbon stabilization and saturation in highly weathered soils? *Catena* 163, 13–23. DOI: 10.1016/j.catena.2017.12.003
Brown, T.T. and Huggins, D.R. (2012) Soil carbon sequestration in thedryland cropping region of the Pacific Northwest. *Journal of Soil and Water Conservation* 67, 406–415. DOI: 10.2489/jswc.67.5.406
Carvalho, J.L.N., Cerri, C.E.P., Feigl, B.J., Piccolo, M.C., Godinho, V.P. *et al.* (2009) Carbon sequestration in agricultural soils in the Cerrado region of the Brazilian Amazon. *Soil and Tillage Research* 103(2), 342–349. DOI: 10.1016/j.still.2008.10.022
Cayuela, M.L., van Zwieten, L., Singh, B.P., Jeffery, S., Roig, A. *et al.* (2014) Biochar's role in mitigating soil nitrous oxide emissions: a review and meta-analysis. *Agriculture, Ecosystems & Environment* 191, 5–16. DOI: 10.1016/j.agee.2013.10.009
Cerri, C.E.P., Sparovek, G., Bernoux, M., Easterling, W.E., Melillo, J.M., Cerri, C.C. (2007) Tropical agriculture and global warming: impacts and mitigation options. *Scientia Agricola* 64, 83–99. DOI: 10.1590/S0103-90162007000100013
Cogo, N.P., Moldenhauer, W.C. and Foster, G.R. (1984) Soil loss reductions from conservation tillage practices. *Soil Science Society of America Journal* 48, 368–373. DOI: 10.2136/sssaj1984.03615995004800020029x
Cogo, N.P., Levien, R. and Schwarz, R.A. (2003) Soil and water losses by rainfall erosion influenced by tillage methods, slope-steepness classes, and soil fertility levels. *Revista Brasileira de Ciência do Solo* 27, 743–753. DOI: 10.1590/S0100-06832003000400019
Conceição, P.C., Amado, T.J.C., Mielniczuk, J. and Spagnollo, E. (2005) Soil organic matter and other attributes as indicators to evaluate soil quality in conservation systems. *Revista Brasileira de Ciência do Solo* 29, 777–788. DOI: 10.1590/S0100-06832005000500013
Corbeels, M., Scopel, E., Cardoso, A., Bernoux, M., Douzet, J. *et al.* (2006) Soil carbon storage potential of direct seeding mulch based cropping systems in the Cerrados of Brazil. *Global Change Biology* 12(9), 1773–1787. DOI: 10.1111/j.1365-2486.2006.01233.x
Crews, T.E. (2005) Perennial crops and endogenous nutrient supplies. *Renewable Agriculture and Food Systems* 20, 25–37. DOI: 10.1079/RAF200497
De Oliveira Ferreira, A., Sá, J.C.M., Briedis, C. and Figueiredo, A.G. (2009) Desempenho de genótipos de milhocultivados com diferentesquantidades de palha de aveia-preta e doses de nitrogênio (Corn genotype performance under black oat crop residues and nitrogen fertilization). *Pesquisa Agropecuária Brasileira* 44(2), 173–179. DOI: 10.1590/S0100-204X2009000200009
De Oliveira Ferreira, A., Sá, J.C.M., Harms, M.G., Miara, S., Briedis, C. *et al.* (2012) Carbon balance and crop residue management in dynamic equilibrium under a no-till system in Campos Gerais. *Revista Brasileira de Ciência do Solo* 36, 1583–1590. DOI: 10.1590/S0100-06832012000500022
De Oliveira Ferreira, A., Amado, T.J.C., Nicoloso, R.S., Sá, J.C.M., Fiorin, J.E. *et al.* (2013) Soil carbon stratification affected by long-term tillage and cropping systems in southern Brazil. *Soil and Tillage Research* 133, 65–74. DOI: 10.1016/j.still.2013.05.011
De Oliveira Ferreira, A., Amado, T.J.C., Rice, C.W., Diaz, D.A.R., Briedis, C. *et al.* (2018a) Driving factors of soil carbon accumulation in Oxisols in long-term no-till systems of South Brazil. *Science of The Total Environment* 622–623, 735–742. DOI: 10.1016/j.scitotenv.2017.12.019
De Oliveira Ferreira, A., Sá, J.C.M., Lal, R., Tivet, F., Briedis, C. *et al.* (2018b) Macroaggregation and soil organic carbon restoration in a highly weathered Brazilian Oxisol after two decades under no-tillage. *Science of the Total Environment* 621, 1559–1567. DOI: 10.1016/j.scitotenv.2017.10.072
De Oliveira Ferreira, A., Sá, J.C.M., Lal, R., Amado, T.J.C., Inagaki, M.T. *et al.* (2021a) Can no-tillage restore soil organic carbon to levels under natural vegetation in a subtropical and tropical typic quartzipisamment? *Land Degradation & Development* 32, 1742–1750. DOI: 10.1002/ldr.3822
De Oliveira Ferreira, A., Amado, T.J.C., Rice, C.W., Goncalves, D.R.P. and Diaz, D.A.R. (2021b) Comparing on-farm and long-term research experiments for soil carbon recovery of conservation agriculture in Southern Brazil. *Land Degradation & Development* 32, 3365–3376. DOI: 10.1002/ldr.4015
Dietzel, R., Jarchow, M.E. and Liebman, M. (2015) Above-and belowground growth, biomass, and nitrogen use in maize and reconstructed prairie cropping systems. *Crop Science* 55(2), 910–923. DOI: 10.2135/cropsci2014.08.0572

DuPont, S.T., Beniston, J., Glover, J.D., Hodson, A., Culman, S.W. *et al.* (2014) Root traits and soil properties in harvested perennial grassland, annual wheat, and never-tilled annual wheat. *Plant Soil* 381(1), 405–420. DOI: 10.1007/s11104-014-2145-2

Ellert, B.H. and Bettany, J.R. (1995) Calculation of organic matter and nutrients stored in soils under contrasting management regimes. *Canadian Journal of Soil Science* 75, 529–538. DOI: 10.4141/cjss95-075

Ernst, O.R., Dogliotti, S., Cadenazzi, M. and Kemanian, A.R. (2018) Shifting crop-pasture rotations to no-till annual cropping reduces soil quality and wheat yield. *Field Crops Research* 217, 180–187. DOI: 10.1016/j.fcr.2017.11.014

Fargione, J.E., Bassett, S., Boucher, T., Bridgham, S.D., Conat, R.T. *et al.* (2018) Natural climate solutions for the United States. *Science Advances* 4(11), eaat1869. DOI: 10.1126/sciadv.aat1869

Franzluebbers, A.J., Sawchik, J. and Taboada, M.A. (2014) Agronomic and environmental impacts of pasture-crop rotations in temperate North and South America. *Agriculture, Ecosystems & Environment* 190, 18–26. DOI: 10.1016/j.agee.2013.09.017.

Fuentes, R.L., Telles, T., Soares Júnior, D., Melo, T., Friedrich, T. *et al.* (2021) Expansion of no-tillage practice in conservation agriculture in Brazil. *Soil and Tillage Research* 208, 104877. DOI: 10.1016/j.still.2020.104877

García-Préchac, F., Ernst, O., Siri-Prieto, G. and Terra, J.A. (2004) Integrating no-till into crop-pasture rotations in Uruguay. *Soil and Tillage Research* 77, 1–13. DOI: 10.1016/j.still.2003.12.002

García-Préchac, F., Salvo, L., Ernst, O., Siri-Prieto, G., Quincke, J.A. *et al.* (2022) Long-term effects of different agricultural soil use and management systems on soil degradation in Uruguay. In: Li, R., Napier, T.L., El-Swaify, S.A., Sabir, M. and Rienzi, E. (eds) *Global Degradation of Soil and Water Resources*. Springer, Singapore, pp. 77–92.

Ghimire, R. and Khanal, B.R. (2020) Soil organic matter dynamics in semi-arid agroecosystems transitioning to dryland. *PeerJ* 8, e10199. DOI: 10.7717/peerj.10199

Ghimire, R., Machado, S. and Rhinhart, K. (2015) Long-term crop residue and nitrogen management effects on soil profile carbon and nitrogen in wheat–fallow systems. *Agronomy Journal* 107, 2230–2240. DOI: 10.2134/agronj14.0601

Ghimire, R., Thapa, V.R., Cano, A. and Acosta-Martinez, V. (2019) Soil organic matter and microbial community responses to semi-arid croplands and grasslands management. *Applied Soil Ecology* 141, 30–37. DOI: 10.1016/j.apsoil.2019.05.002

Ghimire, R., Clay, D., Thapa, S. and Hurd, B. (2022) More carbon per drop to enhance soil carbon sequestration in water-limited environments. *Carbon Management* 13(1), 450–462. DOI: 10.1080/17583004.2022.2117082

Grahmann, K., Rubio, V., Terra, J.A. and Quincke, J.A. (2019) Long-term observations in contrasting crop-pasture rotations over half a century: statistical analysis of chemical soil properties and implications for soil sampling frequency. *Agriculture, Ecosystems & Environment* 287, 106710. DOI: 10.1016/j.agee.2019.106710

Hussain, M.Z., Bhardwaj, A.K., Basso, B., Robertson, G.P. and Hamilton, S.K. (2019) Nitrate leaching from continuous corn, perennial grasses, and poplar in the US Midwest. *Journal of Environmental Quality* 48(6), 1849–1855. DOI: 10.2134/jeq2019.04.0156

Inagaki, T.M., Sá, J.C.M., Caires, E.F. and Gonçalves, D.R.P. (2016) Lime and gypsum application increases biological activity, carbon pools, and agronomic productivity in highly weathered soil. *Agriculture, Ecosystems & Environment* 231, 156–165. DOI: 10.1016/j.agee.2016.06.034

Jantalia, C.P., Resck, D.V.S., Alves, B.J.R., Zotarelli, L., Urquiaga, S. *et al.* (2007) Tillage effect on C stocks of a clayey Oxisol under a soybean-based crop rotation in the Brazilian Cerrado region. *Soil and Tillage Research* 95(1–2), 97–109. DOI: 10.1016/j.still.2006.11.005

Kahlon, M.S., Lal, R. and Ann-Varughese, M. (2013) Twenty-two years of tillage and mulching impacts on soil physical characteristics and carbon sequestration in Central Ohio. *Soil & Tillage Research* 126, 151–158. DOI: 10.1016/j.still.2012.08.001

Kassam, A., Friedrich, T., Shaxson, F. and Pretty, J. (2009) The spread of conservation agriculture: justification, sustainability and uptake. *International Journal of Agricultural Sustainability* 7, 292–320. DOI: 10.3763/ijas.2009.0477

Keesstra, S.D., Bouma, J., Wallinga, J., Tittonell, P., Smith, P. *et al.* (2016) The significance of soils and soil science towards realization of the UN sustainable development goals. *Soil* 2, 111–128. DOI: 10.5194/soil-2-111-2016

Keith, A., Singh, B. and Singh, B.P. (2011) Interactive priming of biochar and labile organic matter mineralization in a smectite-rich soil. *Environmental Science & Technology* 45, 9611–9618. DOI: 10.1021/es202186j

Lal, R. (1976) No-tillage effects on soil properties under different crops in western Nigeria. *Soil Science Society of America Journal* 40, 762–768. DOI: 10.2136/sssaj1976.03615995004000050039x

Lal, R. (2004) Soil carbon sequestration impacts on global climate change and food security. *Science* 304, 1623–1627. DOI: 10.1126/science.1097396

Lal, R. (2006) Enhancing crop yields in the developing countries through restoration of the soil organic carbon pool in agricultural lands. *Land Degradation & Development* 17(2), 197–209. DOI: 10.1002/ldr.696

Lal, R. (2014) Societal value of soil carbon. *Journal of Soil and Water Conservation* 69, 186A–192A. DOI: 10.2489/jswc.69.6.186A

Lal, R. (2015) Sequestering carbon and increasing productivity by conservation agriculture. *Journal of Soil and Water Conservation* 70, 55A–62A. DOI: 10.2489/jswc.70.3.55A

Lal, R. (2019) Rights-of-soil. *Journal of Soil and Water Conservation* 74, 81A–86A. DOI: 10.2489/jswc.74.4.81A

Lal, R., Smith, P., Jungkunst, H.F., Mitsch, W.J., Lehmann, J. *et al.* (2018) The carbon sequestration potential of terrestrial ecosystems. *Journal of Soil and Water Conservation* 73, 145A–152A. DOI: 10.2489/jswc.73.6.145A

Lal, R., Brevik, E.C., Dawson, L., Field, D., Glaser, B. *et al.* (2020) Managing soils for recovering from the COVID-19 pandemic. *Soil Systems* 4, 46. DOI: 10.3390/soilsystems4030046

Lee, J.W., Hawkins, B., Day, D.M. and Reicosky, D.C. (2010) Sustainability: the capacity of smokeless biomass pyrolysis for energy production, global carbon capture and sequestration. *Energy & Environmental Science* 3, 1695–1705. DOI: 10.1039/C004561F

Liu, X., Burras, C.L., Kravchenko, Y.S., Duran, A., Huffman, T. *et al.* (2012) Overview of Mollisols in the world: distribution, land use and management. *Canadian Journal Soil Science* 92, 383–402. DOI: 10.4141/cjss2010-058

Machado, S. (2011) Soil organic carbon dynamics in the Pendleton long-term experiments: implications for biofuel production in Pacific Northwest. *Agronomy Journal* 103, 253–260. DOI: 10.2134/agronj2010.0205s

Mitchell, J.P., Shrestha, A., Mathesius, K., Scow, K.M., Southard, R.J. *et al.* (2017) Cover cropping and no-tillage improve soil health in an arid irrigated cropping system in California's San Joaquin Valley, USA. *Soil and Tillage Research* 165, 325–335. DOI: 10.1016/j.still.2016.09.001

Nicoloso, R.S. and Rice, C.W. (2021) Intensification of no-till agricultural systems: an opportunity for carbon sequestration. *Soil Science Society of America Journal* 85(5), 1395–1409. DOI: 10.1002/saj2.20260

Nielsen, D.C. and Vigil, M.F. (2010) Precipitation storage efficiency during fallow in wheat-fallow systems. *Agronomy Journal* 102(2), 537–543. DOI: 10.2134/agronj2009.0348

Norton, J.B., Mukhwana, E.J. and Norton, U. (2012) Loss and recovery of soil organic carbon and nitrogen in a semi-arid agroecosystem. *Soil Science Society of America Journal* 76, 505–514. DOI: 10.2136/sssaj2011.0284

Oldeman, L.R., Hakkeling, R.T.A., Sombroek, W.G. and Batjes, N. (1991) Global assessment of human-induced soil degradation (GLASOD). *World Map of the Status of Human-induced Soil Degradation.* Winand Staring Centre–ISSSFAO–ITC, Wageningen, Netherlands.

Paruelo, J.M., Jobbágy, E.G. and Sala, O.E. (2001) Current distribution of ecosystem functional types in temperate South America. *Ecosystems* 4, 683–698. DOI: 10.1007/s10021-001-0037-9

Paustian, K., Lehmann, J., Ogle, S., Reay, D., Robertson, G.P. *et al.* (2016) Climate-smart soils. *Nature* 532(7597), 49–57. DOI: 10.1038/nature17174

Paustian, K., Larson, E., Kent, J., Marx, E. and Swan, A. (2019) Soil C sequestration as a biological negative emission strategy. *Frontiers in Climate* 1, 8. DOI: 10.3389/fclim.2019.00008

Paye, W.S., Ghimire, R., Acharya, P., Nilahyane, A., Mesbah, A.O. *et al.* (2022) Cover crop water use and corn silage production in semi-arid irrigated conditions. *Agricultural Water Management* 260, 107275. DOI: 10.1016/j.agwat.2021.107275

Plaza-Bonilla, D., Arrúe, J.L., Cantero-Martínez, C., Fanlo, R., Iglesias, A. *et al.* (2015) Carbon management in dryland agricultural systems:a review. *Agronomy for Sustainable Development* 35, 1319–1334. DOI: 10.1007/s13593-015-0326-x

Rasmussen, P.E., Albrecht, S.L. and Smiley, R.W. (1998) Soil C and N changes under tillage and cropping systems in semi-arid Pacific Northwest agriculture. *Soil & Tillage Research* 47, 197–205. DOI: 10.1016/S0167-1987(98)00106-8

Rosenzweig, S.T., Stromberger, M.E. and Schipanski, M.E. (2018) Intensified dryland crop rotations support greater grain production with fewer inputs. *Agriculture, Ecosystems & Environment* 264, 63–72. DOI: 10.1016/j.agee.2018.05.017

Rubio, V., Diaz-Rossello, R., Quincke, J.A. and van Es, H.M. (2021) Quantifying soil organic carbon's critical role in cereal productivity losses under annualized crop rotations. *Agriculture, Ecosystems & Environment* 321, 107607. DOI: 10.1016/j.agee.2021.107607

Sá, J.C.M. and De Oliveira Ferreira, A. (2018) The soil science in the evolution of no-tillage system in Brazil. *SBCS Bulletim (Brazilian Society of Soil Science)* 44, 54–57.

Sá, J.C.M., Cerri, C.C., Dick, W.A., Lal, R., Solismar Filho, P.V. *et al.* (2001) Organic matter dynamics and carbon sequestration rates for a tillage chronosequence in a Brazilian Oxisol. *Soil Science Society of America Journal* 65, 1486–1499. DOI: 10.2136/sssaj2001.6551486x

Sá, J.C.M., Séguy, L., Gozé, E., Bouzina, S., Husson, O. *et al.* (2006) Carbon sequestration rates in no-tillage soils under intensive cropping systems in tropical agroecozones. *Edafologia* 13, 139–150.

Sá, J.C.M., Cerri, C.C., Lal, R., Dick, W.A., De Cassia Piccolo, M. *et al.* (2009) Soil organic carbon and fertility interactions affected by a tillage chronosequence in a Brazilian Oxisol. *Soil and Tillage Research* 104, 56–64. DOI: 10.1016/j.still.2008.11.007

Sá, J.C.M., Santos, J.B., Lal, R., De Moraes, A., Tivet, M. *et al.* (2013) Soil-specific inventories of landscape carbon and nitrogen stocks under no-tillage and native vegetation to estimate carbon offset in a subtropical ecosystem. *Soil Science Society of America Journal* 77, 2094–2110. DOI: 10.2136/sssaj2013.01.0007

Sá, J.C.M., Tivet, F., Lal, R., Briedis, C., Hartman, D.C. *et al.* (2014) Long-term tillage systems impacts on soil C dynamics, soil resilience and agronomic productivity of a Brazilian Oxisol. *Soil and Tillage Research* 136, 38–50. DOI: 10.1016/j.still.2013.09.010

Sá, J.C.M., Séguy, L., Tivet, F., Lal, R., Bouzinac, S. *et al.* (2015) Carbon depletion by plowing and its restoration by no-tillage cropping systems in oxisols of subtropical and tropical agro-ecoregions in Brazil. *Land Degradation & Development* 26, 531–543. DOI: 10.1002/ldr.2218

Sá, J.C.M., Lal, R., Cerri, C.C., Lorenz, K., Hungria, M. *et al.* (2017) Low-carbon agriculture in South America to mitigate global climate change and advance food security. *Environment International* 98, 102–112. DOI: 10.1016/j.envint.2016.10.020

Sá, J.C.M., Potma Gonçalves, D.R., Ferreira, L.A., Mishra, U., Inagaki, T.M. *et al.* (2018) Soil carbon fractions and biological activity based indices can be used to study the impact of land management and ecological successions. *Ecological Indicators* 84, 96–105. DOI: 10.1016/j.ecolind.2017.08.029

Sá, J.C.M., Lal, R., Briedis, C., De Oliveira Ferreira, A.,Tivet, F. *et al.* (2022) Can C-budget of natural capital be restored through conservation agriculture in a tropical and subtropical environment? *Environmental Pollution* 298, 118817. DOI: 10.1016/j.envpol.2022.118817

Séguy, L., Bouzinac, S. and Husson, O. (2006) Direct-seeded tropical soil systems with permanent soil cover: learning from Brazilian experience. In: Uphoff, N. and Theis, J.E. (eds) *Biological Approach to Sustainable Soil Systems*. CRC Press, Boca Raton, Florida, pp. 323–342.

Shaver, T.M., Peterson, G.A., Ahuja, L.R., Westfall, D.G., Sherrod, L.A. *et al.* (2002) Surface soil physical properties after twelve years of dryland no-tillage management. *Soil Science Society of America Journal* 66, 1296–1303. DOI: 10.2136/sssaj2002.1296

Sherrod, L.A., Peterson, G.A., Westfall, D.G. and Ahuja, L.R. (2003) Cropping intensity enhances soil organic carbon and nitrogen in a no-tillage agroecosystem. *Soil Science Society of America Journal* 67, 1533–1543. DOI: 10.2136/sssaj2003.1533

Sisti, C.P.J., Santos, H.P., Kohhann, R., Alves, B.J.R., Urquiaga, S. *et al.* (2004) Change in carbon and nitrogen stocks in soil under 13 years of conventional or zero tillage in southern Brazil. *Soil and Tillage Research* 76, 39–58. DOI: 10.1016/j.still.2003.08.007

Six, J., Elliott, E.T. and Paustian, K. (2000) Soil macroaggregate turnover and microaggregate formation: a mechanism for C sequestration under no-tillage agriculture. *Soil Biology and Biochemistry* 32(14), 2099–2103. DOI: 10.1016/S0038-0717(00)00179-6

Six, J., Conant, R.T., Paul, E.A. and Paustian, K. (2002a) Stabilization mechanisms of soil organic matter: implications for C-saturation of soils. *Plant and Soil* 241(2), 155–176. DOI: 10.1023/A:1016125726789

Six, J., Feller, C., Denef, K., Ogle, S.M., Sá, J.C.M. *et al.* (2002b) Soil organic matter, biota and aggregation in temperate and tropical soils – effects of no-tillage. *Agronomie* 22, 755–775. DOI: 10.1051/agro:2002043

Skinner, R.H. and Comas, L.H. (2010) Root distribution of temperate forage species subjected to water and nitrogen stress. *Crop Science* 50, 2178–2185. DOI: 10.2135/cropsci2009.08.0461

Spokas, K.A., Cantrell, K.B., Novak, J.M., Archer, D.W., Ippolito, J.A. *et al.* (2012) Biochar: a synthesis of its agronomic impact beyond carbon sequestration. *Journal of Environmental Quality* 41, 973–989. DOI: 10.2134/jeq2011.0069

Thapa, V.R., Ghimire, R., Mikha, M.M., Idowu, O.J. and Marsalis, M.A. (2018) Land use effects on soil health in semi-arid drylands. *Agricultural & Environmental Letters* 3(1), 180022. DOI: 10.2134/ael2018.05.0022

Thapa, V.R., Ghimire, R., Duval, B.D. and Marsalis, M.A. (2019) Conservation systems for positive net ecosystem carbon balance in semi-arid drylands. *Agrosystems, Geosciences & Environment* 2(1), 1–8. DOI: 10.2134/age2019.03.0022

Thapa, V.R., Ghimire, R., VanLeeuwen, D., Acosta-Martínez, V. and Shukla, M. (2022) Response of soil organic matter to cover cropping in water-limited environments. *Geoderma* 406, 115497. DOI: 10.1016/j.geoderma.2021.115497

Thapa, V.R., Ghimire, R., Adhikari, K.P. and Lamichhane, S. (2023) Soil organic carbon sequestration potential of conservation agriculture in arid and semi-arid regions: a review. *Journal of Arid Environments* 217, 105028. DOI: 10.1016/j.jaridenv.2023.105028

Tisdall, J.M. and Oades, J.M. (1982) Organic matter and water-stable aggregates in soils. *Journal of Soil Science* 33, 141–163. DOI: 10.1111/j.1365-2389.1982.tb01755.x

Tivet, F., Sá, J.C.M., Lal, R., Borszowskei, P.R., Briedis, C. *et al.* (2013a) Soil organic carbon fraction losses upon continuous plow-based tillage and its restoration by diverse biomass-C inputs under no-till in sub-tropical and tropical regions of Brazil. *Geoderma* 209–210, 214–225. DOI: 10.1016/j.geoderma.2013.06.008

Tivet, F., Sá, J.C.M., Lal, R., Briedis, C., Borszowskei, P.R. *et al.* (2013b) Aggregate C depletion by plowing and its restoration by diverse biomass-C inputs under no-tillage in sub-tropical and tropical regions of Brazil. *Soil and Tillage Research* 126, 203–218. DOI: 10.1016/j.still.2012.09.004

Verhoeven, E., Pereira, E., Decock, C., Suddick, E., Angst, T. *et al.* (2017) Toward a better assessment of biochar–nitrous oxide mitigation potential at the field scale. *Journal of Environmental Quality* 46, 237. DOI: 10.2134/jeq2016.10.0396

West, T.O. and Post, W.M. (2002) Soil organic carbon sequestration rates by tillage and crop rotation: a global data analysis. *Soil Science Society of America Journal* 66, 1930–1946. DOI: 10.2136/sssaj2002.1930

Wingeyer, A.B., Amado, T.J., Studdert, G.A., Varela, C.H., Garcia, F.O. *et al.* (2015) Soil quality impacts of current south American agricultural practices. *Sustainability* 7(2), 2213–2242. DOI: 10.3390/su7022213

# 10 Central and West Asia and North Africa (CWANA)

**Aly Abousabaa[1,2*], Michael Baum[3*], and Vinay Nangia[4]**
*[1]CGIAR's Regional Director, Central and West Asia & North Africa; [2]Director General, ICARDA; [3]Deputy Director General, ICARDA; [4]Research Team Leader, ICARDA/Morocco*

**Abstract**
The CWANA region is experiencing an increase in the severity and frequency of high temperatures, agricultural and ecological droughts, precipitation droughts, floods, climate variability, and climate change-compounding events, with significant consequences for society and the environment. ICARDA, alongside its national and international partners and donors, promotes diversified cropping systems and delivers its solutions as integrated packages that include innovative technologies, capacity-building programs, instruments for better market access, and the current best genetic and agronomic approaches. These include sustainable, water-smart irrigation, cultivation of drought-tolerant crop varieties, water management approaches such as rainwater harvesting and deficit irrigation, precision farming technologies, and low-water-footprint cropping systems, raised-bed technology, and others. ICARDA breeding programs have focused their efforts on the development of germplasm tolerant to major abiotic and biotic stresses.

## Introduction

The CWANA region covers wide areas across Central Asia, West Asia, and North Africa and contains a great diversity of human cultures and agro-ecological systems (Baum and Al-Zu'bi, 2021). What unifies this area is a relative semi-arid and aridness that is only going to become more extreme with ongoing climate change (CC). The region is expected to experience an increase in the severity and frequency of high temperatures, agricultural and ecological droughts, precipitation droughts, floods, climate variability, and CC-compounding events with significant consequences for society and the environment. Despite the inherent constraints in the CWANA region of limited productive agricultural land (in the MENA region: 33% for crops and pastures, 5% for crops alone), severe water scarcity, and unsustainable surface and groundwater abstraction (for use in agriculture), the predominantly rainfed agri-food systems (70%) are frequently the largest employers of the populace and of women, and significantly contribute to national GDP and local food security (Baum and Al-Zu'bi, 2021). The CWANA regional agri-food systems are, however, vulnerable to overuse of the limited natural resources; the biophysical effects of present and future CC and their interplay with societal pressures, e.g. gender inequalities; youth unemployment; demographic shifts; predominance of small, non-specialized smallholdings; and

*Corresponding author: a.abousabaa@cgiar.org

 *Climate Change and Sustainable Agro-ecology in Global Drylands* (A.S. El-Beltagy, R. Lal and K. Malik eds.)
DOI: 10.1079/9781800624870.0010

institutional and economic pressures. The consequences of experiencing multiple compounding events in places that do not have resilient systems in place to cope with them can be devastating. For instance, the Syrian uprising in 2011 occurred after the most severe drought on record (2007–2010), and the prolonged unsustainable use of freshwater resources for irrigation caused widespread crop failure and mass migration from rural areas to urban centers, and civil unrest.

ICARDA is an international organization, headquartered in the MENA region, undertaking agricultural research for development. ICARDA provides innovative, science-based solutions for countries across dry areas. Its research aims to reduce poverty and enhance food, water, nutritional security, and environmental health in the face of global challenges, including CC. ICARDA addresses the challenges facing the world's dry areas: rising temperatures, critical water scarcity, diminishing natural resources, and an insecure food and nutritional future under often unstable social conditions. Its work closely aligns with the UN Sustainable Development Goals (SDGs). ICARDA has taken action to address the vulnerability of CWANA regional agri-food systems by developing resilient solutions, such as drought-resistant crops and sustainable irrigation practices. These solutions aim to mitigate the impacts of CC, improve food security, and reduce vulnerability to multiple compounding events (Hamwieh and Imtiaz, 2015; Istanbuli *et al.*, 2022). ICARDA innovations help countries and communities strengthen their adaptation to CC and transition towards food and nutritional security. For example, ICARDA's research on drought-resistant crop varieties and sustainable water management practices could have mitigated the effects of the severe drought in Syria and prevented widespread crop failure (Malhotra *et al.*, 2004). By promoting these innovative solutions, ICARDA helps vulnerable communities build resilience to compounding events and achieve long-term food and nutritional security. ICARDA technologies and innovations promote the sustainable use of natural resources and biodiversity, creating greater resilience to CC (Shideed *et al.*, 2007). ICARDA delivers its solutions as integrated packages that include innovative technologies, capacity-building programs, instruments for better market access, and the current best agronomic approaches. These include sustainable, water-smart irrigation; cultivation of drought-tolerant crop varieties; water management approaches such as rainwater harvesting and deficit irrigation; precision farming technologies; and low-water-footprint cropping systems (Solh, 2010).

## Genetic Resources

CWANA hosts four Vavilovian centers of origin for many species of global importance, with wheat-based and mixed-farming systems at the core of its food and nutritional security (Vavilov, 1926; Conservation International, 2016; Maxted and Vincent, 2021). Cereals (durum, bread wheat, barley, rice), potatoes/sweet potatoes, food and forage legumes, dryland fruit trees, and vegetables are critical for farming-system diversification, intensification, and sustainable production in CWANA. Other genetic solutions (breeds or fish) and the livestock value chain are important for the food systems in CWANA and for promoting feed and forage crops. Eight types of farming systems have been identified in the CWANA region (FAO – https://www.fao.org/). The highland, rainfed, and dryland mixed and irrigated farming systems account for the majority of crop, fruit, vegetable, and livestock production in the region and form the basis of meeting the food, nutrition, and livelihood security needs for millions of households.

ICARDA makes extensive use of the germplasm collections in its gene bank to find new sources of tolerance to cope with CC. Since the disruption of the gene bank operations in Aleppo, ICARDA has opened new gene banks in Terbol, Lebanon (2016), and Rabat, Morocco (2022) (Kehel *et al.*, 2023). The ICARDA gene bank is one of the nine gene banks under the umbrella of the GeneBank Initiative (GBI – https://www.cgiar.org/initiative/genebanks/). ICARDA hosts one of the largest collections of cereals and legumes landraces and crop wild relatives in the world, and in the last years has been developing new strategies that couple GIS and genomic-based information to better mine the resources and thus find the proverbial 'needle in the haystack' that are the genes for stress-tolerance (Tsivelikas *et al.*, 2023). Gene banks worldwide hold collections of plant genetic resources for long-term conservation, and maintain crop diversity for current and future use by crop-improvement research, direct use, and training. Most of the gene banks are facing major problems of size and organization. Some collections have grown so large that their main activities, which are the conservation and use of genetic diversity, have become challenging. Another challenging aspect of plant genetic resources conservation is the lack of information about accessions, specifically precise evaluation information. This is mainly due to the challenge of evaluating the entire collection of a gene bank. However, several gene banks maintain and curate passport information. But passport data do not help a user of the gene bank discern which accession in a database potentially contains the trait of interest. The concept of core and mini-core collections has been proposed as a strategy that allows the use of a small portion of a germplasm collection to represent the total collection. Core collections should be dynamic and need to be adjusted when additional germplasm and new information become available. The remaining accessions in the collection should, however, still be conserved as a secondary source of diversity. Concerns about core collections include rendering the reserve collection more vulnerable to loss, lack of representation of rare, endemic alleles, and poor relation to the specific needs of users. To address the latter concern, specialized core collections have been established around a particular trait, region, or type of material. For adaptive traits, core and mini-core collections may not capture the needed diversity.

## The focused identification of the germplasm strategy

An alternative to random selection and core collections is the use of the focused identification of germplasm strategy (FIGS), a trait-based approach, which assists gene bank managers in identifying desired genetic material with a high probability of having the sought trait (Bari *et al.*, 2012). In the last ten years, ICARDA, in collaboration with the Vavilov Institute in Russia and GRDC-Australia, has invested in the development of FIGS, which uses germplasm collection site agro-climatic and edaphic information to two adaptive traits. FIGS is based on the premise that the environment strongly influences natural selection and, consequently, the geographical distribution of organisms. It creates 'best-bet' trait-specific subsets of material by passing accession-level information, especially agro-climatic site information, through a series of filters that increase the chances of finding the adaptive trait of interest.

Accessions from these areas have a higher probability of containing traits and genes of interest. From this calculation, smaller subsets of genetic material are assembled that have a high potential to contain the plant traits that breeders need to develop robust crop varieties – capable of tolerating drought and extreme temperatures and resisting pests and disease. In several hundred searches delivered to date, FIGS has demonstrated that it can identify specific traits for breeders rapidly and precisely. In some cases, it has identified traits that researchers have been looking for, unsuccessfully, for a number of years. At its core is a powerful algorithm that matches plant traits with geographic and agro-climatic information about the places where samples were collected. This allows the rapid searching of thousands of plant samples conserved in gene banks to pinpoint a number of high-potential types that can meet the breeder's strategy.

FIGS has been successfully used to identify sources of resistance for several useful traits for breeding globally such as Sunn pest in wheat in Syria; Russian wheat aphid in bread wheat; abiotic stresses, such as drought adaptation in *Vicia faba* L.; resistance to stem rust in bread and durum wheat; and stem rust and stripe rust in accessions of wheat landraces (El Bouhssini *et al.*, 2009; Khazaei *et al.*, 2013). FIGS is also an efficient tool for linking gene bank accessions to a trait of interest. FIGS has also been used in chickpea to identify accessions with resistance to Fusarium wilt and Ascochyta blight and tolerance to drought and salinity. This has led to the development of new genotypes with improved disease resistance (Istanbuli *et al.*, 2022).

Under GRDC (Grains Research and Development Corporation) investment in chickpea with ICARDA, research has been conducted between 2015 and 2020. A total of 1364 chickpea accessions were collected from 31 countries selected by FIGS and classified into seven subsets based on stress-tolerance traits. These subsets encompassed Ascochyta blight, BGM, cold, drought, FW, salinity, and virus resistance. Through collaborations with partner institutions in the CWANA region, the subsets were evaluated in different locations for biotic and abiotic stress response. Salinity tolerance was evaluated through field and greenhouse experiments using hydroponics in Egypt, resulting in the selection of five salt-tolerant genotypes. Similarly, drought-tolerance screenings in Lebanon and Egypt identified 13 genotypes displaying favorable performance under stress conditions. Cold-tolerance assessments in Turkey and Lebanon revealed 18 genotypes with enhanced tolerance. Additionally, evaluations of chickpea virus subsets highlighted resistance or moderate resistance to specific viruses in controlled greenhouse and field conditions. Five genotypes were resistant (% infection < 10) to BWYV; eight genotypes were resistant (% infection < 10) to CpCSV; and three genotypes were moderately resistant (% infection < 40) to AMV. These findings provide valuable insights for future research and breeding efforts to improve chickpea varieties. FIGS is a valuable tool for plant breeders and researchers seeking to improve crop resilience and sustainability in the face of CC and other challenges.

## ICARDA's Research on Heat- and Drought-tolerant Variety Development

With the inherent and increasing adverse effects of CC, ICARDA breeding programs for cereals and legumes have focused their efforts on the development of germplasm tolerant to major abiotic and biotic stresses. As water is becoming scarce, even in the irrigated areas of the CWANA and SSA regions, ICARDA's germplasm development approach is to identify genotypes with disease resistance, high yield potential and water-use efficiency, so that wheat genotypes targeted for irrigated areas can cope with temporary drought periods. Similarly, this approach minimizes yield losses and maximizes yield gains during drought and good seasons, respectively, for the rainfed production system (Tadesse *et al.*, 2016). ICARDA breeding programs have, since their origin, worked together with the national programs to bring farmers more tolerant varieties to prevalent stresses. ICARDA being a CGIAR center targeting mainly the dry areas, the breeding programs are routinely exposing their breeding lines to hot and dry environments, selecting the ones that were able to cope best with these stresses. In the last decade, and with the advent of the new genotyping and phenotyping technologies, ICARDA breeding programs have been able to better characterize and understand the response of the new breeding lines to these stresses, for instance through the development and use of the new Heat Phenotyping Platform in Sudan and the Heat and Drought Phenotyping Platform in Morocco (Tadesse *et al.*, 2022b).

With this information, together with the new genotyping technologies, breeders have been able to identify the genetic mechanisms behind superior tolerance to heat and drought. Cropping season 2021/22 was one of the harshest of the last 50 years in Morocco. Despite the difficult season, ICARDA was able to phenotype and harvest most of the yield trails. The effect of drought makes phenotypic data collection more difficult and increases the experimental errors and coefficient of variation of yield trails, especially for manually collected data. The use of the high-throughput system substantially increased the precision of data collection with a reduced amount of time. The conventional selection and evaluation approach under typical drought and heat-stress conditions, and the sowing-time gradient approach for heat tolerance, augmented by use of associated physiological traits, were, for a long time, used to select drought- and heat-tolerant germplasm. Capable of screening thousands of plots per day with state-of-the-art sensors (multispectral cameras, Lidar, high-resolution RGB cameras etc.) or high-precision lysimeters, coupled with high-precision sensors, physiologists are able to provide breeders with high-quality and -quantity data as never before.

More than 880 new varieties have been released for cultivation – and this generates benefits worth US$ 850 million every year. Technology 'packages' have been developed – crop management, irrigation, pest control – to maximize yields from the new varieties. In the last five years, more than 120 cereal and legume varieties of ICARDA origin have been released by the national programs in more than 20 countries.

These genotypes offer higher and more stable yields and higher tolerance/resistance to diseases, insect pests, drought, heat, cold, parasitic weeds, and other stress factors. Some varieties also offer large improvements in end-use quality, nutritional value, and other traits. ICARDA breeders were able to identify the genetic mechanisms behind superior tolerance to heat and drought in chickpea, resulting in the release of many new heat- and drought-tolerant varieties in Ethiopia, Kenya, and Sudan (Fikre *et al.*, 2020). New bread wheat varieties are reaching more than 500,000 ha in the heat-stressed Ethiopian lowlands (Tadesse *et al.*, 2022a). ICARDA's efforts focus on developing improved varieties with tolerance to abiotic stresses like drought. The chickpea varieties selected for drought are Sultan (Azerbaijan); Ghab 5, FLIP97-706C, FLIP97-7C (Lebanon); and Sids1 (Egypt), while the lentil drought-tolerant varieties are Extra and Jemmat Shaim in Morocco (Figs 10.1 and 10.2) (Idrissi *et al.*, 2019; Ouji *et al.*, 2023).

Under the Dissemination of Interspecific ICARDA Varieties and Elites through Participatory Research (DIIVA-PR) project, funded by the Crop Trust, ICARDA in Morocco, alongside its partners INRA-Maroc and Benchaib Semences, has recently developed six promising new drought-tolerant varieties of durum wheat (El Haddad *et al.*, 2021a, b) and barley (Sanchez-Garcia, 2021) to help increase production, resilience, and nutritional quality for farmers. With their climate-smart and enhanced food-quality traits, such as heat, drought, and pest resistance, the new varieties aim to strengthen food and nutrition security and improve the livelihoods and resilience of Moroccan farmers. This groundbreaking research aligns with the Generation Green 2020–2030 strategy of the Moroccan Ministry of Agriculture by aiming to increase production while putting farmers at the center of the research. Integrated with other agricultural innovations, the new varieties will also support the country's goal of achieving resilient food sovereignty for major cereal crops.

To ensure the viability of the new varieties, crop breeders from ICARDA and INRA evaluated the new varieties of barley and durum wheat *in situ* across Morocco, exposing them to some of the most severe droughts on record, and identifying the most promising ones. The farmers also tested the varieties for four years across 30 Moroccan farms, engaging more than 200 local farmers in a participatory approach promoted by the DIIVA-PR project.

## Durum wheat varieties

Nachit was released by INRA in 2017 and should be commercially distributed by SONACOS starting in 2025. Its wild emmer origin ensures substantial grains and deep roots, suitable for collecting water from the soil's deepest part. As such, it is very well adapted to the northern plateaus and irrigated areas of Beni Mellal. Thanks to its big grains and high yield potential, some farmers can achieve more than ten tons per ha in irrigated conditions. Jabal was released in 2021 and should become commercially available from Benchaib Semences in 2025. It is a very rustic variety derived from wild goat grass, characterized by a tall stance ideal for producing straw, an appealing long black spike, and a superficial root system suitable for the shallow soils of the Atlas mountains and the southern regions. Jawahir is the latest addition released in 2022 by INRA and has yet to be sold to a seed company. It is the top-performing variety in terms of yield and good semolina color. It is also resistant to a very dangerous insect (Hessian fly) thanks to its parentage from a wild Armenian species and extremely drought-tolerant, making it perfectly suitable for the Settat-Safi region, especially if cultivated via conservation agriculture (CA).

## Barley varieties

Chiffa was released by INRA in 2016. It is a 'food and fodder' barley variety. It has easy thresh-ability (naked), a trait traced back to ancient varieties grown in the Nepalese highlands; it is drought-tolerant, with high β-glucan content, a dietary fiber with many health benefits against diabetes and ischemic diseases, and it has excellent resistance to net blotch disease. Assiya was also released in 2016 by INRA. It is a 'food and fodder' barley variety. It has easy thresh-ability, its grain cover (hull) detaching easily during threshing, a trait that traces back to ancient varieties grown in the Nepalese highlands. It has

**Fig. 10.1.** New drought-tolerant barley elite genotype issued from crosses with gene bank material being tested in farmers' fields in Morocco (*left*); and wheat variety (Sids 15) released in Egypt. (Photo: Dr. W. Tadesse)

**Fig. 10.2.** Two drought-tolerant lentil varieties, Jemaat Shaim and Extra, released in Morocco in 2019. (Photo: Dr. W. Tadesse)

high grain and straw production even under drought conditions and excellent resistance to net blotch disease. Khnata was released by INRA in 2017. It is a 'feed' barley variety with high grain and straw production. It increases farm profitability and resilience by providing straw for the livestock even in dry years and large grain size for a six-row variety with more carbohydrates.

In summary, breeders are now able to better introgress traits for tolerance to heat and drought into the modern elite lines designed to perfectly fit the prevalent crop/livestock farming systems and the agronomic technologies used by the farmers. All these traditional and new approaches have resulted in dozens of new tolerant varieties being co-developed with national programs and released to farmers in the most drought- and heat-prone areas in the region. Whether they are the new heat-tolerant bread wheat varieties being upscaled in Sudan and Ethiopia or the durum wheat ones in Senegal and Morocco or the barley ones in North Africa and Ethiopia, these new technologies are already making an impact in farmers' fields.

## Different Agro-management Techniques to Optimize Water-use Efficiency

### Diversified cropping systems

The challenges faced by small-scale family farmers operating in the world's drylands to produce nutritious and marketable food reliably and in enough quantity are countless. It is in the driest regions of the globe that CC-related shocks can be felt most vividly – temperatures are soaring at unprecedented rates; freshwater supplies from rain, rivers and aquifers are shrinking; and pests and diseases proliferate. All the while, biodiversity is declining due to inappropriate agricultural practices, and yields and production are struggling to keep up with rising demand. The expansion of monoculture farming, where farmers intensively and repeatedly cultivate a single-type crop on the same tired land, while relying on large quantities of agrochemical inputs, has greatly diminished the presence of wild pollinators and natural enemies, making crops increasingly unsuccessful and more vulnerable to pests, while the natural biodiversity of soil declines with every repeated crop to the point of collapse.

### *Description of the solution*

To support dryland farmers' livelihoods, income, and nutritional security as well as biological diversity and environmental sustainability, ICARDA, alongside its national and international partners and donors, promotes diversified cropping systems (DCS) across Africa, Asia, the Middle East, and North Africa (MENA).

DCS go back to basics by merging local knowledge and traditional ways, with ICARDA's cutting-edge scientific innovations to reintroduce diverse, self-sustaining cropping systems with low environmental footprints to produce food, feed, and fiber, deliver crucial ecosystem services to safeguard biodiversity and biochemical nutrient cycles, and better regulate water use. ICARDA promotes different options to diversify existing mono-cropping systems (Figs 10.3–10.5) to improve the productivity, stability, and delivery of ecosystem services at the farm level: crop rotation through introduction of different crops – food and forage legume crops in cereal–cereal rotation; intercropping; and relay cropping. In all three, the secondary crop, often a legume, a forage, or a cash crop – a fruit or vegetable – is integrated into a predominantly cereal-based system, with temporal and spatial differences. In a crop rotation model, for example, once the main cereal crop is harvested, the land is utilized to cultivate the secondary crop until the next cereal season. The nutritious residues it produces enrich the land, improve soil microbiome health, optimize yields, and increase efficiency while conserving water and resources. Since legumes capture nitrogen and fixate it in the soil, farmers' reliance on nitrogen-based fertilizers is reduced, with positive impacts on the agro-ecosystem and local biodiversity. However, in particularly water-scarce environments, land can be left fallow following the cereal crop.

**Fig. 10.3.** Crop rotation: different types of crops are rotated in a set seasonal order.

### *The framework of solution application*

In late 2019, ICARDA scientists quantified the economic advantages of legume–cereal rotations

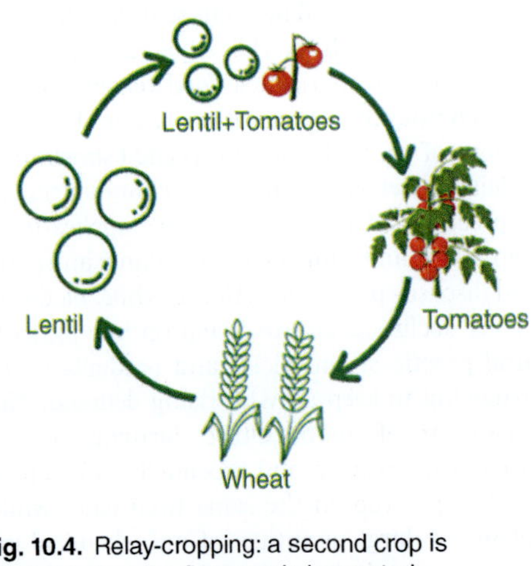


**Fig. 10.4.** Relay-cropping: a second crop is planted before the first crop is harvested.

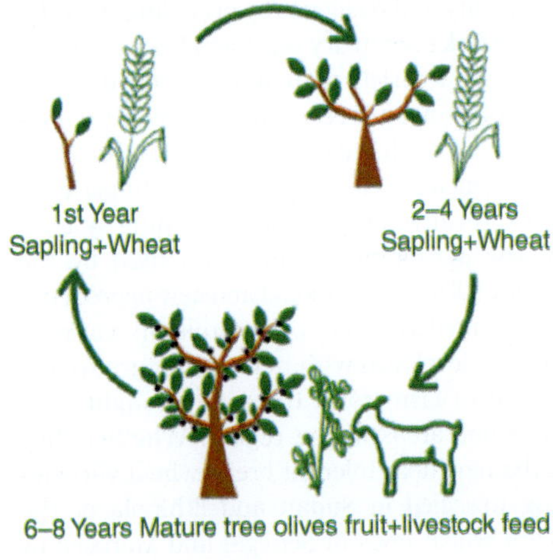


**Fig. 10.5.** Intercropping: two or more crops or trees are grown simultaneously with or without a row arrangement.

over cereal monocropping in dry areas. The study, which took place in Morocco's rainfed areas, involved 1230 family farming households over a period of two years. The adoption of improved faba bean varieties and wheat rotations in Morocco's rainfed areas resulted in a two-year average gross margin 48% higher than wheat monocropping. Diversifying the wheat-based system through relay seeding of low-water-requiring high-value spring crops in an early-maturity lentil variety, combined with supplementary drip-irrigation, increased system productivity, farm income by $410–10,295 ha$^{-1}$, and doubled water-use efficiency over a cereal monocrop in the Mediterranean rainfed environment of Morocco (Devkota and Yigezu, 2020).

A replicated field experiment was conducted in ICARDA's Marchouch research station in Morocco during 2019/20 cropping season. Commercially grown, short-duration lentil variety Bakariya was planted during the second week of December in two different crop geometries. Four different short-duration and high-economic-value spring crops, i.e. bean, chickpea, onion, and quinoa, were planted in between two paired rows a month before the lentil harvest. Irrigation (harvested groundwater applied using drip-irrigation) was applied immediately after seeding to ensure better germination. The amount of supplemental irrigation water and rainfall was measured during the crop growing season. Winter crops, i.e. lentil, was converted to wheat-equivalent yield and spring crops were converted to chickpea-equivalent yield considering the current farmgate price of the respective crops. Total income was calculated considering the grain/bulb yield per ha and multiplying by the current market price.

The main crop growing season had low rainfall (270 mm compared to 440 mm long-term average) with poor distribution, where > 26 mm rain fell during January and February, the major growth period for the winter crop. However, occurrence of late-season rainfall, i.e. March, April, and May, favored growth of the spring crop (Fig. 10.6).

Lentil yield was not affected due to crop geometry and relay seeding of different spring crops. In winter crops, wheat-equivalent yield and total income and economic water-use efficiency ($/mm water) was higher for lentil than in wheat (Fig. 10.7). Higher equivalent yield of lentil than wheat was due to the high price for lentil, and the comparatively low wheat yield could be related to prolonged drought during the active tillering stage. This study found that in low- and poor-distribution rainfall years, growing lentil is more economical than wheat. All

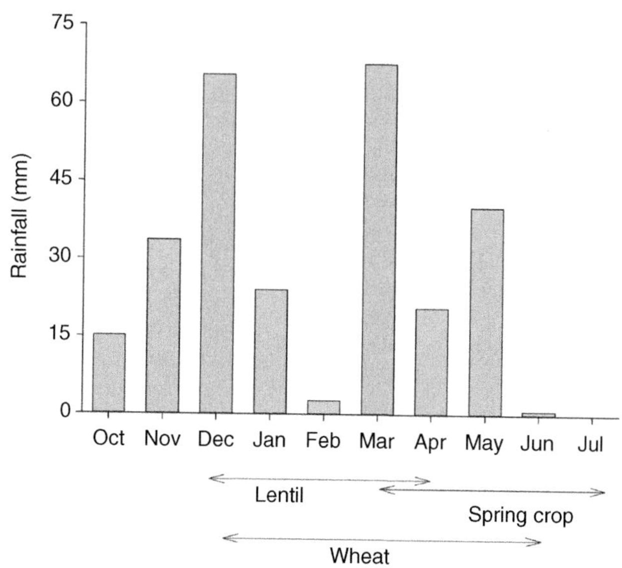


**Fig. 10.6.** Rainfall amount (mm) during 2019/20 crop growing season (Oct–July) in Marchouch research station, Morocco. (From Devkota and Yigezu, 2020)

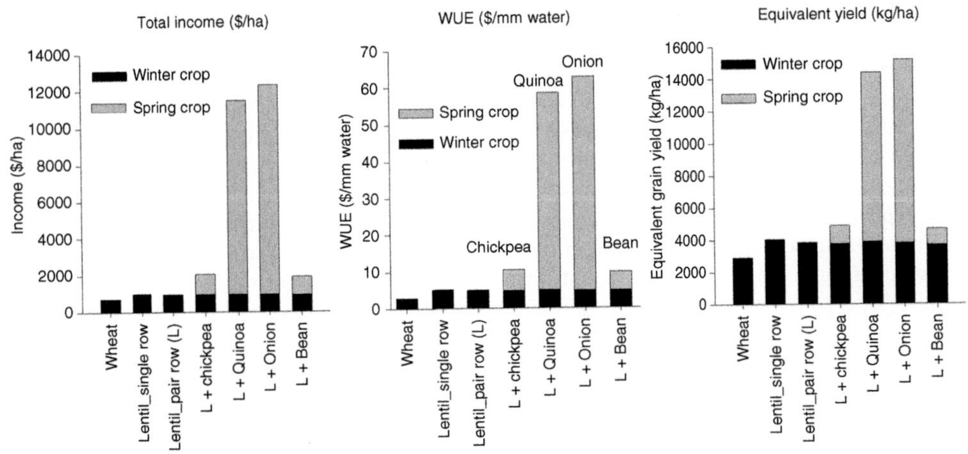


**Fig. 10.7.** Wheat- (for winter) and chickpea- (for spring season) equivalent yield, total income and water-use efficiency (WUE) of winter- and spring-planted crops. (From Devkota and Yigezu, 2020)

evaluated spring crops performed well in relay seeding with lentil, and produced additional income of US$ 900–11,000 ha$^{-1}$ (Fig. 10.7). Among the spring crops evaluated, onion produced significantly higher yield and income (US $11,378 ha$^{-1}$) followed by quinoa (US$ 10,523 ha$^{-1}$), chickpea (US$ 1,111 ha$^{-1}$), and bean (US$ 982 ha$^{-1}$). Similarly, water- (mostly rainwater) use efficiency was significantly higher with inclusion of a spring crop than sole lentil and wheat crop (Fig. 10.7), especially in the year with occurrence of late rainfall.

In low rainfall years, low-water-requiring crops, such as lentil, are more economical than wheat. In years with late rainfall or with supplementary irrigation, relay seeding of onion followed by quinoa, chickpea, and bean with winter-seeded lentil, have the highest economic water-use efficiency compared to a sole crop of wheat in the rainfed Mediterranean region. This practice leads to sustainable intensification of rainfed production systems, helping to generate higher income opportunities for farmers.

The other key benefit of intercropping, particularly in the dry areas, is the higher crop-per-drop ratio compared to monocultures. However, harvesting diversified fields is more labor-intensive than when one single crop is cultivated, which can lead to higher costs to farmers. DCS have myriad beneficial impacts for soil quality, yields, nutrition, income, and biodiversity health.

BENEFITS

- improved soil quality by rotations and/or mix of trees, cereals, legumes, roots, tubers, and vegetables;
- mitigation of heatwaves by providing shelter for crops and animals when trees are planted;
- diversification of farmers' crops and the food on their plates, and improvement of nutrition security at household level;
- enrichment of the soil microbiome, which is the best natural defense against soil-borne pests and viruses;
- creation of additional jobs to manage diversified fields and landscapes not suited to large and automated machinery;
- sustaining pollinators and natural enemies to create higher net income per surface;
- attraction of better market prices and creation of opportunities to develop value chains that support agro-ecological products;
- reduction of rural outmigration by creating better incomes, opportunities, and livelihoods, even on small areas of land.

## Supplemental irrigation

Supplemental irrigation (SI) refers to giving rainfed crops limited amounts of additional water during dry spells in order to improve and stabilize yields. SI is presented to farmers as a simple but highly effective package that allows them to plant and manage their crops while adapting to CC. The solution provides farmers with a cheap, robust, and portable irrigation system; information on when and how much to water crops; and advice on appropriate fertilization and crop diversity. When precipitation is insufficient for normal crop growth, farmers can provide additional water by conventional surface irrigation or by sprinkling or drip, when possible. SI allows farmers to plant their crops early, increasing yields by mitigating the damage done by low-moisture environments. With SI, farmers can react to changing climate patterns by planting and growing earlier.

### *The framework of solution application*

Wheat is the main staple food crop grown in the Mediterranean rainfed regions in Morocco, but production is unable to meet demand. Average wheat yield is lower than neighboring countries with similar climatic conditions. The existence of a large attainable yield gap shows the potential to increase wheat yield in the region. However, high variability in inter- and intra-seasonal rainfall presents a big risk for the farmers to invest in best-management practices. Optimizing genotype selection, seeding time, and crop demand-based water management practices have the potential to alleviate the risk associated with seasonal rainfall variability and enhance the stability of yield of this crop. The objective of the study presented here was to explore the best combination of genotype and management practices to enhance yield and water productivity (WP) of wheat in the Mediterranean rainfed production environment. Ten advanced durum wheat genotypes were evaluated at two seeding dates (timely – 12 November, and late – 25 December) and two water management regimes – rainfed and crop demand-based supplemental irrigation for three growing seasons (2016/17–2018/19) with contrasting rainfall amounts and distribution patterns. Our results indicate that combining appropriate genotypes and management is more effective than solely relying on the choice of the genotypes. The effect of seeding time on grain yield and WP was influenced by rainfall amount and its distribution. In low rainfall years (280 mm vs. 440 mm long-term average), early sowing enhanced yield and water productivity by 45–100% compared to late seeding, but no advantage was observed in wet years. Similarly, in a dry year, the application of crop demand-based supplemental irrigation increased yield by 2.1–2.5 t $ha^{-1}$ (72–90%). On average, more than 50% yield variability was explained by yearly variation in rainfall amount and distribution followed by supplemental irrigation (by 14%) and seeding time (by 12%). The contribution of the genotype on yield variability was insignificant. The response of genotype on grain yield and water productivity varied with seeding time and rainfall pattern. Selection of appropriate genotype, adjustment in seeding time, and application of crop demand-based supplemental irrigation help to enhance the stability of yield and WP of wheat under a Mediterranean rainfed environment.

Additionally, supplemental irrigation benefited millions of farmers and provided water and food security.

- **Water-use efficiency.** Using SI, farmers increased their water-use efficiency from 0.50 $kg/m^3$ to 0.99 $kg/m^3$. On-farm water productivity was 2.5 $kg/m^3$ under SI, compared to 0.3–1.0 $kg/m^3$ under rainfed conditions and 0.75 $kg/m^3$ under full irrigation.
- **Increased productivity.** SI increased the productivity of rainfed crops. Indeed, field trials have shown massive increases in wheat and barley yields with small quantities of SI. For example, wheat production in Syria increased from 2.4 million tons to over 4 million tons. In Yemen, SI increased the productivity of sesame by 103–120%.
- **Risk aversion.** The practice is highly adaptive, and it provides a high return that helps mitigate risk of crop failure. A cost-benefit analysis showed that the gross margin sustained a positive momentum over ten years, with supplemental irrigation responsible for about one third.
- **Increase in income.** The high adoption of this practice and resulting productivity gains have increased farmers' per-ha incomes by 35–50%.
- **Increase yields.** Effectively, millions of farmers benefited from SI. Yields of rainfed crop increased by over 400%. Using a $m^3$ of water at the optimum time and with SI's management package, a farmer could

produce 2.0–3.5 more kg of grain than by rainfed production alone.

### *Supplemental irrigation for early sowing*

In the lowlands of the Mediterranean region and rainfed south-Asian countries, farmers usually sow/seed their land only when a sufficient amount of rain has fallen. The date by which this amount of rain has fallen is usually called the 'onset rainfall', meaning the beginning of the rainy season.

Here, continuous cropping prevails as pure cereal or cereal–legume rotations. Every week's delay after this time results in a yield decrease of 200–250 kg/ha. If the onset of the seasonal rain is delayed, early sowing can be realized with the help of an SI system. With SI, it is possible to decide on the sowing date of the basically rainfed crops without needing to wait for the onset of the seasonal rain. This results in a longer growing season, better yield, and an earlier maturity that helps crops to escape terminal drought.

In the highlands, frost conditions occur between December and March and field crops remain dormant. Usually, the first rainfall, sufficient to germinate the seeds (the onset rain), comes late and results in a small crop stand when the frost occurs. As a result, rainfed yields are much lower than when the crop stand, pre-frost, is good. Ensuring a good crop stand before frost can be achieved by early sowing and applying 50–70 mm of SI. Supplemental irrigation, given at early sowing, dramatically increases wheat yield and water productivity. In the highlands of Turkey, applying 50 mm of SI to wheat sown early has increased grain yield by more than 60%, adding more than 2 t/ha to the average rainfed yield of 3.2 t/ha (Ilbeyi *et al.*, 2006). WP reached 4.4 kg/m$^3$ of consumed water compared to WP values for wheat from 1 kg/m$^3$ to 2 kg/m$^3$ under traditional practices. Similar results were found in the Iran highlands for wheat and barley (Tavakoli *et al.*, 2010).

### *Supplemental irrigation to alleviate moisture stress*

Unlike full irrigation, the timing for SI cannot be determined in advance. This is because the basic source of water to rainfed crops is rainfall which, being variable in amount and distribution, is difficult to predict. Since SI water is best given when the soil moisture drops to a critical level, the time for irrigation can best be determined by measuring the soil moisture on a regular basis. Supplemental irrigation is even referred to as life-saving irrigation in some south-Asian countries. Usually, however, one to three supplemental irrigation applications of not more than 100 mm each, annually, appear sufficient, depending on the rainfall amount and distribution.

SI provides farmers with benefits such as higher incomes, lower risk, increased water productivity, and encouragement to adopt modern technology like seeds and fertilizer. However, it is crucial that farmers understand how to apply SI to maximize water-use efficiency and productivity. The most important step is determining the best time to irrigate and the right amount of water to use. Integrating cultural practices and improved varieties, in which extension and human capacity-building services can play major roles, is also important for achieving optimal results. SI was developed for semi-arid environments in sub-Saharan Africa, including Burkina Faso and Niger, where rainfall comes in short bursts and dry spells are frequent. ICARDA and partners, including farming organizations, developed and optimized SI packages (including training) for rainfed cropping systems in Syria, Tunisia, Morocco, Iran, Iraq, Jordan, Lebanon, Yemen, Turkey, China, and Pakistan. As policies prioritize rainfed agro-ecosystems, SI's potential will grow.

### *Productivity and profitability*

Risk weighs on the daily lives of poor rainfed farmers, and investment packages have to help reduce that risk. Risks include not only climatic and limited access to reliable technology and water, but also unstable land tenure and poorly functioning product and credit markets. Investing in SI can have a significant impact on justified returns from dry-farming systems. Biophysical returns on water with SI are higher than those under conventional irrigation and are highest with deficit applications. An integrated investment package including water harnessing and irrigation technology, irrigation scheduling, training, and cropping and fertilizing guidance is probably the best. Combined soil and water management investments can also have a high return.

The key requirements for successful investment in SI include:

- determining the most appropriate scheduling, crops and cropping patterns, and socio-economic feasibility;
- strong water-user associations with incentives for local communities to use water efficiently;
- managing the economic and environmental consequences of using water in SI; and
- developing policies that foster an enabling environment for the adoption of water-efficient technologies.

A three-year study by ICARDA, in collaboration with GCSAR (General Commission for Scientific and Agricultural Research) of Syria, investigated the effects of adopting supplemental irrigation for olive productivity. The research experiment was carried out at Jillin Agricultural Research Station, located 25 km north of Dara'a City with an average annual rainfall of 400 mm. The experiment included three water treatments: irrigation to 100% of crop water requirement (CWR), irrigation to 50% CWR, and rainfed treatment (control). The Sorani and Jlout cultivars of olive were, on average, 30 years old and spaced 8 x 8 m apart. Soils have a high content of clay (between 67% and 71 %), field capacity (volumetric) ranged between 45% and 46%, and bulk density between 1.2 g $cm^{-3}$ and 1.4 g $cm^{-3}$. In the 2012 growing season, the application of deficit supplemental irrigation at 50% CWR (353 mm rainfall + 162 mm irrigation) increased fruit yield by 2460 kg $ha^{-1}$ (33%) for Sorani and 3900 kg $ha^{-1}$ (96%) for Jlout, compared with the rainfed treatment. At 100% CWR (353 mm rainfall + 325 mm irrigation), fruit yield increased by a further 2030 kg $ha^{-1}$ (25%) for Sorani and 1250 kg $ha^{-1}$ (13%) for Jlout. In terms of percentage oil extracted from fruit, Sorani produced the highest values when grown under 50% CWR (24%). In the 2011 growing season, for 50% CWR treatment, fruit yields increased by as much as 113% (4375 kg $ha^{-1}$ vs. 9360 kg $ha^{-1}$) when 115 mm of irrigation was applied (in addition to 316 mm rainfall) to rainfed Sorani. A further increase of 2652 kg/ha (28%) over 113% achieved with 50% CWR treatment, was achieved by applying an additional 115 mm of irrigation (100% CWR) in addition to the rainfall. The year 2013 was wetter and hotter than 2012 and 2011. Yields of 50% CWR treatment increased by as much as 77% (4836 kg $ha^{-1}$ vs. 8580 kg $ha^{-1}$) when 142 mm of irrigation was applied (in addition to 416 mm rainfall) to rainfed Sorani. For the same cultivar, a further increase of 1404 kg $ha^{-1}$ (16%) was achieved by applying an additional 142 mm of irrigation (100% CWR) in addition to the rainfall. Presently, 695,711 ha are planted to rainfed olives in Syria. Based on the above results, if 25% of this area applies supplemental irrigation, yields will increase by approx. 90% for Jlout and 100% for Sorani. At US$ 0.70/kg olives, this translates to about US$ 646 million additional income for farmers growing Jlout and US$ 670 million for farmers growing Sorani.

A study by ICARDA scientists (Bader *et al.*, 2011) in different provinces of Syria recorded distribution of farmers who have heard about and adopted supplemental irrigation. It was reported that, generally, it is not necessary for the farmer to know what supplemental irrigation means before they adopt it. In fact, most farmers apply supplemental irrigation at a time when they do not know exactly what it means. About 83% of the sampled farmers who own water sources had heard about supplemental irrigation, and 72% of them had adopted it. However, only 21% of farmers were aware of supplemental irrigation in its proper form, and received information through communication channels such as extension, public media, neighbors, and ICARDA scientists. Moreover, adopters of supplemental irrigation who used groundwater gained 20,823 SL/ha (SL = Syrian pound) net return, compared to non-adopters, who used more water but whose gain was only 15,386 SL/ha (Table 10.1). WP, the other indicator of the benefit of supplemental irrigation, is defined by Shideed *et al.* (2005) as the ratio of crop production (kg) to the unit of water used ($m^3$). Water productivity was calculated as a ratio of production to total water applied, including rainfall. Supplemental irrigation resulted in an increase in water productivity for adopters compared to non-adopters (0.94 kg/$m^3$ vs. 0.8 kg/$m^3$), a highly significant difference ($p < 1\%$) as shown in Table 10.1.

Supplemental irrigation can be a useful water management tool to increase water productivity

**Table 10.1.** Impact of supplemental irrigation on net returns and water productivity in wheat farming based on groundwater. (From ICARDA)

| | Net return SL/ha | WP kg/m$^3$ |
|---|---|---|
| Adopters | 20,823 | 0.94 |
| Non-adopters | 15,386 | 0.80 |

Source: Field surveys. SL: Syrian pound; WP: Water productivity

from rainfed forage and sheep production from pastures in water-deprived areas. A three-year study (Ates *et al.*, 2013) by scientists at ICARDA and the Bahri Dagdas International Agricultural Research Institute, Konya, compared sheep production from permanently sown pastures with four levels of irrigation in Konya, Central Anatolia region of Turkey. Pastures were established in 2007 with red fescue (*Festuca rubra* L.), Kentucky bluegrass (*Poa pratensis* L.), perennial ryegrass (*Lolium perenne* L.), white clover (*Trifolium repens* L.) and bird's foot trefoil (*Lotus corniculatus* L.) and irrigated at 100%, 75%, 50% or 25% of field capacity (FC). Established pastures were grazed rotationally by flocks of weaned lambs between 2008 and 2010. Each pasture treatment produced similar liveweight gains in lambs (g head$^{-1}$ d$^{-1}$) ($p = 0.35$), but the total animal production (kg ha$^{-1}$ d$^{-1}$) differed ($p < 0.01$) consistently in each year of the study. Pastures irrigated at FC 100% and FC 75% produced approximately 2 kg ha$^{-1}$ d$^{-1}$ liveweight. Lower levels of irrigation reduced ($p < 0.01$) liveweight gain per ha to less than 1 kg ha$^{-1}$ d$^{-1}$ in pastures irrigated at FC 25%. The average annual liveweight production was 498, 445, 380, and 198 kg ha$^{-1}$ for FC 100%, 75%, 50%, and 25%, respectively. Water productivity per unit of meat produced was low with full irrigation, particularly during summer months, but increased for each unit of meat produced with deficit irrigation.

### *Potential for scaling out*

The ICARDA multi-disciplinary team conducting suitability analysis for supplemental irrigation in West Asia and North Africa (Ziadat *et al.*, 2014) suggested that site selection criteria be:

- rainfall: the maximum rainfall amount of 500 mm since areas that receive more than 500 mm may not need supplementary irrigation;
- slope: no more than 5%, to avoid soil erosion; and
- crops: winter crops are chosen since supplemental irrigation is a method that can be used to compensate for the deficiency of rainfall during certain periods of winter.

Figure 10.8 shows areas where supplemental irrigation can be applied. To identify more precisely those areas with a high potential for supplemental irrigation, the water resources were superimposed on the similarity map for supplemental irrigation. Areas close to irrigation water sources should be targeted by local governments for implementing supplemental irrigation. The locations of dams in the WANA region that are close to those areas identified by the rainfed benchmark/supplementary irrigation results are shown in Fig. 10.8.

## Mechanized raised bed (MRB) technology package

MRB technology raises the seedbed simultaneously seeding wheat, consequently creating furrows, of which the length depends on the dimensions of the farm-field. The inter-furrow spacing, and the width depends on the type of crop planted and on the soil type. The field/terrain may not exceed a slope of 5%. This allows water to infiltrate, reach the end of the furrow and for safe runoff, preventing waterlogging. So, to implement MRB technology, the field might be leveled prior to implementation, but this is often not the case as MRB is designed to local conditions, and most fields already have a slope of less than 5%. The practice of the machine requires a loose soil, so it is required that the field is ploughed prior to seeding, and therefore MRB is not seen as a type of plowing. If the field preparation is done, an MRB machine can start seeding and raising the seedbed. Specific characteristics for

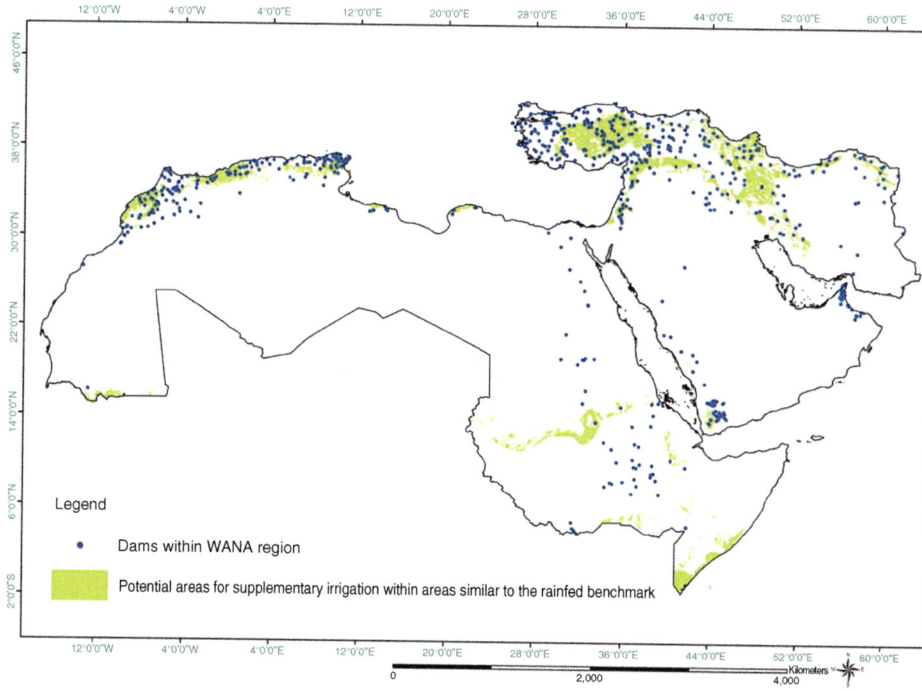


**Fig. 10.8.** Potential areas for supplementary irrigation within West Asia and North Africa. (From Ziadat *et al*., 2014)

MRB technology with respect to conventional seeding are that seeding and raising seedbeds are done simultaneously. Raising seedbeds mechanically saves 80% of the workload with respect to manually raising seedbeds. The after-harvest practice depends on the farmer's preference, as some farmers prefer letting the stubble be grazed, while others clear the field. The after-harvest practices are thus independent with respect to MRB.

In addition, to complement the MRB, high-quality seeds are offered. These are beneficial as they yield higher than the regular wheat seeds. Nevertheless, the machine can also be used with regular wheat seeds, if the high-quality seeds are found too expensive by the farmer. The high-quality seeds are hybrid and need to be replaced after three years to ensure high quality is maintained.

### *Description of the solution*

To build livelihood resilience, ICARDA and national partners in Egypt and Sudan, along with financial support from AFESD, have developed a unique community-based innovative package. The mechanized raised-bed technology is adapted to dryland conditions and improves water-use efficiency, improving farmers' livelihoods through increased farm productivity with fewer inputs. Conventional irrigated agriculture usually involves flood irrigation in flat land or furrow irrigation, which often results in uneven water distribution in the field, saturating some crops, leaving others too dry, and wasting valuable water resources. MRB machines are attached to standard farm tractors, and quickly and easily create raised soil beds, on which they simultaneously sow crops, while creating adequate furrows along which optimum water travels easily, vastly increasing water efficiency (Fig. 10.9).

The technology has significant positive impacts for local farmers as applied water is saved by 25%, water pumping costs decreased by 25%, seed rate reduced by 50%, farming costs decreased by 30%, fertilizer use efficiency increased by 30%, and crop yield increased by 15–30% with respect to conventional farming. Overall, it is estimated that farming under MRB is about a third of the cost with respect to conventional farming. Manually raising seedbeds was considered too

**Fig. 10.9.** ICARDA mechanized raised bed planter being used for land preparation and sowing in Egypt, and comparison between flat planting (FP) and mechanized raised bed planting (MRB). (From Swelam, 2016)

expensive due to the required amount of labor. Thus, regular flood irrigation was practiced. Practically, this results in full-surface flooding of the field. This has significantly higher evaporation, hence increasing salinization, as opposed to furrow irrigation through raised seedbeds. Also, because water is well distributed over the field due to the furrows (reducing water stress and waterlogging), there is less leaching of the nutrient, hence increased nutrient efficiency. The raised seedbeds allow excess water, in case of a heavy rainfall event or over-irrigating, to safely run off. These features of better water disposal and reduced evaporation make MRB technology well suited with respect to CC, which leads to more concentrated rainfall events and increased temperature, hence increased evaporation. In addition, as MRB prevents, e.g., waterlogging, it prevents land degradation (e.g. salinization).

### *The framework of solution application*

MRB technology has been applied by ICARDA and national partners in Egypt. Egypt is a water-scarce country and 95% of its water comes from beyond its borders. In addition, Egypt is highly dependent on the import of wheat; 50% of the wheat demand is met by import. In Egypt, water scarcity and mediocre yield are two issues that keep the majority of people working in the agricultural sector in poverty. Water is scarce as the annual precipitation is less than 250 mm, leading most farming to be irrigated. The water comes from the Nile river. However, available irrigation water per farmer is rather low due to population growth. Thus, due to clay soil and the use of flood irrigation, waterlogging and uneven water distribution over the field lead to salinization of the soil, harming the farmer's yields. Furthermore, yield is affected by the lack of water, insufficient use of fertilizers, and the use of relatively low-quality seeds. In particular, fertilizers are expensive, leading to insufficient and poor application.

In effect, the core objective of the MRB technology package is to fit within this context and realize more output with less input, consequently improving livelihoods. Indeed, MRB improves farmers' resilience with increased water and nutrient efficiency. The adopters of MRB receive benefits such as improvement in their livelihoods, a decreased workload, increased yields, and more efficient use of resources (water, fertilizer, and seeds). The first-stage research and designing of MRB technology was done in 2003. Introductions and pilots of the technology were designed together with regular farmers in the Nile Delta area from 2010 until 2013. In 2015, MRB technology was proven beneficial, permitting outscaling. Thanks to the shown potential of MRB technology and Egypt's reliance on foreign countries for water and wheat, MRB technology has become a strong component of Egypt's national wheat campaign. In Al-Sharkia, for instance, land devoted to raised-bed-sown wheat increased from 950 ha in 2009/10 to 33,600 ha in 2014/15. A similar trend has been noted nationwide across different parts of Egypt. Raised-bed planting now covers some 105,000 ha – the result of government efforts to replicate successes in Al-Sharkia. The Egyptian government aims to cultivate 2 million acres of wheat under MRB technology by end of 2022.

### *Scale of adoption*

Therefore, the livelihoods of farmers who have adopted this technology have been greatly improved. Farmers who have adopted MRB agree that it is affordable, easy to apply, improves production, and is cost-saving. Furthermore, since this technology increases irrigation efficiency, it can mitigate existing upstream-downstream issues in terms of availability, as there is more available irrigation water. Also, as MRB technology is currently outscaled, it creates employment opportunities since MRB machines are locally produced from scratch.

- The technology has transformed irrigated wheat-based producing systems across Egypt, Sudan, and beyond, achieving significantly higher crop yields of 20–30% while ensuring 25% savings in irrigation water and reducing farming costs by about 20%.
- Scaling-out activities have led to substantial increases in the wheat area cultivated by MRB throughout Egypt with production expected to increase from 8 million tons in 2017 to over 14 million tons by 2025. This could result in a total irrigation water saving of more than 1.5 billion m$^3$/year by 2025, benefiting an additional 200,000 ha.
- ICARDA's unique research-for-development model differs from other global MRB in its close development partnership with NARS and the farming communities it targets so that a machine suitable for real-world agro-ecosystems is developed.
- ICARDA also delivers an integrated approach of inputs, demonstrations, and training field days, involving stakeholders in decision making. Policies in Egypt, Sudan, and Nigeria have also enabled a suitable environment for the adoption of MRB technology.
- Numerous agreements and funding have been deployed to out-scale the MRB outside of Egypt through donors such as the Bill & Melinda Gates Foundation, the EU, and FAO, among many others.

### *Productivity and profitability*

Impacts in Egypt's Al-Sharkia Governorate, where the innovation was disseminated, have been particularly impressive. The results, obtained during the initiative's first phase, compared the raised-bed technique with conventional flat-surface irrigation. They include:

- 25% average saving in applied irrigation water;
- 30% average increase in grain yield;
- an average 73% increase in water-use efficiency; and
- 30–50% saving in the quantity of seed used for planting.

In addition, raised-bed planting brought savings in energy and labor. The average time needed to pump water to irrigate 1 ha of wheat planted on a raised bed was 29.4 hours, compared to 43.9 hours on a flat field. The subsequent reduction in the costs of labor and fuel – approximately 33% – contributed to a rise in farmers' incomes.

In Sudan, raised-bed planting was also tested, and demonstrated that wheat yields could be increased by an average of 15%. In Morocco, water-use efficiency increased by 13% compared to basin irrigation.

### *Potential for scaling out*

Scaling-out activities have led to substantial increases in the wheat area cultivated by MRB throughout Egypt with production expected to increase from 8 million tons in 2017 to over 14 million tons by 2025. This could result in a total irrigation water saving of more than 1.5 billion m$^3$/year by 2025, benefiting an additional 200,000 ha.

## MRB technology package

The biggest challenge in dryland agriculture is low and erratic rainfall. New technology packages are available that can help build effective, low-cost systems for harvesting rainwater, delivering larger and more stable water supplies for crops and livestock, even in very dry areas.

### *Description of the solution*

Water harvesting is based on the principle of depriving part of the land of its share of rain, which is usually small and non-productive, and adding it to the share of another part. This brings the amount of water available to the latter area closer to crop water requirements and thereby permits economic agricultural production. Water harvesting may occur naturally or by intervention. Natural water harvesting can be observed after heavy storms, when water flows to depressions, providing areas for farmers to cultivate. Water harvesting by intervention involves inducing runoff and either collecting or directing it, or both, to a target area for use. Besides being applied to agriculture, water harvesting may be developed to provide drinking water for humans and animals as well as for domestic and environmental purposes.

### *The framework of solution application*

**Case study – Iraq**

Wheat is one of the most important food crops in areas of west Asia such as in the Kurdistan region of Iraq (IKR). Bread (*Triticum aestivum* L.) and durum (*Triticum durum* L.) wheat are cultivated in Kurdistan where consumption is high, and yields are low despite the potential for increased production. One of the reasons for low productivity in Kurdistan is the dependence on rain for watering the crop. Most of the wheat in the region is rainfed and moisture stress adversely affects yields. The region is, however, blessed with hills and valleys which generate ample runoff during the wet season which could be captured for use during dry spells.

ICARDA, the General Directorate of Agriculture Research & Extension of IKR, and Japanese International Cooperation Agency (JICA) implemented a study in Zurgaziraw near Erbil City in IKR in 2014–15 on the use of water-harvesting techniques to provide supplemental irrigation to wheat crops. The site has an average annual rainfall of 250 mm and is located at the foothill of a series of mountains which allows ample runoff from the mountains to flow through the site. Simeto wheat cultivar was sown on 21 November 2014 at a rate of 128 kg ha$^{-1}$. Due to the very low rainfall, three treatments were tested: (i) 0.25 ha (plot A) with 200 mm supplemental irrigation; (ii) 0.25 ha (plot B) with 100 mm supplemental irrigation; and (iii) 0.25 ha (plot C) under rainfed (control).

An important constraint to SI is the availability of water resources when irrigation is needed. Groundwater or surface water may be available in some places but often not. In these situations, rainwater harvesting can be used to collect runoff water and store it for use as SI. In Iraq, a six-step process (Fig. 10.10) was followed: (1) selection of a suitable site with steep, sloped, barren land generating non-beneficial runoff; (2) design and excavation of a water-harvesting reservoir; (3) reservoir lined to prevent seepage and reinforced with a spillway and rocks to prevent rupture; (4) reservoir monitored for water collection and maintenance during the wet season; (5) sprinkler system designed and installed in the field towards the end of the wet season; and (6) soil moisture monitored using a measuring device with supplemental irrigation applied to both treatment plots (plot A: 200 mm SI; plot B: 100 mm SI) each time the moisture level dropped below 30% of field capacity. At maturity, the crop was harvested from three 1 m$^2$ plots in each treatment (nine plots in total) to calculate yield, spikes, and grains per unit area.

Ten supplemental irrigation doses were applied to both treatment plots (A and B). Since the total amounts were set at 200 mm and 100 mm for plots A and B, respectively, the amount received in each irrigation application operation for plot B was half of plot A. Crop yield increased, on average, by 91% (0.75 t ha$^{-1}$ under rainfed vs. 1.43 t ha$^{-1}$ under 100 mm SI treatment) with the addition of 100 mm supplemental irrigation, compared with rainfed-only plots. With 200 mm of supplemental irrigation, crop yield increased, on average, by a further 112% (3.03 t ha$^{-1}$ under 200 mm SI treatment vs. 1.43 t ha$^{-1}$ under 100 mm SI treatment) on top of the 91% increase achieved by switching from rainfed to 100 mm SI (Table 14.2).

**Table 10.2.** Yield and yield components measured at the three plots during the 2014–15 growing season at Zurgaziraw area near Erbil City in Iraqi Kurdistan region (IKR). (From ICARDA)

| Location | Spikes (per m$^2$) | Grain per spike | 1000-grain weight (g) | Grain yield (t ha$^{-1}$) |
|---|---|---|---|---|
| Plot A (200 mm supplemental irrigation) | 320a | 21a | 45.47a | 3.03a |
| Plot B (100 mm supplemental irrigation) | 151b | 21a | 44.03b | 1.43b |
| Plot C (rainfed) | 192ab | 12a | 31.40c | 0.75c |

Values with same letter do not differ significantly ($p$ > 0.05) between irrigation treatments.

**Fig. 10.10.** A six-step process to implement water harvesting for supplemental irrigation.

**Case study – Jordan rangelands (Badia)**

In a project on Community-based Optimization of the Management of Scarce Water Resources in Agriculture in West Asia and North Africa, which started in 2004 with funding support from Arab Fund for Economic and Social Development (AFESD), International Fund for Agricultural Development (IFAD), and OPEC Fund for International Development (OFID), ICARDA established several benchmark research sites in different agro-ecosystems including the Badia in Jordan for the implementation of an integrated watershed management approach. GIS-based land suitability analysis was performed to integrate biophysical and socio-economic factors into a comprehensive site-characterization map. In the two communities of Mharib and Al-Majidyya, micro-catchment water-harvesting techniques, like contour ridges and furrows, runoff strips, and bunds were designed and established. The Vallerani machine, a plow with a hydraulic arm for intermittent pit excavation along the natural contour of the slope, was used for excavating furrows (Fig. 10.11).

Scientific research led to a low-cost Contour Laser Guiding (CLG) system, based on GPS guidance technology, which was adapted and implemented to the Vallerani unit. For the sustainability of the project, communities were actively engaged, and they agreed not to graze the newly planted shrub for at least two years after planting, a time needed for the establishment of drought-tolerant Atriplex species as a source of fodder shrubs. This system reduced the pressure on rangelands and also provided sources of nutritious feed from the drought-tolerant Atriplex species. The integrated approach that was

*Continued*

Continued.
used in the Badia recovery to sustain plant cover and conserve soil from degradation has resulted in community resilience to harsh environments and to the severe effects of CC. The use of the mechanized Vallerani system has allowed the implementation of rainwater harvesting packages at a large scale. The main beneficiaries are the herders and rural communities of marginal lands where sheep and goats are the main source of income for better livelihoods. This system has enabled them to grow fodder shrubs and reduce pressure on rangelands for livestock grazing in areas where rainwater is not enough to support stable and sustainable production and better livelihoods.

Beneficiary farmers enjoy more than double the yield for barley and 1.6 times for rangeland and forage shrubs production compared to those grown without rainwater harvesting. Highly degraded lands have been rehabilitated by this improved vegetation, thus mitigating land degradation. The GPS-guided micro-catchment rainfall system has reduced cost and time required to delineate contours for the plow to follow and tripled the construction capacity (up to 30 ha/day), improved efficiency and precision, and substantially reduced the cost of establishing micro-catchments.

In partnership with Jordan's National Agricultural Research Center (NARC), USDA-Agricultural Research Service, Texas A&M University, Jordanian Ministry of Agriculture, and University of Jordan, another project on Water and Livelihood Initiative, funded by the United States Agency for International Development, focused on improving livelihoods of participating rural communities and reducing land degradation at pilot sites using the Vallerani rainwater harvesting package. Under the Badia Restoration Program, Jordan's Ministry of Environment has purchased three new Vallerani machines for the rehabilitation of rangelands in the Badia region. It has substantially facilitated wide dissemination of this technology. In Jordan, the Vallerani rainwater harvesting package developed through scientific research has been implemented on over 3864 ha of rangeland so far, with adoption rates tripling since the start of the project. Biodiversity has increased by capturing 67–73% of available seeds in the harvested area compared to 3–6% in the control sites. Around 40–50% of precipitation is prevented from being lost by evaporation at the intervention sites.

## Feasibility of using mechanized rainwater harvesting technique

Planting barley using the mechanized rainwater harvesting technique:

- 17% economic internal rate of return (EIRR) as compared to 7.8% in traditional way of planting

Planting shrubs using mechanized rainwater harvesting technique was more feasible than planting shrubs in traditional way:

- 13% EIRR as compared to 7.4% in traditional way of planting
- The contribution of environmental benefits in the calculation of return on investment for rainwater harvesting techniques have led to increased EIRR of 17% compared to 13%.

### *End-user's profile*

One key limitation is that the technical aspects of water-harvesting structures – often complex – are usually implemented by unskilled labor. Laying ridges or contour lines is essential to the proper functioning of a water-harvesting system, but this is a complicated procedure and requires special training. Proper site selection is also required. Failure to adequately tailor a method to site characteristics – topography, soil type, vegetation cover etc. – will result in failure.

In addition, the contextual environment in the drylands is increasingly unfavorable. The breakdown in collective conservation systems, subsidized feed, and a corresponding increase in animal populations and overgrazing means that unless new legislation is introduced, and existing institutions are reformed, dry ecosystem restoration schemes will have limited success.

### *Productivity and profitability*

A PRACTICAL AND COST-EFFECTIVE APPROACH. In an effort to overcome these constraints, ICARDA scientists worked with two communities in Jordan's Badia – Mhareb and Majdieh – to design, test, and promote a practical rainwater harvesting package. The package combines indigenous knowledge with mechanization and a contour laser guiding system to enhance the accuracy of ridges and bunds.

**Fig. 10.11.** Field of Vallerani contour bunds after construction. (From Oweis)

Efforts were also taken to improve the selection of restoration sites, design appropriate structures, select the right shrubs, and, most importantly, implement sustainable grazing strategies and ensure ongoing maintenance.

With support from Jordan's National Center for Agricultural Research and Extension (NCARE), 80% and 90% of farmers in Mhareb and Majdieh, respectively, used the package. Jordan's Ministry of Environment also adopted it, allocating funds for its implementation across 2000 ha so far – an area the Ministry is planning to extend even further.

The result: rapid vegetation growth, more animal feed, less soil erosion, and enhanced biodiversity. The package is also cost-effective; it costs a mere US\$ 32 ha$^{-1}$, which includes the production, planting, and maintenance of shrub seedlings (US\$ 11 ha$^{-1}$) and the economic internal rate of return is estimated at some 13%.

Other experts calculated a benefit-cost ratio (BCR) of > 1.5 for revegetation of degraded Badia areas through mechanized rainwater harvesting and saltbush. However, costs vary based on the seedling costs and availability of trained labor.

### *Potential for scaling out*

GIS-based land suitability analysis conducted by the ICARDA team suggests that a total area of 2.7 million ha in Jordan has the potential for adoption of mechanized rainwater harvesting packages, with further prospects of scaling out on an area of 300 million ha in MENA countries. IFAD contributed to these efforts through the Agriculture Resource Management Project (ARMP) II, a development project, where tenders for the purchase of three Vallerani machines were made in 2015. The valuation and assessment of environmental benefits associated with implementing rainwater harvesting techniques provide encouraging justification for public investment in scaling out of these techniques in the dry areas of Jordan. The scaling-out efforts will achieve higher success if water harvesting interventions are included in plans for integrated land and water resources development taking into consideration all the necessary technical, agricultural, socio-economic, and institutional aspects and inputs.

## References

Ates, S., Isik, S., Keles, G., Aktas, A.H., Louhaichi, M. *et al.* (2013) Evaluation of deficit irrigation for efficient sheep production from permanent sown pastures in a dry continental climate. *Agricultural Water Management* 119(2013), 135–143. DOI: 10.1016/j.agwat.2012.12.017

Bader, A., Jouni, N. and Shideed, K. (2011) Adoption and impact of supplemental irrigation in wheat-based systems in Syria. *Water Benchmarks of CWANA – Improving Water and Land Productivities in Rainfed Systems – Number* 8. University of Aleppo, Syrian Arab Republic. Available at: https://hdl.handle.net/20.500.11766/8207 (accessed 16 July 2024).

Bari, A., Street, K., Mackay, M., Endresen, D.T.F., De Pauw, E. and Amri, A. (2012) Focused identification of germplasm strategy (FIGS) detects wheat stem rust resistance linked to environmental variables. *Genetic Resources and Crop Evolution* 59, 1465–1481. DOI: 10.1007/s10722-011-9775-5

Baum, M. and Al-Zu'bi, M. (2021) *From Fragility to Resilience in Central and West Asia and North Africa.* CGIAR System Organization, Montpellier, France. Available at: https://hdl.handle.net/10568/121104 (accessed 16 July 2024).

Conservation International (2016) *Impact Report 2015*. Available at: https://conservation.org.gy/cg/cc/media/CI_2015_ImpactReport.pdf (accessed 13 April 2023).

Devkota, M. and Yigezu, Y.A. (2020) Explaining yield and gross margin gaps for sustainable intensification of the wheat-based systems in a Mediterranean climate. *Agricultural Systems* 185, 102946. DOI: 10.1016/j.agsy.2020.102946

El Bouhssini, M., Street, K., Joubi, A., Ibrahim, Z. and Rihawi, F. (2009) Sources of wheat resistance to Sunn pest, *Eurygaster integriceps* Puton, in Syria. *Genetic Resources and Crop Evolution* 56, 1065–1069. DOI: 10.1007/s10722-009-9427-1

El Haddad, N., Kabbaj, H., Zaïm, M., El Hassouni, K., Sall, A.T. *et al.* (2021a) Crop wild relatives in durum wheat breeding: Drift or thrift? *Crop Science* 61(1), 37–54. DOI: 10.1002/csc2.20223

El Haddad, N., Sanchez-Garcia, M., Visioni, A., Jilal, A., El Amil, R. *et al.* (2021b) Crop wild relatives crosses: multi-location assessment in durum wheat, barley, and lentil. *Agronomy* 11, 2283. DOI: 10.3390/agronomy11112283

Fikre, A., Desmae, H. and Ahmed, S. (2020) Tapping the economic potential of chickpea in sub-Saharan Africa. *Agronomy* 10(11), 1707. DOI: 10.3390/agronomy10111707.

Hamwieh, A. and Imtiaz, M. (2015) Identifying water-responsive and drought-tolerant chickpea genotypes. *Crop and Pasture Science* 66(10), 1003–1011. DOI: 10.1071/CP14225

Idrissi, O., Sahri, A., Houasli, C. and Nsarellah, N. (2019) Breeding progress, adaptation and stability for grain yield in Moroccan lentil improved varieties. *Crop Science* 59(3), 925–936. DOI: 10.2135/cropsci2018.07.0431

Ilbeyi, A., Ustun, H., Oweis T., Pala, M. and Benli, B. (2006) Wheat water productivity in a cool highland environment: effect of early sowing with supplemental irrigation. *Agricultural Water Management* 82, 399–410. DOI: 10.1016/j.agwat.2005.08.005

Istanbuli, T., Abu Assar, A., Tawkaz, S., Kumar, T., Alsamman, A.M. *et al.* (2022) The interaction between drought stress and nodule formation under multiple environments in chickpea. *PLOS ONE* 17(10), e0276732. DOI: 10.1371/journal.pone.0276732

Kehel, Z., Yazbek, M., Tsivelikas, A. and Atoui, M. (2023) *ICARDA 2030 Genetic Resources Conservation and Use Strategy*. ICARDA, Beirut.

Khazaei, H., Street, K., Santanen, A., Bari, A. and Stoddard, F.L. (2013) Do faba bean (*Vicia faba* L.) accessions from environments with contrasting seasonal moisture availabilities differ in stomatal characteristics and related traits? *Genetic Resources and Crop Evolution* 60, 2343–2357. DOI: 10.1007/s10722-013-0002-4

Malhotra, R.S., Sarker, A. and Saxena, M.C. (2004) Drought tolerance in chickpea and lentil – present status and future strategies. *Challenges and Strategies of Dryland Agriculture* 32, 257–273. DOI: 10.2135/cssaspecpub32.c17

Maxted, N. and Vincent, H. (2021) Review of congruence between global crop wild relative hotspots and centres of crop origin/diversity. *Genetic Resources and Crop Evolution* 68(4), 1283–1297. DOI: 10.1007/s10722-021-01114-7

Ouji, A., Amri, M., Bouhadida, M., Omri, N., Khamassi, K. *et al.* (2023) Registration of 'Krib' new lentil variety in Tunisia. *Genetic Resources and Crop Evolution* 70, 711–715. DOI: 10.1007/s10722-023-01540-9

Sanchez-Garcia, M. (2021) Food, feed, forage and malt: barley is the ultimate multipurpose crop for nutrition and livelihood security in the MENA drylands. In: Amsallem, I. and Hubert, B. (eds) *Agroecological Transformation for Sustainable Food Systems*. Agropolis International, Montpellier, France. DOI: 10.23708/fdi:010082500

Shideed, K., Oweis, T., Gabr, M. and Osman, M. (2005) *Assessing On-Farm Water Use Efficiency: A New Approach*. ICARDA, Aleppo, Syria.

Shideed, K., Alary, V., Laamari, A., Nefzaoui, A. and El-Mourid, M. (2007) Ex post impact assessment of natural resource management technologies in crop-livestock systems in dry areas of Morocco and Tunisia. In: Waibel, H. and Zilberman, D. (eds) *International Research on Natural Resource Management: Advances in Impact Assessment*. ICARDA, pp. 169–195. DOI: 10.1079/9781845932831.0169

Solh, M. (2010) Ensuring food security in a changing climate: How can science and technology help? In: Sohl, M. and Sexena, M.C. (eds) *Food Security and Climate Change in Dry Areas*. Proceedings of an international conference, 1–4 Febuary, Amman, Jordan. ICARDA, Aleppo, Syria.

Swelam, A. (2016) *Raised-Bed Planting in Egypt: An Affordable Technology to Rationalize Water Use and Enhance Water Productivity*. ICARDA, Amman, Jordan. Available at: https://repo.mel.cgiar.org/handle/20.500.11766/5900 (accessed 16 July 2024).

Tadesse, W., Miloudi, M., Abdalla, O. and Rajaram, S. (2016) Wheat breeding at ICARDA: achievements and prospects in the CWANA region. In: Bonjean, A., Angus, B. and van Ginkel, M. (eds) *World Wheat Book, Vol.* 3. Limagrain.

Tadesse, W., Debele, T., Kachala, M., Rachdad, F., Turaki, Z. *et al.* (2022a) Innovation platform: approaches and success stories for development and out-scaling of heat tolerant wheat varieties in Africa. *Agricultural Research and Technology* 27(2). Available at: https://juniperpublishers.com/artoaj/pdf/ARTOAJ.MS.ID.556361.pdf (accessed 16 July 2024).

Tadesse, W., Zegeye, H., Tolesa, D., Kassa, D., Shiferaw, W. *et al.* (2022b) Wheat breeding in Ethiopia: retrospects and prospects. *Crop Breeding, Genetics and Genomics* 4(3). DOI: 10.20900/cbgg20220003

Tavakoli, A., Oweis, T., Farahani, H., Ashrafi, S., Asadi, H.A. *et al.* (2010) *Improving Rainwater Productivity with Supplemental Irrigation in Upper Karkheh River Basin of Iran*. ICARDA, Aleppo, Syria. Available at: https://hdl.handle.net/20.500.11766/8846 (accessed 16 July 2024).

Tsivelikas, A., Yazbek, M. and Kehel, Z. (2023) *Conservation of Genetic Resources at ICARDA*. ICARDA Publications. Available at: https://hdl.handle.net/10568/136149 (accessed 16 July 2024).

Vavilov, N.I. (1926) Tzentry proiskhozhdeniya kulturnykh rastenii [*The Centers of Origin of Cultivated Plants*]. *Works of Applied Botany and Plant Breeding* 16, 1–248. (in Russian, English)

Ziadat, F.M., Mazahreh, S., Haddad, M., Benabdelouahab, T. and Attaher, S. *et al.* (2014) *Similarity and Suitability Analysis to Assist the Out-Scaling of Sustainable Water and Land Management Practices in West Asia and North Africa*. ICARDA, Beirut. Available at: https://repo.mel.cgiar.org/handle/20.500.11766/8856 (accessed 16 July 2024).

# 11 Innovations, Strategies, and Policies for Building Resilience of India's Dryland Farming to Climate Change

**M.L. Jat[1], Ch. Srinivasa Rao[2]*, V. Padmaja-Karanam[1], and Rajbir Singh[3]**
[1]*ICRISAT, Hyderabad, India;* [2]*NAARM, Hyderabad, India;* [3]*Natural Resource Management Division, ICAR, New Delhi, India*

**Abstract**

Climate change is a real and undeniable threat to the entire civilization. India is among the most severely impacted countries with extreme weather events felt intensely on agriculture (decline in yields by 3–9%), water resources, and ecosystems, threatening food security and rural livelihoods. To address these challenges, India has been actively pursuing new programs and innovative strategies for climate-proofing agriculture through its National Action Plan on Climate Change (NAPCC); major initiatives being National Mission on Sustainable Agriculture (NMSA), National Innovations on Climate Resilient Agriculture (NICRA), and international collaborative programs: conservation agriculture (CA), landscape approach, etc., thus promoting climate-resilient agriculture through resilient crop varieties, integrated natural resources management, digital agriculture, capacity building of farmers, and supporting community-based adaptation initiatives. Results from these initiatives such as CA have shown the potential to produce more food (10–15%), increase farmers' income (25–50%), and lower carbon footprint by 25–30%. Through a multifaceted approach, embracing innovative technologies, adaptive strategies, and supportive policies such as the National Adaptation Fund for Climate Change (NAFCC), and crop insurance schemes, India's agriculture sector is enhancing its resilience to climate change.

## Introduction

A growing climate crisis, rapidly degenerating soils, drying aquifers, extinction of agrobiodiversity, and highly volatile markets have raised stern fears for achieving the Sustainable Development Goals, when only seven annual harvests remain until 2030. Anthropogenic climate change (CC) has already slowed down agricultural productivity growth by 21% in the last 50 years (Crippa *et al.*, 2021). Climate change, driven primarily by human activities such as the burning of fossil fuels, deforestation, industrial processes, and agricultural activities, releases greenhouse gases (GHG) and fluorinated gases which trap heat in the Earth's atmosphere, leading to the greenhouse effect and global warming. In 1896, the Nobel Laureate Svante Arrhenius predicted that an increase in atmospheric $CO_2$ from burning fossil fuels would lead to global warming, and doubling of $CO_2$ concentrations (then about 260 ppm) would raise the temperatures of the Earth by about 2–6°C. At present, the atmospheric $CO_2$ concentration is over 420 ppm. All this supports the estimate by the Intergovernmental Panel on Climate Change

*Corresponding author: cherukumalli2011@gmail.com

 *Climate Change and Sustainable Agro-ecology in Global Drylands* (A.S. El-Beltagy, R. Lal and K. Malik eds.)
DOI: 10.1079/9781800624870.0011

(IPCC) that the Earth will be warmer by 1.4–5.8°C during the current century (IPCC, 2021).

Climate variability and change are threatening agri-food systems, particularly in the drylands where climate-induced risks are more intense. These risks further compound existing vulnerability, with high levels of poverty and inequity, food and nutritional insecurity, degrading natural resources, and poor-resource governance. Agri-food systems' vulnerability reflects multiple overlapping crises. There is good awareness and intention among the major stakeholders on the importance of building resilience, as was focused on during COP27. However, actions are limited due to lack of science–policy interface, unified approaches, and tools for designing and implementing context-specific and data-driven holistic solutions for supporting climate resilience and sustainability of agri-food systems. This has posed a serious threat to the food and nutritional security and livelihoods of billions, especially in drylands. Without long-term corrective measures, the challenge will further compound, given the agri-food systems' contribution of a third of global anthropogenic GHG (Ortiz-Bobea *et al.*, 2021) and rapidly deteriorating soil health and drying aquifers.

## Climate Change Impacts on Indian Agriculture

The impacts of CC are global, but countries like India are highly vulnerable because a large population depends on agriculture. There has been a significant increase in the frequency of extreme weather events in recent years affecting farm productivity and adversely impacting food security. As per the latest reports, the global average temperature rise is 0.99°C since pre-industrial times (1850). Sixteen of the 17 warmest years over the last 136 years, have occurred since 2001 and 2016 was the warmest. India's average mean annual temperature during 2022 was +0.51°C above the 1981–2010 average (Fig. 11.1), thus making 2022 the fifth warmest year on record since 1901 (IMD, 2022). With reference to the rate of warming based on the 21st century emissions pathway, it is estimated that by the end of the century, average temperature in India will increase by 1.1–4.1°C over the 1986–2005 baseline (World Bank Group, 2021).

Worldwide, agriculture both contributes to and is sensitive to CC impacts and is threatened by it. Agriculture is one of the largest contributors to the GDP in South Asia, besides being the main source of employment for a labor force of millions. As more than 50% of the cropped area in India is rainfed, the agricultural economy is centered on the reliability of the monsoon. Global warming is projected to have significant impact on conditions affecting agriculture, including temperature, $CO_2$, glacier runoff, precipitation, and interaction of these elements. These conditions in turn determine the carrying capacities of different ecosystems to produce enough biomass including food for the human population and domesticated animals. The overall effect of CC on agriculture and related sectors will depend on the balance of these effects. Climate change is predicted to have severe consequences on agriculture and the rural poor (Fig. 11.2). This is critical because increased frequency of extreme weather events like droughts and floods can have severe impacts (IPCC, 2007a). Recent research indicated that monsoon rainfall in India became

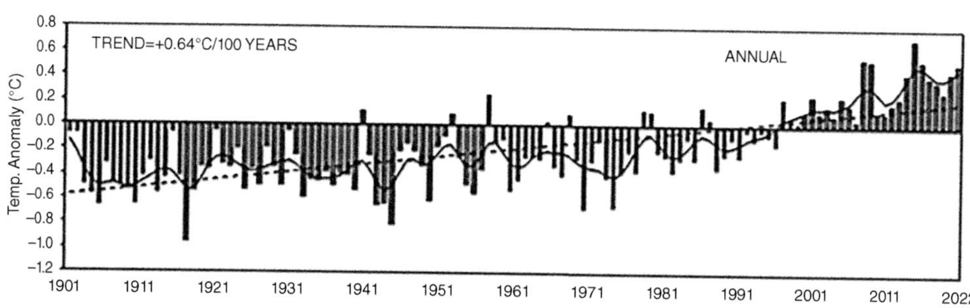


**Fig. 11.1.** Annual mean land-surface air temperature anomalies averaged over India for the period 1901– 2022. (From IMD, 2022)

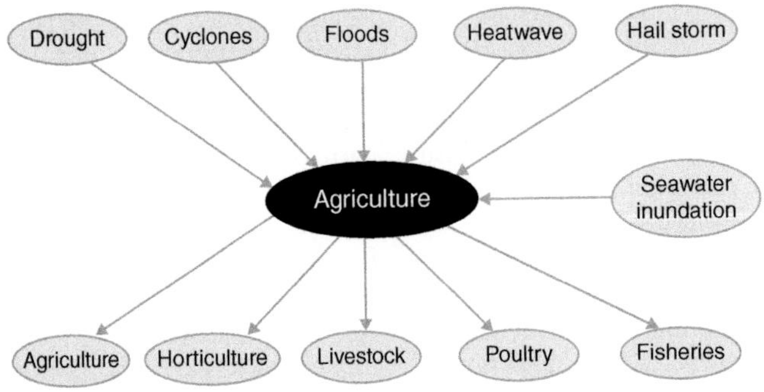


**Fig. 11.2.** Climate change impacts in different sub-sectors of agriculture. (From Srinivasarao *et al.*, 2016c)

more erratic with intense rainfall events and reduced number of rainy days (Krishnan *et al.*, 2020). Given the dominance of rainfed agriculture, the onset, duration, spatial extent, and total precipitation during the monsoon is the most critical factor in determining the livelihoods of a large majority of the population in rural areas. Moreover, resilience is typically low in rural areas as the existing asset base is limited, and services are often insufficient. Global warming is likely to affect all these factors (Sterrett, 2011).

Crop growth and yield are largely determined by weather during the growing season. With slight variations from normal weather, the effectiveness of externally supplied inputs decreases, and consequently yields are affected. Increasing $CO_2$ concentrations in the atmosphere and predicted higher atmospheric temperature are also expected to influence crop production through changes in plant growth and transpiration (Mall *et al.*, 2006). Increased temperature reduces the grain yield due to faster development, increased respiration losses and decreased time of grain growth in food crops. Yield reduction in wheat, rice, and corn due to temperature rise during winter is expected (Challinor *et al.*, 2007; Challinor and Wheeler, 2008; Srinivasarao *et al.*, 2020).

Key impacts of CC would be an increase in temperature, particularly during post-monsoon and winter, increased frequency of floods during the monsoon, decreases in winter precipitation, and lower number of rainy days, all affecting crop growth and production.

The IPCC reports, and a few other studies, indicate a probability of a 10–40% loss in crop production in India with increase in temperature by 2080–2100 (Rosenzweig and Parry, 1994; Fischer *et al.*, 2002; Rosenzweig *et al.*, 2002; IPCC, 2007b). Studies conducted in India (Aggarwal and Sinha, 1993; Rao and Sinha, 1994; Lal *et al.*, 1998; Saseendran *et al.*, 2000; Mall and Aggarwal, 2002; Aggarwal, 2008) have confirmed similar declining trends in agricultural productivity due to CC. For every one degree increase in temperature, the yields of wheat, soybean, mustard, groundnut, and potato are expected to decline by 3–7% (Aggarwal, 2009; Mohanty *et al.*, 2017) and in rice by 6% (Saseendran *et al.*, 2000; Bouman *et al.*, 2007). Projections indicate the potential loss of 4–5 MT of wheat with every rise of 1°C temperature throughout the growing period with current land use in India alone (Aggarwal, 2008). Losses are also significant in other crops, such as mustard, peas, tomatoes, onion, garlic, and fruit crops (Samra and Singh, 2004; Aggarwal, 2009). Sorghum yields are predicted to vary from +18 to −22 % depending on a rise of 2–4°C in temperature and increase by 20–40% of precipitation (Mall *et al.*, 2006). Rainfed areas are likely to be more vulnerable in terms of extreme events (Mall *et al.*, 2006). Aberrations in the south–west monsoon could include a delay in the onset of the monsoon, long dry spells, and early withdrawal, etc., adversely affect the productivity (Lal, 2001). This increase in variability could make it more difficult for resource-poor farmers to take decisions on investing in inputs and new technologies (Pandey *et al.*, 2000; Bal *et al.*, 2021).

Climate change is also expected to accelerate the hydrological cycle. In India, winter precipitation is projected to decline, and this is likely to result in a greater demand for water during the winter season. During the monsoon season, the intensity of rains is projected to increase, which will imply more frequent and severe floods and a lesser recharge of groundwater (Jain, 2012). Further, water for agriculture is becoming increasingly scarce, and CC-induced higher temperatures will increase crops' water requirements, so shortages will become more serious in the coming decades. More than 50% of India is under severe water stress already. By 2025, 15–20 million of the world's 79 million ha of irrigated rice lowlands, which provide three-quarters of the world's rice supply, are expected to suffer some degree of water scarcity (Bouman *et al.*, 2007). It is also estimated that to eliminate hunger and undernourishment of the world's population, the additional water requirements may be equivalent to all freshwater resources withdrawn and used today for agricultural, industrial, and domestic purposes (SIWI *et al.*, 2005).

Increases in frequency and magnitude of heatwaves, droughts, inland and coastal flooding, sea-level rise, and tropical cyclone etc., would decrease food stability and increase risk of food insecurity (Bhattacharyya *et al.*, 2020). The latest IPCC AR-6 Report has highlighted that the major challenges for sustainable agriculture in the coming years will be increase in rainfall, high inter-annual variability, intense and frequent heatwaves, temperature likely to increase by 1.5–4.0°C and rise in sea level by 0.3 m (IPCC, 2021). Studies in India showed significant negative impacts of CC, predicted to reduce yields by 4.5–9.0%, depending on the magnitude and distribution of warming (Srinivasa Rao *et al.*, 2016; Kumar *et al.*, 2011). The National Institute of Disaster Management suggests the economic damage inflicted due to floods and heavy rains in the last 65 years – between 1953 and 2018 – is approximately INR four trillion (US$ 50 billion). According to a recent World Bank study, India will lose 2.8% of its GDP because of CC impact by 2050 and this will lead to significant reductions in living standards (Mani *et al.*, 2018). India is on track to meet its commitment in its nationally determined contributions (NDCs) under the Paris Agreement to reduce its emissions intensity of GDP by 33–35% below 2005 levels well before 2030 (MoEFCC, 2021). But developing and targeting a portfolio of component technologies through bundling on a system basis and defining their recommendation domains are critical for the needed investment decisions for impact at scale (Jat, 2020; Singh *et al.*, 2021). Hence, it is important to accelerate CC research towards developing resilient technologies and their scaling.

## Recent Climate Risks in Indian Agriculture

Indian agriculture continues to face several climate-related risks that have significant implications for food security, rural livelihoods, and the overall economy. Climate change and increased extreme weather events in recent decades as well as uncertainty in prediction of those events further add to the woes of the farmers causing widespread losses of agricultural output. Western Rajasthan, parts of Haryana, Uttar Pradesh, Maharashtra, southern Bihar, Madhya Pradesh, southern Gujarat, northern parts of Andhra Pradesh, and Karnataka are regularly facing dryness, and these regions are highly vulnerable to drought (Bhadwal *et al.*, 2007). The super-cyclone in 1999 that hit Orissa took over a million lives besides property loss in the coastal areas (Ahluwalia and Malhotra, 2006). The coastal state of Andhra Pradesh was similarly impacted during the *Hudhud* cyclone in 2014. According to the *State of the Climate in Asia 2021* report, published by the World Meteorological Organization (WMO), floods in 2021 inflicted economic losses to the tune of $3.2 billion (WMO, 2022). In 2022, extreme heat gripped large parts of India, affecting hundreds of millions of people and agriculture in particular (Bal *et al.*, 2022). Some of the recent climate risks that have been observed are:

- erratic monsoons: in recent years, monsoon patterns have become increasingly unpredictable, with variations in timing, intensity, and duration. Delayed or inadequate monsoons are leading to water scarcity, crop failures, and reduced agricultural productivity;
- extreme weather events: a rise in extreme weather events such as cyclones, floods, and droughts causes massive damage to crops, infrastructure, and livestock, leading to significant economic losses for farmers;
- heatwaves and extreme temperatures: increasing temperatures due to CC have led to more frequent and intense heatwaves in

different parts of the country. High temperatures during critical stages of crop growth can adversely affect yields and reduce agricultural output;

- water scarcity and depletion of groundwater resources: CC, coupled with unsustainable water management practices, has led to water scarcity in many regions. Overextraction of groundwater for irrigation has resulted in the depletion of aquifers, further exacerbating the water crisis for agriculture;
- pests and diseases: CC can alter the geographical distribution and prevalence of pests and diseases in crops, making them more difficult to control. Outbreaks of new or invasive pests are leading to significant crop losses and additional financial burdens for farmers;
- shift in crop suitability zones: changing climate patterns can affect the suitability of certain crops in specific regions. Some areas may become less favorable for traditional crops, while others might open up new possibilities for different crops;
- livestock vulnerability: extreme heat events and changes in the availability of water and fodder impact livestock health and productivity. This can adversely affect the livelihoods of pastoral communities and dairy farmers; and
- adaptation challenges: small and marginal farmers, who constitute a significant portion of India's agricultural workforce, often lack access to resources, information, and technologies needed to adapt to CC effectively.

## Need for Climate-resilient Agriculture

Sustainable intensification, regenerating soils, and building resilience through low-emission and nutrition-sensitive innovations relevant for smallholder agriculture are critical for building resilience. This, however, would require a holistic approach around genetic innovations, soil, climate, and resource-informed planning of farming systems coupled with integrated and efficient use of production inputs (both organic and inorganic). Transforming agri-food systems, therefore, must focus on transdisciplinary, multi-stakeholder innovations combined with informed knowledge intensification which is so critical for regenerative agriculture aiming at the concept of 'one health' (soil, plant, animal, human).

Agri-food systems are responsible for a third of the global anthropogenic GHG emissions. It is estimated that about three-quarters of the total emissions from agriculture and land use originate in the developing countries. Agriculture accounts for about 408 Tg of $CO_2e$, out of which, 222.7 Tg (54.6%) of carbon dioxide equivalent ($CO_2e$) are emitted from livestock rearing. The remaining 185.2 Tg $CO_2e$ GHG emissions (45.4%) are from croplands. Emissions from crop agriculture include 17.5% from rice cultivation, 19.1% from fertilizer applied to agricultural soils (mainly urea), 6.7% from manure management, and 2.2% due to field burning of agricultural residues (MoEFCC, 2021). The recent IPCC report highlights that South Asia is among the most vulnerable to CC and climate variability. In the absence of higher adaptation and mitigation, CC can have major consequences for food security and development in the decades to come.

India is already experiencing the impacts of CC on its agriculture sector. To build resilience and adapt to these changes, various innovations and strategies are being explored and implemented in India to address CC impacts on agriculture. In this context, India has been promoting climate-resilient agriculture (CRA), encompassing a combination of sustainable and regenerative agricultural techniques. These practices include conservation agriculture, use of climate-resilient crop varieties, agroforestry, measures to manage water resources effectively during both droughts and floods, crop rotation and diversification, and precision farming. Alongside efforts in establishing climate information services, facilitating community-based adaptation strategies, developing better storage facilities and post-harvest infrastructure, and capacity-building programs are helping in implementing climate-resilient practices and scaling innovative solutions.

## National Initiatives on Climate-Proofing in Indian Agriculture

India is an agrarian economy and highly vulnerable to climate change. Recognizing the significant impact of CC on food security, rural livelihoods, and overall economic stability, several national initiatives and programs are being implemented to climate-proof its agriculture sector. The Indian government launched the National Action Plan on Climate Change (NAPCC) in 2008, outlining eight missions that target specific areas of CC mitigation and adaptation, covering various sectors to address CC, including the National Water Mission, National Mission for Sustainable Agriculture, National Solar Mission, National Mission for Enhanced Energy Efficiency, and others. These missions aim to provide a framework for different stakeholders to collaborate and implement climate-proofing measures to enhance resilience and climate-proofing across different sectors of the economy. These initiatives aim to promote CRA practices, enhance water management, and support farmers in adapting to changing climate patterns. Some of the key national initiatives on climate-proofing in Indian agriculture include those listed below.

## National Innovations in Climate-resilient Agriculture (NICRA)

To meet the challenges of sustaining domestic food production in the face of changing climate and to generate information on adaptation and mitigation in agriculture to contribute to global platforms like United Nations Framework Convention on Climate Change (UNFCCC), Indian Council of Agricultural Research (ICAR), and Ministry of Agriculture and Farmers Welfare, the government of India launched a flagship project, NICRA, in February 2011. It focuses on developing and promoting climate-resilient technologies and practices suited to different agro-ecological regions. NICRA is operated with these objectives: (i) to undertake strategic and applied research on CC adaptation and mitigation; (ii) to validate, demonstrate, and assess the impact of climate-resilient technologies on farmers' fields; and (iii) to strengthen the capacity of scientists, farmers, and other stakeholders on CRA. The outputs of the project are helping the districts and regions prone to extreme weather conditions like droughts, floods, frost, and heatwaves cope with such extremes. The project is implemented through components – strategic research, technology demonstration, and capacity building in 151 clusters of villages in each one of the identified climatically vulnerable districts to demonstrate proven technologies and enhance adaptive capacity of farmers (Srinivasarao *et al.*, 2016a).

In the strategic research component, both short- and long-term research programs with a national perspective have been taken up to evolve adaptation and mitigation strategies in agriculture crops, horticulture, natural resources, livestock, fisheries, and poultry. About 41 ICAR institutes, state agriculture universities, NGOs, and others are involved in strategic research. State-of-the-art research infrastructure was developed like high-throughput phenotyping platforms, free air-temperature elevation systems in open fields, carbon dioxide temperature gradient chambers, a network of 100 automatic weather stations, environmental growth chambers with $CO_2$ and temperature controls, and a special calorimetric system to study livestock response to heat stress (Venkateswarlu *et al.*, 2012).

Assessing vulnerability to CC and variability is the foremost step in evolving appropriate adaptation strategies to changing climate. Such analysis also helps in targeting adaptation investments to regions that are more vulnerable. A highly vulnerable system may not suffer risk due to CC or a less vulnerable system may face risk if it is placed where severe hazard incidence is possible. Vulnerability is assessed by using the indicator method. The recent study on risk and vulnerability assessment of agriculture to CC by ICAR showed that out of 573 districts, 109 were very risk-prone and 201 are risk-prone (Rama Rao *et al.*, 2019). This is useful to policymakers and research managers for prioritization of resources related to CC action plans, thus helping in planning for adaptation for vulnerability and risk reduction.

Contingency planning is one of the major strategies of preparedness for tackling aberrant weather events. In the face of increasing climate variability, adoption and implementation of the district-level contingency plans is a priority for many state governments (Srinivasarao *et al.*, 2020). ICAR has prepared district agricultural contingency plans (DACPs) for 650 districts of the country recommending location-specific climate-resilient crops and varieties, and management practices for use by the state departments of agriculture and farmers. These agricultural contingency plans cover weather aberrations like drought in rainfed and irrigated areas, floods, unseasonal rains, and extreme weather events such as heatwaves, cold waves, untimely and high-intensity rainfall, frost, hailstorms, cyclones, pests, and disease outbreaks etc. The two-step implementation of district plans involves preparedness and real-time activation of the contingency measures. Based on the advanced seasonal forecast before the commencement of the rainy season, DACP interface meetings are organized in the presence of senior district officials to ensure necessary policy interventions. Contingency crops/varieties – horsegram, castor, foxtail millet, pearl millet, cluster bean, rapeseed, and blackgram have been introduced and adopted by farmers at various locations. Thus, an agriculture contingency plan is evolving as a future strategy to manage aberrant weather situations for ensuring food security and livelihoods. Development of agricultural contingency plans and experiences to manage weather aberrations and extreme climatic events with the implementation of these plans were shared at UNFCCC global CC negotiations.

Significant achievements of the project include studies on the impact of elevated temperature and $CO_2$ on crops (rice, wheat, maize, onion), livestock, fisheries, soil carbon, pests, and diseases using integrated simulation-modeling framework; development of climate-resilient crop varieties tolerant to abiotic stresses for cultivation in vulnerable regions of the country; location-specific GHG inventory for different cropping systems and production systems and quantified carbon sequestration potential through agroforestry systems and major cropping systems; and climate-induced pest and disease outbreaks in the country studied in nine crops (ten insects and 29 diseases) to build pest and disease forewarning models. Web-enabled and mobile apps for pest forewarning; forecasted area expansion of black pepper (97–133 districts) and cardamom (24–104 districts) using climate analogs for habitat suitability in the future changing climate; and an environmentally safe spray formulation for synchronization of flowering in mango have all been developed. Efforts are made to characterize heat-shock proteins and heat stress-induced genes in native breeds that can be used as biomarkers to induce resilience.

Soil-health management aims to promote crop- and location-specific sustainable soil-health management that includes residue management, organic farming practices, nutrient management, balanced application of fertilizers, and minimizing the soil erosion/degradation. Soil Quality Index (SQI) is the most appropriate for quantitative assessment of soil quality. In this direction, an effort was made to develop a software 'SQI CAL' for rapid calculation of SQI. Soil aggregates and water stability improved under conservation treatments. Assessment of biochar on productivity, nutrient-use efficiency and C sequestration potential of maize-based cropping systems in the north-eastern hill region indicated a higher SMBC, DHA activity, and SOC with application of biochar @ 5.0 Mg/ha along with 75% RDF + 4 Mg/ha FYM, while exchangeable aluminum and exchangeable acidity were reduced. Balanced fertilization and integrated nutrient management systems as evidenced from various nutrient management field experiments under rainfed conditions (Srinivasarao *et al.*, 2015b) are key to sustainable crop production in terms of building up of C, structural stability and crop yield, and climate mitigation.

Participatory technology demonstration on site-specific technology interventions for coping with climate variability in climatically vulnerable districts of NICRA is the flagship outreach program. This helps to generate awareness and build capacity of farmers and other stakeholders on climate-resilient agriculture. In addition, this also supports evolving innovative institutional mechanisms at village level that enable the communities to respond to climate stresses in a continuing mechanism beyond the project cycle. The technology demonstrations aim at taking the resilient technologies to the farmers' doorsteps for reducing the vulnerability and enhancing the adaptive capacity of the farming community at village level and developing climate-resilient model villages (Fig. 11.3) to upscale further. One village or a cluster of villages from each of the 151 selected districts were selected for this purpose. Village-level institutional mechanisms such as Village Level Climate Risk Management Committees (VCRMC) and custom hiring centers are established for managing the infrastructure created and to improve the timeliness of operations during the limited windows of moisture availability in rainfed areas and to promote small-farm mechanization for adoption of climate-resilient

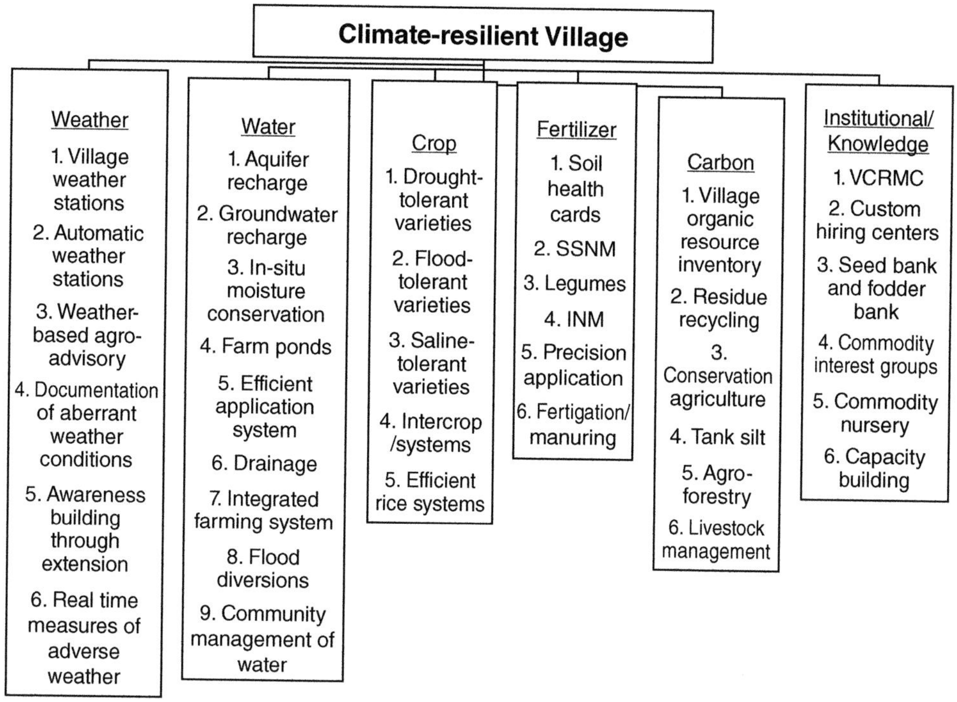


**Fig. 11.3.** Framework of climate-resilient village in India of NICRA. SSNM, site-specific nutrient management; INM, integrated nutrient management; VCRMC, village level climate risk management committee. (From Srinivasarao *et al.*, 2016a)

practices. These interventions helped farmers to reduce yield losses and enhanced their adaptive capacity against climatic variability.

A large-scale capacity-building program on climate-resilient agriculture was undertaken with more than 1200 scientists, 874 research scholars, and 160 doctoral and postgraduate students who were involved in CC research and dissemination of climate-resilient technology across the country. These resilient practices are being adopted by communities and spreading beyond NICRA villages. During one decade, over 17,000 training programs were conducted across India under NICRA to educate stakeholders on various aspects of CC and resilient technologies, covering > 0.5 million stakeholders to enable wider adoption of climate-resilient technologies and increase in yields (Srinivasarao *et al.*, 2015a).

ICAR, together with its partners, has developed 2279 crop varieties since 2014, out of which 1888 are climate-resilient. Out of these, 407 are abiotic stress-tolerant and 1481 are biotic stress-tolerant varieties. Heat-tolerant wheat varieties are currently used in up to 50% of the area, to reduce the impact of heat stress in northwestern India. Innovative technological and policy inputs to manage risks posed by CC are essential to retain farmers in farming and reduce farm distress. Convincing farmers to adopt the technology and upscaling across climatically vulnerable areas is still a huge challenge.

Large-scale adoption of resilient technologies across the country has resulted in substantial reduction of emissions from the agriculture sector in the country. India's Biennial Update reports indicate a reduction in emissions contributions from agriculture from 18% in 2010 to 16% in 2014 and 14.45% in the latest report. Some of the technologies, like direct-seeded rice (DSR), micro-irrigation (drip and sprinkler), balanced rations for livestock, avoidance of crop residue burning, neem-coated urea, solar pumps, agroforestry etc., have been widely demonstrated across the country in the NICRA 151 clusters.

### National Mission for Sustainable Agriculture (NMSA)

Under the National Action Plan on Climate Change (NAPCC), NMSA focuses on promoting climate-resilient agriculture through the adoption of climate-resilient practices, sustainable water management, and integrated farming systems. NMSA, made operational from 2014, aims at making agriculture more productive, sustainable, remunerative, and climate-resilient by promoting location-specific integrated/composite farming systems; soil and moisture conservation measures; comprehensive soil-health management; efficient water management practices, and mainstreaming rainfed technologies and crop diversification. Important components of the NMSA program are:

- **On-farm water management (OFWM).** OFWM has a focus on enhancing water-use efficiency by promoting technological interventions like drip and sprinkler technologies, efficient water application and distribution system, secondary storage etc. Thereafter, these activities have been subsumed under the 'Per Drop More Crop' (PDMC) component of Pradhan Mantri Krishi Sinchayee Yojana (PMKSY) launched during 2015–16.
- **Rainfed area development program.** This focuses on an integrated farming system for enhancing productivity and minimizing risks associated with climatic variability. Under this program, crops are integrated with activities like horticulture, livestock, fisheries, vermi-organic compost, green manuring, apiculture etc. to enable farmers to maximize farm returns for sustained livelihoods and mitigate the impacts of drought, flood, or other extreme weather events with an income opportunity from allied activities. Since the inception of the scheme, an area of 0.342 mha has been brought under the integrated farming system.
- **Soil-health management (SHM).** SHM aims to promote integrated nutrient management (INM) through judicious use of chemical fertilizers including secondary and micronutrients in conjunction with organic manures and biofertilizers for improving soil health and its productivity, and strengthening of soil and fertilizer testing facilities to improve soil test-based recommendations to farmers for improving soil fertility (Fig. 11.4). The activities include training and demonstrations on balanced use of fertilizers. A soil-health card scheme has been implemented in the country since February 2015 and provides a soil-health card to all farmers in the country. The card provides information to farmers on nutrients status of their soil and recommendations on appropriate dosage of nutrients to be applied for improving soil health and its fertility.

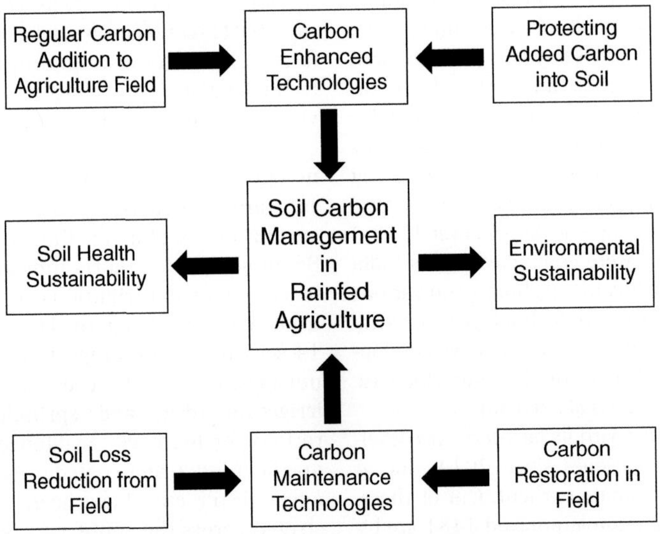


**Fig. 11.4.** Carbon management strategies in rainfed agriculture. (From Srinivasarao and Srinivas, 2023)

- **Pradhan Mantri Krishi Sinchai Yojana (PMKSY).** PMKSY aims to improve water-use efficiency in agriculture by providing financial assistance for the development of irrigation infrastructure, water harvesting, and water distribution systems. The PMKSY-PDMC mainly focuses on water-use efficiency at farm level through precision/micro-irrigation (drip and sprinkler irrigation). Besides promoting precision irrigation and better on-farm water management practices to optimize the use of available water resources, the component also supports micro-level water storage or water conservation/management activities to supplement source creation. Since 2015/16, an area of 3.1 million ha has been brought under micro-irrigation.
- **Paramparagat Krishi Vikas Yojana (PKVY).** PKVY promotes organic farming practices that enhance soil health and resilience to CC by conserving water and reducing GHGs. It encourages farmers to adopt organic inputs, conserve traditional seeds, and implement integrated pest management techniques.
- **Rashtriya Krishi Vikas Yojana (RKVY).** RKVY supports state governments in implementing agricultural development projects, including those that focus on climate-resilient agriculture, crop diversification, and capacity building for farmers.
- **Pradhan Mantri Fasal Bima Yojana (PMFBY).** PMFBY is a crop insurance scheme that provides financial support to farmers in case of crop losses due to adverse weather events or other calamities, reducing the financial risks associated with climate variability.

## Green India Mission

Green India Mission was launched in February 2014 under the NAPCC, with the objective to safeguard the biological resources of the nation and associated livelihoods against the perils of adverse CC and to recognize the vital impact of forestry on ecological sustainability, biodiversity conservation, and food-, water-, and livelihood-security. It aims at protecting, restoring, and enhancing India's diminishing forest cover and responding to CC through adaptation and mitigation measures. It envisages a holistic view of greening that extends beyond tree planting. GIM focuses on multiple ecosystem services such as biodiversity, water, biomass, preserving mangroves, wetlands, critical habitats, etc. along with carbon sequestration and hydrological services, provisioning services like fuel, fodder, timber and non-timber forest produce, and increased forest-based livelihood income of households living in and around forests. Traditional ecological knowledge of communities, along with forestry science and state-of-the-art technology, would improve mission interventions.

## Climate finance

This was established in 2015 to address the vulnerabilities arising from changing climate. It plays a crucial role by focusing on financing projects and programs of the state and Union Territories of India that enhance adaptive capacities and resilience to CC. The National Bank for Agriculture and Rural Development (NABARD) is designated a National Implementing Entity (NIE) given its presence across the country and considering its arrangement for Adaptation Fund (AF) under the Kyoto Protocol. It facilitates the implementation of NAFCC of project areas/concepts identified under the State Action Plan on Climate Change (SAPCC) and the relevant missions from SAPCC by channeling funds and supporting climate-resilient initiatives and promoting sustainable practices, particularly in the agriculture and rural sectors. It also facilitates developing projects, their appraisal, disbursement of funds, monitoring and evaluation of the projects, and capacity building of state government and relevant stakeholders (www.nabard.org). So far, around 30 projects have been sanctioned to different states in India with a budget outlay of US$ 100 million.

## National Food Security Mission

The NFSM works to increase food production and productivity in various crops through the promotion of climate-resilient technologies and practices.

### Mission for Integrated Development of Horticulture

This mission focuses on developing climate-resilient horticulture practices, including protected cultivation, precision irrigation, and post-harvest management.

### National Bamboo Mission

NBM promotes bamboo cultivation and utilization, which contributes to climate resilience by providing an alternative source of income for farmers and helping in soil conservation and carbon sequestration.

### National Livestock Mission

NLM aims to enhance livestock productivity and climate resilience by providing support for breed improvement, feed management, and veterinary services for livestock farmers.

These initiatives demonstrate India's commitment to climate-proofing its agriculture sector by promoting sustainable practices, efficient water management, and farmer-centric policies.

## International initiatives and programs on climate-resilient agriculture in India

To increase agricultural productivity, enhance climate resilience, and reduce GHG emissions, several international collaborative programs and initiatives have been implemented in India to facilitate research, and transfer of knowledge and technology for climate adaptation and mitigation in agriculture. These programs involve collaboration between India, international organizations, and foreign governments to access climate-resilient agricultural technologies and best practices in order to address CC challenges in agriculture. India collaborates with CGIAR centers such as International Crops Research Institute for the Semi-Arid Tropics (ICRISAT), International Maize and Wheat Improvement Center (CIMMYT), and others, to access resilient crop varieties and agricultural technologies to improve agricultural productivity and resilience in the face of CC.

### CGIAR research program on climate change, agriculture, and food security (CCAFS)

The CCAFS research program was launched by CGIAR in 2010 aimed at developing new approaches and innovative solutions to promote more adaptable and resilient agriculture and food systems; mainstream climate variability and CC issues into development strategies and institutional agendas; enhance people's understanding of CC issues; and ensure that decisions, policies and actions are based on the best information and data available (CCAFS, 2019). With this vision, significant efforts were made to generate research-based knowledge and enhanced capacity for scaling CRA solutions at national and sub-national levels in India. Some of the research activities include innovative institutional mechanisms to scale out CRA through participatory evaluation of selected CRA technologies, developing an ICT-based climate information system (CIS), developing new insurance products, and capacity building engagements. Some of the success stories are:

- A climate-resilient village (CRV) agricultural research for development (AR4D) approach is being used by several government and non-government institutions in India to identify models of CRA portfolios, to be brought to scale.
- A solar irrigation cooperative pilot in Gujarat state stimulated the announcement of around US$ 21 billion investment towards the Kisan Urja Suraksha evam Utthaan Mahaabhiyan (KUSUM) scheme for solar irrigation in India.
- An innovative use of models for determining triggers for site-specific rainfall thresholds led to a win-win insurance product for farmers, industry, and the government, which was used by more than a million farmers in the state of Maharashtra.
- Flood mapping and flood insurance products are being piloted in Bihar.
- Many climate-resilient agricultural practices are being promoted via India's National

Agroforestry Policy to make agricultural landscapes more climate-resilient.

### Sustainable and resilient farming systems intensification in the Eastern Gangetic Plains

A successful international multi-partnership program on Sustainable and Resilient Farming System Intensification in the Eastern Gangetic Plains (SRFSI), is a regional project run between 2014 and 2021 funded via Australian Centre for International Agricultural Research (ACIAR). The project was led by CIMMYT together with South-Asian NARS and aimed to reduce poverty in the Eastern Gangetic Plains (Bihar and West Bengal) by establishing an evidence base by which conservation agriculture-based sustainable intensification (CASI) systems could improve productivity, profitability, and sustainability benefits to smallholder farmers in the Eastern Gangetic Plains while safeguarding the environment.

Overall, the research outputs provided a strong justification that the CASI-based system can offer multiple benefits to smallholder farmers across the EGP. The findings indicated a 10% yield increase, 17% reduction in water use, 44% reduction in labor use, 62% reduction in energy use, 16% cut in emissions, and 56% increase in profits. SRFSI was tasked with supporting partners to institutionalize CASI and promote its widespread adoption by smallholder farmers for their benefit. Focus was placed on capacity development as the pathway to CASI institutionalization, where more than 60,000 people received some form of training through the SRFSI project (with approximately 30% women).

SRFSI focused on creating self-sustaining enabling environments for the sustainability of CASI scaling beyond the life of the project itself. Focus was placed on ensuring integration of CASI through convergence with local and governmental programs. This included the commissioning of the regional Centre of Excellence for Conservation Agriculture (CECA) in West Bengal. Likewise, huge efforts have been put into development of the SRFSI online digital repository (https://srfsi.cimmyt.org/ [accessed 18/7/24]) to make the learnings of the project accessible to different stakeholders in local languages) and to ensure that such knowledge can continue to be used and institutionalized. A clear indication of the culmination of these efforts is the convergence evident in West Bengal where state government policy has incorporated CASI in their various operational guidelines and schemes, including the mandatory requirement for three CASI machines in any government-supported custom hire centers.

International programs on CRA in India play a crucial role in providing financial support, technical expertise, and knowledge exchange to enhance the country's efforts in CRA. Also, these programs reveal that global networks and partnerships are extremely important in catalyzing adoption of new technologies, mainly through knowledge sharing and enhanced capacity development of all stakeholders (Paroda, 2018). By leveraging international partnerships and resources, India is strengthening its agricultural sector's capacity to adapt to CC impacts while ensuring food security and sustainable rural development.

### Landscape approach for scaling climate-resilient, regenerative, and carbon-neutral agricultural systems

Soil degradation is one of the major global challenges, especially in drylands. The incremental population pressure and rapidly changing food habits have led to significant land-use changes, a key contributor to climate change. A large part of boundaries, which are forest cover, has been converted into grazing and agriculture lands in search of various provisioning services and facilitating urbanization. These changes have been accompanied by negative externalities such as CC, loss of biodiversity, poor retention ability of the landscape, and accelerated soil degradation (IPCC, 2019). These alterations have negatively impacted available natural resources, sustainability, and productivity of the landscape at local, regional, and global scales. In addition, these landscapes largely belong to marginal and small landholders and coincide with high poverty and malnutrition.

The landscape resource-conservation approach is a promising solution for regenerating desolate landscapes into productive areas that can address the interlinked challenges of food insecurity, malnutrition, and poverty along with CC. This approach supports building groundwater resilience, moisture retention ability, and improves baseflow and cropping intensification with carbon neutrality. In addition, regenerative landscapes with increased resource availability hold huge opportunity for crop intensification and system productivity. Scaling such good practices will also generate employment opportunities locally catering to the need of landless, marginal, and vulnerable farming families. This not only enhances household income but also controls outmigration, reduces drudgery, and improves the overall quality of life. Scaling regenerative landscapes with CRA technologies is the way forward to address the UN Sustainable Development Goals (UN-SDGs) including social equality, peace, and harmony. ICRISAT, together with NARES in India, has been intensively working on a landscape approach for building resilience of the drylands (ICRISAT, 2023).

## Conservation agriculture initiative

A conservation agriculture (CA) system is characterized by three interlinked core principles: (i) minimum mechanical disturbance of soil or direct seeding; (ii) permanent organic mulch (crop residues, cover crops) on the soil; and (iii) diversified and efficient crop rotations, along with other complementary good agronomic management practices to optimize soil health and resilience to CC (FAO and ITPS, 2015; Farooq and Siddique, 2015; Kar *et al.*, 2021). CA is currently adopted over 205 million ha globally (Kassam *et al.*, 2022), and has shown promise to address the soil health and CC mitigation through soil carbon sequestration and GHG mitigation (Jat *et al.*, 2019c, 2022; Gathala *et al.*, 2020). Over the past two decades, CA has been promoted on a larger scale in India, which demonstrates its potential for building resilience, reducing GHG emissions, and sequestering carbon. A meta-analysis of CA-based practices in South Asia shows a mean yield advantage of 5.8% with 12.6% higher water-use efficiency, an increase in net economic return of 25.9% and a reduction of 12–33% in global warming potential (Jat *et al.*, 2020). Further, there has been a significant advancement in various CA-based interventions for a food- and water-secure world. The recent research innovations on bundling of CA with innovative agronomic practices such as laser leveling, soil-moisture tension-based irrigation scheduling, sub-surface drip-fertigation, automated irrigation etc. have shown promise in this respect (Sidhu *et al.*, 2021).

Science-based evidence on CA has demonstrated that system-targeted CA-based management bundles have the potential to produce more food (10–15%) from less water (20–75%) and energy (20–45%) while increasing farmers' income (25–50%) in an environmentally responsible manner through lowering carbon footprints by 25–30% (Jat *et al.*, 2019c). System-based research by Jat *et al.* (2009) showed complementing effects of layering laser land leveling and CA on crop and water productivity and economic returns. Studies conducted by Kakraliya *et al.* (2018) in a rice–wheat (RW) rotation of western IGP revealed that layering of various management practices over CA in a RW rotation improved system productivity and profitability by 6% and 19%, respectively, saved 22% of irrigation water, and improved water productivity index (WPI) and total water productivity ($WP_{I+R}$) by 37% and 26% on a three-year mean basis, respectively, while reducing global warming potential (GWP) by 40% compared to farmers' practice. In addition, the bundled deployment of these practices was found more adapted to climate risks rather than conventional practices.

Sandhu *et al.* (2019) reported that layering drip-irrigation with residue retention in a no-till maize–wheat system, saved 25.5 ha-cm water in a sandy loam soil and increased water productivity by 66% and 259% in wheat and maize, respectively, compared to the conventional tillage-based furrow irrigated system. Results from another set of research showed that CA coupled with sub-surface drip-irrigation (SDI) in a rice–wheat system saved about 50% water and improved yields by 10–20% by eliminating surface-water evaporation and reducing the incidence of weeds (Sidhu *et al.*, 2019). The amount of energy required to pump groundwater is about 50% less with an SDI system; the improvement in irrigation water-use efficiency resulted in higher net returns. In a recent study by Jat *et al.* (2019c), bundling of CA with sub-surface drip-irrigation termed as CA+ were compared with CA alone and conventional tillage-based flood-irrigated RW rotation. In contrast to conventional till RW rotation, which consumed 1889 mm/ha irrigation water, CA+ system saved 58.4% and 95.5% irrigation water in RW and maize–wheat (MW) rotations, respectively. On a system basis, CA+ practices saved 46.7% and 44.7% irrigation water under RW and MW systems respectively, compared to their respective CA-based systems with flood irrigation. CA+ in RW system recorded 11.2% higher crop productivity and improved irrigation water productivity by 145% and profitability by 29.2% compared to conventional farmers' practice. A synthesis of the potential of CA practices for CC adaptation and mitigation is given in Table 11.1.

## Policies on Climate-Proofing

India has taken several policy initiatives and measures to address climate change and climate-proofing. Apart from the programs mentioned earlier, some of the relevant policy initiatives are summarized below.

- **Nationally determined contributions (NDCs).** As part of the global efforts under the Paris Agreement, India submitted its intended NDCs in 2015, which later became its nationally determined contributions (NDCs). India pledged to reduce its GHG emissions intensity by 33% of GDP, increase the share of non-fossil fuel-based energy, and enhance carbon sink capacity, among with other measures.
- **State action plans on climate change (SAPCC).** Most states in India have formulated SAPCC to address their specific CC challenges and develop region-specific adaptation and mitigation strategies.
- **National water policy.** India's national water policy emphasizes sustainable water management, water conservation, and climate-resilient water resource planning.
- **National Disaster Management Plan.** India has a comprehensive National Disaster Management Plan supported by state disaster management plans that include measures to manage and respond to climate-induced disasters effectively.

## Community of Practitioners (CoPs) and Networks on Climate-Proofing in India

In India, individuals and communities are making efforts to ward off the adverse effects of CC and adapt through appropriate solutions. They are embracing several climate-resilient practices and using their traditional wisdom. Agriculture being a state subject under the Indian constitution, planning and policy implementation falls substantially within the purview of respective states and local institutions, with the central government providing the broad policy framework and guidelines (NMSA). The development of the State Action Plan on Climate Change (SAPCC) needs to be considered as a dynamic document. It is important to periodically review, update, and integrate the agriculture, forestry, and land-use components in the SAPCCs. Many states have either revised or are in the process of revising SAPCCs in synergy with the post-2020 NDC goals and the SDGs. For mainstreaming of climate considerations across the development programs in the agriculture sector, spatial planning, natural, socio-economic, and demographic conditions need to be considered. India has well-developed institutional arrangements for disaster management at the national, state, and district levels.

There are several CoPs and networks in India focusing on climate-proofing and addressing CC challenges. These initiatives aim to foster collaboration and knowledge sharing among various stakeholders, including government agencies, non-governmental organizations, academic institutions, and local communities. Some examples of potential CoPs and networks on climate-proofing in India are:

- **Indian Network for Climate Change Assessment (INCCA).** INCCA is an initiative under the Ministry of Environment, Forest, and Climate Change (MoEFCC) in India. It aims to assess CC impacts and vulnerabilities in different regions of the country and facilitate the development of climate-resilient strategies.
- **India Climate Collaborative (ICC).** The ICC is a platform that brings together various actors, including civil society organizations, philanthropic foundations, and

**Table 11.1.** Evidence of climate-resilient agricultural practices in India. (From Pathak *et al.*, 2021)

| Climate-resilient technology | Crop/ cropping system | Productivity gain (kg/ha) | Profitability (US$/ ha) | Water saving (%) | Adaptation to climatic risks | GHG mitigation (kg $CO_2$ eq/ ha)[&] | References |
|---|---|---|---|---|---|---|---|
| Laser land leveling (LLL) | Rice-wheat; maize-wheat | 350–645 | 150–220 | 15–20 | * | 400–600 | Jat *et al.* (2015); Kakraliya *et al.* (2018) |
| Zero tillage without residue | Wheat | 300–450 | 110–160 | | * | 1000–1500 | Yadvinder-Singh *et al.* (2015); Thind *et al.* (2019) |
| Zero tillage with residue | Wheat | 410–660 | 153–232 | 6–10 | * # | 1500–2000 | Gathala *et al.* (2014); Aryal *et al.* (2016); Choudhary *et al.* (2018); Kakraliya *et al.* (2018) |
| Direct seeded rice (DSR) | Rice | ±150–220 | 120–155 | 12–18 | * # | 1000–2250 | Gathala *et al.* (2014); Chakraborty *et al.* (2017); Kakraliya *et al.* (2018) |
| Permanent beds | Maize-wheat | 450–950 | 220–350 | 65–70 | * # | 800–1000 | Jat *et al.* (2013, 2016); Parihar *et al.* (2017a, b); Das *et al.* (2018); Jat *et al.* (2019c) |
| | Maize-mustard | 400–1120 | 280–460 | 70–85 | * | | Gora *et al.* (2022) |
| | Pigeonpea-wheat | 400–900 | | 15–25 | | | Das *et al.* (2016) |
| Site-specific nutrient management (SSNM) | Rice-wheat; maize-wheat | 350–650 | 160–220 | | # | 500–1000 | Singh *et al.* (2015); Sapkota *et al.* (2017) |
| Sub-surface drip-irrigation | Rice-wheat | 280–550 | 150–220 | 40–50 | *# | 500–1500 | Jat *et al.* (2019c); Sidhu *et al.* (2019) |
| | Maize-wheat | 500–1020 | 320–580 | 70–85 | * | | Jat *et al.* (2019c); Sidhu *et al.* (2019); Patra *et al.* (2021) |
| Legumes inclusion (Mungbean) | Rice-wheat; maize-wheat | 320–350 | 180–250 | | # | 65–120 | Gathala *et al.* (2014); Choudhary *et al.* (2018); Jat *et al.* (2018); Thind *et al.* (2019) |
| Crop diversification | Wheat- and mustard-based systems | 400–1050 | 250–450 | 15–75 | * # | Significant but varied widely depending on different crops | Gora *et al.* (2022) |
| Portfolios of climate-resilient practices (LLL + ZT + Mulch + SSNM) | Rice-wheat | 110–220 | 280–360 | 15–20 | *# | 2800–3000 | Kakraliya *et al.* (2018) |

*Adaptation to extreme climate events such as excess rainfall during season

#Adaptation to extreme climate events such as terminal heat stress during season

&These are both life-cycle analysis as well as soil respiration under various studies.

research institutions, to collectively work on CC issues in India. It focuses on promoting innovative solutions and supporting climate-resilience efforts.

- **State-specific initiatives.** Several Indian states have their own CoPs and networks dedicated to CC adaptation and mitigation. For example, the Climate Change Cells in Kerala, Odisha, and Karnataka work on climate-resilience planning at the regional level.
- **United Nations Development Program (UNDP) projects.** UNDP often collaborates with the Indian government and local organizations to implement climate-related projects aimed at building resilience and promoting sustainable development.
- **Civil society organizations and NGOs.** Numerous NGOs and civil society organizations in India work on climate-proofing and community-based adaptation projects. Some examples include the Center for Science and Environment (CSE), WWF-India, and Oxfam India.

Given the localized nature of climate impacts and the adaptation needs, it is essential to further downscale it to the district or village level based on scientific evidence generated through collaborative research to mobilize communities to adapt to changing climate. However, to build climate resilience at community level, it is important to mobilize collective action which could be possible through collaborative actions with civil society, non-profit organizations, and local government. These initiatives often collaborate and engage in knowledge exchange through workshops, conferences, and online platforms to foster collective action and share best practices in climate-proofing measures.

## Potential Opportunities for Scaling CRA

Although agriculture is one of the significant sources of GHGs, it is also a target for large-scale GHG mitigation because of its potential to sequester carbon in vegetation and soils (Lal, 2015) or reduce GHG emissions by adopting regenerative agricultural practices. It is the most cost-effective and scalable solution. Offsetting or reducing the net $CO_2$ emissions to the atmosphere will be the key to mitigating the major factor of global warming and climate change.

India's efforts to build resilience to CC in agriculture are ongoing, and new innovations and initiatives are emerging. Integration of sustainable agricultural practices that promote CC adaptation, reduce GHG emissions, and enhance food security and resilience are important. Further scaling CRA is crucial for addressing CC impacts on agriculture and ensuring sustainable food production. Here are some potential opportunities for scaling climate-resilient agriculture:

- **Policy and financial support.** Governments and international organizations can play a vital role by implementing policies and providing financial incentives to promote CRA practices. This could include repurposing subsidies for climate-resilient crops, investments in research and development, and tax breaks for adopting sustainable farming techniques.
- **Knowledge sharing and capacity development.** Educating farmers about CRA practices and providing technical support can increase their adoption. Training programs, workshops, and demonstration farms can help farmers understand the benefits of CRA and how to implement these practices effectively.
- **Climate-resilient technologies.** Advancements in technology, such as precision agriculture, sensor-based irrigation systems, and climate-resilient crop varieties can significantly enhance agricultural productivity and reduce resource wastage, making them critical components for scaling CRA.
- **Climate finance mechanisms.** Development banks, climate funds, and private investors can channel financial resources towards CRA projects. This can include funding for climate-resilient infrastructure, sustainable farming practices, and climate-risk insurance for farmers.
- **Carbon credits.** Carbon credits can be a potential instrument for incentivizing farmers for adoption of CRA. Also, under the Paris Agreement, India has committed to reducing GHG emission intensity of GDP at 33–35% by 2030 compared to 2005 levels. Regenerative agricultural practices that help

sequester carbon in the soil and reduce GHGs can be incentivized by quantifying these activities as carbon credits. Linking agriculture to carbon marketplace would serve as a strong motivation for the farmers to transition to sustainable agriculture practices promoted by the government. Carbon credits can encourage practices that mitigate emissions, such as adoption of sustainable agricultural practices, thereby helping sequester carbon in the soil, reduce emissions from agricultural activities, and enhance overall ecosystem health.

- **Public-private partnerships.** Collaboration between governments, NGOs, and the private sector can accelerate the adoption of CRA practices. Private companies can bring in innovation, technology, and market linkages, while governments and NGOs can provide support and ensure equitable distribution of benefits.
- **Farmer cooperatives and networks.** Encouraging farmers to work together through cooperatives or networks can facilitate the exchange of knowledge, resources, and access to markets. These collaborative efforts can also attract more significant investment and support for CRA initiatives.
- **Climate-resilient certification and labeling.** Creating certification and labeling schemes for climate-resilient products can incentivize consumers to choose sustainable agricultural products. This can drive demand for CRA practices and encourage farmers to adopt them.
- **Climate information and advisory services.** Access to accurate and timely climate information is essential for farmers to plan their activities and make informed decisions. Setting up climate information centers and advisory services can help disseminate climate-related data and guidance to farmers.
- **Integration of CRA into agricultural policies and programs.** Embedding CRA principles into national agricultural policies and programs ensures long-term commitment and mainstreaming of climate-resilient practices.
- **International cooperation and knowledge exchange.** Encouraging collaboration between countries and facilitating knowledge sharing can help identify successful CRA practices that can be adapted and scaled up in different regions and contexts.

Scaling climate-resilient agriculture is a multifaceted endeavor that requires concerted efforts from governments, researchers, private sector, farmers, civil society, and consumers. By addressing challenges and seizing these opportunities, CRA adoption can be advanced and contribute to building a resilient and sustainable agricultural sector in the face of climate change.

## Conclusions and Way Forward

Building resilience to climate change is a pressing challenge for India, given its vulnerability to climate impacts and the vastness of its population and ecosystems. Addressing CC requires global cooperation and collective action to reduce GHG emissions, transition to cleaner energy sources, enhance climate resilience, and implement sustainable practices. Over the years, India has taken significant steps to address CC through various innovations, strategies, and policies. These efforts have involved multiple sectors and stakeholders, from renewable energy promotion to disaster management and sustainable agriculture practices.

The adaptation and mitigation practices evaluated under various initiatives are being upscaled by several state governments across the country. Resilient practices and technologies have been documented for upscaling (Prasad *et al.*, 2014). A strategic approach of both central and state programs on CC mitigation is crucial to progress the adaptive ability by addressing the existing climate risks and future vulnerabilities. An intensive and comprehensive program is essential with collective responsibilities and climate actions under NAPCC and SAPCC (Srinivasa Rao *et al.*, 2019).

Innovations such as CRA, green infrastructure development, and climate information services have shown promising results in enhancing India's ability to cope with climate-related risks. Moreover, policies that promote community-based adaptation, renewable energy integration, and ecosystem-based approaches have contributed to a more comprehensive and inclusive climate-resilience agenda.

By integrating climate resilience into agricultural planning and implementation, India aims to enhance food security, protect livelihoods, and build a more resilient and sustainable agricultural system in the face of CC. To further strengthen climate resilience in India, a holistic and coordinated approach is essential. Some key steps on the way forward include:

- **Scaling up successful initiatives.** Identify and scale up successful climate-resilience initiatives to reach a broader population and expand their impact across different regions.
- **Enhancing climate research and data.** Invest in climate research and data collection to better understand regional climate patterns and emerging risks, facilitating evidence-based decision-making.
- **Mainstreaming climate into development plans.** Integrate climate considerations into national and regional development plans across sectors to ensure climate resilience becomes an integral part of India's growth trajectory.
- **Empower local communities.** Empower local communities to play an active role in CC adaptation and resilience-building efforts. Engaging communities in decision-making processes fosters ownership and ensures that solutions are contextually relevant.
- **Private sector engagement.** Encourage private sector participation in climate-resilience projects by offering incentives and creating an enabling business environment for green investments.
- **Capacity building and training.** Invest in capacity building and training programs to equip government officials, professionals, and communities with the knowledge and skills needed to implement climate-resilient strategies effectively.
- **Incentivizing climate-resilient agriculture through carbon credits.** Carbon credits can encourage practices that mitigate emissions and promote adoption of sustainable agricultural practices.
- **Innovative financing mechanisms.** Develop innovative financing mechanisms, such as green bonds and climate funds, to mobilize resources for climate-resilient projects and initiatives.
- **Public awareness behavioral change.** Raise public awareness about CC and the importance of individual actions in building resilience. Encouraging sustainable behaviors and lifestyles can have a positive impact on climate resilience.
- **Collaboration and international cooperation.** Collaborate with other countries and international organizations to access climate finance, technology transfer, and expertise in climate adaptation and resilience building.
- **Review and adaptation.** Continually review and update climate policies and strategies based on new information, technological advancements, and changing climate trends.

Addressing CC is a complex and multifaceted challenge and building resilience to CC is a long-term endeavor that requires sustained commitment from all stakeholders. By combining innovative approaches, well-designed policies, and coordinated efforts with active participation from government, private sector, civil society, communities, and individuals, India can continue its path towards a more climate-resilient and sustainable future. As the challenges of CC continue to evolve, India's proactive and adaptive approach on sustainable development, incentivizing adoption of climate-resilient agricultural practices and ecosystem services through carbon credits, clean-energy technologies, and responsible environmental stewardship will play a crucial role in ensuring a resilient and sustainable future for the environment, its people, and generations to come.

## References

Aggarwal, P.K. (2008) Global climate and Indian agriculture: impacts, adaptation and mitigation. *Indian Journal of Agricultural Sciences* 78(10), 911–919. Available at: http://www.indiaenvironmentportal.org.in/files/Global%20Climate%20Change.pdf (accessed 18 July 2024).

Aggarwal, P.K. (2009) *Global Climate Changes and Indian Agriculture: Case Studies from ICAR Network Project.* Indian Council of Agricultural Research, New Delhi.

Aggarwal, P.K. and Sinha, S.K. (1993) Effect of probable increase in carbon dioxide and temperature on wheat yields in India. *Journal of Agricultural Meteorology* 48(5), 811–814. DOI: 10.2480/agrmet.48.811

Ahluwalia, V.K. and Malhotra, S. (2006) *Environmental Science.* Ane Books India, Delhi, Chennai, Kolkata, Bangalore.

Aryal, J.P., Sapkota, T.B., Stirling, C.M., Jat, M.L., Jat, H.S. *et al.* (2016) Conservation agriculture-based wheat production better copes with extreme climate events than conventional tillage-based systems – a case of untimely excess rainfall in Haryana, India. *Agriculture, Ecosystems and Environment* 233, 325–335. DOI: 10.1016/j.agee.2016.09.013

Bal, S.K., Sandeep, V.M., Kumar, P.V., Subba Rao, A.V.M., Pramod, V.P. *et al.* (2021) Assessing impact of dry spells on the principal rainfed crops in major dryland regions of India. *Agricultural and Forest Meteorology* 313(2022), 108768. DOI: 10.1016/j.agrformet.2021.108768

Bal, S.K., Prasad, J.V.N.S. and Singh, V.K. (2022) *Heat Wave 2022 Causes, Impacts and Way Forward for Indian Agriculture*. Technical Bulletin No. ICAR/CRIDA/TB/01/2022, ICAR-Central Research Institute for Dryland Agriculture, Hyderabad, Telangana, India.

Bhadwal, S., Kelkar, U. and Bhandari, P.M. (2007) Impact on agriculture help reduce vulnerability. *The Hindu* survey of the environment. *The Hindu* (Special Issue).

Bhattacharyya, P., Pathak, H. and Pal, S. (2020) Climate smart agriculture: concepts, challenges and opportunities. *Green Energy and Technology*. Springer Nature. DOI: 10.1007/978-981-15-9132-7

Bouman, B.A.M., Barker, R., Humphreys, E., Tuong, T.P., Atlin, G.N. *et al.* (2007) Rice: feeding the billions. In: Molden, D. (ed.) *Water for Food, Water for Life: A Comprehensive Assessment of Water Management in Agriculture*. IWMI, Colombo, Sri Lanka, pp. 515–549. Available at: https://www.iwmi.cgiar.org/assessment/Water%20for%20Food%20Water%20for%20Life/Chapters/Chapter%2014%20Rice.pdf (accessed 18 July 2024).

CCAFS (2019) CCAFS in the South Asian Region. CGIAR Research Program on Climate Change, Agriculture and Food Security (CCAFS), Delhi. Available at: https://hdl.handle.net/10568/105602 (accessed 18 July 2024).

Chakraborty, D., Ladha, J.K., Rana, D.S., Jat, M.L., Gathala, M.K. *et al.* (2017) A global analysis of alternative tillage and crop establishment practices for economically and environmentally efficient rice production. *Nature Scientific Reports* 7, 9342. DOI: 10.1038/s41598-017-09742-9

Challinor, A.J. and Wheeler, T.R. (2008) Use of a crop model ensemble to quantify $CO_2$ stimulation of water-stressed and well-watered crops. *Agricultural and Forest Meteorology* 148(6–7), 1062–1077. DOI: 10.1016/j.agrformet.2008.02.006

Challinor, A.J., Wheeler, T.R., Craufurd, P.Q., Ferro, C.A.T. and Stephenson, D.B. (2007) Adaptation of crops to climate change through genotypic responses to mean and extreme temperatures. *Agriculture, Ecosystems and Environment* 119(1–2), 190–204. DOI: 10.1016/j.agee.2006.07.009

Choudhary, K.M., Jat, H.S., Nandal, D.P., Bishnoi, D.K., Sutaliya, J.M. *et al.* (2018) Evaluating alternatives to rice-wheat system in western Indo-Gangetic Plains: crop yields, water productivity and economic profitability. *Field Crops Research* 218, 1–10. DOI: 10.1016/j.fcr.2017.12.023

Crippa, M., Solazzo, E., Guizzardi, D., Monforti-Ferrario, M., Tubiello, F.N. *et al.* (2021) Food systems are responsible for a third of global anthropogenic GHG emissions. *Nature Food* 2, 198–209. DOI: 10.1038/s43016-021-00225-9

Das, T.K., Bandyopadhyay, K.K., Bhattacharyya, R., Sudhishri, S., Sharma, A.R. *et al.* (2016) Effects of conservation agriculture on crop productivity and water-use efficiency under an irrigated pigeonpea-wheat cropping system in the western Indo-Gangetic Plains. *Journal of Agricultural Science* 154(8), 1327–1342. DOI: 10.1017/S0021859615001264

Das, T.K., Saharawat, Y.S., Bhattacharyya, R., Sudhishri, S., Bandyopadhyay, K.K. *et al.* (2018) Conservation agriculture effects on crop and water productivity, profitability and soil organic carbon accumulation under a maize-wheat cropping system in the north-western Indo-Gangetic Plains. *Field Crops Research* 215, 222–231. DOI: 10.1016/j.fcr.2017.10.021

FAO and ITPS (2015) *Status of the World's Soil Resources (SWSR) – Main Report*. Food and Agriculture Organization of the United Nations and Intergovernmental Technical Panel on Soils, Rome. Available at: https://www.fao.org/3/a-i5199e.pdf (accessed 18 July 2024).

Farooq, M. and Siddique, K.H.M. (2015) Conservation agriculture: concepts, brief history, and impacts on agricultural systems. In: Farooq, M. and Siddique, K.H. (eds) *Conservation Agriculture*. Springer International Publishing, Switzerland, pp. 3–17. DOI: 10.1007/978-3-319-11620-4_1

Fischer, G., Shah, M. and Velthuizen, H. (2002) *Climate Change and Agricultural Vulnerability.* International Institute for Applied Systems Analysis, Laxenburg, Austria. Available at: https://pure.iiasa.ac.at/id/eprint/6670/1/XO-02-001.pdf (accessed 18 July 2024).

Gathala, M.K., Kumar, V., Sharma, P.C., Saharawat, Y.S., Jat, H.S. *et al.* (2014) Optimizing intensive cereal-based cropping systems addressing current and future drivers of agricultural change in the northwestern Indo-Gangetic Plains of India. *Agriculture, Ecosystems and Environment* 187, 33–46. DOI: 10.1016/j.agee.2013.06.002

Gathala, M.K., Laing, A.M., Tiwari, T.P., Timsina, J., Islam, S. *et al.* (2020) Energy-efficient, sustainable crop production practices benefit smallholder farmers and the environment across three countries in the Eastern Gangetic Plains, South Asia. *Journal of Cleaner Production* 246, 118982. DOI: 10.1016/j.jclepro.2019.118982

Gora, M.K., Kumar, S., Jat, H.S., Kakraliya, S.K., Choudhary, M. *et al.* (2022) Scalable diversification options delivers sustainable and nutritious food in Indo-Gangetic plains. *Nature Scientific Reports* 12, 14371. DOI: 10.1038/s41598-022-18156-1

ICRISAT (2023) Innovations to transform drylands towards inclusive, resilient and sustainable agri-food systems. In: Kumar, A., Kumar, S., Mandapati, J. and Peachey, R. (eds) *International Crops Research Institute for the Semi-Arid Tropics, Patancheru, Telangana, India.*

IMD (2022) *Statement on Climate of India during 2022.* Climate Research and Services (CRS), Indian Meteorological Department, Ministry of Earth Sciences (MoES), Government of India. Available at: https://internal.imd.gov.in/press_release/20230106_pr_2096.pdf (accessed 18 July 2024).

IPCC (2007a) *Climate Change 2007: The Physical Science Basis.* Contribution of Working Group I to the Fourth Assessment Report of the Intergovernmental Panel on Climate Change, Cambridge University Press, Cambridge, UK.

IPCC (2007b) *Climate Change 2007: Impacts, Adaptation and Vulnerability.* Contribution of Working Group II to the Fourth Assessment Report of the Intergovernmental Panel on Climate Change, Cambridge University Press, Cambridge, UK.

IPCC (2019) Summary for policymakers. In: Shukla, P.R., Skea, J., Calvo Buendia, E., Masson-Delmotte, V., Pörtner, H.-O. *et al.* (eds) *Climate Change and Land: An IPCC Special Report on Climate Change, Desertification, Land Degradation, Sustainable Land Management, Food Security, and Greenhouse Gas Fluxes in Terrestrial Ecosystems.* Cambridge University Press, Cambridge, UK.

IPCC (2021) Climate Change 2021: the physical science basis. In: Masson-Delmotte, V., Zhai, P., Pirani, A., Connors, S.L., Péan, C. *et al.* (eds) *Contribution of Working Group I to the Sixth Assessment Report of the Intergovernmental Panel on Climate Change.* Cambridge University Press, Cambridge, UK and New York. DOI:10.1017/9781009157896

Jain, S.K. (2012) Sustainable water management in India considering likely climate and other changes. *Current Science* 102(2), 177–188. Available at: http://www.jstor.org/stable/24083846 (accessed 18 July 2024).

Jat, H.S., Datta, A., Choudhary, M., Yadav, A.K., Choudhary, V. *et al.* (2019a) Effects of tillage, crop establishment and diversification on soil organic carbon, aggregation, aggregate associated carbon and productivity in cereal systems of semi-arid northwest India. *Soil and Tillage Research* 190, 128–138. DOI: 10.1016/j.still.2019.03.005

Jat, H.S., Sharma, P.C., Datta, A., Choudhary, M., Kakraliya, S.K. *et al.* (2019b) Re-designing irrigated intensive cereal systems through bundling precision agronomic innovations for transitioning towards agricultural sustainability in north-west India. *Nature Scientific Reports* 9, 17929. DOI: 10.1038/s41598-019-54086-1

Jat, H.S., Choudhary, M., Kakraliya, S.K., Gora, M.K., Kakraliya, M. *et al.* (2022) A decade of climate-smart agriculture in major agri-food systems: earthworm abundance and soil physico-biochemical properties. *Agronomy* 12(3), 658, 1–24. DOI: 10.3390/agronomy12030658

Jat, M.L. (2020) Tech tools for the modern farm. In: Vembu, V. and Madhavan, N. (eds) *Handbook of Indian Agriculture 2020: Charting the Way Out of the Farm Crisis.* The Hindu Group, India, pp. 149–153.

Jat, M.L., Gathala, M.K., Ladha, J.K., Saharawat, Y.S., Jat, A.S. *et al.* (2009) Evaluation of precision land leveling and double zero-till systems in rice-wheat rotation: water use, productivity, profitability and soil physical properties. *Soil and Tillage Research* 105, 112–121. DOI: 10.1016/j.still.2009.06.003

Jat, M.L., Gathala, M.K., Saharawat, Y.S., Tetarwal, J.P., Gupta, R. *et al.* (2013) Double no-till and permanent raised beds in maize–wheat rotation of north-western Indo-Gangetic plains of India: effects on crop yields, water productivity, profitability and soil. *Field Crops Research* 149, 291–299. DOI: 10.1016/j.fcr.2013.04.024

Jat, M.L., Singh, Y., Gill, G., Sidhu, H.S., Aryal, J.P. *et al.* (2015) Laser-assisted precision land leveling impacts in irrigated intensive production systems of south Asia. In: Lal, R. and Stewart, B.A. (eds)

*Advances in Soil Science, Soil Specific Farming: Precision Agriculture*. CRC Press, Boca Raton, Florida, pp. 323–352. DOI: 10.1201/b18759-14

Jat, M.L., Dagar, J.C., Sapkota, T.B., Yadvinder-Singh, Govaerts, B. *et al.* (2016) Climate change and agriculture: adaptation strategies and mitigation opportunities for food security in South Asia and Latin America. *Advances in Agronomy* 137, 127–236. DOI: 10.1016/bs.agron.2015.12.005

Jat, M.L., Chakraborty, D., Ladha, J.K., Rana, D.S., Gathala, M.K. *et al.* (2020) Conservation agriculture for sustainable intensification in South Asia. *Nature Sustainability* 3(4), 336–343. DOI: 10.1038/s41893-020-0500-2

Jat, S.L., Parihar, C.M., Singh, A.K., Nayak, H.S., Meena, B.R. *et al.* (2019c) Differential response from nitrogen sources with and without residue management under conservation agriculture on crop yields, water-use and economics in maize-based rotations. *Field Crops Research* 236, 96–110. DOI: 10.1016/j.fcr.2019.03.017

Kakraliya, S.K., Jat, H.S., Singh, I., Sapkota, T.B., Singh, L.K. *et al.* (2018) Performance of portfolios of climate smart agriculture practices in a rice-wheat system of western Indo-Gangetic plains. *Agricultural Water Management* 202, 122–132. DOI: 10.1016/j.agwat.2018.02.020

Kar, S., Pramanick, B., Brahmachari, K., Saha, G., Mahapatra, B.S. *et al.* (2021) Exploring the best tillage option in rice based diversified cropping systems in alluvial soil of eastern India. *Soil and Tillage Research* 205, 104761. DOI: 10.1016/j.still.2020.104761

Kassam, A., Friedrich, T. and Derpsch, R. (2022) Successful experiences and lessons from conservation agriculture worldwide. *Agronomy* 12(4), 769. DOI: 10.3390/agronomy12040769

Krishnan, R., Sanjay, J., Gnanaseelan, C., Majumdar, M., Kulkarni, A. *et al.* (2020) *Assessment of Climate Change Over the Indian Region.* A Report of the Ministry of Earth Sciences, Government of India. ISBN 978-981-15-4326-5, Springer Nature, Singapore.

Kumar, S.N., Aggarwal, P.K., Rani, S., Jain, S., Saxena, R. *et al.* (2011) Impact of climate change on crop productivity in Western Ghats, coastal and northeastern regions in India. *Current Science* 101(3), 332–341. Available at: https://www.jstor.org/stable/24078512

Lal, M., Singh, K.K., Rathore, L.S., Srinivasan, G. and Saseendran, S.A. (1998) Vulnerability of rice and wheat yields in NW India to future changes in climate. *Agricultural and Forest Meteorology* 89(2), 101–114. DOI: 10.1016/S0168-1923(97)00064-6

Lal, R. (2001) Soil degradation by erosion. *Land Degradation and Development* 12(6), 519–539. DOI: 10.1002/ldr.472

Lal, R. (2015) Restoring soil quality to mitigate soil degradation. *Sustainability* 7(5), 5875–5895. DOI: 10.3390/su7055875

Mall, R.K. and Aggarwal, P.K. (2002) Climate change and rice yields in diverse agro-environments of India. I. Evaluation of impact assessment models. *Climatic Change* 52, 315–330. DOI: 10.1023/A:1013702105870

Mall, R.K., Singh, R., Gupta, A., Srinivasan, G. and Rathore, L. (2006) Impact of climate change on Indian agriculture: a review. *Climatic Change* 78, 445–478. DOI: 10.1007/s10584-005-9042-x

Mani, M., Bandyopadhyay, S., Chonabayashi, S., Markandya, A. and Mosier, T. (2018) *South Asia's Hotspots: The Impact of Temperature and Precipitation Changes on Living Standards*. The World Bank Group, Washington, DC. Available at: http://hdl.handle.net/10986/28723 (accessed 18 July 2024).

MoEFCC (2021) *India: Third Biennial Update Report to the United Nations Framework Convention on Climate Change.* Ministry of Environment, Forest and Climate Change, Government of India.

Mohanty, M., Nishant, K.S., Sonali, P.M., Chaudhary, R.S., Reddy, K.S. *et al.* (2017) Climate change impacts vis-à-vis productivity of soybean in vertisol of Madhya Pradesh. *Journal of Agrometeorology* 19(1), 10–16. DOI: 10.54386/jam.v19i1.749

Ortiz-Bobea, A., Ault, T.R., Carrillo, C.M., Chambers, R.G. and Lobell, D.B. (2021) Anthropogenic climate change has slowed global agricultural productivity growth. *Nature Climate Change* 11, 306–312. DOI: 10.1038/s41558-021-01000-1

Pandey, S., Behura, D., Villano, R. and Naik, D. (2000) Economic cost of drought and farmers' coping mechanisms: a study of rainfed rice systems in eastern India, No. 287599, IRRI Discussion Papers, International Rice Research Institute, Laguna, Philippines. DOI: 10.22004/ag.econ.287599

Parihar, C.M., Jat, S.L., Singh, A.K., Ghosh, A., Rathore, N.S. *et al.* (2017a) Effects of precision conservation agriculture in a maize-wheat-mungbean rotation on crop yield, water-use and radiation conversion under a semiarid agro-ecosystem. *Agricultural Water Management* 192, 306–319. DOI: 10.1016/j.agwat.2017.07.021

Parihar, C.M., Jat, S.L., Singh, A.K., Majumdar, K., Jat, M.L. *et al.* (2017b) Bio-energy, water-use efficiency and economics of maize-wheat-mungbean system under precision-conservation agriculture in semi-arid agro-ecosystem. *Energy* 119, 245–256. DOI: 10.1016/j.energy.2016.12.068

Paroda, R.S. (2018) *Reorienting Indian Agriculture: Challenges and Opportunities.* CAB International, Wallingford, UK.

Pathak, H., Srinivasarao, Ch. and Jat, M.L. (2021) Conservation agriculture for climate change adaptation and mitigation in India. *Journal of Agricultural Physics* 21(1), 182–196.

Patra, K., Parihar, C.M., Nayak, H.S., Rana, B., Singh, V.K. *et al.* (2021) Water budgeting in conservation agriculture-based sub-surface drip irrigation in tropical maize using HYDRUS-2D in south Asia. *Nature Scientific Reports* 11, 16770. DOI: 10.1038/s41598-021-93866-6

Prasad, Y.G., Maheswari, M., Dixit, S., Srinivasarao, C., Sikka, A.K. *et al.* (2014) *Smart Practices and Technologies for Climate Resilient Agriculture.* Central Research Institute for Dryland Agriculture (ICAR), Hyderabad, India.

Rama Rao, C.A., Raju, B.M.K., Islam, A., Subba Rao, A.V.M., Rao, K.V. *et al.* (2019) *Risk and Vulnerability Assessment of Indian Agriculture to Climate Change.* ICAR-Central Research Institute for Dryland Agriculture, Hyderabad, India.

Rao, G.D. and Sinha, S.K. (1994) Impact of climatic change on simulated wheat production in India. In: Rosenzweig, C. and Iglesias, I. (eds) *Implications of Climate Change for International Agriculture: Crop Modelling Study.* US Environmental Protection Agency, Washington, DC (Section 6: Asia-India), pp. 1–17.

Rosenzweig, C. and Parry, M.L. (1994) Potential impact of climate change on world food supply. *Nature* 367, 133–138. DOI: 10.1038/367133a0

Rosenzweig, C., Tubiello, F.N., Goldberg, R., Mills, E. and Bloomfield, J. (2002) Increased crop damage in the US from excess precipitation under climate change. *Global Environmental Change* 12(3), 197–202. DOI: 10.1016/S0959-3780(02)00008-0

Samra, J.S. and Singh, G. (2004) *Heat Wave of March 2004: Impact on Agriculture.* Indian Council of Agricultural Research, New Delhi.

Sandhu, O.P., Gupta, R.K., Thind, H.S., Jat, M.L., Sidhu, H.S. *et al.* (2019) Drip irrigation and nitrogen management for improving crop yields, nitrogen use efficiency and water productivity of maize-wheat system on permanent beds in north-west India. *Agricultural Water Management* 219, 19–26. DOI: 10.1016/j.agwat.2019.03.040

Sapkota, T.B., Jat, R.K., Singh, R.G., Jat, M.L., Stirling, C.M. *et al.* (2017) Soil organic carbon changes after seven years of conservation agriculture in a rice–wheat system of the eastern Indo-Gangetic Plains. *Soil Use and Management* 33, 81–89. DOI: 10.1111/sum.12331

Saseendran, S.A., Singh, K.K., Rathore, L.S., Singh, S.V. and Sinha, S.K. (2000) Effects of climate change on rice production in the tropical humid climate of Kerala, India. *Climatic Change* 44, 495–514. DOI: 10.1023/A:1005542414134

Sidhu, H.S., Jat, M.L., Singh, Y., Sidhu, R.K., Gupta, N. *et al.* (2019) Sub-surface drip fertigation with conservation agriculture in a rice-wheat system: a breakthrough for addressing water and nitrogen use efficiency. *Agricultural Water Management* 216, 273–283. DOI: 10.1016/j.agwat.2019.02.019

Sidhu, R.K., Kumar, R., Rana, P.S. and Jat, M.L. (2021) Automation in drip irrigation for enhancing water use efficiency in cereal systems of South Asia: status and prospects. *Advances in Agronomy* 167, 247–300. DOI: 10.1016/bs.agron.2021.01.002

Singh, B., Singh, V., Purba, J., Sharma, R.K., Jat, M.L. *et al.* (2015) Site-specific fertilizer nitrogen management in irrigated transplanted rice (*Oryza sativa*) using an optical sensor. *Precision Agriculture* 16, 455–745. DOI: 10.1007/s11119-015-9389-6

Singh, N.P., Anand, B., Singh, S., Srivastava, S.K., Srinivasarao, C. *et al.* (2021) Synergies and trade-offs for climate-resilient agriculture in India: an agro-climatic zone assessment. *Climatic Change* 164, 11. DOI: 10.1007/s10584-021-02969-6

SIWI, IFPRI, IUCN and IWMI (2005) *Let it Reign: The New Water Paradigm for Global Food Security*. Final Report to Commission on Sustainable Development (CSD)-13. Stockholm International Water Institute, Stockholm.

SRFSI-ACIAR (2021) *Sustainable and Resilient Farming Systems Intensification in the Eastern Gangetic Plains (SRFSI). Final Report*. ACIAR, Canberra. Available at: https://www.aciar.gov.au/project/cse-2011-077 (accessed 18 July 2024).

Srinivasa Rao, Ch., Prabhakar, M., Maheswari, M., Srinivasa Rao, M., Sharma, K.L. *et al.* (2016) *National Innovations in Climate Resilient Agriculture (NICRA). Research Highlights 2015–16*. Central Research

Institute for Dryland Agriculture, Hyderabad, India. Available at: http://www.nicra-icar.in/nicrarevised/images/publications/NICRA%20Res%20Highlights%202015-16.pdf (accessed 18 July 2024).

Srinivasa Rao, Ch., Prasad, R.S. and Mohapatra, T. (2019) *Climate Change and Indian Agriculture: Impacts, Coping Strategies, Programmes and Policy*. Indian Council of Agricultural Research, Ministry of Agriculture and Farmers' Welfare and Ministry of Environment, Forestry and Climate Change, Government of India, New Delhi.

Srinivasarao, Ch. and Srinivas, K. (2023) Carbon management for sustainable soil health and environment in rainfed agroecosystems. *Indian Journal of Fertilizers* 19(14), 276–299. Available at: https://www.researchgate.net/publication/370301372_Carbon_Management_for_Sustainable_Soil_Health_and_Environment_in_Rainfed_Agriculture (accessed 18 July 2024).

Srinivasarao, Ch., Maheswari, M., Srinivasa Rao, M., Sharma, K.L., Vanaja, M. *et al.* (2015a) *National Innovations on Climate Resilient Agriculture (NICRA). Research Highlights 2014–15.* Central Research Institute for Dryland Agriculture, Hyderabad, India.

Srinivasarao, Ch., Lal, R., Prasad, J.V.N.S., Gopinath, K.A., Singh, R. *et al.* (2015b) Potential and challenges of rainfed farming in India. *Advances in Agronomy* 133, 113–181. DOI: 10.1016/bs.agron.2015.05.004

Srinivasarao, Ch., Gopinath, K.A., Prasad, J.V.N.S. and Singh, A.K. (2016a) Climate resilient villages for sustainable food security in tropical India: concept, process, technologies, institutions, and impacts. *Advances in Agronomy* 140(3), 101–214. DOI: 10.1016/bs.agron.2016.06.003

Srinivasarao, Ch., Rao, K.V., Desai, S., Subba Rao, A.V.M., Osman, Md. *et al.* (2016b) *Preparedness for Agriculture Contingencies – Kharif, 2016 Summary of Interface Meetings and Way Forward*. Technical Bulletin 2/2016. Central Research Institute for Dryland Agriculture, Indian Council of Agricultural Reserarch. Hyderabad Telangana, India.

Srinivasarao, Ch., Rao, K.V. and Saxena, C. (2016c) Drought proofing through implementation of district agriculture contingency plans in India. In: *Twelfth International Dryland Development Conference*, 21–24 August, Alexandria, Egypt. International Dryland Development Commission (IDDC), pp. 17–18.

Srinivasarao, Ch., Rao, K.V., Gopinath, K.A., Prasad, Y.G., Arunachalam, A. *et al.* (2020) Agriculture contingency plans for managing weather aberrations and extreme climatic events: development, implementation and impacts in India. *Advances in Agronomy* 159, 35–91. DOI: 10.1016/bs.agron.2019.08.002

Sterrett, C. (2011) *Review of Climate Change Adaptation Practices in South Asia*. Oxfam Research Reports. Oxford, UK. Available at: https://www.oxfam.org/en/research/review-climate-change-adaptation-practices-south-asia (accessed 18 July 2024).

Thind, H.S., Sharma, S., Yadvinder-Singh and Sidhu, H.S. (2019) Rice-wheat productivity and profitability with residue, tillage and green manure management. *Nutrient Cycling in Agroecosystems* 113(2), 113–125. DOI: 10.1007/s10705-018-09967-8

Venkateswarlu, B., Kokate, K.D., Gopinath, K.A., Srinivasarao, C., Anuradha, B. *et al.* (eds) (2012) *Coping with Climate Variability: Technology Demonstration on Farmers' Fields in Vulnerable Districts*. Central Research Institute for Dryland Agriculture, Hyderabad, India. Available at: http://nicra-icar.in/nicrarevised/images/Books/NICRA-TDC-Coping%20with%20Climate%20Variability.pdf (accessed 18 July 2024).

WMO (2022) *State of the Climate in Asia 2021* (WMO-No. 1303). World Meteorological Organization (WMO), Geneva, Switzerland. Available at: https://public.wmo.int/en/our-mandate/climate/wmo-statement-state-of-global-climate/asia (accessed 18 July 2024).

World Bank Group (2021) *Climate Risk Country Profile: India*. World Bank Group, Washington, DC. Available at: https://climateknowledgeportal.worldbank.org/sites/default/files/country-profiles/15503-WB_India%20Country%20Profile-WEB.pdf (accessed 18 July 2024).

Yadvinder-Singh, Singh, M., Sidhu, H.S., Humphreys, E., Thind, H.S. *et al.* (2015) Nitrogen management for zero till wheat with surface retention of rice residues in north-west India. *Field Crops Research* 184, 183–191. DOI: 10.1016/j.fcr.2015.03.025

# 12 Climate Change and Sustainable Agro-ecology in China's Dryland: A Case Study

**Bai-Jian Lin[1], Yu-Xin Wei[2], Zhe Chen[2], Cong He[2], Olatunde Pelumi Oladele[2], Jian-Fu Xue[3], Xing Wang[4], Wei Bai[5], Yuan-Feng Sun[6], Ren-Hua Sun[6]*, Xin Zhao[2], and Hai-Lin Zhang[2]**

*[1]Northeast Agricultural University, Heilongjiang, China; [2]China Agricultural University, Beijing, China; [3]Shanxi Agricultural University, Shanxi, China; [4]Northwest A&F University, Shaanxi, China; [5]Liaoning Academy of Agricultural Science, Liaoning, China; [6]Rural Energy and Environment Agency, Ministry of Agriculture and Rural Affairs (MOARA), Beijing, China*

**Abstract**

Global climate change is constraining the practices of agricultural production due to which developing climate-smart agriculture (CSA) faces enormous challenges in mitigating greenhouse gas (GHG) emissions and ensuring food security. Currently, few studies have systematically reviewed the progress of CSA in developing dryland agriculture in China. This chapter reviews the problems and challenges in CSA practices of China's dryland farming. The application of CSA management practices in China's dryland can increase the amount of soil-organic carbon (SOC) and maintain or increase crop yield, but GHG emissions remain controversial. Adaptation to climate change can be addressed by adopting conservation tillage, optimizing fertilizer application, and improving technologies, such as straw conversion into biochar. Future research should consider the combined plant-soil-climate system when developing CSA management practices in dryland agricultural systems to prioritize the achievement of synergies among soil-quality improvement, increase crop yield, and facilitate climate adaptation.

## Introduction

Climate change is one of the most serious threats to agricultural systems and human survival. Recent research has highlighted that dryland ecosystems are more sensitive to climate change and vulnerable to global warming than traditional agricultural systems (Huang *et al.*, 2017; Yao *et al.*, 2021). In the past three decades, increasing GHG emissions have caused extreme weather events, such as severe droughts and heatwaves, which have seriously affected dryland agricultural production (Leisner, 2020). Thus, dryland agriculture faces more challenges than ever before, which require solutions to improve agricultural technology and agronomic measures from regional to global level (Qiao *et al.*, 2022). Increasing global food yield and sequestering carbon emissions should be achieved synergistically, which is the core goal of dryland

*Corresponding author: sunrh_abc@163.com


DOI: 10.1079/9781800624870.0012

agriculture in mitigating and adapting to climate change.

The total amount of arable land in China ranges from 130 to 171.3 Mha, and approximately 50% of this land is composed of dryland fields. In recent decades, agronomists have attempted to adopt climate-smart agriculture (CSA) management practices as a systematic approach to ensure the goals of sustainable agricultural development (FAO, 2013). Improvements in nutrient management and the use of water-saving agricultural technology, conservation agriculture, crop diversification, plastic film management, integrated management practices that facilitate soil fertility, and other innovative technologies are within the scope of CSA (Zhao *et al.*, 2023). However, the feasibility of the technology, costs, and potential benefits are critical to determine whether to implement CSA in fields or regions.

Currently, researchers have attempted to use various methods to evaluate the comprehensive benefits of CSA management practices, including literature surveys, long-term field experiments, and modeling, and a large amount of positive or negative evidence has been accumulated. For example, inorganic fertilizer combined with manure can achieve the preferable synergistic benefit of improving crop yield and reducing the emissions of ammonia ($NH_3$). However, the excessive input of inorganic fertilizers increases GHG emissions and environmental problems (Xu *et al.*, 2022). Conservation and organic agriculture can improve soil nutrients, reduce environmental costs, and maintain soil health, but the benefits for yield are controversial. Similarly, irrigation can also help to reduce GHG emissions and improve crop productivity by altering the soil-moisture dynamics in dryland agriculture, but this also depends on irrigation methods, such as flooding, and furrow or sprinkler irrigation (Emde *et al.*, 2021).

Furthermore, some new technological applications can improve the ecological environment and economic benefits of dryland agriculture (Wolfert and Isakhanyan, 2022). For example, combining Big Data and artificial intelligence (AI) technology makes it possible to improve the systematic optimization of farming systems based on site-specific conditions. However, the existing CSA management practices primarily focus on a single technology or the entire farmland system, and few studies have systematically reviewed the progress and challenges of CSA in drylands.

It is necessary to conduct a comprehensive review of the development and challenges of CSA to better understand its impact when applied to dryland agriculture in China. This case study will be described from the following aspects based on a literature survey method: (i) What has happened to the changes in the management technology of dryland farmland in China?; (ii) What are the current constraints on developing dryland farmland management technology in China? What other challenges are involved?; and (iii) What are the optimal management practices for dryland farmland in China to address climate change? The answers to these questions should provide a deeper understanding of the methods of dryland farmland management in China and provide new strategies to advance CSA.

## Climate-smart Agriculture and Sustainable Intensification Practices in China

### Conservation tillage

Research on and the application of conservation tillage in the dryland regions of northern China have been developed since the 1970s (Fig. 12.1). Conservation tillage has been promoted in the dryland of China since 2002, and the area of application exceeded 8.6 million ha in 2014 (Li *et al.*, 2017). Conservation tillage in China refers to a sustainable agricultural technology that can promote the integrated development of ecological, economic, and social benefits. This approach aims to reduce soil erosion, improve precipitation utilization, improve soil fertility, and increase crop yield (Zhang *et al.*, 2005).

The responses of crop yield to conservation tillage are controversial and debatable in China. Some studies have demonstrated that conservation tillage can increase grain yield and the efficiency of using precipitation by optimizing water use (Peng *et al.*, 2020). In addition, no-till with residue return can maintain a high photosynthetic rate in winter wheat (*Triticum aestivum* L.), which is conducive to the accumulation of dry matter and improvements in yield (Zhang *et al.*, 2017). However, some studies have found that conservation tillage reduced crop yield, possibly owing to increases in soil compaction, reduced crop stands, and poor seedling emergence (Zhang

**Fig. 12.1.** Conservation tillage in arid areas in China.

*et al.*, 2014). As a result, the reduction in yield under conservation tillage has restricted its popularization and implementation in many regions (Zhang *et al.*, 2014; Pittelkow *et al.*, 2015). However, adopting conservation tillage under appropriate site-specific conditions may improve the stability of yield and alleviate constraints related to the soil. Therefore, appropriate strategies that alleviate yield limitations and harness the benefits of conservation tillage are needed to improve production.

The significance of SOC sequestration for climate change (CC) mitigation has attracted increased attention, owing to the dual role of soil as both a substantial source and sink of atmospheric carbon dioxide ($CO_2$) (Lal, 2004). Soil organic carbon is an essential indicator of soil quality and crop productivity (Lal, 2015). Numerous studies have shown that conservation tillage has the potential to enhance the accumulation of SOC in the upper soil layers (Zhao *et al.*, 2015; Pu *et al.*, 2019; Kan *et al.*, 2020). Residue return was believed to primarily enhance the amount of SOC owing to the richness in sources of the soil carbon pool, while the benefits of no-till were primarily owing to the improvement in soil structure and enhanced physical protection of the soil through the formation of aggregates (Six *et al.*, 2002). Furthermore, higher particulate organic carbon under conservation tillage contributed to the formation of macro-aggregates and their stability at the upper soil depths, thus protecting the SOC from mineralization (Kan *et al.*, 2020). However, conservation tillage increased the contents of SOC in the upper soil layers but did not increase the storage of SOC in the whole soil profile (Du *et al.*, 2010). Therefore, we can continue to investigate the mechanism of SOC sequestration under conservation tillage and how to coordinate the distribution of SOC in the soil layer.

The anthropogenic emission of GHG, such as $CO_2$, methane ($CH_4$), and nitrous oxide ($N_2O$), is a major contributor to global warming. Upland soils are usually small sinks of $CH_4$ (Zhao *et al.*, 2016). A meta-analysis of Chinese farmland showed that the use of no-till for a long-term period and with fewer inputs of nitrogen fertilization could decrease the emissions of $N_2O$ in alkaline soil (Zhao *et al.*, 2016). A four-year observation showed that no-till clearly reduced the emissions of $N_2O$ and $CO_2$ but greatly increased the uptake of $CH_4$ compared with plow tillage (Wang *et al.*, 2022). Furthermore, a field study showed that no-till with straw return emphatically increased crop yield without significantly increasing the emissions of soil $N_2O$ (Yuan *et al.*, 2022). However, the retention of crop residues increased the emissions of $N_2O$ compared with the practice of residue removal in winter wheat and summer maize (*Zea mays*) seasons, but the increased uptake of $CH_4$ and crop yield finally resulted in a lack of significant increase in total emissions (Yao *et al.*, 2017). The primary pathways for the production of $N_2O$ in agricultural soil are nitrification and denitrification processes. Environmental factors, such as temperature, moisture, and pH, among others, and soil-carbon and nitrogen components are the key factors that affect the emissions of $N_2O$.

The impact of conservation tillage on dryland extends beyond the earlier-mentioned areas. Soybean-based crop rotation could offset the decline in the carbon sustainability index under no-till. Moreover, conservation agriculture has the potential to optimize the trade-offs between economic advantages and environmental considerations in the North China Plain, as demonstrated by Liu *et al.* (2023a). Another study conducted in the same region revealed that retaining crop residues led to a decrease in the carbon footprint per unit area, yield, and overall income throughout the year. This reduction was attributed to improved SOC stocks, enhanced crop yield, and increased net income (He *et al.*, 2022). No-till could be a preferred 'climate-resilient' technology to lower the carbon footprint and enhance the ecosystem services of climate regulation for the winter wheat–summer maize system in the North China Plain (Zhang *et al.*, 2016). Compared with plow tillage, no-till for maize can reduce soil evaporation and enhance water-use efficiency (Zhang *et al.*, 2002).

## Optimizing Chemical Fertilizer Input

Integrated nutrient management is a nutrient regulatory strategy to maintain optimal ecosystem productivity and sustainability for fertilizer, a practice that has been widely used in Chinese dryland farming. Integrated nutrient management emphasizes nutrient cycling on regional and national scales. Nutrients are supplemented based on quantitative environmental nutrient status, and fertilizer application is limited by the amount of nutrients the ecosystem can tolerate. Ultimately, this approach aims to align production, ecological, environmental, and economic advantages with other agronomic strategies (He and Lv, 2008).

The productivity of dryland crops in northern China has been increasing due to the optimal coordination of various policy, social, and technological factors. Among these factors, technological innovation, particularly integrated nutrition management, has consistently played the most indispensable role. Generally, the primary constraints on crop productivity in dryland farmland are infertile land and the scarcity of water, which can be effectively mitigated through integrated nutrient management. To establish a rational and efficient structure for dryland farming, it is important to merge effective management strategies with fertility optimization techniques. Key technical measures include crop rotation systems such as grain–oilseed–legume, grain–legume, and grain–grass. Additionally, adopting intercropping patterns such as wheat–potato and maize–peanut can maximize the utilization of natural resources such as sunlight, heat, and water. These practices collectively contribute to achieving higher crop yield and increased income for farmers. Fertilizer inputs are increased by combining organic and mineral fertilizers, and timely inputs are combined with patterns of crop nutrient demands to improve the scientific application of fertilizer. The practice has shown that drylands in northern China can achieve twice as much yield as water-filled land through the judicious application of organic fertilizers. The practice has demonstrated that in the arid regions of northern China, judicious utilization of organic fertilizers can lead to a twofold increase in yield compared to water-irrigated land. Moreover, the combination of organic and synthetic nitrogen fertilizers (OR-N) frequently offers a nutrient-rich source that complements intermediate nutrients (such as calcium, magnesium, and sulfur) and micronutrients (including molybdenum, boron, iron, and manganese). This supplementation contributes to the enhancement of crop yield, as observed in studies by Redding *et al.* (2016).

Furthermore, organic fertilizers enhance the microbial load and enzymatic activity associated with the soil carbon and nitrogen cycle (Lazcano *et al.*, 2013). This results in the release of more active carbon and active nitrogen fractions through mineralization and decomposition, leading to an increase in soil nutrient content (Ma *et al.*, 2022). Gong *et al.* (2009) concluded that crop yield could be increased by applying NPK mineral fertilizers, which lead to increased SOC stocks through inputs of crop residues and root biomass. Through long-term trials spanning 40–50 years of dryland farming, it was discovered that the application of mineral fertilizers for several decades resulted in a net accumulation of soil carbon (C) surpassing the removal of nitrogen (N) from grains by 60–190%. This phenomenon held even with the significant return of crop residues. Notably, the losses were more pronounced in the subsoil or the entire soil profile (0–46 cm) compared to the surface or plow layer. In conclusion, although the lack of comparable baseline data and the lack of prolonged field trials may confound the results of trials caused by mineral fertilizers, there is a pressing necessity to develop comprehensive integrated nutrient management systems for dryland agricultural soils in China. These systems should be designed to tackle the challenges posed by climate change while prioritizing ecological sustainability.

Long-term experiments based in Gansu Province, China, found that inputs of organic fertilizer, which included manure, increased the formation of soil-forming carbonates and reduced the loss of dissolved primary carbonates that increased the amount of inorganic carbon stocks by 8.9–11.0 Mg C $ha^{-1}$; this accounted for 4.8–17.0% of the initial stock at a depth of 120 cm (Liu *et al.*, 2023b). In general, reliance on mineral fertilizers as a standalone approach is never a sustainable option for nutrient management. The combined application of organic and mineral fertilizers for 26 years significantly reduced the soil capacity and increased the total concentration of organic carbon, organic carbon stocks, and labile fractions of organic C fractions. In contrast, applications of inorganic fertilizers have often produced contradictory effects on the concentrations of SOC and its fractions, which affects the microbial biomass carbon. Wu *et al.* (2004) found that the sole use of nitrogen fertilizer was ineffective at maintaining or restoring the organic matter content of the Loess Plateau. Indeed, SOC is subject to multiple influences from climate, soil resource status, and other field management practices, and frequently does not respond promptly to changes in mineral fertilizers.

Mechanized straw return and crop residue mulching is an economical and convenient way to efficiently utilize nutrients. Particularly in the dryland soils of northern China, crop residue mulching is an important agricultural measure that conserves water. Straw can also improve soil structure, promote water infiltration, and increase soil-water storage, which, in turn, promotes root development and increases crop yield (Kan *et al.*, 2020). Extensive long-term trials have indeed validated these sustainable field practices mentioned. However, when it comes to field management, it becomes crucial to consider the optimal productivity of the entire ecosystem. This involves thoroughly acknowledging the interconnected and complex impacts of climatic conditions and soil fertility levels. By doing so, it becomes possible to formulate an all-encompassing nutrient management system, fostering effective resource utilization and nutrient cycling.

The application of mineral N fertilizer is the most important factor that influences the emissions of soil $N_2O$ and offers direct substrates for nitrifying and denitrifying microorganisms, which then produce the intermediate substrate $N_2O$ (Liu *et al.*, 2022). Integrated nutrient management is essential to combat CC and reduce the emissions of $N_2O$ from dryland soils. In recent years, China has implemented the Action Plan for Zero Growth in Fertilizer Use by 2020. A series of integrated nutrient management measures for the optimization of nitrogen fertilizer have been implemented, including the 4Rs (the right time to apply fertilizer, the right place to apply fertilizer, the right amount of fertilizer, and the right rate to apply fertilizer) as effectively as the Integrated Soil-Crop System Management (ISSM). Xia *et al.* (2017) showed that developing an integrated fertilizer strategy enabled farmers to significantly increase the yield of China's staple crops by 1.3% and the utilization of N fertilizer by 48.2%, and significantly reduce the emissions of $N_2O$ from the soil by 31.2%. In addition, a comprehensive assessment of the agronomic and environmental impacts of having a comprehensive system to develop an integrated nutrient management system for virtually any N management practice in different environments is needed and crucial to enhance crop yield and reduce GHG emissions from dryland farmlands in China.

### Water-saving agriculture

Water-saving agriculture aims to maximize the use of water and fertilizer resources, and several technical options, such as conservation tillage, limited irrigation, drought-resistant chemical control, and the cultivation of water-saving varieties, are proposed for the characteristics of semi-arid regions in China (Deng *et al.*, 2006). Water-saving agriculture aims to maximize the use of water and fertilizer resources. Various technical approaches, including conservation tillage, controlled irrigation, chemical management for drought resistance, and the cultivation of water-saving crop varieties, have been suggested to suit the unique conditions of semi-arid regions in China (Deng *et al.*, 2006). The development of water-saving agriculture holds a significant strategic position in ensuring Chinese food security and the green and high-quality development of agriculture (Li *et al.*, 2020).

Water-saving irrigation measures, such as sprinklers and drip-irrigation, are currently widely used. These types of measures, including micro-sprinklers and drip-irrigation, can improve the growth environment of crops, which results in higher root vigor and microbial activity and an increase in the rate of soil respiration (Zhang *et al.*, 2012; Guo *et al.*, 2016). In wheat research, water-saving irrigation promoted the development of tillers and the formation of spikes (Feng *et al.*, 2022) and obtained higher yield of grain and water-use efficiency (Zhang *et al.*, 2003; Feng *et al.*, 2022). Film-mulched drip-irrigation significantly increased the yield of maize and water and nitrogen-use efficiency (Sui *et al.*, 2018; Geng *et al.*, 2021). A higher accumulated temperature of topsoil, the increase in soil nutrients in the root zone, and the increase in net radiation intercepted were the three primary reasons for the increase in yield (Sui *et al.*, 2018). However, in the semi-arid areas of the Loess Plateau where soil moisture conditions are unpredictable, the yield-enhancing effects of plastic film mulching are somewhat limited and even lead to reduced yield and harvests. For example, studies on spring wheat covered with mulch found that the yield was reduced when there was low soil moisture before sowing (Li *et al.*, 2010). Drought-resistant chemical control focuses on regulating two key factors for crop growth, including water and fertilizer. The rational application of chemical agents can improve the soil structure to reduce soil erosion, increase the infiltration of water and fertilizer, promote physiological growth, and improve crop quality (Yang *et al.*, 2013). In studies on winter wheat, the application of highly absorbent polymers was found to promote wheat growth by such factors as improving root vigor, increasing wheat dry matter weight and tiller number, and improving soil physicochemical properties and water-use efficiency, which ultimately lead to increases in yield (Wei *et al.*, 2023).

Conservation tillage is also one of the leading technologies for water-saving agriculture in drylands. Long-term no-till can increase water-use efficiency by reducing the distribution and uptake of water by the roots in the deep soil layer and increasing the concentration of SOC on the soil surface. Furthermore, conservation tillage improved the soil organic matter, available phosphorus (P), and available potassium, and enhanced grain yield and water-use efficiency in rainfed regions in the North China Plain (Shao *et al.*, 2016).

In the semi-arid areas of China, plastic mulch did not affect the concentration of organic carbon and its stock in the topsoil (Wang *et al.*, 2016b). However, full plastic film mulching could negatively affect the storage of soil carbon and control of GHG emissions (Zhang *et al.*, 2015a, b). Furthermore, short-term continuous cropping could improve the total content of organic carbon in the soil in long-term drip-irrigation under the membrane. In contrast, with the extension of continuous cropping years, the content of SOC significantly decreased; the soil carbon bank management index decreased; the soil quality decreased, and the soil fertility was affected (Gao *et al.*, 2021). Some studies on the emissions of $N_2O$ from the soil showed that mulching could increase the fluxes of these emissions owing to the possible promotion of denitrification by mulching on the one hand and the close relationship between mulching and soil temperature, soil moisture, and nutrients on the other (Meng *et al.*, 2022). However, the application of plastic film effectively reduced the emissions of $N_2O$ from the soil, while promoting yield under suitable fertilizer rates and irrigation (Yu *et al.*, 2017). Micro-sprinkler and drip-irrigation water-saving methods significantly increased the emissions of $N_2O$ from the soil compared with diffuse irrigation. Micro-sprinkler irrigation increased the content of soil microbial carbon, promoted nitrification and denitrification, and increased the fluxes of $N_2O$ emissions (Zhang *et al.*, 2012), while drip-irrigation reduced the leaching of nitrogen. The total content of mineral nitrogen in the soil was higher than in diffuse irrigation; thus, the emissions of $N_2O$ from the soil were higher under drip-irrigation than under diffuse irrigation (Wang *et al.*, 2016a).

## Biochar input

In recent years, biochar has been initiated as a soil conditioner to compensate for the low efficiency of carbon conversion from the traditional straw return to dryland agricultural soils in northern China. Biochar generally comprises carbon (C), hydrogen (H), oxygen (O), nitrogen (N), sulfur (S), phosphorus (P), potassium (K), calcium (Ca), magnesium (Mg), sodium (Na), silicon (Si) etc., with C normally reaching more than 60%. Biochar is generally porous and has a large specific surface area, which can improve the soil's physical structure and thus adsorb different types and forms of nutrients (Yuan *et al.*, 2016). As the promotion of inorganic carbon continues, there is no denying biochar's scientific, advanced, and practical nature in agricultural applications. Nonetheless, it is accompanied by some debate and disagreement (Chen *et al.*, 2019b). The direct or indirect application of biochar to the soil affects its microcosmic environment. Generally, biochar particles are less dense and more porous than soil, which helps the soil to retain more air and water, thus reducing the soil bulk density and stimulating root growth and the activity of soil microorganisms.

Studies have found that the application of biochar (100 t $ha^{-1}$) reduces the tensile strength of the soil by 18 kPa, which contributes to the ability of mycorrhizae to extract nutrients and improves seed germination and crop yield (Hussain *et al.*, 2017). Typically, biochar exhibits a slow rate of decomposition. However, it works quite differently from organic carbon, as it reduces nutrient loss. It accomplishes this by aggregating and retaining nutrients through functional groups, thereby enriching the soil's fertility. In addition, biochar is mostly alkaline or possesses a large lime equivalent value. This property allows it to serve as a substitute for lime, enhancing the efficiency of specific nutrients in acidic soils by increasing the saturation of soil bases, lowering the level of exchangeable aluminum, depleting soil protons, and increasing the pH of acidic soils (Ding *et al.*, 2013). As previously explained, the efficacy of biochar in soil enrichment is based on appropriate biochar technology. This means that field management practices must fully account for soil type, basic fertility, and the type and amount of biochar used as a feedstock. Wang (2018) discovered that biochar is more effective in fertilizing degraded land with poor soil fertility.

In saline soils, biochar can significantly improve the soil pH, electrical conductivity, soil nutrients, and associated enzyme activity for soil remediation. Moreover, biochar is an important strategy to combat soil desertification in the drylands of northwest China. Studies have found that the application of biochar can control soil sanding by increasing the ability to adsorb sodium ($Na^+$), potassium ($K^+$), calcium ($Ca^{2+}$), magnesium ($Mg^{2+}$), and $NH_4^+$, and improving the utilization of water in sandy soils (Wang *et al.*, 2018). In addition, management practices in China have proved that biochar and conventional fertilizers provide mutual positive feedback. Song *et al.* (2023) revealed that a 15% reduction in nitrogen fertilizer combined with biochar promoted photosynthetic production, dry matter accumulation and translocation, and the enhancement of yield in spring wheat. A northwest dryland soil experiment showed that the application of biochar enhanced the content of water-stable macro-agglomerates in the soil and the stability of agglomerates, which increase the total yield of crops during the wheat and maize seasons.

Biochar is difficult to degrade because of its stable carbon chain structure. Typically, it can form a carbon sink directly in the soil and profoundly impacts the soil carbon and nitrogen cycles (Lu *et al.*, 2021). Dong *et al.* (2019) demonstrated that biochar accelerated the turnover of soil inorganic carbon and increased the production of soil-forming carbonate and total inorganic carbon in the 0–40 cm soil layer. Moreover, since biochar improved the soil pH and the abundance of microbial communities, it indirectly influenced the percentage of $N_2O$ emissions produced by nitrification owing to the process of nitrification and reduced the amount of GHG emissions from dryland farmland. A life-cycle assessment and national data integration study showed that integrated biochar solution management systems could improve the net environmental and economic benefits by 36.2% and contribute to China's 2060 target for carbon neutrality (Xia *et al.*, 2023).

## Plastic film mulch

Plastic film mulching is a major path to increasing crop yield in the semi-arid areas of arid zones (Fig. 12.2). It tends to have a variety of functions, including reducing soil erosion, increasing soil temperature while inhibiting soil water evaporation, and reducing weed elimination and other effects. With CC and increasing demands for food, the use of plastic film and the area covered by it have increased dramatically, and the resulting ecological effects have raised concerns.

**Fig. 12.2.** Plastic film mulching cultivation in arid areas.

Plastic mulch usually reduces the net radiation income of dryland soils, which weakens the sensible heat exchange and brings the latent heat exchange to zero. This results in significant warming effects on soils in the 0–5 cm tillage layer. Research has indicated that this warming effect follows a U-shaped trend throughout the crop reproductive period. In the early stages of fertility, mulch can increase the ground temperature of the soil by more than 5°C owing to the low vegetation cover, which is highly beneficial for the emergence and development of seedlings. As the crop matures and the vegetation cover increases, the warming effect of the mulch gradually decreases. However, later in the crop season, as the leaves dry out, the warming effect of the mulch increases further. Notably, higher soil temperatures in the later stages of fertility can be detrimental to nutrient uptake by the crop and can also increase the loss of mineralization of soil nutrients. In addition, the mulch film, owing to its physical isolation, will prevent water vapor loss, thus acting as a barrier to drought and moisture (Wang *et al.*, 2003). Case studies in the Gansu and Shanxi provinces have consistently concluded that mulching can increase the soil-water content of the cultivated layer by 1–4% (Bu *et al.*, 2006). However, in years with less rainfall, the mulch, while increasing the water content of the cultivated layer, reduced the subsoil-water content, thus showing no difference in soil-water storage between 0–2 m (Li *et al.*, 2003).

Since changes in the temperature of soil water directly affect the growth and development of crops, plastic mulching has an impact on the promotion of early sprouting, seedling emergence, and the growth of stems with leaves, specifically the root system, as well as on the effectiveness of soil nutrients and their uptake by crops. National-scale surveys have shown that plastic film mulching technology has improved the water-use efficiency of crops and increased crop yield in China's drylands by 32.7–58%. However, the inappropriate use of mulch can also have a range of negative effects on crop growth (Liu *et al.*, 2014a). Currently, the polyethene film used in China is only 4–8 μm, which results in easy damage and resistance to removal. Over the years, the residual plastic in the soil accumulates and can damage the soil ecology and affect crop growth. Yan *et al.* (2014) revealed that when residual plastic in the soil exceeded 37.5 kg $ha^{-1}$, plant growth was affected, and the seedling emergence of winter wheat decreased by 25% and the tiller number by 17%. Cotton (*Gossypium hirsutum*) yield decreased by 4%, 8%, 12%, and 19% when there were 80, 170, 280, and 370 kg $ha^{-1}$ of residual plastic in the soil, respectively. Through the extended and irrational use of mulch over the long term, the initial gains in yield can be outweighed by sustained reductions in yield (Bi *et al.*, 2008). Model simulations to assess the combined effect of mulching technology in Hubei province found that when mulching technology was applied for 36 years, the rate of yield reduction due to mulch residue was greater than the rate of yield increase due to mulching. Furthermore, the influence of mulch residue would persist, leading to ongoing declines in crop yield.

Several studies have also found that conventional plastic mulching is not climate-friendly. Mulching also affects the biological properties of the soil and alters the soil mechanism of the carbon and nitrogen cycles. Studies have shown that plastic mulching increases the water storage and temperature of the soil, which enhances the activities of microorganisms that mineralize nitrogen and soil peroxidase and urease, and promotes crop growth while increasing the volatilization of $N_2O$ and $NH_3$. Inadequate management of water and nitrogen under mulching conditions further exacerbates this process. Some studies have shown that mulching does not significantly affect the stocks of SOC (Luo *et al.*, 2015). Liu *et al.* (2014b) showed that mulching increases the yield and subsoil biomass but also stimulates the mineralization of SOC and enzyme activity and enhances the carbon source function of the soil. In the long term, the current stage of mulching technology still has some drawbacks and does not seem to be a sustainable management practice. Looking ahead, it becomes imperative to enhance mulching technology, striving to create and implement low-cost, environmentally friendly, and biodegradable mulching methods.

## Case Studies of Sustainable Dryland Farming in China

### Northeast China

Focusing on major national needs and the frontiers of dryland agricultural science and technology, several research institutes and higher education institutions in northeast China have innovated a series of major key technologies. These include resource-efficient planting techniques, land conservation practices, efficient water and fertilizer utilization methods, and technology to combat farmland erosion. These innovations have been developed in response to both national needs and the frontiers of dryland agricultural science and technology.

Resource-efficient planting techniques combine the use of arable land with the use of nutrients, providing robust scientific and technological support for sustainable production. Through years of field research, northeast China has devised planting strategies founded on rational annual rotations of maize and soybean, as well as optimized configurations for maize fields. This has led to the establishment of resource-efficient planting systems tailored to various dryland zones in northeast China.

Conservation tillage technology has been tailored to the four major soil types of northeast China, and its viability has been verified and enhanced through multiple rotation cycles. For the first time, the standard parameter thresholds for high-yield tillage layers based on soil physical and chemical properties have been systematically proposed, and the supporting tillage method for constructing high-yield tillage layers has been determined, filling the gap in the quantitative research of reasonable tillage layer construction parameters. On this basis, specific details regarding the method, timing, residue retention quantities, and nitrogen application levels have been elucidated for three distinct soil types: black soil, brown soil, and cinnamon soil. Additionally, research on organic–inorganic application techniques was carried out to establish a model for high-yield fertilization of maize in black soil, brown soil, and cinnamon soil.

Years of field research have led researchers to innovate key technologies, such as crop intercropping, dryland autumn mulching, and monopoly film sowing over the straw to improve the efficiency of water and fertilizer use in dryland crops in northeast China. The dryland autumn mulching technology is used to prepare and mulch the land after the autumn crop harvest with the objective of 'using the autumn rain in spring and preserving the spring moisture in autumn' to achieve the cross-seasonal regulation of water resources in dryland farming by reducing the ineffective evaporation of soil moisture during the dormant period of the crop in autumn, winter, and spring. The monopoly film furrow sowing and residue mulching technology combines monopoly furrow planting and monopoly film mulching to suppress evaporation and collect rainfall, and wide and narrow row planting technology, which retains and concentrates the limited precipitation in the furrow planting area as much as possible. The farmland erosion control technology constructs a synergistic and integrated system of above-ground, surface and below-ground erosion control. For example, intercropping, film or residue mulching, and conservation tillage change the micro-topography and reduce soil erosion.

### Northwest China

Agriculture and food production is critical to human survival and development, particularly in arid regions with scarce water resources. CC is expected to exacerbate these challenges, with potential impacts on crop yield in arid regions in China. To begin with, changes in temperature and precipitation patterns are likely to affect crop yield and the timing of planting and harvesting. For example, elevated temperatures may increase the frequency of heat stress on crops, leading to decreased yield (IPCC, 2014). In addition, extreme weather events may also lead to soil erosion and land degradation, reducing the productivity of agricultural land (Kong *et al.*, 2002). In recent years, arid regions in China have experienced changes in precipitation patterns, including decreased rainfall and heightened variability. According to a study by Zhang *et al.* (2020), the average annual precipitation in northwest China decreased by 7.4% from 1960 to 2018, with more frequent droughts and heavy rainfall events. In general, the population in arid regions relies heavily on agriculture and animal husbandry for their livelihoods. However, the impact of CC on agriculture and food production, as well as water resources, poses significant challenges to the population's livelihood in these regions. In conclusion, the population in arid regions of China faces significant livelihood challenges, including low income, water scarcity, and CC impacts.

*Continued*

Continued.

Many water-saving agricultural measures have been adopted to address this issue and ensure the sustainable development of agricultural production in northwest China. Drip-irrigation and micro-sprinkler irrigation are two popular irrigation methods used in agriculture. Previous studies have shown that drip-irrigation can enhance 45.5% of water-use efficiency and increase crop yield by 13.4% compared with traditional irrigation methods (Huang *et al.*, 2022). Rainwater harvesting and irrigation is a system that collects and stores rainwater for later use in irrigation. The collection and storage of rainwater reduces the demand for traditional sources of water, such as rivers and groundwater. For example, a study by Ren *et al.* (2016) found that the use of rainwater harvesting and irrigation in the Loess Plateau region of China increased crop yield by up to 32.3% and improved the content of soil moisture. Mulching cultivation is an agricultural technique. Mulching cultivation in the arid region of northwest China typically involves using organic materials, such as straw, leaves, or crop residues; or inorganic materials, such as plastic films or stones to cover the soil surface. A study by Hu *et al.* (2019) found that the use of straw mulching in the arid region of northwest China increased the content of soil moisture, reduced fluctuations in soil temperature, and improved maize yield by up to 30%. In the arid region of northwest China, where water resources are scarce, conservation tillage has become an essential strategy to increase crop yield and improve soil health. For example, a study by Qi *et al.* (2022) found that conservation tillage increased water-use efficiency by 48.2% and moisture storage by 5.7% compared with conventional tillage, which ultimately improved crop production. Crop rotation and fallowing are effective agronomic practices to improve soil fertility, crop yield, and water-use efficiency in the arid region of northwest China. Crop rotation increases crop yield by disrupting the cycles of diseases and pests, improving the ability of soil to retain water, reducing soil-borne diseases, and improving water-use efficiency. Drought is a common natural disaster in the arid region of northwest China, which seriously affects crop growth and development, and reduces crop yield. Breeding drought-tolerant crop varieties is an important strategy to mitigate the impact of drought on crop production. For example, the drought-tolerant maize variety 'Zhengdan 958' has gained widespread adoption in northwest China's arid belt. Remarkably, its yield under drought conditions surpasses traditional variants by 24.7% (Chen *et al.*, 2019a).

**North China**

Climate change has a profound impact on crop yield in the dry farming region in north China. It is necessary to clarify the effect of CC on agriculture production, to innovate drought-resistant technology, and to analyze the mechanism of increasing grain production in dryland regions. The implementation of suitable farming practices can effectively adapt to and mitigate the adverse effects of future CC. This involves enhancing the efficiency of precipitation utilization, improving soil structure and fertility, and bolstering the crop-soil system's resilience to environmental changes. Soil-mulching management techniques, e.g. plastic mulching, straw mulching, and gravel-sand mulching, increase crop yield by preserving soil moisture, reducing soil evaporation and erosion, relegating soil temperature, and reducing weed infestation. Soil mulching with ridge-furrow planting is one of the most viable practices to maximize rainfall usage in arid and semi-arid regions. In addition, the adoption of conservation tillage, soil testing, and formulated fertilization technology improve soil fertility and optimize soil structure, which can enhance the adaptive capacity of the crop-soil system to address the challenges posed by CC.

Dryland wheat is important in agricultural production in Shanxi province in north China. Numerous environmental issues, such as an increase in the frequency of drought and large fluctuations in yield, have been issues that affect the production of dryland winter wheat under CC on the Loess Plateau (Fig. 12.3). The water-saving technology of 'Three in Advance' in dryland wheat fields, with deep plowing/subsoiling, residue incorporation (or mulching), and the application of organic fertilizer during the summer fallow season as the core, is a dry-farming technology with distinctive regional characteristics. Compared with no-till during the summer fallow season, the adoption of deep plowing/subsoiling during the summer fallow season can facilitate the incorporation of summer precipitation deeper into the soil layer. This can result in an increase of 81.0/58.7% in the efficiency of water and 7.8/6.0% in the storage of soil water in a 0–300 cm soil profile before sowing, by 30.8/25.8% in the annual average yield and 12.1/11.1% in water-use efficiency during the production of dryland winter wheat (Ren *et al.*, 2019).

*Continued*

Continued.

**Fig. 12.3.** Dryland farming in the Loess Plateau.

## Challenges/Opportunities for Future Development

Human activities have further exacerbated the challenge of CC to global food security in dryland cropland. The optimization of CSA practices to address CC and achieve Sustainable Development Goals has become a major international challenge (Zhao *et al.*, 2023). However, some controversies exist in the efforts to balance the comprehensive effects of dryland farmland CSA using practical techniques. For example, protective agriculture has achieved the principles of soil carbon sequestration and grain yield promotion but has increased the emissions of $NH_3$ (Xu *et al.*, 2022). Recently, the conversion of crop straw into biochar has substantial potential for carbon sequestration and the reduction of emissions in dryland soils. However, technology and funding still limit the conversion process (Xia *et al.*, 2023). Although inputs of organic material increase crop yield and reduce GHG emissions compared with mineral fertilizers, their effectiveness depends on various factors, such as crops, climate, and soil (Xu *et al.*, 2022). In addition, the application of manure can lead to the accumulation of heavy metals such as lead, chromium, and cadmium in soil and plants, and the problem of excessive phosphorus is a particular problem (Jin *et al.*, 2020). Water resources are the foundation of food security in dryland agriculture, and the transformation of irrigation methods may improve water efficiency and reduce the emissions of $CO_2$ related to infrastructure (Emde *et al.*, 2021). However, irrigation methods still face challenges balancing the conservation of water, reduction in emissions, and promotion of yield.

Currently, the optimization of CSA management practices based on onsite conditions offers opportunities for the precise regulation of dryland agricultural systems. For example, the 4Rs principle (right source, time, rate, and placement) of precise fertilization technology has played an important role in achieving a collaborative reduction in emissions and an increase in crop yield (Bowles *et al.*, 2018). Furthermore, improving the collection of manure and storage devices for it and their application technologies is crucial to address the issues of increasing the reduction in emissions in dryland farmlands in the future. Conservation agriculture and cultivation technology can improve the soil structure, strengthen nutrient circulation, and maintain the stability of soil moisture in dryland agriculture, which are regarded as sustainable agricultural technologies (Tao *et al.*, 2019). Biochar has a substantial potential to repair soil, improve its quality, and enhance crop productivity; thus, an integrated biomass pyrolysis and power generation system is necessary to convert straw into biochar (Xia *et al.*, 2023). Agriculture is bound to shift toward mechanization and intelligence and fully integrate internet technology, climate services, and farmers' needs, to more effectively apply CSA management techniques to achieve sustainable agricultural development goals (Zhao *et al.*, 2023).

Owing to the complex interrelationships among agronomic management, GHG emissions, SOC sequestration, and crop yield, future research should combine plants, soil, and climate with studying the systems used by crops to adapt and mitigate capabilities to CC. In particular, 'internet plus weather' services and CSA management are applied to achieve synergy between climate adaptation and crop yield. Furthermore, exploring how to optimize crop systems to adapt to CSA management and their interactions with the environment plays a key role in ensuring water–carbon–food–energy security and achieving the sustainable development of CSA. Therefore, the urgent task is to develop CSA management practices, achieve soil carbon sequestration and reduce emissions, increase grain production, and increase the adaptability of crops to CC.

## Conclusions

Global CC severely restricts the production of dryland farming and human survival. This case study summarized the progress and challenges of applying CSA management to dryland farming systems to adapt to CC and showed that the impact of current CC on dryland farming primarily includes food security, water resources, and people's livelihoods. The CSA practices advocated in agriculture can address these challenges, but their effectiveness also depends on the specific situation. The CSA practices that are widely used in dryland farming in China include irrigation, mulching cultivation, conservation tillage, crop rotation, biochar, and fallow breeding of drought-tolerant crop varieties. However, CSA management still faces significant challenges in balancing the relationship between organic carbon sequestration, food security, and GHG emissions based on local conditions. Identifying and prioritizing appropriate CSA practices is necessary to help establish sustainable agricultural production systems. In the future, the combination of internet technology and CSA management is considered to be the primary direction of development for dryland farming in China.

## References

Bi, J.Y., Wang, X.F. and Zhu, D.L. (2008) Effect of plastic-film mulch on crop yield. *Transactions of the Chinese Society of Agricultural Engineering* 172–175. (in Chinese)

Bowles, T.M., Atallah, S.S., Campbell, E.E., Gaudin, A., Wieder, W.R. *et al.* (2018) Addressing agricultural nitrogen losses in a changing climate. *Nature Sustainability* 1(8), 399–408. DOI: 10.1038/s41893-018-0106-0

Bu, Y.S., Miao, G.Y., Zhou, N.J., Shao, H.L. and Wang, J.C. (2006) Analysis and comparison of the effects of plastic film mulching and straw mulching on soil fertility. *Scientia Agricultura Sinica* 39,1069–1075. (in Chinese) Available at: http://journals.caass.org.cn/zgnykx/EN/10.3864/j.issn.0578-1752.at-2005-5566 (accessed 20 July 2024).

Chen, L., Xin, X.L., Zhang, J.B., Marc, R.G., Nie, G.S. *et al.* (2019a) Soil characteristics overwhelm cultivar effects on the structure and assembly of root-associated microbiomes of modern maize. *Pedosphere* 29(3), 360–373. DOI: 10.1016/S1002-0160(17)60370-9

Chen, W.F., Meng, J., Han, X.R., Lan, Y. and Zhang, W.M. (2019b) Past, present, and future of biochar. *Biochar* 1, 75–87. DOI: 10.1007/s42773-019-00008-3

Deng, X.P., Shan, L., Zhang, H.P. and Turner, N.C. (2006) Improving agricultural water use efficiency in arid and semiarid areas of China. *Agricultural Water Management* 80, 23–40. DOI: 10.1016/j.agwat.2005.07.021

Ding, Y.L., Liu, J. and Wang, Y.Y. (2013) Effects of biochar on microbial ecology in agriculture soil: a review. *Chinese Journal of Applied Ecology* 24, 3311–3317. DOI: 10.13287/j.1001-9332.2013.0560 (in Chinese)

Dong, X.L., Singh, B.P., Li, G.T., Lin, Q.M. and Zhao, X.R. (2019) Biochar increased field soil inorganic carbon content five years after application. *Soil & Tillage Research* 186, 36–41. DOI: 10.1016/j.still.2018.09.013

Du, Z.L., Ren, T.S. and Hu, C.S. (2010) Tillage and residue removal effects on soil carbon and nitrogen storage in the North China Plain. *Soil Science Society of America Journal* 74, 196–202. DOI: 10.2136/sssaj2009.0048

Emde, D., Hannam, K.D., Most, I., Nelson, L.M. and Jones, M.D. (2021) Soil organic carbon in irrigated agricultural systems: a meta-analysis. *Global Change Biology* 27(16), 3898–3910. DOI: 10.1111/gcb.15680

FAO (2013) *Climate-Smart Agriculture: Sourcebook.* Food and Agriculture Organization of the United Nations (FAO), Rome.

Feng, L.J., Yu, Z.W., Zhang, Y.L. and Shi, Y. (2022) Effects of irrigation on tiller occurrence, photo-assimilates production and distribution in different stem and tillers and spike formation in wheat. *Acta Agronomica Sinica* (in Chinese). Available at: https://zwxb.chinacrops.org/EN/10.3724/SP.J.1006.2022.21045 (accessed 20 July 2024).

Gao, W.C., Yang, W.J., Shi, C.L. and Chen, L. (2021) Study on soil organic carbon and its change of activity in continuous cropping cotton field under mulched drip irrigation. *Xinjiang Agricultural Sciences* 58, 1603–1609. (in Chinese) DOI: 10.6048/j.issn.1001-4330.2021.09.005

Geng, Y.H., Cao, G.J.L., Wang, C., Wang, M. and Huang, J.X. (2021) Can drip irrigation under mulch be replaced with shallow-buried drip irrigation in spring maize production systems in semiarid areas of northern China? *Journal of the Science of Food and Agriculture* 101, 1926–1934. DOI: 10.1002/jsfa.10808

Gong, W., Yan, X.Y., Wang, J.Y., Hu, T.X. and Gong, Y.B. (2009) Long-term manuring and fertilization effects on soil organic carbon pools under a wheat-maize cropping system in North China Plain. *Plant and Soil* 314, 67–76. DOI: 10.1007/s11104-008-9705-2

Guo, S.F., Qi, Y.C., Yin, F.H., Peng, Q., Dong, Y.S. *et al.* (2016) Effect of irrigation patterns on soil $CO_2$ and $N_2O$ emissions from winter wheat field in North China Plain. *Environmental Science* 37, 1880–1890. (in Chinese) DOI: 10.13227/j.hjkx.2016.05.036

He, C., Wang, Y.Q., Yu, W.B., Kou, Y.H., Yves, B.N. *et al.* (2022) Comprehensive analysis of resource utilization efficiency under different tillage systems in North China Plain. *Journal of Cleaner Production* 347. DOI: 10.1016/j.jclepro.2022.131289

He, Q. and Lv, S.H. (2008) Integrated management of nutrient resources – the inevitable choice of agricultural development in our country. *Sichuan Agricultural Science and Technology* 245, 16–17. (in Chinese)

Hu, Y., Ma, P., Zhang, B., Hill, R.L., Wu, S. *et al.* (2019) Exploring optimal soil mulching for the wheat-maize cropping system in sub-humid drought-prone regions in China. *Agricultural Water Management* 219, 59–71. DOI: 10.1016/j.agwat.2019.04.004

Huang, J., Li, Y., Fu, C., Chen, F., Fu, Q. *et al.* (2017) Dryland climate change: recent progress and challenges. *Reviews of Geophysics* 55(3), 719–778. DOI: 10.1002/2016RG000550

Huang, Y., Li, S., Hu, D., Cong, X. and Zhang, R.Q. (2022) Effects of drip irrigation on water use efficiency and yield of maize based on Meta-Analysis. *Journal of China Agricultural University* 27(5), 96–105. (in Chinese) Available at: https://webofscience.clarivate.cn/wos/alldb/full-record/CSCD:7220695 (accessed 20 July 2024).

Hussain, M., Farooq, M., Nawaz, A., Al-Sadi, A.M., Solaiman, Z.M. *et al.* (2017) Biochar for crop production: potential benefits and risks. *Journal of Soils and Sediments* 17, 685–716. DOI: 10.1007/s11368-016-1360-2

IPCC (Intergovernmental Panel on Climate Change) (2014) *Climate Change 2014: Impacts, Adaptation, and Vulnerability*. Contribution of Working Group II to the Fifth Assessment Report of the Intergovernmental Panel on Climate Change (ed. Field, C.B., Barros, V.R., Dokken, D.J., Mach, K.J., Mastrandrea, M.D. *et al.*). Cambridge University Press, Cambridge, UK and New York.

Jin, X.P., Bai, Z.H., Oenema, O., Winiwarter, W., Velthof, G. *et al.* (2020) Spatial planning needed to drastically reduce nitrogen and phosphorus surpluses in China's agriculture. *Environmental Science & Technology* 54(19), 11894–11904. DOI: 10.1021/acs.est.0c00781

Kan, Z.R., He, C., Liu, Q.Y., Qi, J.Y., Dang, Y.P. *et al.* (2020) Characteristics of carbon mineralization and accumulation under long-term conservation tillage. *Catena* 193. DOI: 10.1016/j.catena.2020.104636

Kong, Z.H., Zhang, X.S. and Zhou, G.S. (2002) Agricultural sustainability in a sensitive environment – a case analysis of Loess Plateau in China. *Journal of Environmental Sciences* 14(3), 357–366. Available at: https://webofscience.clarivate.cn/wos/alldb/full-record/WOS:000177001000012 (accessed 20 July 2024).

Lal, R. (2004) Soil carbon sequestration impacts on global climate change and food security. *Science* 304, 1623–1627. DOI: 10.1126/science.1097396

Lal, R. (2015) Sequestering carbon and increasing productivity by conservation agriculture. *Journal of Soil and Water Conservation* 70, 55A–62A. DOI: 10.2489/jswc.70.3.55A

Lazcano, C., Gomez-Brandon, M., Revilla, P. and Dominguez, J. (2013) Short-term effects of organic and inorganic fertilizers on soil microbial community structure and function. *Biology and Fertility of Soils* 49, 723–733. DOI: 10.1007/s00374-012-0761-7

Leisner, C.P. (2020) Review: climate change impacts on food security – focus on perennial cropping systems and nutritional value. *Plant Science* 293. DOI: 10.1016/j.plantsci.2020.110412

Li, S.Q., Li, D.F., Li, F.M., Bai, H.Y., Ling, L. *et al.* (2003) Soil ecological effects of plastic film mulching in semiarid agro-ecological system. *Journal of Northwest A&F University (Natural Science Edition)*, 21–29. (in Chinese) DOI: 10.13207/j.cnki.jnwafu.2003.05.006

Li, S.Z., Wang, Y., Fan, T.L., Wang, L.M., Zhao, G. *et al.* (2010) Effects of different plastic film mulching modes on soil moisture, temperature and yield of dryland maize. *Scientia Agricultura Sinica* 43, 922–931. (in Chinese) Available at: https://www.cabdirect.org/cabdirect/abstract/20103112857 (accessed 20 July 2024).

Li, W., Xue, C.X., Yao, S.B. and Zhu, R.X. (2017) The adoption behavior of households conservation tillage technology: an empirical analysis based on data collected from 476 households on the Loess Plateau. *Chinese Rural Economy* 44–57, 94–95. (in Chinese)

Li, X.R., Jiang, W.L., Guan, X. and Li, M.B. (2020) An overview of water conservancy green development in China. *Journal of China Agricultural Resources and Regional Planning* 41, 49–55. (in Chinese) Available at: https://webofscience.clarivate.cn/wos/alldb/full-record/CSCD:6878285 (accessed 20 July 2024).

Liu, E.K., He, W.Q. and Yan, C.R. (2014a) 'White revolution' to 'white pollution': agricultural plastic film mulch in China. *Environmental Research Letters* 9. DOI:10.1088/1748-9326/9/9/091001

Liu, E.K., Zhou, J., Yang, X., Jin, T., Zhao, B.Q. *et al.* (2023a) Long-term organic fertilizer-induced carbonate neoformation increases carbon sequestration in soil. *Environmental Chemistry Letters*. DOI: 10.1007/s10311-023-01568-4

Liu, W.X., Liu, W.S., Yang, M.Y., Wei, Y.X., Chen, Z. *et al.* (2023b) Effects of tillage and cropping sequences on crop production and environmental benefits in the North China Plain. *Environment Science and Pollution Research* 30, 17629–17643. DOI: 10.1007/s11356-022-23371-4

Liu, X.E., Li, X.G., Hai, L., Wang, Y.P., Fu, T.T. *et al.* (2014b) Film-mulched ridge-furrow management increases maize productivity and sustains soil organic carbon in a dryland cropping system. *Soil Science Society of America Journal* 78, 1434–1441. DOI: 10.2136/sssaj2014.04.0121

Liu, X.J., Cui, Z.L., Hao, T.X., Yuan, L.X., Zhang, Y. *et al.* (2022) A new approach to holistic nitrogen management in China. *Agricultural Science Engineering* 9, 490–510. DOI: 10.15302/J-FASE-2022453

Lu, T.P., Wang, X.J., Du, Z.L. and Wu, L.P. (2021) Impacts of continuous biochar application on major carbon fractions in soil profile of North China Plain's cropland: in comparison with straw incorporation. *Agriculture Ecosystems & Environment* 315. DOI: 10.1016/j.agee.2021.107445

Luo, S.S., Zhu, L., Liu, J.L., Bu, L.D., Yue, S.C. *et al.* (2015) Mulching effects on labile soil organic nitrogen pools under a spring maize cropping system in semi-arid farmland. *Agronomy Journal* 107, 1465–1472. DOI: 10.2134/agronj14.0643

Ma, L., Wang, S.T., Shi, L., Zhang, R., Wang, K. *et al.* (2022) Effects of manure combined with nitrogen fertilizer on yield and quality of winter wheat and soil biological characteristics in drylands of Northwest China. *Chinese Journal of Applied Ecology* 33, 2718–2724. (in Chinese) DOI: 10.13287/j.1001-9332.202210.014

Meng, C.B., Zhao, J.Y., Wang, N., Yang, K.J. and Wang, F.X. (2022) Black plastic film mulching increases soil nitrous oxide emissions in arid potato fields. *International Journal of Environmental Research and Public Health* 19. DOI: 10.3390/ijerph192316030

Peng, Z.K., Wang, L.L., Xie, J.H., Li, L.L., Coulter, J.A. *et al.* (2020) Conservation tillage increases yield and precipitation use efficiency of wheat on the semi-arid Loess Plateau of China. *Agricultural Water Management* 231. DOI: 10.1016/j.agwat.2020.106024

Pittelkow, C.M., Liang, X.Q., Linquist, B.A., van Groenigen, K.J., Lee, J. *et al.* (2015) Productivity limits and potentials of the principles of conservation agriculture. *Nature* 517, 365–U482. DOI: 10.1038/nature13809

Pu, C., Kan, Z.R., Liu, P., Ma, S.T., Qi, J.Y. *et al.* (2019) Residue management induced changes in soil organic carbon and total nitrogen under different tillage practices in the North China Plain. *Journal of Integrative Agriculture* 18, 1337–1347. DOI: 10.1016/S2095-3119(18)62079-9

Qi, X.P., Li, G., Yuan, J.Y. and Chang, H.G. (2022) Effects of conservation tillage on the water storage, moisture conservation and yield of dry-land wheat fields of central Gansu Province. *Arid Zone Research* 39(1). (in Chinese) DOI: 10.13866/j.azr.2022.01.30

Qiao, L., Wang, X.H., Smith, P., Fan, J.L., Lu, Y.L. *et al.* (2022) Soil quality both increases crop production and improves resilience to climate change. *Nature Climate Change* 12(6), 574. DOI:10.1038/s41558-022-01376-8

Redding, M.R., Lewis, R., Kearton, T. and Smith, O. (2016) Manure and sorbent fertilizers increase on-going nutrient availability relative to conventional fertilizers. *Science of the Total Environment* 569, 927–936. DOI: 10.1016/j.scitotenv.2016.05.068

Ren, A., Sun, M., Xue, L., Deng, Y., Wang, P. *et al.* (2019) Spatio-temporal dynamics in soil water storage reveals effects of nitrogen inputs on soil water consumption at different growth stages of winter wheat. *Agricultural Water Management* 216, 379–389. DOI: 10.1016/j.agwat.2019.01.023

Ren, X., Zhang, P., Liu, X., Ali, S., Chen, X. *et al.* (2016) Impacts of different mulching patterns in rainfall-harvesting planting on soil water and spring corn growth development in semihumid regions of China. *Soil Research* 55(3), 285–295. DOI: 10.1071/SR16127

Shao, Y.H., Xie, Y.X., Wang, C.Y., Yue, J.Q., Yao, Y.Q. *et al.* (2016) Effects of different soil conservation tillage approaches on soil nutrients, water use and wheat-maize yield in rainfed dry-land regions of North China. *European Journal of Agronomy* 81, 37–45. DOI: 10.1016/j.eja.2016.08.014

Six, J., Feller, C., Denef, K., Ogle, S.M., Sa, J.C.D. *et al.* (2002) Soil organic matter, biota and aggregation in temperate and tropical soils – effects of no-tillage. *Agronomie* 22, 755–775. DOI: 10.1051/agro:2002043

Song, S.L., Yang, W.J., Chen, Y.X., Jia, Y.H., Li, D.P. *et al.* (2023) Effect of reduced nitrogen fertilizer combined with biochar on photosynthetic characteristics, dry matter accumulation and distribution, and yield of spring wheat in irrigated area of northern Xinjiang. *Journal of Triticeae Crops* 43(3), 311–320. Available at: https://webofscience.clarivate.cn/wos/alldb/full-record/CABI:20230215901 (accessed 20 July 2024).

Sui, J., Wang, J.D., Gong, S.H., Xu, D., Zhang, Y.Q. *et al.* (2018) Assessment of maize yield-increasing potential and optimum N level under mulched drip irrigation in the Northeast of China. *Field Crops Research* 215, 132–139. DOI: 10.1016/j.fcr.2017.10.009

Tao, F.L., Palosuo, T., Valkama, E. and Makipaa, R. (2019) Cropland soils in China have a large potential for carbon sequestration based It on literature survey. *Soil & Tillage Research* 186, 70–78. DOI: 10.1016/j.still.2018.10.009

Wang, G., Liang, Y., Zhang, Q., Jha, S.K., Gao, Y. *et al.* (2016a) Mitigated $CH_4$ and $N_2O$ emissions and improved irrigation water use efficiency in winter wheat field with surface drip irrigation in the North China Plain. *Agricultural Water Management* 163, 403–407. DOI: 10.1016/j.agwat.2015.10.012

Wang, J., Li, F.M., Song, Q.H. and Li, S.Q. (2003) Effects of plastic film mulching on soil temperature and moisture and on yield formation of spring wheat. *Chinese Journal of Applied Ecology* 205–210. (in Chinese)

Wang, L.C., Li, Y.K., Guo, W.Z., Qu, J.S., Ma, L. *et al.* (2018) Influence of biochar on growth of cabbage mustard in arid area of northwestern China and its water-saving effect. *Jiangsu Agricultural Sciences* 46, 97–100. (in Chinese) DOI: 10.15889/j.issn.1002-1302.2018.06.025

Wang, W.Y., Zhang, H.P., Mo, F., Liao, Y.C. and Wen, X.X. (2022) Reducing greenhouse gas emissions and improving net ecosystem economic benefit through long-term conservation tillage in a wheat-maize multiple cropping system in the Loess Plateau, China. *European Journal of Agronomy* 141. DOI: 10.1016/j.eja.2022.126619

Wang, Y.P., Li, X.G., Fu, T.T., Wang, L., Turner, N.C. *et al.* (2016b) Multi-site assessment of the effects of plastic-film mulch on the soil organic carbon balance in semiarid areas of China. *Agricultural and Forest Meteorology* 228, 42–51. DOI: 10.1016/j.agrformet.2016.06.016

Wei, X.H., Zhang, Z.W., Li, S.D., Luo, Z.M., Xia, Y.H. *et al.* (2023) Effects of water retaining agent on growth and yield of winter wheat in dryland. *Journal of Anhui Agricultural Sciences* 51(1), 33–35. (in Chinese) DOI: 10.3969/j.issn.0517-6611.2023.01.007

Wolfert, S. and Isakhanyan, G. (2022) Sustainable agriculture by the Internet of Things – a practitioner's approach to monitor sustainability progress. *Computers and Electronics in Agriculture* 200. DOI: 10.1016/j.compag.2022.107226

Wu, T.Y., Schoenau, J.J., Li, F.M., Qian, P.Y., Malhi, S.S. *et al.* (2004) Influence of cultivation and fertilization on total organic carbon and carbon fractions in soils from the Loess Plateau of China. *Soil & Tillage Research* 77, 59–68. DOI: 10.1016/j.still.2003.10.002

Xia, L.L., Lam, S.K., Chen, D.L., Wang, J.Y., Tang, Q. *et al.* (2017) Can knowledge-based N management produce more staple grain with lower greenhouse gas emission and reactive nitrogen pollution? A meta-analysis. *Global Change Biology* 23, 1917–1925. DOI: 10.1111/gcb.13455

Xia, L.L., Cao, L., Yang, Y., Ti, C.P., Liu, Y.Z. *et al.* (2023) Integrated biochar solutions can achieve carbon-neutral staple crop production. *Nature Food* 4(3), 236–246. DOI: 10.1038/s43016-023-00694-0

Xu, P., Li, G., Houlton, B.Z., Ma, L., Ai, D. *et al.* (2022) Role of organic and conservation agriculture in ammonia emissions and crop productivity in China. *Environmental Science & Technology* 56(5), 2977–2989. DOI: 10.1021/acs.est.1c07518

Yan, C.R., Liu, E.K., Shu, F., Liu, Q., Liu, S. *et al.* (2014) Review of agricultural plastic mulching and its residual pollution and prevention measures in China. *Journal of Agricultural Resources and Environment* 31(2), 95–102. (in Chinese) Available at: https://webofscience.clarivate.cn/wos/alldb/full-record/CABI:20153275907 (accessed 20 July 2024).

Yang, P.L., Liao, R.K., Ren, S.M. and Li, Y.K. (2013) Application of chemical regulating technology for utilization of water and fertilizer in dry-land agriculture. *Transactions of the Chinese Society for Agricultural Machinery* 44, 100–109. (in Chinese) DOI: 10.6041/j.issn.1000-1298.2013.06.019

Yao, Y., Liu, Y., Wang, Y. and Fu, B. (2021) Greater increases in China's dryland ecosystem vulnerability in drier conditions than in wetter conditions. *Journal of Environmental Management* 291, 112689. DOI: 10.1016/j.jenvman.2021.112689

Yao, Z.S., Yan, G.G., Zheng, X.H., Wang, R., Liu, C.Y. *et al.* (2017) Straw return reduces yield-scaled N2O plus NO emissions from annual winter wheat-based cropping systems in the North China Plain. *Science of the Total Environment* 590, 174–185. DOI: 10.1016/j.scitotenv.2017.02.194

Yu, Y.X., Tao, H., Jia, H.T. and Zhao, C.Y. (2017) Impact of plastic mulching on nitrous oxide emissions in China's arid agricultural region under climate change conditions. *Atmospheric Environment* 158, 76–84. DOI: 10.1016/j.atmosenv.2017.03.020

Yuan, J.Y., Yan, L.J., Li, G., Sadiq, M., Rahim, N. *et al.* (2022) Effects of conservation tillage strategies on soil physicochemical indicators and N2O emission under spring wheat monocropping system conditions. *Science Reports* 12. DOI: 10.1038/s41598-022-11391-6

Yuan, S., Zhao, L.X., Meng, H.B. and Shen, Y.J. (2016) The main types of biochar and their properties and expectative researches. *Journal of Plant Nutrition and Fertilizer* 22, 1402–1417. DOI: 10.11674/zwyf.14539

Zhang, F., Li, M., Qi, J.H., Li, F.M. and Sun, G.J. (2015b) Plastic film mulching increases soil respiration in ridge-furrow maize management. *Arid Land Research and Management* 29, 432–453. DOI: 10.1080/15324982.2015.1018456

Zhang, H.L., Chen, F., Qin, Y.D. and Zhu, W.S. (2002) Water consumption characteristics for summer corn under no-tillage with mulch. *Transactions of the Chinese Society of Agricultural Engineering* 36–40. (in Chinese) Available at: http://www.tcsae.org/en/article/id/20020210 (accessed 20 July 2024).

Zhang, H.L., Gao, W.S., Chen, F. and Zhu, W.S. (2005) Prospect and present situation of conservation tillage. *Journal of China Agricultural University* 16–20. (in Chinese)

Zhang, H.L., Lal, R., Zhao, X., Xue, J.F. and Chen, F. (2014) Opportunities and challenges of soil carbon sequestration by conservation agriculture in China. *Advances in Agronomy* 124. DOI: 10.1016/B978-0-12-800138-7.00001-2

Zhang, Q.B., Yang, L., Sun, B., Zhang, W.F., Luo, H.H. *et al.* (2012) Respiration characteristics of cotton soil under irrigation and fertilization measures in arid region. *Transactions of the Chinese Society of Agricultural Engineering* 28, 77–84. (in Chinese) Available at: http://www.tcsae.org/en/article/id/20121413 (accessed 20 July 2024).

Zhang, Q.Q., Song, Y.F., Wu, Z., Yan, X.Y., Gunina, A. *et al.* (2020) Effects of six-year biochar amendment on soil aggregation, crop growth, and nitrogen and phosphorus use efficiencies in a rice-wheat rotation. *Journal of Cleaner Production* 242. DOI: 10.1016/j.jclepro.2019.118435

Zhang, X., Davidson, E.A., Mauzerall, D.L., Searchinger, T.D., Dumas, P. *et al.* (2015a) Managing nitrogen for sustainable development. *Nature* 528, 51–59. DOI: 10.1038/nature15743

Zhang, X.Q., Pu, C., Zhao, X., Xue, J.F., Zhang, R. *et al.* (2016) Tillage effects on carbon footprint and ecosystem services of climate regulation in a winter wheat-summer maize cropping system of the North China Plain. *Ecological Indicators* 67, 821–829. DOI: 10.1016/j.ecolind.2016.03.046

Zhang, X.Q., Zhao, X.L., Wang, Y.Q., Pu, C., Baoerjiang, M. *et al.* (2017) Effects of tillage practices on photosynthetic performance diurnal variation during filling stage and grain yield of winter wheat. *The Journal of Applied Ecology* 28, 885–893 (in Chinese). DOI: 10.13287/j.1001-9332.201703.020

Zhang, X.Y., Pei, D. and Hu, C.S. (2003) Conserving groundwater for irrigation in the North China Plain. *Irrigation Science* 21, 159–166. DOI: 10.1007/s00271-002-0059-x

Zhao, J.F., Liu, D.S. and Huang, R.X. (2023) A review of climate-smart agriculture: recent advancements, challenges, and future directions. *Sustainability* 15(4). DOI: 10.3390/su15043404

Zhao, X., Zhang, R., Xue, J.F., Pu, C., Zhang, X.Q. *et al.* (2015) Management-induced changes to soil organic carbon in China: a meta-analysis. *Advances in Agronomy* 134, 1–50. DOI: 10.1016/bs.agron.2015.06.002

Zhao, X., Liu, S.L., Pu, C., Zhang, X.Q., Xue, J.F. *et al.* (2016) Methane and nitrous oxide emissions under no-till farming in China: a meta-analysis. *Global Change Biology* 22, 1372–1384. DOI: 10.1111/gcb.13185

# 13 Dryland Agriculture in Pakistan under Changing Climate: A Case Study

**Muhammad Sarwar Khan[1], Iqrar Ahmad Khan[1], Syed Aftab Wajid[1], and Kauser Abdulla Malik[2]***

*[1]University of Agriculture, Faisalabad, Pakistan; [2]Forman Christian College (A Chartered University), Lahore, Pakistan*

**Abstract**

Drylands are the largest biome on Earth, and most of this area exists in developing countries. Pakistan has a predominantly arid and semi-arid climate and has substantial drylands which are subject to various hazards of climate change. Agriculture in Pakistan is significantly affected by short-term climate variability and long-term climate change. Periods of severe droughts followed by devastating floods are common in the country and have contributed to the low crop yields. Development of climate-resilient crop varieties, appropriate agronomic practices, interventions in the irrigation system, crop diversification, and conservation of water resources can contribute to the betterment of dryland agriculture, thus helping to uplift food and feed demands in the country.

## Introduction

Pakistan has a predominantly arid to semi-arid climate, where 5 mha of the cultivable land is rainfed. Droughts have severe repercussions in the country, including food shortage, animal losses, crop failure, and water scarcity. These effects may result in social and economic instability, unemployment, and social conflicts. The country relies on its water resources to support its agricultural economy. With 37.4% of the labor force employed in agriculture, and 22.7% of GDP contributed by agriculture, it is the primary industry in Pakistan (Government of Pakistan, 2022). In Pakistan, drylands, arid lands, and/or rainfed areas are used to designate lands that solely depend on rainfall and are not irrigated by canals or tube well water. Most parts of Sindh, Baluchistan, southern parts of Punjab, and Khyber Pakhtunkhwa receive less than 250 mm of annual rainfall and are considered drylands. The Potohar Plateau covers an area of 1.8 million ha and is an important region of the drylands of Pakistan.

Dryland agriculture depends on livestock raised on native vegetation and crops grown with rainwater. Major dryland crops include wheat, chickpea, sorghum, millet, barley, maize, lentil, peanut, rapeseed-mustard, and guar seeds (Rashid and Ryan, 2007). The agricultural sector, particularly in dryland regions, faces enormous challenges like reduced crop yields, soil degradation, and increased incidence of insects, pests, and diseases due to climate change (CC). To address these challenges, there is a need for developing and adapting scientific interventions to uplift dryland agriculture. Adaptation strategies can lessen farmers' susceptibility to the

*Corresponding author: kausermalik@fccollege.edu.pk

 *Climate Change and Sustainable Agro-ecology in Global Drylands* (A.S. El-Beltagy, R. Lal and K. Malik eds.)
DOI: 10.1079/9781800624870.0013

effects of CC and increase the resilience of their farming operations. These interventions, on the other hand, can help to improve soil health, reduce soil erosion, and increase crop productivity. In this chapter, we have highlighted the impact of CC on Pakistan's dryland agriculture and discussed the development and adaptation of scientific interventions that can be implemented to reduce vulnerability and build resilience to these impacts.

## Agro-ecological Zones of Punjab, Pakistan

Pakistan is an agricultural economy with diversified climatic conditions. Punjab is the second-largest province, accounting for 26% of the country's geographical area and about 60% of its population. Punjab has various ecological conditions for crop production, with a cotton–wheat cropping system in the south, mixed cropping in central Punjab, and a rice–wheat cropping system in upper Punjab, while having an arid area with a single wheat crop per year and variable rainfall within the province ranging from 150 mm to 500 mm. Punjab province is Pakistan's food basket that always ensures national food security. FAO defined agro-ecological zones (AEZs) as regions having similar climatic and edaphic conditions for the characterization of agricultural productivity (Rosenzweig *et al.*, 2014). Pakistan was divided into ten agro-ecological zones for the first time in 1980 by the Pakistan Agriculture Research Council in collaboration with FAO based on water availability. Despite continuous ecological variations, there was no re-zoning until 2001. After 2001, abrupt changes in climatic parameters changed the available soil and ecological resources and demanded redefining these zones based on newly established ecological conditions triggered by changing climate. The University of Agriculture Faisalabad, with the collaboration of FAO and other agricultural organizations, redefined the agro-ecological zones of Punjab-Pakistan in 2019 for restructuring the agricultural landscape in Punjab (FAO, 2019). The basic objective of redefining the zones of Punjab was the optimal use of land resources synchronized with climatic resources. These zones result from the interaction of numerous environmental factors, including topography, geology, rainfall, temperature, and soil characteristics. The rice–wheat, cotton–wheat, central mixed, high-rainfall, low-rainfall, barani, and desert regions of the Cholistan, Thal, and Suleiman mountains are among the agro-ecological zones in Punjab, Pakistan. Sustainable agriculture and rural livelihood growth depend on understanding Punjab's agro-ecological zones. It enables the development of strategies for reducing the effects of CC and environmental degradation on agriculture, as well as the identification of the crops and farming methods that are best suited for each zone. Additionally, it offers chances for agricultural production diversification, a boost in productivity and profitability, and an uplift of living standards in rural areas.

## Dryland Areas of Punjab

Drylands are categorized into three sub-groups depending on the annual precipitation and evapotranspiration: hyper-arid, arid, and semi-arid (Spinoni *et al.*, 2015). Drylands in Punjab cover most of the state's geographical area. These regions are primarily situated in the western and southern parts of the state. Despite their ecological vulnerability, drylands support a diverse array of flora and fauna and provide habitat to millions of people (Ahmad *et al.*, 2019). Dryland agriculture in Punjab is a high-risk, low-input enterprise with small and fragmented landholdings (Rashid *et al.*, 2004). The uncertainty in precipitation and frequent droughts are the primary causes of low crop productivity in drylands.

### Soil resources and farming systems of drylands in Punjab

The productivity of drylands in Punjab is low, mainly due to climatic conditions rather than soil health. The prominent soil types in drylands are Aridsols, Alfisols, Inceptisols, and Entisols, which indicate the arid and semi-arid conditions of the regions, leading to low crop productivity (Rafiq *et al.*, 1996). However, adapting proper soil and management practices such as smart irrigation, fertilization, and conservation tillage can help enhance the productivity of drylands. The rainfed agriculture (i.e. non-irrigated) techniques in drylands are categorized into three categories: Barani, Sailaba, and Khushkaba. Barani agriculture, which relies solely on rainfall, is commonly used in the Potohar Plateau and northern mountains of Punjab. The Sailaba farming technique utilizes soil moisture gained through summer floods and rains and is practiced in areas near major rivers prone to seasonal flooding. Khushkaba, which involves the diversion of runoff from large barren lands to irrigate crops, is a common farming practice in the highlands. Another practical use of drylands is as rangelands, which play a critical role in supporting the state's livestock sector (FAO, 2016). It is estimated that rangelands in Punjab cover about 8.2 million ha of drylands and support 13 million head of livestock. While the forage production rate ranges from 400 to 4500 kg $ha^{-1}$, the rangelands are under tremendous pressure due to overgrazing, leading to land degradation (i.e. desertification) and biodiversity loss (FAO, 2016).

## Challenges of dryland agriculture

The challenges faced by dryland agriculture are categorized into biophysical and socio-economic. High-intensity monsoon rains in July and August cause soil erosion, flooding, and high evaporation rates, leading to soil degradation. The exploitation of overgrazed rangelands by impoverished residents for fuel exacerbates the problem. Even though dryland crop yields are often low and uncertain, it is noteworthy that the average yields of farmers are significantly below the potential. Due to unstable incomes and low production efficiency, low priority is given to dryland agriculture. Consequently, without secure agro-based livelihoods, local populations must migrate in search of other income resources. These challenges make dryland agriculture less profitable and highly prone to climatic events, laying the foundation for socio-economic challenges such as low-priority businesses and, thereby, less investment.

## Impact of climate change on migration and displacement

Climate change is not only negatively impacting the environment and productivity in the drylands, but the food insecurity, declining income, and limited options for livelihood are also forcing the communities to migrate towards cities. Several studies conducted in the drylands of Punjab and Khyber Pakhtunkhwa have indicated that CC and heat stress have significantly affected the well-being of people in rural areas, especially in the dry areas, by reducing crop productivity, declining farm income, and loss of animals to natural catastrophes (Umar and Fahad, 2018). Rising temperatures increase vulnerability to extreme weather events, increasing hot days and physical stress. Migration is an adaptation strategy, but its relationship with improving thermal comfort and heat exposure is weak. The study found that migration in rural Pakistan improves livelihood resilience and adaptation to CC. Migrant families have better living standards, access to resources, and the potential to use remittances for productive purposes. They are more income-diverse, engaged in progressive farming, and have lower dependency ratios. Migrant households have better anticipatory capacities and access to social networks, information, and financial resources. However, understanding resilience in societies with varying social, economic, and cultural values is challenging. Overall, migration benefits all income groups in rural Pakistan, improving livelihoods and rural development.

## Government initiatives for sustainable dryland agriculture

The Punjab government has taken several initiatives to address the above-mentioned challenges to promote sustainable dryland agriculture practices. Some of these are:

**1.** solar-powered tube wells on a subsidy basis to empower farmers to irrigate crops in a timely way and without relying on grid electricity or diesel-powered pumps, which are both expensive and unreliable;
**2.** subsidized water conservation practices such as drip/sprinkler irrigation systems, on-farm water storage facilities, and mulch, which are helpful in mitigating water scarcity in drylands;
**3.** provision of quality seeds: high-quality seed of drought-resistant varieties are supplied to farmers to improve crop productivity;
**4.** high-tech laboratories at the district level to test soil and water health. Such information is

critical for determining correct fertilizer type and quantity, to save investment and improve soil health;

**5.** rehabilitation programs to restore rangelands, include reseeding, water conservation, and fencing to improve drylands' vegetation cover and grazing conditions; and

**6.** investment in research and development to improve seed quality so that it is more tolerant to heat, pests, and other diseases in drylands.

## Redefined agro-ecological zones under changing climate

Over the past few decades, significant changes in the agro-climate and land-use patterns have been observed, indicating the influence of CC. Ten agro-ecological zones have been identified in Pakistan, including the western dry mountains (19.17%), western dry plateau (14.4%), sandy desert (14%), northern irrigated plain (11.3%), and coastal belt (9.9%) (Fig. 13.1). Water scarcity in various ecologies calls for interventions such as rainwater harvesting, water-efficient technologies, and surface storage for sustainable agriculture. Collaboration with international partners can facilitate agricultural economic growth and support rural communities. Building awareness and capacity among local communities and institutions is crucial for efficient water use and climate-smart agriculture. Monitoring water resources is necessary for sustainable agriculture in the face of changing climate conditions.

## Integrated framework for agro-ecological zones (AEZ)

Agro-ecological zones were devised based on crop norms, soil, climate, and water availability using R software. Data were collected from various departments, but the information was not in the same grid size, making it challenging to determine crop suitability. The data were converted to the same grid size of 0.5 × 0.5 km to address this issue, allowing for soil, climate, and water information at each point. A high-resolution database for every input parameter within the specific grid in the geo-statistical section, with size explained in detail, was developed, as given in Fig. 13.2.

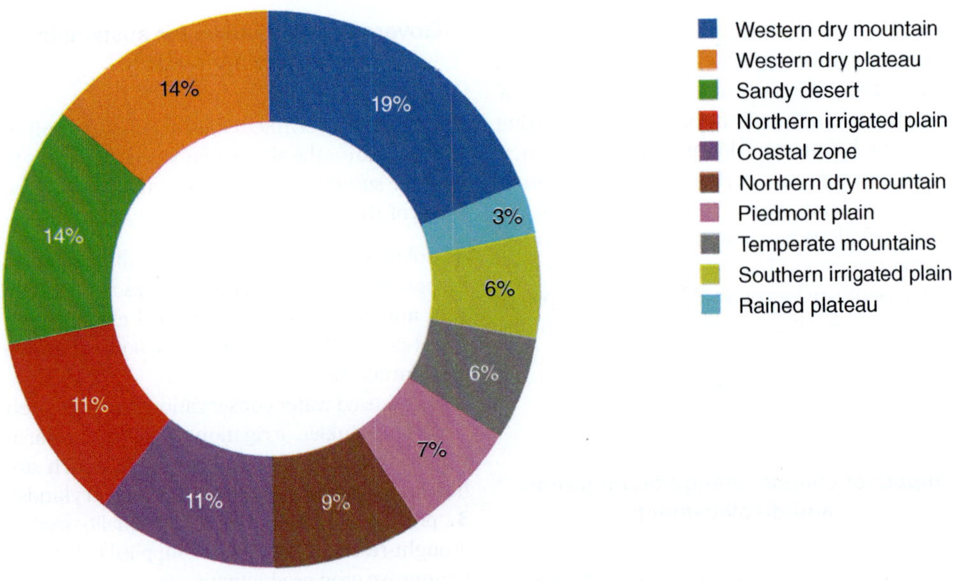


**Fig. 13.1.** Donut plot representing agro-ecological zones of Pakistan. (Drawn using Plotly Library in R software)

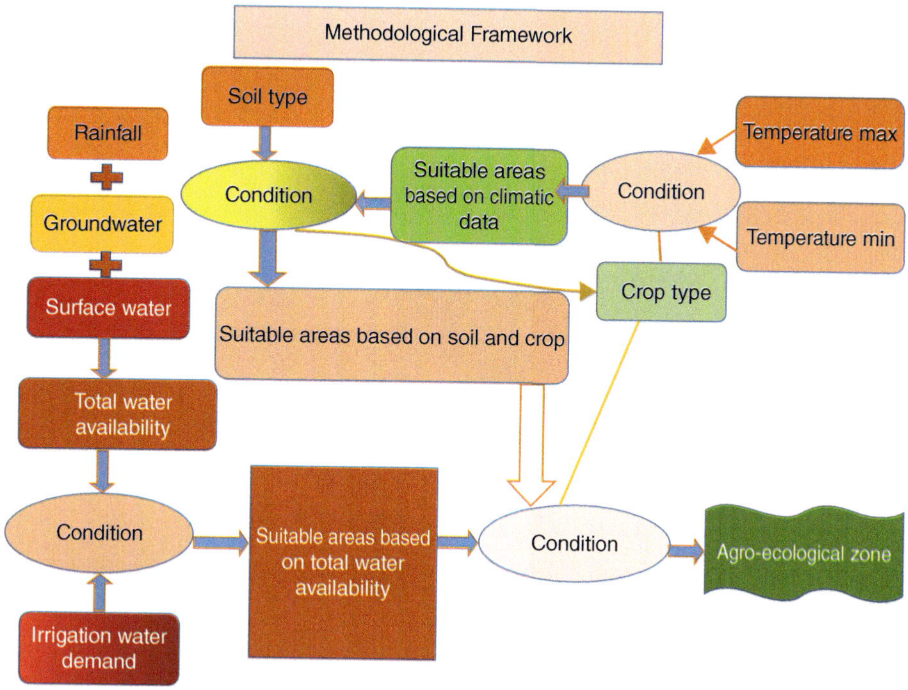


**Fig. 13.2.** Methodological framework to establish the AEZ.

## Geo-statistical analysis

We had datasets in point format, but these point formats have some missing data and are at very low resolution. To improve the resolution, the interpolation was performed at the scale of 0.005°. To quantify spatial dependence, the semi-variogram is used, which mathematically describes the mean square variability between two neighboring points of distance *h*, as shown in the equation:

$$\gamma(h)=\frac{1}{2N(h)}\sum_{i=1}^{N(h)}\left[z(x_i+h)-z(x_i)\right]^2$$

The semi variogram, $\gamma(h)$ is expressed as a separate vector or function of the lag distance, *h*, between two points, $z(x_i)$ is the random variable at location $x_i$, where $N(h)$ represents the total number of observations pairs separated by distance *h*. To determine three parameters, namely the sill (c), the range (A0), and nugget (c0), the experimental semi-variogram, $\gamma(h)$, is fitted to a theoretical model, as defined by Isaaks and Srivastava (1989). These improved interpolated maps were then considered in GIS for the extraction of point data at a resolution of 0.005°.

## Climate layer

**(i)** Temperature layer: the difference in the maximum temperature was significant with the average minimum and maximum temperatures of 18.5–20°C to 39–40.5°C in different ecological zones of Punjab.
**(ii)** Water layers: interpolated surface water supply map for Punjab, indicating surface water, groundwater, and rainfall received during the year, is given in Fig. 13.3.
**(iii)** Soil layer: it defines the different crop-growing areas in Punjab along with other factors.

## Crop norms

Crop norms are the data on the sowing requirement of the specific crop including soil, climate, water, and other input use till harvesting for better crop production (Singh *et al.*, 2008). Crop norms of newly defined agro-ecological zones of Punjab are depicted in Fig. 13.4.

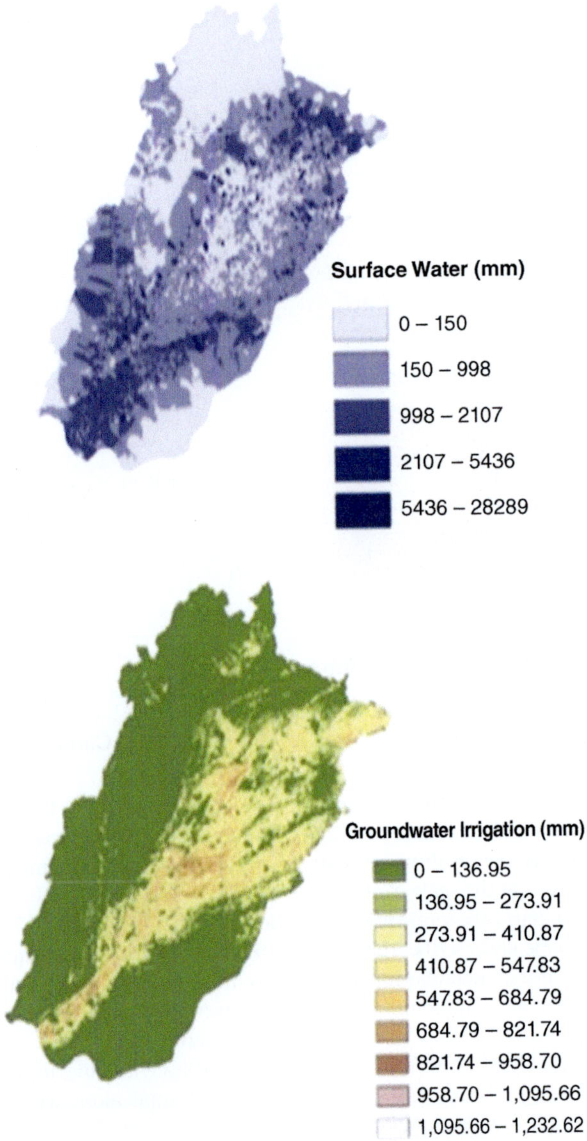


**Fig. 13.3.** Variation in surface water and groundwater availability in Punjab.

## CC Impact on Water Availability

Climate change has a significant impact on the availability of water around the world. Due to the increase in temperature, changes in precipitation trends, snowpack, and evaporation rates are altering the availability and distribution of water resources. Changing patterns of precipitation lead to drought and flooding simultaneously in different regions.

### Impact of CC on glaciers

Like many other parts of the world, CC is a major threat to freshwater resources in Pakistan (Lutz *et al.*, 2014). Several studies have reported that glaciers in Pakistan are receding, particularly at lower altitudes of the HKH, due to higher temperatures, diminishing snowfall, and heavier summer rainstorm events (Qasim *et al.*, 2023).

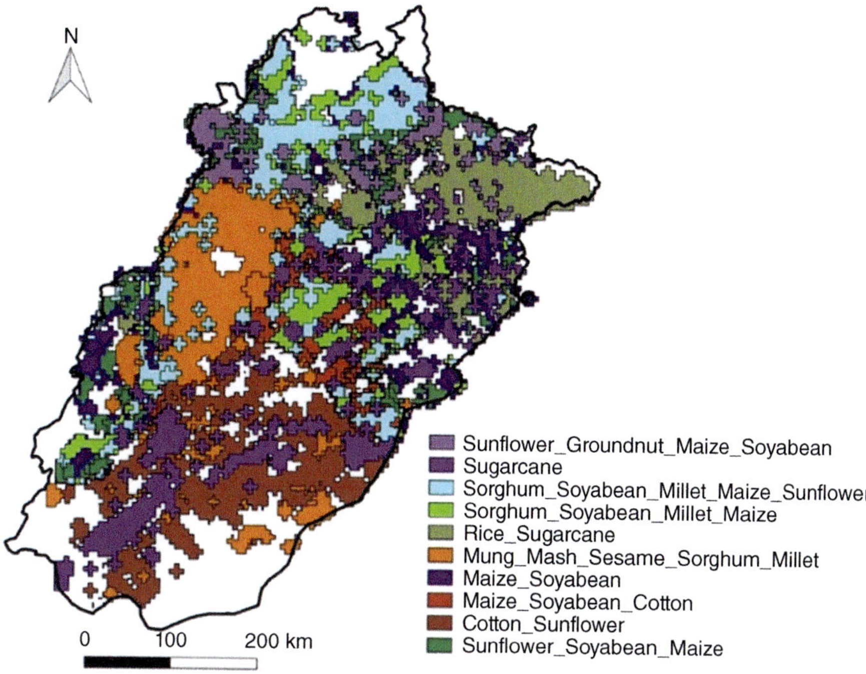


**Fig. 13.4.** Newly defined agro-ecological zones of Punjab.

The rate at which glaciers are melting has significantly increased during the past few decades, causing changes in the hydrological regime of the area (Latif *et al.*, 2021). For instance, an analysis of the Upper Indus basin, Jhelum, and Kabul river basins from 1961 to 2013 showed a higher increase in summertime streamflow compared to other seasons. This has increased the risk of summer flooding, exacerbated by glacial lake outbursts causing flooding downstream. It is reported that the outburst of glacial lakes played a major role in the historic flooding in Pakistan in 2022 (Mallapaty, 2022). Moreover, such CC impacts are expected to worsen in the future.

## Contribution of precipitation to freshwater resources

### *Precipitation contribution to freshwater*

Precipitation plays an important role in managing the water resources of Pakistan and its variability has triggered serious threats to the Pakistani economy. The precipitation in the country is mainly dependent on western and monsoon winds, with uneven distribution throughout the year. The average annual precipitation varies from 70 mm in the south with the driest and hottest climate to > 2000 mm in the upper reaches of the Indus basin. The upper reaches of the Indus basin receive snow, which plays an important role in supplying water to the country and fulfilling its needs for agriculture and socio-economic activities in the dry season (Ahmad *et al.*, 2018).

## Impact of precipitation across agro-ecological zones

The streamflow in an area mainly depends on the amount and pattern of precipitation. To check the mean annual precipitation distribution and the number of rainy days across agro-ecological zones of Pakistan, the precipitation data from meteorological stations installed in these zones were collected from the Pakistan Meteorological Department. The temporal extent of the data was up to 54 years. The number of rainy days and average annual precipitation help quantify the availability of water in the area for better management. Frequent precipitation with the highest number of rainy days increases the annual precipitation amount, which can cause floods in the area. Figure 13.5 presents the number of rainy days and mean annual precipitation across agro-ecological zones in Punjab. Most of the stations located in the agro-ecological zones of upper Punjab (zone VII to zone XIV maize–wheat mixed cropping to high-rainfall zones) depict the highest number of rainy days between 88 mm and 214 mm with mean annual precipitation amount between 312 mm and 1739 mm. The number of rainy days and mean annual precipitation amount decreases from north (zone VII) to south (zone I) towards the dry and hot zones (Cholistan desert). As the station is located in zone VII (cotton mix cropping), zone VI (mix cropping), zone V (semi-desert irrigated), zone IV (Rod-i-Kohi), zone III (cotton–sugarcane), zone II (arid irrigated), and zone I (Cholistan desert) depict very few numbers of rainy days with mean annual precipitation amount between 28 mm and 310 mm, showing the reduction in mean annual precipitation and number of rainy days from north to south of the country (Fig. 13.5).

To study the change in precipitation patterns and their extreme behaviors across agro-ecological zones in Punjab, Pakistan, the Concentration Index (CI), which was developed by Martin-Vide (2004), was used to examine the precipitation concentration on daily and monthly scales. The CI for the annual precipitation records and for the precipitation amounts greater than 1 mm (named here as wet days) was analyzed and presented in Fig. 13.6. The CI for wet days and annual precipitation depicted a latitudinal gradient from lower values in the northwest (zone XIV (high rainfall)) to higher in the south (zone I (Cholistan desert)), in general, shown in Fig. 13.6 (a and b).

Moreover, the high rainfall zone (zone XIV) depicts the lowest values of wet days CI and annual CI (0.592–0.625), with an increase in their values towards the Cholistan desert, except for the arid irrigated zone at the most southern part of the province, which shows the lowest value of CI. The agro-ecological zones in the most northern part of the country, from zone XIV to zone VIII (maize–wheat mix cropping to high-rainfall zones), have maximum mean annual precipitation and number of rainy days, a north-to-south increase also appears in annual CI and wet days CI, in general. The precipitation stations located in low-precipitation zones (zones I–VII, Cholistan desert to cotton mix cropping), show higher values of the CI (0.672–0.723 for annual and wet days CI). Overall, CI values of wet days and annual precipitation exhibit a similar mix pattern, with lower values at most of the agro-ecological zones in the north with higher altitudes and highest values at lower elevations near the dry and arid zones, indicating a lesser number of rainfall spells with higher concentrations. The most northern extreme shows the minimum value of annual CI values throughout the country. While describing the relationship of CI with elevation, it was observed that the highest values of CI were at lower altitudes, except at some elevations where there was a mixed trend.

In general, a high value of CI can be related to torrential and heavy precipitations for a lesser period that can cause extreme events in the form of floods, as recently witnessed (August 2022) in the provinces of Baluchistan and Sindh.

## Impact of drought on the water resources of Pakistan

Pakistan is one of the most water-stressed countries in the world, with a per capita water availability less than half the global average. Droughts are a persistent issue in many areas of Pakistan, notably in the arid and semi-arid regions of Baluchistan, Sindh, and southern Punjab. Pakistan has recently seen several devastating droughts. For instance, in 2018, drought led to food instability, crop failures, and water shortages, which in some places resulted in famine and starvation. The Sindh and Baluchistan provinces were severely impacted. Many causes, such as less rainfall, excessive temperatures, and ineffective water management, contributed to the drought. In Baluchistan, the drought affected 4.3 million animals and almost 6.6 million people for over two years (USAID, 2019). The drought also brought on malnutrition, severe food insecurity, and relocation.

Similarly, during 2015, drought affected over 2 million people and lasted several months.

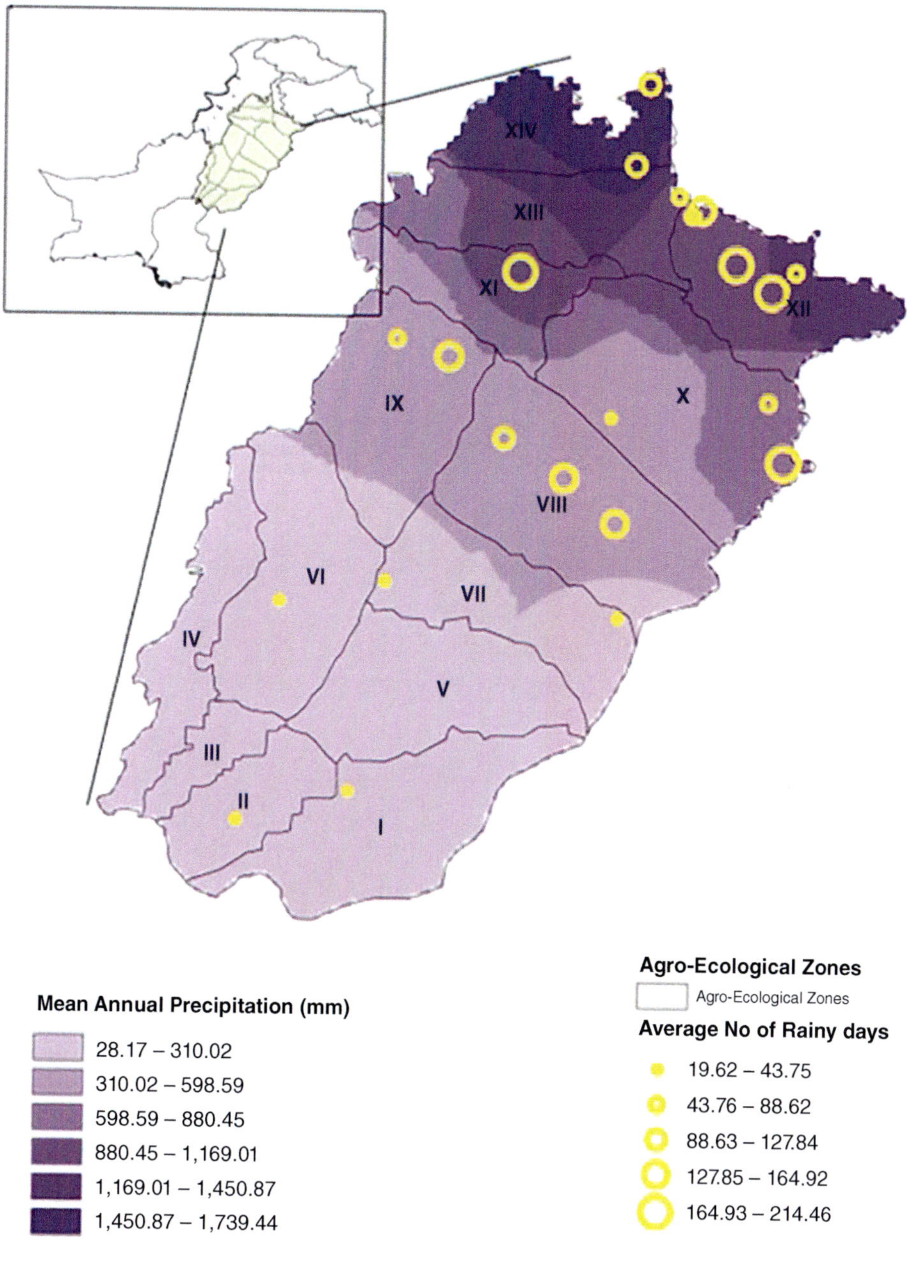


**Fig. 13.5.** Mean average precipitation and number of rainy days across agro-ecological zones of Punjab.

The impact of drought on the country's resources was significant and its major rivers including Chenab, Indus, and Jhelum experienced lower flow rates, which led to a decline in water availability for drinking and irrigation for agriculture purposes, leading to declining agricultural productivity. Similarly, the country faced three major droughts from 1998 to 2018 according to the study conducted by Pakistan's Meteorological Department. During these droughts, agricultural

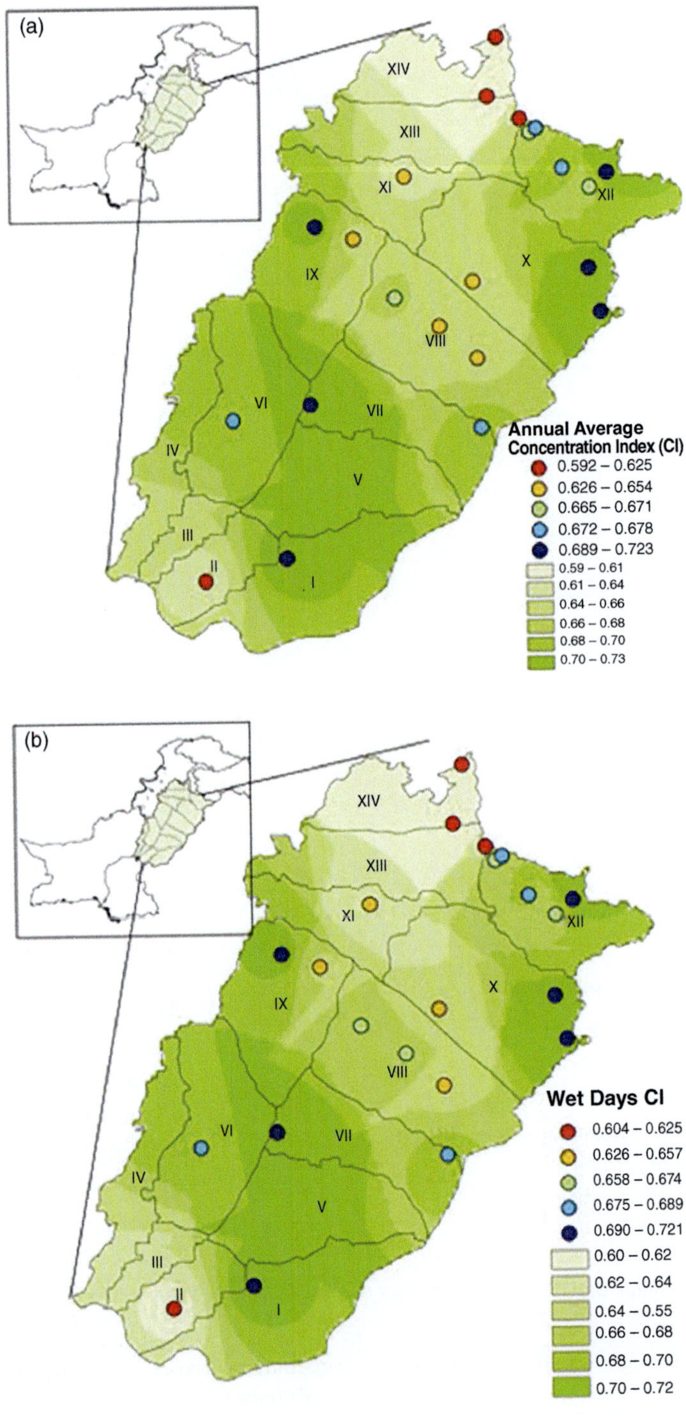


**Fig. 13.6.** Concentration Index (CI) of annual (a) and wet (b) days with spatial distribution of each agro-ecological zone.

production declined by approximately 40% (Siddiqui *et al.*, 2016).

### Mitigation strategies for water reservoir management

#### *Short-term strategies*

It is necessary to build a reasonable number of micro-ponds (at the farm level) and efficient water application systems by using pipes. Moreover, mini- and small dams should be built annually, along with rainwater harvesting at the household level, making it a legal obligation for citizens.

#### *Medium-term strategies*

Irrigation canals should be lined in poor groundwater quality zones to ensure wastewater treatment at the industrial level and through wetlands. Low-delta crops should be introduced.

#### *Long-term strategies*

Multi-purpose large dams should be built with at least one mega-dam every ten years. Link canals should be lined to avoid waterlogging, flood canals created for groundwater recharge in floodplains and desert areas, and line river reaches to promote river navigation.

## Present and Future Agriculture of Different Agro-ecological Zones in Pakistan: A Modeling Approach

### AgMIP (Agricultural Model Intercomparison and Improvement Project) Consortium

AgMIP Consortium started working in early 2012 for the improvement of crop models, CC impact assessment, use of climate models for future downscaled data, and as an economic model for improving the livelihood of farmers and other stakeholders. Mitigation and adaptation strategies were developed for early (2012–2039), mid- (2040–2069), and late-century (2070–2099) scenarios regarding crop yields and their economics for the livelihood of the stakeholders.

#### *Concept and selection of baseline data*

To create future climate scenarios, there was a need for a baseline period using observed climate and variability data recorded from 1961 to 1990 due to GHG (Karl *et al.*, 1996) and climatological reasons may favor a defined baseline period for which scientists are working (IPCC, 1994).

#### *Climate data for baseline accuracy and downscaled data*

Numerous gridded climate data sets showed the average surface climate around the world (Leemans and Cramer, 1991), and in individual countries (Kittel *et al.*, 1995), for the yearly swing in weather and daily gridded baseline climate data, both at regional and global levels, they were required to generate more climate scenarios (Widmann and Bretherton, 2000). Meteorological data were accessible for 1981–2010 with daily weather series from the AgMIP dataset by Modern-Era Retrospective Analysis for Research & Applications (AGMERRA; https://gmao. Gsfc.nasa.gov/reanalysis/MERRA [accessed 21/7/24]). This database provides daily temperature, rainfall, and radiation from NASA/GEWEX SRB to cover the 1980–2010 period (Ruane *et al.*, 2015).

#### *Surveyed data and CC trends*

The surveyed data from 155 farmers were collected from the study area of the rice–wheat cropping zone of Punjab. Five already mentioned climate models can deal with future climate based on dry/wet, cool/wet, hot/wet, hot/dry, and middle qualities (Rosenzweig *et al.*, 2014).

#### *Representative Agricultural Pathways (RAPs) for climatic interventions*

The future agricultural production systems are translated into RAPs. These were developed through continuous engagement and meetings with scientists and stakeholders, with farmer practices at the field level regarding resources used and information on inputs available from literature. Changes in key national- and regional-level drivers were evaluated and inputs from global models, such as population and economic growth rate, were also used as RAPs. These portray potential future agricultural production systems with technological advancement, soil health, the genetic potential of the crop, etc. Changes in biophysical resources, institutional policies, technological advancements, and socio-economic conditions were also taken as bases of RAPs. It has been observed that an increase in production is a function of inputs and technological advancements in the agriculture production system. Mechanized farming, increasing crop intensity, and ecosystem disturbances are also destroying the natural resources of production systems.

## Adaptation and Agronomic Interventions under Climatic Change

### The impacts of CC on agriculture

Overall, CC may have significant social, economic, and environmental impacts, particularly for countries like Pakistan, where agriculture is a crucial sector of the economy and a source of livelihood for millions of people. The significant impacts are:

1. Rising temperatures can affect crop growth, development, and yield. High temperatures can also increase water demand, leading to water stress for crops and reducing their productivity.
2. Changes in precipitation patterns are changing due to CC, leading to more frequent and intense droughts and floods. This can lead to crop failure, soil erosion, and loss of soil fertility.
3. Pest attack and disease incidence may aggravate as changing climatic conditions can create favorable conditions for them.
4. Increased intensity of rainfall events due to CC can lead to soil erosion, which can cause soil degradation, reduced soil fertility, and reduced crop yields.
5. CC is altering the availability of water for agriculture. Droughts and reduced water availability can lead to crop failure and reduced yields, while floods can cause soil erosion and damage to crops.

### Adaptation measures for agriculture

CC adaptation strategies for the agricultural sector are crucial. Adaptation measures for agriculture can help farmers cope with the challenges of CC and build resilience in their farming systems. However, the success of these measures depends on a range of factors, including access to finance, technology, supportive policy frameworks, and effective partnerships between farmers, governments, and other stakeholders.

Here are some examples of adaptation measures that can be implemented in agriculture:

1. **Crop diversification.** Farmers can diversify their crops by growing a variety of crops that are better suited to changing climate conditions, crop rotation, and intercropping.
2. **Water management.** Using efficient irrigation systems, rainwater harvesting, and agriculture practices for water conservation can help to increase water-use efficiency, reduce soil erosion because of runoff, and increase soil moisture to improve soil health.
3. **Improved soil health.** Improving soil health with practices like conservation agriculture, cover cropping, and agroforestry can help to enhance soil organic matter, lessen soil erosion, and promote soil fertility, thus improving crop productivity and resilience.
4. **Use of climate-smart technologies.** Farmers can use climate-smart technologies such as solar-powered irrigation systems, improved seeds, and precision farming technologies to increase productivity and reduce vulnerability to climate impacts.
5. **Early warning systems.** Implementing early warning systems for extreme weather events, pests, and diseases can help farmers to take appropriate measures to reduce damage to their crops and lessen their susceptibility to the effects of the climate.
6. **Improved seeds and technologies.** Selecting drought-tolerant crops together with modern agriculture practices such as precision farming (e.g., drip irrigation and mechanization) can significantly increase dryland productivity.
7. **Training and education.** Providing farmers and stakeholders with training and education on climate-smart agriculture practices can help them to adopt water-efficient irrigation, crop management, and soil conservation, which is critical to improving dryland farming.

### Agronomic interventions for sustainable dryland agriculture

Agronomic interventions can help to reduce the environmental impact on agriculture, improve soil health and fertility, and increase productivity and profitability, leading to more sustainable farming systems. Agronomic interventions that can be implemented for sustainable agriculture are given below:

**1. Integrated Pest Management (IPM).** IPM is a technique for eradicating pests that integrates multiple pest control methods and minimizes the use of chemical pesticides. This method emphasizes the use of natural predators and biological controls, crop rotation, and the use of resistant varieties.
**2. Conservation agriculture.** Conservation agriculture involves practices such as minimum tillage, crop rotation, and cover cropping, which help to maintain soil health, prevent soil erosion, and improve water retention.
**3. Precision agriculture.** Precision agriculture involves the use of advanced technologies such as sensors and GPS to monitor crop growth and optimize inputs such as fertilizers and water usage, leading to reduced environmental impact and improved yields.
**4. Agroforestry.** Agroforestry involves the integration of trees into farming systems. It has the potential to offer numerous advantages, including the preservation of biodiversity, carbon sequestration, and soil.
**5. Organic farming.** Organic farming practices involve the use of natural inputs such as compost, manure, and biological pest control methods to improve soil health and reduce environmental impacts.
**6. Crop diversification.** Agricultural diversification is raising a variety of crops to lower the risk of crop failure brought on by pests, diseases, and CC. Diversification can also provide farmers with a range of income streams and improve the nutritional quality of diets.

## Soil-management practices for improved crop yield in dryland zones

Major crops grown in dryland areas include groundnut (50% of total area), pearl millet (80% of total area), chickpea (90% of total area), and pigeon pea (95% of total area). Dryland agriculture is achieved by raising livestock on native vegetation and conversion of the natural ecosystem into an agro-ecosystem for crop production through tillage practices. Dryland agriculture mainly emphasizes sustainable crop yields by conserving water and using limited fertilizer and other inputs, under wind- and water-erosion constraints.

The drylands may contribute to most of the world's food-grain production in future but it is necessary to cover the yield gaps compared with humid and sub-humid regions (Parr *et al.*, 1990). Although drylands exhibit restrictive climatic conditions, including low moisture availability, these lands can yet play an important role in controlling multiple ecosystem functions and services.

## Features and Health Assessment of Dryland Region Soils in Pakistan

In Pakistan, drylands, arid lands, and/or rainfed areas are used to designate lands that solely depend on rainfall and are not irrigated by canals or tube well water. Most parts of Sindh, Baluchistan, southern parts of Punjab, and Khyber Pakhtunkhwa receive less than 250 mm of annual rainfall and are considered drylands. The Potohar Plateau covers an area of 1.8 million ha and is an important region of drylands in Pakistan.

In these regions, the soils are mainly developed from loess, alluvium, and colluvium parent material (Nizami *et al.*, 2004). Soils are often coarse-textured with low soil organic matter and fertility and are easily susceptible to wind and water erosion (Reynolds *et al.*, 2007; Khan and Mahmood-ul-Hasan, 2019). These soil properties are seriously influenced by climatic conditions, which determine soil development and productivity (Koutroulis, 2019).

## Biological approaches for economic utilization of saline/drylands

Dryland soils have several characteristics in common, including: (i) low soil fertility; (ii) low organic matter; (iii) alkaline pH; (iv) calcium carbonate-rich subsoil; (v) sandy and coarse-textured soils; and (vi) soil salinization (Wale and Dejenie, 2013). In dryland soils, water-holding capacity and the ability to supply nutrients are the major soil characteristics mainly responsible for crop productivity, and these properties are controlled by soil depth. But soil depth in drylands is often limited due to the hardpan layer, which restricts the soil's water-holding capacity. Low soil fertility in drylands is mainly characterized by the leaching of nutrients and the weathering of minerals (Nizami *et al.*, 2004). In Pakistan, according to an estimate, there are nearly 6 mha of salt-affected wastelands, which are predominantly in the dry areas of the country. To utilize these lands, there has been R&D program going on since 1970 which developed into Biosaline Agriculture Technology, which aims to economically utilize salt-affected land and saline (brackish) underground water through profitable and integrated use of genetic resources (plants, livestock, fish, insects, microbes) and improved agronomic practices. This biological approach was demonstrated successfully at various locations in Pakistan. A pilot project namely Saline Agriculture Farmers Participatory Project (SAFPDP), funded by the government, was launched for rehabilitation of 10,000 ha of saline drylands using this technology with farmers' participation. This technology has also been demonstrated on field stations in nine countries of the Middle East and North Africa under an IAEA-funded Inter-regional Model Project INT/5144 led by Nuclear Institute for Agriculture and Biology (NIAB), Faisalabad.

## Cholistan: a model dryland desert in Pakistan

Cholistan is a vast sandy desert spread over about 26,000 km$^2$, covering two-thirds of the total area of the Bahawalpur division (Fig. 13.7). In the east, it is bounded by the Bahawalnagar, Bahawalpur, and Rahim Yar Khan districts of Punjab adjacent to the Indo-Pak international boundary extending along the east and south. The word cholistan is a distorted form of the Turkish word *chilistan* meaning waterless wasteland.

The diversified topography of this desert region has resulted in rich flora and fauna. Prominent but sparse vegetation in Cholistan includes tree species, small shrubs, and medicinal herbs.

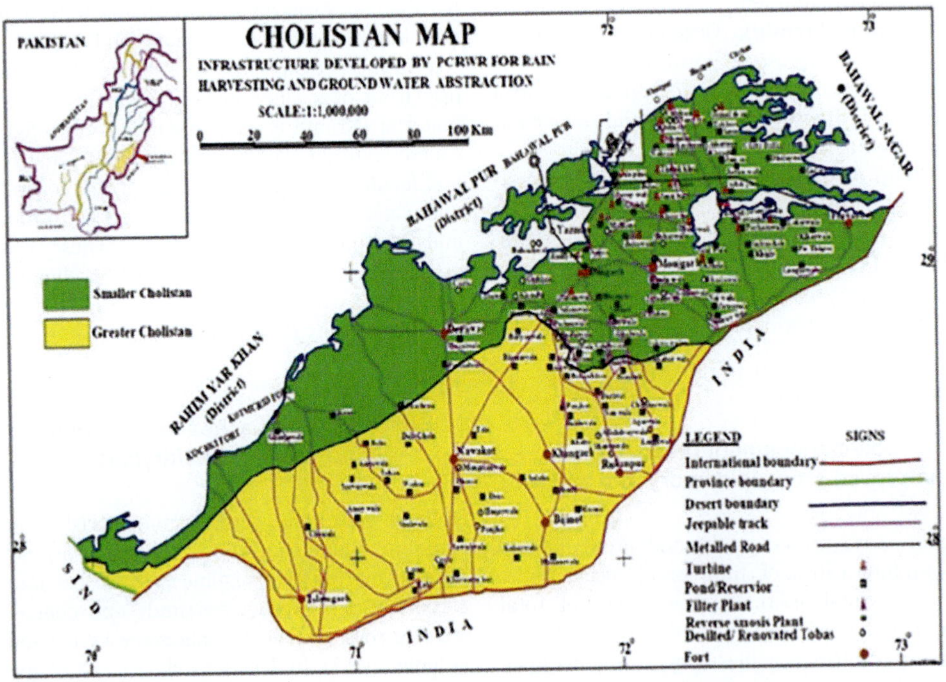


**Fig. 13.7.** Map of Cholistan.

The desert also serves as a habitat for animals and birds including some endangered species of wildlife, such as the blackbuck, Indian gazelle, Indian wild ass, and the great Indian bustard. Agriculture and animal husbandry are the two principal occupations of the region. In view of its biodiversity, it is an ideal area to try out all the biological approaches for land utilization.

There are three sources of water: irrigation supplies through canals, groundwater, and harvested rainwater. Groundwater is mostly saline with fewer pockets of freshwater in scattered spots which may serve as a source of drinking water supplies for the human population. Rainwater harvesting is also a traditional method known to local people. Communities generally develop tobas, ponds or kunds to harvest the rare but substantial rainfall events.

The Pakistan Council of Research in Water Resources (PCRWR) has executed several flagship projects in Cholistan especially for rainwater harvesting by constructing scientifically designed rainwater-harvesting ponds, which has given a boost to the various agricultural activities related to crops and livestock.

### Impact of CC on soil health of dryland areas

In dryland agriculture, CC is one of the major detrimental factors, negatively impacting crop and livestock production and exacerbating hunger in these areas. Frequent droughts and floods in rural areas of the drylands of Pakistan due to changing climate are serious threats to the assets of poor families (Bantilan and Aupama, 2006).

CC is accelerating land degradation, resulting in decreased biological or economic productivity of rainfed cropland. Due to CC in rainfed areas, uncertain rainfalls make soil moisture a major limiting factor for crop production. At the same time, flooding due to heavy rainfall in a changing climate enhances soil erosion, which decreases crop productivity.

The climatic variables, such as elevated temperature, uncertain rainfall, and elevated $CO_2$, severely influence cropping systems. Elevated temperatures cause a decrease in crop productivity by shortening crop cycles, hastening anthesis, and inducing floral aberrations (Zampieri *et al.*, 2020). Similarly, increased temperature during grain filling can cause low grain quality. Drought stress during grain filling results in earlier senescence (Asseng *et al.*, 2019). Meanwhile, elevated $CO_2$ may positively affect crop biomass production, but too-high concentrations can lead to an elevation of tissue temperature. However, the interactive effects of the above-mentioned climatic factors on crop production might be distinctive. Due to futuristic climate changes, there are further chances of soil organic carbon (SOC) declining because of higher soil respiration rates. Soil desertification due to drought and heatwave events can further aggravate SOC losses through soil erosion. The loss of SOC is intensified due to erosion since it specifically displaces the finer soil fractions, and SOC is largely bound to these fine fractions (Quiroga *et al.*, 1996).

## Future Implications and Strategies for Carbon Sequestration and CC Mitigations in Drylands

Carbon sequestration is one of the utmost imperative opportunities for reducing GHG and $CO_2$ production. By increasing the carbon quantity in the soil with the help of sustainable crop management practices, we can diminish the emission of $CO_2$ to the atmosphere. The global potential of soil C sequestration in drylands is estimated to range from 0.5–1.4 Pg C year$^{-1}$ (Lal, 2019) and their SOC stock is estimated to be 241 Pg, about 27% of the global SOC stock. As drylands are vulnerable to climatic changes, it will likely be compounded by the consequences of global CC. At the same time, a high amount of atmospheric $CO_2$ might decrease CC impacts by boosting plant water-use efficiency (WUE) and productivity (Tietjen *et al.*, 2010). But this increased production may not show any significant increase in the SOC stock since it accelerates the soil respiration by stimulating microbial activities.

The effect of erosion on SOC loss from surface soil is obvious as they have high SOC concentrations. But with that, soil erosion can expose carbonates and/or other minerals in the soil to climatic elements, which may cause their weathering and subsequent $CO_2$ release and/or sequestration through the process of silicate weathering. Maintaining vegetative cover via planting or adding crop residues can play an important role in reducing soil erosion in drylands (Delgado-Baquerizo *et al.*, 2013) and may also enhance weathering processes. This process should be counteracted by management and human intervention. Management of extreme events (drought, heatwaves, and wildfires) is an important loophole to preserve the SOC stock of drylands. SOC storage in drylands may be an important negative feedback in the global C cycle since they faced a historic loss of SOC, and therefore have a great sink capacity. If eroded and degraded drylands were rehabilitated, they could potentially sequester up to 0.4–0.6 Gt C $yr^{-1}$ (Lal, 2001).

Dryland areas of Pakistan are diverse regarding their soils, climate, resource base, production systems, and demographic features. Thus, C sequestration potential cannot be grouped broadly for strategy planning and its implementation. In fact, even within a defined agro-ecological zone, the strategies must be tailored to suit local situations. To achieve C sequestration, introduction of trees is very important as C input from trees is more resistant to mineralization than that from crops. Similarly, zero tillage along with green manuring and addition of farmyard manure can help to enhance C sequestration annual rates (0.1–0.25 t/ha).

## Adaptation and modification of crop plants for sustainable agriculture in drylands

Dryland regions frequently have salinized soil and nutritional deficits, reducing crop production (Varshney *et al.*, 2021). Some plant species have developed morphological, physiological, and biochemical systems that enable them to live, develop, and reproduce in drylands. Plants can use various coping mechanisms to deal with water stress, such as tolerance, avoidance, and escape (Seleiman *et al.*, 2021). Depending on the species and the water stress's duration, intensity, and timing, these approaches may serve diverse purposes and are not mutually exclusive. Contrary to these natural adaptation strategies, plants may be modified to cope with the stressful environment of drylands. These modifications include ideotype breeding, molecular breeding, genetic engineering, and genome editing.

## Current Scenario of Crop Diversity and Genetic Variation in Dryland Areas of Pakistan

The development of modern cultivars exhibiting high productivity and better-quality results from a limited genetic base. The purpose of developing these cultivars was to fulfill yield requirements under ideal field conditions with an adequate supply of various inputs, like water, space, and fertilizer (Khan *et al.*, 2022). Hence, development efforts focus on reproduction rather than protection and resistance mechanisms. Crop development resulted in a loss of phenotypic and genotypic diversity (Renzi *et al.*, 2022). New crops are needed in these areas to increase crop diversity. Dryland agriculture in Pakistan is a high-risk, low-input industry for resource-poor farmers who usually employ sub-par seed, insufficient and unbalanced fertilizers, and sub-par crop management techniques. Crop yields are far below their demonstrated possible potential as a result (Rashid and Ryan, 2007).

### Genetic approaches for crop improvement in dryland areas

#### *Utilization of wild species for abiotic stress tolerance*

Wild plants have evolved to survive under extreme abiotic stress, such as high temperature ranges, evaporation, salt, light intensity, and low levels of precipitation and soil moisture (Yates *et al.*, 2014). Their genetic components can be exploited through breeding and genetic engineering methods to engineer plants that can withstand abiotic stress (Shinozaki and Yamaguchi-Shinozaki, 2007).

Numerous plants, such as *Cynanchum komarovii* and *Prunus mongolica* Maxim, native to drylands, have shown differential expression of more than 3000 genes under abiotic stress conditions (Wang *et al.*, 2015). Most of those genes were found to be involved in coding proteins controlling two types of processes: (i) signal transduction for transcription factors, protein kinases, and ABA biosynthesis; and (ii) synthesis of water and ion channels, scavenging molecules, defense-related proteins, and proteins controlling osmolyte biosynthesis (Seki *et al.*, 2003). Gene-coding proteins responsible for these functions have been found in several models and cultivated plants (Mizoi *et al.*, 2013). Their analogs in plants of dryland origin may be exploited as authoritative means for developing abiotic stress-tolerant plants through breeding and transgenic approaches. For instance, a novel DREB2B (EsDREB2B) gene has been identified in a dryland legume, *Eremosparton songoricum*. Similarly, some other novel DREBs have been isolated from a dehydration-tolerant moss (*Syntrichia caninervis* Mitt.). These DREB genes were initially transformed in yeast and conferred high abiotic stress tolerance to yeast (Li *et al.*, 2016). In another effort, the gene of the WRKY21 transcription factor has been isolated from the desert shrub *Larrea tridentate* (Zou *et al.*, 2004). Similarly, another DREB gene (PeDREB2L) was isolated from a dryland tree, *Populus euphratica* Oliva, and transformed into *Arabidopsis thaliana* that then exhibited a high level of drought and low temperate zone stress tolerance (Chen *et al.*, 2011). There is a need to use genetic resources in modern genetic engineering programs to develop abiotic stress-tolerant cultivars for sustainable dryland agriculture in Pakistan.

### *Ideotype breeding for drought tolerance and improved water-use efficiency*

CC has made traits selection and objectives of breeding programs more challenging. Now, breeders must consider reconstructing the whole plant architecture by incorporating the ideal features, which is called ideotype breeding. The drylands have diverse and dynamic environmental conditions. To address the challenges of harsh environmental conditions, ideotype breeding in drylands focuses on selecting traits relevant to the specific conditions in these regions. For dryland areas, the plant architecture should be designed for short growth duration, efficient root system, drought tolerance, and altered leaf morphology for better photosynthesis. The ideotype breeding for drylands is generally focused on water-use efficiency, root system architecture, and carbon sequestration (Varshney *et al.*, 2021).

Each ecological zone has its own growth conditions and requirements; hence, confining the plant ideotype to a single model in any crop would not be an appropriate approach. Since the cropping scheme of drylands differs from irrigated areas, their crops have different ideotype features. For monocots and dicots, steep, deep, and cheap root ideotype concepts have been hypothesized for optimized water and nitrogen acquisition. This model suggests that long thick lateral roots, developed at steep angles, provide access to soil moisture and nitrates at depths (Lilley and Kirkegaard, 2007). But the trade-off for longer and deeper roots with upper-layer roots can limit access to immobile nutrients phosphorus, potassium, and iron, usually present in the top soil layers. The root architecture traits for water-stress adaptation in dry areas are root angle, root diameter, rooting depth, lateral root intensity, and root length distribution (Liu *et al.*, 2011).

Regarding vegetative traits, the genotypes in dry areas require fewer and smaller leaves to reduce water loss from transpiration. The reduced stomatal density conserves subsoil moisture for later phases of the crop cycle and provides moisture during the grain-filling stage (Blessing *et al.*, 2018). Reduced leaf area in cereals like wheat, barley, and sorghum maintains leaf growth and photosynthesis under severe drought stress (Zhao *et al.*, 2019).

Ideotype breeding is a moving goal, and the key traits change according to agronomic practices, climate, market, and government policies. The strategy for ideotype breeding for drylands starts from developing an optimal ideotype model based on the values of morphological and physiological traits. Then the base material is selected from the available genetic diversity and the desirable features are assembled in a single genotype. Depending upon the nature of the traits' breeding procedure like the single cross, three-way cross, multiple crosses, and backcross, the composite crossing is adopted for hybridization. A practical ideotype model requires minimum resources per unit of dry matter produced with maximum efficiency for water and nutrients for a high harvest index (Reynolds *et al.*, 2017). There is considerable genetic diversity in the primary and secondary gene pools for cereals, legumes, and forages available to utilize for developing cultivars suitable for dry areas.

### *Molecular breeding and genetic engineering for climate-resilient crops*

Modern biotechnological interventions, like molecular/speed breeding, genome editing, engineering multigenic traits, and exploring beneficial microbial associations, are crucial to increasing crop production. Molecular breeding has emerged as a valuable tool to resolve existing bottlenecks in classical breeding, i.e. sexual compatibility, linkage drag, and the requirement of too much labor and time. DNA-based MAS (marker-assisted selection) is an effective method to develop germplasm for dryland agriculture. Growth-stage independent selection can be carried out with the effect of environmental conditions, so it is very effective for selecting and screening potential genotypes at an early growth stage. Speed breeding has further helped speed up the breeding program. Recent interventions in GAB (genome-assisted breeding) paved the way for constructing highly annotated pan-genomes of different crops, thus exploring the landscape of genetic diversity across a wide range of crop plants. These pan-genomes can exploit unique genetic variations leading to the development of CC-resistant crop varieties. In Pakistan, though promising achievements have been made, research is limited to genetic diversity analysis and characterization of genes involved in abiotic stress tolerance (Anwar *et al.*, 2019). Wheat has been chosen for molecular breeding, emphasizing dryland agriculture. Ahmad *et al.* (2014) worked out QTL mapping for drought tolerance, whereas Rashed *et al.* (2010) developed RAPD markers associated with drought tolerance in wheat. These molecular markers have been employed to develop efficient crop varieties that can perform in drylands (Muhammad *et al.*, 2016). Three wheat varieties (Siran, Atta-Habib, and Ghanimat-e-IBGE) have been developed through molecular breeding and approved for commercial cultivation. Alleles associated with drought tolerance were studied in 153 Pakistani hexaploid wheat cultivars and diploid and tetraploid wheat accessions (Ur Rehman *et al.*, 2021) to devise molecular markers for dryland breeding. Muhammad *et al.* (2016) tested 52 wheat varieties for their potential against drought by employing morphological parameters and SSR markers.

Genetic transformation or genome editing is a more precise technique to develop abiotic stress-tolerant crops. In Pakistan, numerous biotech research institutes are working on the genetic transformation of crop plants. Different crop plants (cotton, wheat, sugarcane, brassicas, and potatoes) have been transformed for different traits and several cases are waiting for approval from the National Biosafety Committee. According to the ISAAA (International Services for the Acquisition of Agri-Biotech Applications) GM Approval Database 2022, six transformation events have been approved (two for cotton and four for maize) in the country. These events are either for insect resistance or herbicide tolerance; none has been developed indigenously. The complex regulatory process for the approval and release of transgenic crops has discouraged research on the genetic transformation of food crops. Genome editing is a comparatively acceptable approach because genome-edited plants are not classified as transgenic in some countries and do not require approval from regulatory bodies. Several research labs are working on genome editing of different crop plants to tailor them for desired traits, but the research is still in its infancy. It will hopefully come up with some valuable achievements in the coming years.

## Conclusion

Climate change is one of the most pressing challenges facing agriculture in Pakistan. The impacts of CC, such as droughts, floods, and changing weather, are already being felt by farmers nationwide, particularly in the dryland regions. To address these challenges and build more resilient and sustainable farming systems, there is a need to adapt scientific interventions that consider the local context and the needs of farmers. Adaptation measures such as developing CC-resilient genotypes, crop diversification, conservation agriculture, and water management can help farmers cope with the impacts of CC and improve their livelihoods. Though the scientific community is playing its role, more focused research needs to be initiated. Policy decisions and legislation for adopting new technologies must be sensitized and revised. Likewise, re-zoning is also equally important to uplift agriculture in the drylands. Certain initiatives taken by the government are encouraging the promotion of water conservation and sustainable agriculture practices, thus helping to mitigate these challenges. Hence, there is a dire need to devise short-term, medium-term, and long-term policies to uplift dryland agriculture in the country.

## References


Ahmad, A., Khan, M., Shah, S.H.H., Kamran, M., Wajid, S.A. *et al.* (2019) Agro-ecological zones of Punjab, Pakistan. *Food and Agriculture Organization of the United Nations*, Rome.

Ahmad, I., Yu, Y., Stewart, J.P., Podger, G.M. and Kirby, J.M. (2018) Spatiotemporal analysis of precipitation variability in annual, seasonal, and extreme values over upper Indus River basin. *Atmospheric Research* 213, 346–360. DOI: 10.1016/J.ATMOSRES.2018.06.019

Ahmad, M.A., Khan, S.H., Khan, A.S., Kazi, A.M. and Basra, S.M.A. (2014) Identification of QTLs drought tolerance traits on wheat chromosome 2A using association mapping. *International Journal of Agriculture and Biology* 16, 862–870.

Anwar, S., Mustafa, G., Khan, M.S. and Shahid, M. (2019) Differential expression of stress responsive Scdr1 gene in indigenous sugarcane genotypes. *International Journal of Agriculture and Biology* 22, 1560–8530. DOI: 10.17957/IJAB/15.1165

Asseng, S., Martre, P., Maiorano, A., Rötter, R.P., O'Leary, G.J. *et al.* (2019) Climate change impact and adaptation for wheat protein. *Global Change Biology* 25, 155–173. DOI: 10.1111/gcb.14481

Bantilan, M. and Aupama, K. (2006) Vulnerability and adaptation in dryland agriculture in India's sat: experiences from ICRISAT'S village-level studies. *Journal of SAT Agricultural Research* 2, 1–14.

Blessing, C.H., Mariette, A., Kaloki, P. and Bramley, H. (2018) Profligate and conservative: water use strategies in grain legumes. *Journal of Experimental Botany* 69(3), 349–369. DOI: 10.1093/jxb/erx415

Chen, J., Xia, X. and Yin, W. (2011) A poplar DRE-binding protein gene, PeDREB2L, is involved in regulation of defense response against abiotic stress. *Gene* 483, 36–42. DOI: 10.1016/j.gene.2011.05.010

Delgado-Baquerizo, M., Maestre, F.T., Gallardo, A., Bowker, M.A., Wallenstein, M.D. *et al.* (2013) Decoupling of soil nutrient cycles as a function of aridity in global drylands. *Nature* 502, 672–676. DOI: 10.1038/nature12670

FAO (Food and Agriculture Organization of the United Nations) (2016) *Rangelands of Pakistan: Current Status, Threats and Potential*. FAO, Rome. Available at: https://data.landportal.info/library/resources/faodocrepaf354e5b-3aab-4be5-8742-f95430e90f79/rangelands-pakistan-current-status (accessed 21 July 2024).

FAO (Food and Agriculture Organization of the United Nations) (2019) *Agro-Ecological Zones of Punjab, Pakistan*. FAO, Rome. Available at: https://www.fao.org/documents/card/en/c/ca6938en/ (accessed 21 July 2024).

Government of Pakistan (2022) *Economic Survey of Pakistan 2021–2022*. Ministry of Finance, Economic Affairs, Revenue and Statistics, Islamabad. Available at: https://www.finance.gov.pk/survey/chapter_22/Economic%20Survey%202021-22.pdf (accessed 21 July 2024).

IPCC (Intergovernmental Panel on Climate Change) (1994) Informal working documents, IPCC Special Workshop: Article 2 of the United Nations Framework Convention on Climate Change, 17–21 October, Fortaleza, Brazil.

Isaaks, E.H. and Srivastava, R.M. (1989) *An Introduction to Applied Geostatistics*. Oxford University Press, New York.

Karl, T.R., Knight, R.W., Easterling, D.R. and Quayle, R.G. (1996) Indices of climate change for the United States. *Bulletin of the American Meteorological Society* 77(2), 279–292. DOI: 10.1175/1520-0477(1996)077<0279: IOCCFT>2.0.CO;2

Khan, M., Ali, S., Manghwar, H., Saqib, S., Ullah, F., Ayaz, A. and Zaman, W. (2022) Melatonin function and crosstalk with other phytohormones under normal and stressful conditions. *Genes* 13, 1699. DOI: 10.3390/genes13101699

Khan, S. and Mahmood-ul-Hasan, M.A.K. (2019) Adaptation to climate change and mitigation of its effects in the arid region of Pakistan (1961–2015). *International Journal of Economic and Environmental Geology*, 7–15.

Kittel, T.G.F., Rosenbloom, N.A., Painter, T.H. and Schimel, D.S. (1995) The VEMAP integrated database for modelling United States ecosystem/vegetation sensitivity to climate change. *Journal of Biogeography* 22(4/5), 857. DOI: 10.2307/2845986

Koutroulis, A.G. (2019) Dryland changes under different levels of global warming. *Science of the Total Environment* 655, 482–511. DOI: 10.1016/j.scitotenv.2018.11.215

Lal, R. (2001) Potential of desertification control to sequester carbon and mitigate the greenhouse effect. *Climate Change* 51, 35–72. DOI: 10.1023/A:1017529816140

Lal, R. (2019) Carbon cycling in global drylands. *Current Climate Change Reports* 5, 221–232. DOI: 10.1007/s40641-019-00132-z

Latif, Y., Ma, Y. and Ma, W. (2021) Climatic trends variability and concerning flow regime of Upper Indus Basin, Jehlum, and Kabul River basins Pakistan. *Theoretical and Applied Climatology* 144, 447–468. DOI: 10.1007/s00704-021-03529-9

Leemans, R. and Cramer, W.P. (1991) *The IIASA Database for Mean Monthly Values of Temperature, Precipitation, and Cloudiness on a Global Terrestrial Grid*. International Institute for Applied Systems Analysis, Luxemburg.

Li, H., Zhang, D., Li, X., Guan, K. and Yang, H. (2016) Novel DREB A-5 subgroup transcription factors from desert moss (*Syntrichia caninervis*) confers multiple abiotic stress tolerance to yeast. *Journal of Plant Physiology* 194, 45–53. DOI: 10.1016/j.jplph.2016.02.015

Lilley, J.M. and Kirkegaard, J.A. (2007) Seasonal variation in the value of subsoil water to wheat: simulation studies in Southern New South Wales. *Australian Journal of Agricultural Research* 58(12), 1115–1128. DOI: 10.1071/AR07046

Liu, L., Gan, Y., Bueckert, R. and Rees, K.V. (2011) Rooting systems of oilseed and pulse crops. II: Vertical distribution patterns across the soil profile. *Field Crops Research* 122(3), 248–255. DOI: 10.1016/j.fcr.2011.04.003

Lutz, A.F., Immerzeel, W.W., Shrestha, A.B. and Bierkens, M.F.P. (2014) Consistent increase in High Asia's runoff due to increasing glacier melt and precipitation. *Nature Climate Change* 4, 587–592. DOI: 10.1038/nclimate2237

Mallapaty, S. (2022) Why are Pakistan's floods so extreme this year? *Nature*, 2 September. DOI: 10.1038/d41586-022-02813-6

Martin-Vide, J. (2004) Spatial distribution of a daily precipitation concentration index in peninsular Spain. *International Journal of Climatology* 24(8), 959–971. DOI: 10.1002/JOC.1030

Mizoi, J., Ohori, T., Moriwaki, T., Kidokoro, S., Todaka, D. and Maruyama, K. (2013) GmDREB2A;2, a canonical dehydration-responsive element-binding protein2-type transcription factor in soybean, is post translationally regulated and mediates dehydration-responsive element-dependent gene expression. *Plant Physiology* 161(1), 346–361. DOI: 10.1104/pp.112.204875

Muhammad, I., Inamullah, Ahmad, H., Alam, B., Khan, I.A., Ahmad, I. *et al.* (2016) Marker assisted screening of wheat *(Triticum aestivum* L.) cultivars for drought tolerance and yield improvement. *International Journal of Biosciences* 8(3), 44–52. Available at: https://www.researchgate.net/profile/Sahib_Afridi2/publication/301341944_Marker_assisted_screening_of_wheat_Triticum_aestivum_L_cultivars_for_drought_tolerance_and_yield_improvement/links/571f487c08aed056fa22e8e1.pdf (accessed 21 July 2024).

Nizami, M., Shafiq, M., Rashid, A. and Aslam, M. (2004) *The Soils and Their Agricultural Development Potential in Pothwar*. NARC, Islamabad.

Parr, J., Stewart, B., Hornick, S. and Singh, R. (1990) Improving the sustainability of dryland farming systems: a global perspective. *Advances in Soil Science: Dryland Agriculture: Strategies for Sustainability* 13, 1–8. DOI: 10.1007/978-1-4613-8982-8_1

Qasim, M., Khan, M. and Rashid, W. (2023) Spatial and temporal analyses of land use changes with special focus on seasonal variation in snow cover in District Chitral, a Hindu Kush Mountain region of Pakistan. *Remote Sensing Applications: Society and Environment* 29, 100902. DOI: 10.1016/j.rsase.2022.100902

Quiroga, A., Buschiazzo, D. and Peinemann, N. (1996) Soil organic matter particle size fractions in soils of the semiarid Argentinian pampas. *Soil Science* 161, 104–108.

Rafiq, M., Rashid, A. and Memon, K.S. (1996) Soil resources of Pakistan. *Soil Science* 161, 439–470.

Rashed, M.A., Sabry, S.B.S., Atta, A.H. and Mostafa, A.M. (2010) Development of RAPD markers associated drought tolerance in bread wheat *(Triticum aestivum* L.*). Egyptian Journal of Genetics and Cytology* 39, 131–142.

Rashid, A. and Ryan, J. (2007) Challenges and strategies for dryland agriculture in Pakistan. In: Rao, S.C. and Ryan, J. (eds) *Challenges and Strategies of Dryland Agriculture.* John Wiley, and Sons, Ltd, Chichester, UK, pp. 359–371. DOI: 10.2135/cssaspecpub32.c22

Rashid, A., Ryan, J. and Chaudhry, M.A. (2004) Challenges and strategies for dryland agriculture in Pakistan. *Challenges and Strategies of Dryland Agriculture* 32, 359–371.

Renzi, J.P., Coyne, C.J., Berger, J., von Wettberg, E., Nelson, M. *et al.* (2022) How could the use of crop wild relatives in breeding increase the adaptation of crops to marginal environments? *Frontiers in Plant Science* 13, 886162. DOI: 10.3389/fpls.2022.886162

Reynolds, J.F., Smith, D.M.S., Lambin, E.F., Turner, B., Mortimore, M. *et al.* (2007) Global desertification: building a science for dryland development. *Science* 316(5826), 847–851. DOI: 10.1126/science.1131634

Reynolds, M.P., Alistair, J.D., William, P.J.E., Hoppitt, K., Sonder, S. *et al.* (2017) Strategic crossing of biomass and harvest index – source and sink – achieves genetic gains in wheat. *Euphytica* 213(11), 257. DOI: 10.1007/s10681-017-2040-z

Rosenzweig, C., Elliott, J., Deryng, D., Ruane, A.C., Müller, C. *et al* (2014) Assessing agricultural risks of climate change in the 21st century in a global gridded crop model intercomparison. *Proceedings of the National Academy of Sciences* 111(9), 3268–3273. DOI: 10.1073/pnas.1222463110

Ruane, A.C., Goldberg, R. and Chryssanthacopoulos, J. (2015) Climate forcing datasets for agricultural modeling: merged products for gap-filling and historical climate series estimation. *Agricultural and Forest Meteorology.* DOI: 10.1016/j.agrformet.2014.09.016

Seki, M., Kamei, A., Yamaguchi-Shinozaki, K. and Shinozaki, K. (2003) Molecular responses to drought, salinity and frost: common and different paths for plant protection. *Current Opinion in Biotechnology* 14, 194–199. DOI: 10.1016/S0958-1669(03)00030-2

Seleiman, M.F., Al-Suhaibani, N., Ali, N., Akmal, M., Alotaibi, M. *et al.* (2021) Drought stress impacts on plants and different approaches to alleviate its adverse effects. *Plants (Basel)* 10(2), 259. DOI: 10.3390/plants10020259

Shinozaki, K. and Yamaguchi-Shinozaki, K. (2007) Gene networks involved in drought stress response and tolerance. *Journal of Experimental Botany* 58, 221–227. DOI: 10.1093/jxb/erl164

Siddiqui, A.R., Nazeer, S., Piracha, M.A., Saleem, M.M., Siddiqi, I., Shahzad, S.M. and Sarwar, G. (2016) The production of biochar and its possible effects on soil properties and phosphate solubilizing bacteria. *Journal of Applied Agriculture and Biotechnology* 1(1), 27–40. Available at: https://www.researchgate.net/profile/Ghulam-Sarwar-6/publication/304246322_The_production_of_biochar_and_its_possible_effects_on_soil_properties_and_phosphate_solubilizing_bacteria/links/576a47e008ae3bf53d344dd0/The-production-of-biochar-and-its-possible-effects-on-soil-properties-and-phosphate-solubilizing-bacteria.pdf (accessed 21 July 2024).

Singh, A.K., Tripathy, R. and Chopra, U.K. (2008) Evaluation of CERES-wheat and CropSyst models for water–nitrogen interactions in wheat crop. *Agricultural Water Management* 95, 776–786. DOI: 10.1016/j.agwat.2008.02.006

Spinoni, J., Vogt, J., Naumann, G., Carrao, H. and Barbosa, P. (2015) Towards identifying areas at climatological risk of desertification using the Köppen–Geiger classification and FAO aridity index. *International Journal of Climatology* 35, 2210–2222. DOI: 10.1002/joc.4124

Tietjen, B., Jeltsch, F., Zehe, E., Classen, N., Groengroeft, A. *et al.* (2010) Effects of climate change on the coupled dynamics of water and vegetation in drylands. *Ecohydrology & Hydrobiology* 3, 226–237. DOI: 10.1002/eco.70

Umar, M.A. and Fahad, S. (2018) Role of heat stress in migration decision: a case study of Faisalabad. *Sustainable Development Policy Institute.* Available at: https://idl-bnc-idrc.dspacedirect.org/items/05c63d10-841d-4c82-9685-0cd35100309f (accessed 21 July 2024).

Ur Rehman, S., Sher, A.M., Saddique, M.A.B., Ali, Z., Khan, M.A. *et al.* (2021) Development and exploitation of KASP assays for genes underpinning drought tolerance among wheat cultivars from Pakistan. *Frontiers in Genetics* 12, 684–702. DOI: fgene.2021.684702

USAID (2019) Pakistan complex emergency and drought. Available at: https://2017-2020.usaid.gov/sites/default/files/documents/1866/09.30.19_-_USAID-DCHA_Pakistan_Complex_Emergency_and_Drought_Fact_Sheet_2.pdf (accessed 21 July 2024).

Varshney, R.K., Barmukh, R., Roorkiwal, M., Qi, Y., Kholova, J. *et al.* (2021) Breeding custom-designed crops for improved drought adaptation. *Advanced Genetics* 2(3), e202100017. DOI: 10.1002/ggn2.202100017

Wale, H.A. and Dejenie, T. (2013) Dryland ecosystems: their features, constraints, potentials and managements. *Research Journal of Agricultural and Environmental Management* 2, 277–288.

Wang, J., Zheng, R., Bai, S., Gao, X., Liu, M. and Yan, W. (2015) Mongolian almond (*Prunus mongolica maxim*): the morpho-physiological, biochemical, and transcriptomic response to drought stress. *PLOS ONE* 10(4), e0124442. DOI: 10.1371/journal.pone.0124442

Widmann, M. and Bretherton, C.S. (2000) Validation of mesoscale precipitation in the NCEP reanalysis using a new gridcell dataset for the Northwestern United States. *Journal of Climate* 13(11), 1936–1950. DOI: 10.1175/1520-0442(2000)013%3C1936:VOMPIT%3E2.0.CO;2

Yates, S.A., Chernukhin, I., Alvarez-Fernandez, R., Bechtold, U., Baeshen, M. *et al.* (2014) The temporal foliar transcriptome of the perennial C3 desert plant *Rhazya stricta* in its natural environment. *BMC Plant Biology* 14(2). DOI: 10.1186/1471-2229-14-2

Zampieri, M., Weissteiner, C.J., Grizzetti, B., Toreti, A., van den Berg, M. *et al.* (2020) Estimating resilience of crop production systems: from theory to practice. *Science of the Total Environment* 735, 139378. DOI: 10.1016/j.scitotenv.2020.139378

Zhao, Z., Rebetzke, G.J., Zheng, B., Chapman, S.C. and Wang, E. (2019) Modelling impact of early vigor on wheat yield in dryland regions. *Journal of Experimental Botany* 70(9), 2535–2548. DOI: 10.1093/jxb/erz069

Zou, X., Seemann, J.R., Neuman, D. and Shen, Q.J. (2004) A WRKY gene from creosote bush encodes an activator of the abscisic acid signaling pathway. *The Journal of Biological Chemistry* 279(53), 55770–55779. DOI: 10.1074/jbc.M408536

# 14 Resilience and Adaptation of Egyptian Agriculture

**Adel S. El-Beltagy[1,2*], Magdy Madkour[1], and Ayman Abou Hadid[1]**
[1]*Chair, International Dryland Development Commission (IDDC);* [2]*Arid Land Agricultural Graduate Studies & Research Institute (ALARI), Ain Shams University, Egypt*

**Abstract**

This chapter explores Egyptian agriculture's challenges, focusing on limited resources, sea-level rise, dwindling water supplies, and population growth. It highlights the need for transformation to combat climate change's impact on agriculture, emphasizing the Sustainable Agriculture Development Strategy towards 2030. This strategy incorporates advances in science and technology, like smart agriculture, and stresses human resources development, institutional support, policies, and legislation. Key national programs are discussed, aimed at optimizing water use in irrigation, introducing genetically improved crops and livestock, and enhancing market linkages for farmers. Emphasis is made on linking the impact of climate change to Sustainable Development Goals and the utilization of opportunities of the outcome of COP27 (Loss and Damage). The modalities involving green transition, the adaptation fund, carbon tax, and other mechanisms on the livelihood of Egyptian inhabitants are addressed.

## Particularities of Egyptian Agriculture

Egypt is highly vulnerable to the impacts of climate change (CC), including drought, sea-level rise (SLR), soil salinity, disrupted precipitation patterns, and desertification. These challenges can have devastating effects on the economy, food security, migration patterns, public health, and livelihoods.

### Limited Natural Resources

Egypt is characterized by its very limited arable land and scarce water resources. The total arable land spans nearly 9.3 million acres, which must support a continuously growing population. Water supply is predominantly sourced from the Nile, accounting for 90% and totaling 55 billion $m^3$. The country also experiences erratic rainfall along the Mediterranean coast, varying between 80 mm and 150 mm. CC is projected to decrease agricultural productivity by 22–50% by 2060, accompanied by a reduction in agriculture-related employment of up to 40%. Such challenges underscore the urgency for a transformation in Egyptian agriculture. This transformation should be grounded in science and technology, supported by robust human resource development, and bolstered by adaptive institutions and legislation to enhance resilience and adaptability. While Egypt contributes only 0.6% of annual global carbon dioxide ($CO_2$) emissions, as reported

*Corresponding author: elbeltagy@drylanddevelop.org

 *Climate Change and Sustainable Agro-ecology in Global Drylands* (A.S. El-Beltagy, R. Lal and K. Malik eds.)
DOI: 10.1079/9781800624870.0014

in the Global Carbon Project (2021), it is increasingly becoming one of the countries most affected by CC and ranks prominently among the most vulnerable nations.

## Serious Vulnerability to SLR

CC is intrinsically linked to rising sea levels, which pose a significant threat to Egypt's ecological balance. A case in point is the Nile river delta, where erosion rates, in certain areas, reach up to 10 m annually, leading to a reduction in available agricultural land. This decline in arable farmland complicates the task for Egyptian farmers striving to feed a burgeoning population. Alexandria, Egypt's second-largest city, stands at the forefront of this challenge. Even a moderate surge in sea levels could decimate its beaches and flood its low-lying districts. The encroaching saltwater not only endangers agriculture but can also compromise the structural stability of buildings. The ensuing climate-related displacement confines Egypt's expanding populace to an ever-shrinking living space. Tragically, it is often the nation's poorest, who reside in these vulnerable regions, that suffer the harshest consequences of these CC repercussions. SLR in the delta will cause saline seawater intrusion which will affect 30% of the most fertile delta soil (Pearce, 2010; Michaelson, 2018). An integrated assessment for SLR on the lower Nile delta has shown the adverse impact of water-related security issues (Susnik *et al.*, 2014; Solyman and Abdel Monem, 2020).

According to a report by UNEP/GRID-Arendal (Simonett, 2005), two potential scenarios were outlined concerning the displacement of residents in the Nile delta due to SLR. With a sea-level increase of approximately 50 cm, around 1800 km$^2$ of cropland in the Nile delta could be submerged, affecting the livelihoods of an estimated 3.8 million inhabitants. Conversely, if the rise were to reach 1 m, as some models predict, about 4500 km$^2$ of cropland might be inundated, influencing the lives of approximately 6.1 million people.

This situation would exacerbate the strain on Egypt's food security, which is already challenged by a growing population and diminishing water supplies for agriculture. Overlaying these significant concerns is the added impact of temperature increases due to CC. This temperature rise is predicted to reduce the productivity of crops, livestock, fisheries, fauna, and microflora.

Risk-aversion strategies to counter the impacts of SLR are essential not only for Egypt but also for all the Mediterranean countries. The renowned ecologist, the late Professor El Kassas (Barakat and Hegazy, 1997), former president of the International Union for the Conservation of Nature (IUCN) proposed in a personal communication with Dr. Adel El-Beltagy in the early 1980s that we should consider the option proposed by the German architect Herman Sörgel who introduced the concept of a hydroelectric dam across the Strait of Gibraltar, known as the Atlantropa Project (Fig. 14.1). This proposal aimed not just at generating power (Sörgel, 1932; Cathcart, 1985) but could also be modified to offer potential physical adaptations to address the issue of SLR.

This concept was subsequently expanded upon and discussed in detail, culminating in a proposal that was submitted to Egyptian authorities in August 2015 to be considered for future implementation in cooperation with all Mediterranean countries.

The adverse effects of SLR on the Nile delta are profound. Beyond the already mentioned submergence due to rising sea waters, saline seawater intrusion is poised to disrupt the livelihoods of over 25 million residents. This will have ramifications on agriculture, agro-industry, industrial facilities, and urban settlements close to the coastline.

## The Threat of Diminishing Water Supply from the River Nile

From Pharaonic times and over the subsequent centuries, the Nile delta evolved to become Egypt's most fertile land and primary food production hub. The age-old adage that 'Egypt is the gift of the Nile' remains as relevant today. However, the effects of CC, coupled with a lack of comprehensive and well-studied projects in the Nile basin, will adversely affect Nile water-resource supplies for downstream countries like Sudan and Egypt. This will, in turn, diminish food production and negatively impact the livelihoods of the Egyptian population.

Furthermore, ongoing disputes concerning Nile river rights could potentially disrupt Egypt's water resources. Ethiopia is constructing the

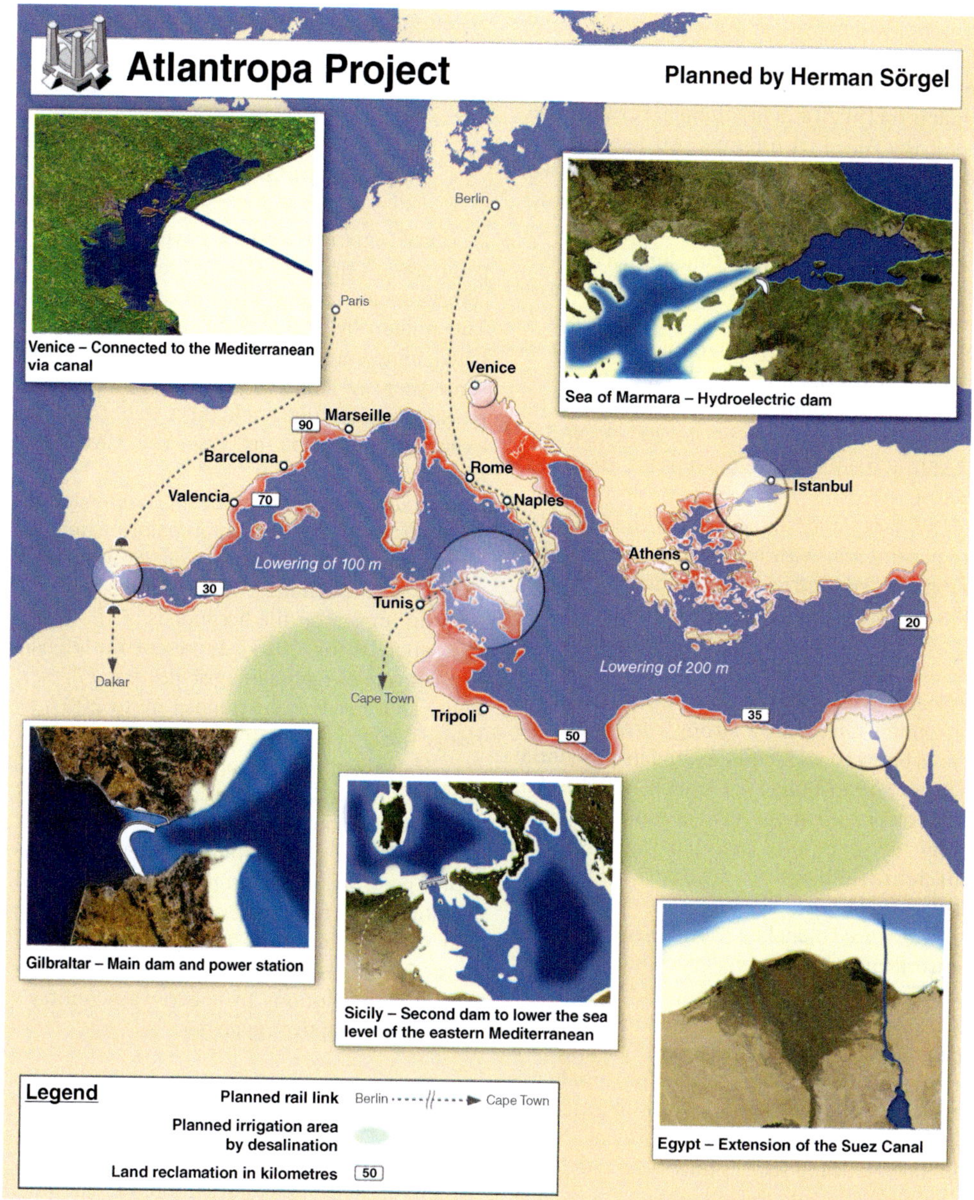


**Fig. 14.1.** An outline map of the various hydroelectricity and land reclamation projects combined in Atlantropa. (Devilm25 (map), VulcanTrekkie45 (translation), CC BY 3.0 –https://creativecommons.org/licenses/by/3.0)

Grand Ethiopian Renaissance Dam (GERD), touted as Africa's most ambitious dam project. Any alterations in the Nile's flow due to this dam could exacerbate scarcity, especially as Egypt's population continues to grow. The consequences of a disrupted water supply, which could hamper agricultural production, are alarming. Indeed, any interference with Egypt's agricultural supply chain could have deeply destabilizing effects.

In addition to the ramifications of the GERD's construction, CC is expected to cause temperatures to rise, exacerbating both flooding and drought, which will in turn influence water runoff in the Blue Nile basin. Research has been conducted to evaluate CC's hydrological impacts on Lake Tana in Ethiopia. For instance, under the Allied Irish Banks (AIB) scenario, a declining pattern in Lake Tana's water storage was identified, attributed to reduced inflows surpassing lake precipitation and surface water inflow over several future timelines. The over-lake evaporation pattern is projected to rise throughout these timeframes (2010–2022) (Nigatu, 2013). Modern techniques, such as machine learning, have also been employed to predict water runoff in the context of CC for Lake Tana (Chattrairat *et al.*, 2022). NASA's satellite systems continually monitor the GERD (NASA, 2022).

Water quality and runoff in the Blue Nile are paramount, with both volume and quality being crucial metrics. Studies by Wondyfraw *et al.*, (2019) and Abegaz *et al.* (2023) noted significant changes due to urbanization and socio-economic developments. Their findings indicate that Lake Tana is currently in a eutrophic state, with a strong correlation observed between water quality and its influential factors. For instance, an uptick in lake skin temperature escalates evaporation, reducing water volume and augmenting insoluble nutrients and water hyacinth (Kebedew *et al.*, 2023).

Advanced modeling techniques, like neural network models, can predict changes in Lake Tana's water level up to 6–7 months in advance, aiding water resource management (Yasuda *et al.*, 2022). However, there exists a significant margin of uncertainty regarding potential variations in Nile flow volumes (Conway, 2005; Beyene *et al.*, 2010; Kingston and Taylor, 2010). Multiple modeling studies have been conducted to gain insights into the impacts on water, food, and economic security (Susnik *et al.*, 2013). Furthermore, a comprehensive modeling approach has been employed to study the water-agricultural system in the Rosetta region of the Nile Delta (Susnik *et al.*, 2013).

Experts in the delta emphasize the importance of adaptive decision making that is receptive to new models, analyses, and insights concerning CC impacts (Cloern *et al.*, 2011; Dettinger *et al.*, 2016). Such decisions should be anchored in integrated monitoring systems, crucial for observing and responding to ecological shifts, comprehensive modeling, and vulnerability assessments.

## Population Growth in Egypt

A critical factor to consider is Egypt's burgeoning population. Currently standing at 113 million (Worldometer, 2023), it is projected to soar to 160 million by 2050. Compounded by anticipated temperature increases of 2–3°C by the same year, agricultural productivity – encompassing crops, livestock, and fisheries – is expected to decline by 25–35%. Given the limited arable land of 9.3 million acres, Egypt's food production capacity falls short of achieving self-sufficiency in key commodities such as wheat, maize, and vegetable oil. Wong stated that 2023, the hottest year recorded, would have an impact on the livelihoods of the world population (Wong, 2023). At present, the intensification rate of Egyptian agriculture is 177%, a figure that needs to be ramped up to 200% to bridge the food production gap.

These four major components of vulnerability to Egyptian agriculture (limited natural resources; serious vulnerability to SLR; threat of diminishing supply from the Nile; and population growth), if not properly addressed, will inevitably lead to a challenging socio-economic and political landscape for Egypt. Addressing these challenges necessitates innovative, knowledge-driven solutions to bolster the country's coping and adaptive capacities.

## Agriculture Strategies

To reposition Egyptian agriculture in relation to dynamic changes, regionally and globally, Egypt has developed several agricultural strategies from the 1980s to 2030: Agricultural Strategy of the 1980s (Egypt Ministry of Agriculture); Agricultural Strategy for the 1990s (World Bank, 1993); Sustainable Agricultural Development Strategy (SADS) towards 2030 (Egypt Ministry of Agriculture, 2009); Business Plan for SADS (2011–2017) (Egypt Ministry of Agriculture, 2010; El-Betalgy *et al.*, 2010); and Business Plan for SADS (2015–2018) (Egypt Ministry of

Agriculture and Ministry of Scientific Research, 2015). These strategies were a result of an interactive participatory consultation process with stakeholders and a bottom-up assessment of development constraints and potential of Egypt's identified five agro-ecological zones, namely: Upper Egypt, Middle Egypt, East Delta, Mid-Delta, and West Delta (Fig. 14.2).

The main framework of the Business Plan was based on SADS' Vision and Mission, which are as follows:

> Vision: To achieve comprehensive economic and social development based on a dynamic agricultural sector capable of sustained and rapid growth, while paying special attention to helping the underprivileged social groups and reducing rural poverty.
> Mission: Modernizing Egyptian agriculture based on achieving food security and improving the livelihood of the rural inhabitants, through the efficient use of development resources, and utilization of geopolitical and environmental comparative advantages of the different agroecological regions.

**The overall objectives of the strategy:**

- promoting sustainable use of natural agricultural resources;
- increasing the productivity of both land and water per unit;
- raising the degree of food security of strategic food commodities;
- increasing the competitiveness of agricultural products in local and international markets and minimizing post-harvest losses;
- improving the climate for agricultural investment; and
- improving the livelihood of the rural inhabitants and reducing poverty rates in the rural areas.

## Increased Sustainable Productivity for Crops and Animals

- Bridging yield and production gaps between the actual and potential production of strategic crops and livestock.

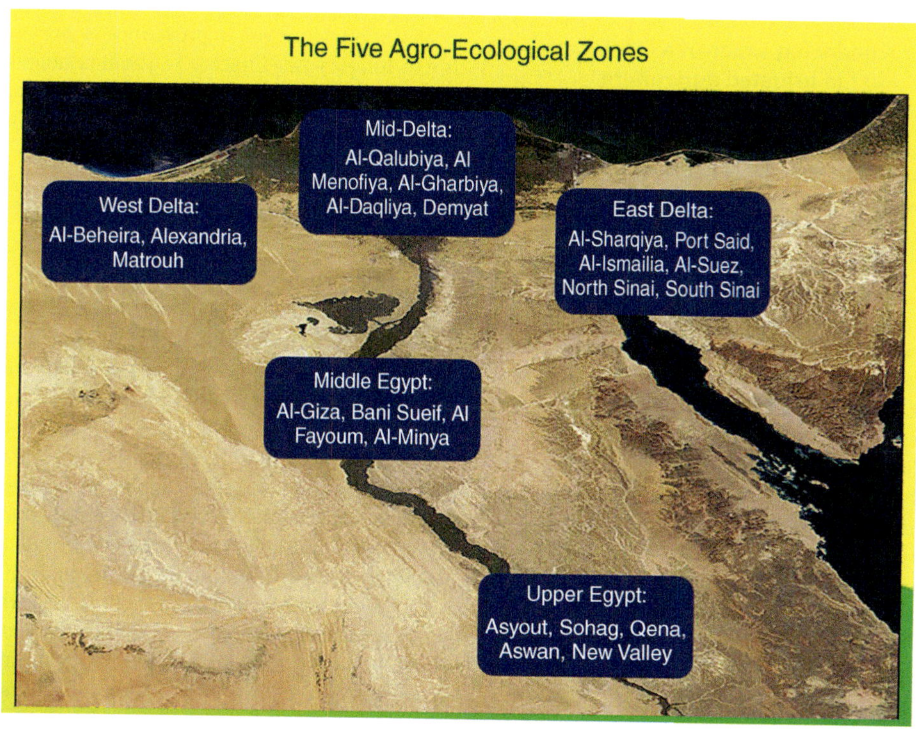


**Fig. 14.2.** Egypt identified five agro-ecological zones, namely: Upper Egypt, Middle Egypt, East Delta, Mid-Delta, and West Delta.

- Improving milk production (quantitatively and qualitatively).
- Promoting intensive fish farming/advanced marine fisheries.
- Addressing anticipated climate change dynamic impacts on agro-ecological zones and their relation to Egyptian food security.
- The use of precision farming techniques, Agriculture 4.0, precision agriculture (El-Beltagy *et al.*, 2021), including AI tools, i.e. machine learning, image processing, neural networks, IoT, Blockchain, Bioinformatic, remote sensing, modeling, etc.) (El-Beltagy *et al.*, 2021), Agriculture 4.0.
- Use of advanced frontiers of genomics, gene editing, and nanotechnology in agriculture applications.
- Advance agro-industry and viable food chains.
- Minimizing pre- and post-harvest food loss.

## Sustainable Use of Agricultural Natural Resources

### Enhancing water-use efficiency in irrigated agriculture

Despite water scarcity and the fact that Egypt's share in the Nile waters is predetermined, water-use efficiency is low, due to high water losses. Water conveyance efficiency is estimated at 70%, and the mean efficiency of field irrigation systems is estimated at only 50%. Hence, one of the main components of the agricultural development strategy is to achieve a gradual improvement of the efficiency of irrigation systems to reach 80% in an area of 8 million acres, and to reduce the areas planted to rice from 1.673 million acres (El-Beltagy and Abou-Hadid, 2008) to 1.3 million acres by 2030 in order to save an estimated 12.4 billion m$^3$ of water.

### Sustainable expansion of reclaimed areas

Expanding reclaimed areas is one of the main pillars of the strategy. This can be achieved by using the water saved through improving field irrigation conveyance systems in reclaiming additional new areas estimated at 1.25 million acres as stage 1 and about 3.1 million acres by 2030 (El-Beltagy *et al.*, 2017).

### Sustainable development of land and water productivity

The strategy pays considerable attention to this component. It is expected that: (i) the cropping area will be increased to 23 million acres by 2030, with an estimated crop intensification rate of 199%, and would increase the economic efficiency of the water unit by 119% by 2030; and (ii) the economic efficiency of the land unit will be raised by 74% by the year 2030.

### Maximizing the sustainable returns of rainfed agriculture

Despite the low rates of rainfall in most of the regions, annual rainfall rates are 100–150 mm on the north coast. Effective rainfall is lower due to evaporation. However, improved water harvesting techniques (see Chapter 5) would be adapted to maximize rainfed water use and supplementary irrigation from groundwater sources.

### Maintaining and protecting agricultural land

Agricultural land in the delta and the Nile valley regions suffers from two important problems: continued encroachment on agricultural land diverting it from agricultural to urban and non-agricultural uses at an annual rate of 33,000 acres until 2010 (Egyptian Ministry of Agriculture and Land Reclamation, 2009); and continued degradation of soil fertility in so many agricultural areas. Rapid urbanization threatens fertile agriculture land and soil carbon in the delta (Abd-Elmabod *et al.*, 2019, 2022). For the problem of encroachment, legislation for agricultural land use should be issued. To maintain soil fertility, we need to have continuous assessment and periodical soil surveys as a basis to establish fertilizer rates, and continued restoration and maintenance of agricultural drainage systems.

### Human resources development and creation of job opportunities, particularly for the young

The strategy development programs and projects are expected to create job opportunities, particularly for the younger generation, in agricultural and related activities, estimated at 4 million jobs by 2030, through (i) reclamation of new areas; (ii) improvement of the irrigation system in the old areas; (iii) adoption of labor-intensive technologies; and (iv) expansion of agricultural support activities in the fields of producing and marketing agricultural inputs and agro-industries. With an estimated family size of five members in the rural areas, the new labor force would provide for an improved livelihood for some 20 million inhabitants. The strategy emphasizes human resource development to provide the needed skills for different development programs, particularly in research and extension activities.

## Institutional Reform

### Agricultural sector institutional reform

The institutional structure of the agricultural sector is highly complex and characterized by duplicative, overlapping duties and responsibilities in some cases and the absence of an institutional structure in others. In addition, some institutional frameworks lack the appropriate mechanisms for conducting the assigned tasks, while some other entities conduct tasks incompatible with their structure and basic functions. The strategy has stressed the importance of the institutional reform of all the partners of Research for Development (R4D), i.e. the Ministry of Agriculture and Land Reclamation (MALR) and its institutions, the cooperative sector, universities and research institutions, civil society organizations (NGOs), agri-business, and private sector active in the field of agri-business, farm machinery, manufacturing of agricultural inputs, etc.

### General objectives of institutional reform:

- delineating the functions of the MALR and related institutions in the fields of research, extension, policy planning and follow-up, providing information and data, developing agricultural resources, planning and monitoring infrastructure development in the newly reclaimed areas, and ensuring availability of agricultural inputs. The ministry would also phase out its role in commercial production;
- merging institutional units with similar functions under one strong entity with defined terms of reference;
- consolidating the agricultural law and related laws, in order to keep abreast of socio-economic developments and international treaties;
- restructuring the agricultural extension service to serve newly emerging functions;
- merging and strengthening entities responsible for information technology, analysis, and dissemination, as well as the entities responsible for human resource training and development;
- assessing the need to establish a new institutional entity to be responsible for the elaboration of standards of agricultural inputs;
- fostering cooperation between agricultural research centers and Egyptian universities in the various domains of frontier sciences and research (national and international);
- establishing mechanisms for safeguarding researchers' royalties, as well as safeguarding their intellectual property rights;
- reviewing the institutional structure of the different entities and units of the MALR, and defining their strengths and weaknesses in relation to other entities, as well as defining the tasks compatible with their functions;
- reforming and supporting civil society organizations active in the field of rural development; enacting a unified law to regulate the establishment of special-purpose associations, instead of enacting a special law for each special-purpose association to be established; and
- the MALR should provide technical support to all institutions and organizations and consider them as principal partners with agricultural research for development (R4D).

## Reviewing and Developing Different Agricultural Policies

Agricultural policies constitute the most important tools and mechanisms for improving the management of the agricultural sector for the general good of society, without sacrificing farmers' interests. Generally, the mix of the proposed agricultural policies aims at meeting the aforementioned six strategic objectives based on the following criteria (Fig. 14.3):

- steering the agricultural sector to be in tandem with national objectives: improving growth rates, raising income levels, promoting exports, and creating more job opportunities for reducing unemployment, especially among the youth;
- achieving the highest possible rates of self-sufficiency in strategic commodities, without compromising the principle of agricultural resources efficiency, especially land and water;
- improving agricultural sector growth rates through achieving substantial progress in land reclamation and development, combined with the application of improved advanced agricultural technologies;
- improving the performance of the wide base of small farmers and encouraging them to interact with the requirements of local and foreign markets, farming cooperatives, and agri-business companies;
- encouraging the private sector to participate actively in the agricultural development effort, through direct agricultural activities, agro-industry, and/or other related and complementary activities; and
- strengthening the efficiency of agricultural research institutions and technology dissemination and enhancing its cooperation with national universities and international technical institutions.

As indicated, the list of proposed policies is grouped into five, related to each other by common objectives of the strategy. The elements and components of these policies are as follows:

1. rationalization of water resources use policy;
2. protection of agricultural land policy;
3. agricultural land maintenance policy;
4. developing inland fisheries and sea aquaculture policy; and
5. agricultural land expansion and settlement policy.

## Capitalizing on Agro-ecological Advantage

Previous agricultural development strategies did not pay specific attention to the endowment of Egypt's agro-climatic regions. Such strategies adopted a generic approach primarily linked to the Central National Plan. This strategy adopts a

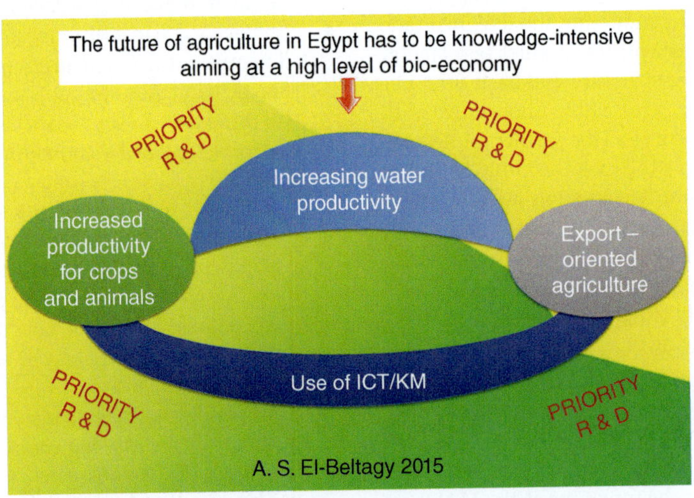


**Fig. 14.3.** Future of agriculture in Egypt with high level of knowledge.

bottom-up approach based on participatory methods (through surveys and meetings) at local level. This approach stems from the concern to improve and optimize the use of natural resources, particularly water for irrigation on farms, and expand land reclamation. To address this situation, it would be necessary to know the attributes of the different regions, as well as the problems and constraints facing future development opportunities in the five geographical regions of Egypt (agro-ecological zones (AEZs)). This requires a dynamic assessment as the baseline will be shifted because of CC.

## Sustainable Agricultural Development Strategy Towards 2030 (Programs and Projects)

The set of the strategic national programs and sub-programs proposed in the strategy constitutes a general reference framework for the detailed major programs and sub-programs to be included in the successive execution plans during the period of the SADS towards 2030. The general framework of the major national programs to be implemented is as follows:

**1.** to rationalize and upgrade the efficiency of water use in agriculture;
**2.** to maintain and upgrade the productive capacity of agricultural land;
**3.** to develop field crops;
**4.** to develop horticultural crops including medicinal and ornamental plants;
**5.** to develop and modernize animal and fisheries production;
**6.** to emphasize the socio-economic development of rural areas;
**7.** to modernize marketing and agro-industries (food chain);
**8.** to optimize agricultural research, extension service, and technology transfer;
**9.** to encourage investment in agriculture and agriculture-related projects;
**10.** to increase the competitiveness of Egyptian agricultural products in local and foreign markets;
**11.** to enhance capacity building related to human resources in agriculture; and
**12.** to promote the role of communications and information technology in agricultural development.

## Effect of Climate Change on Egypt

The increase in Egypt's average temperature has accelerated during the past two decades. Egypt will experience a higher level of warming than the world average by 2100 and face a significant increase in energy demand due to more frequent extreme heat events coupled with urbanization and population growth. Temperatures in 2081–2100 could be around 2.5°C higher in low-emissions scenarios and around 6°C in high emissions scenarios (IPCC climate scenarios ssp1-2.6 55p5-8.5) (IPCC, 2022). The increase in SLR will impact electricity installation and refiners in the delta and north coast (IEA, 2023).

## Strengthening Coping and Adaptive Capacity through Science and Technology (S&T)

The major elements for coping and adaptive capacity are:

- new genetic make-up
- new agro-management techniques

This will require intensive knowledge and enhanced capacity building through an intensive program for human resources development (Fig. 14.4).

Leveraging existing assets, the focus is on pioneering new frontiers with greater emphasis, augmented partnerships, and an elevated level of integration. This approach encourages transdisciplinary and interdisciplinary pathways, fostering knowledge exchange and dissemination.

Enhancing collaboration with top-tier universities across the USA, Europe, Asia, and developing nations will be crucial, particularly in the realm of R4D.

The IPCC defines CC as a change in the mean value and/or variability of the climate's properties that persists for an extended period, usually for decades or longer. Agriculture is extremely vulnerable to CC: increasing temperatures and declining precipitation over arid and semi-arid regions are likely to reduce yields for several primary crops in the next two decades (Kummu *et al.*, 2021; Ortiz-Bobea *et al.*, 2021).

The intensity and distribution of pests and disease outbreaks may also become increasingly

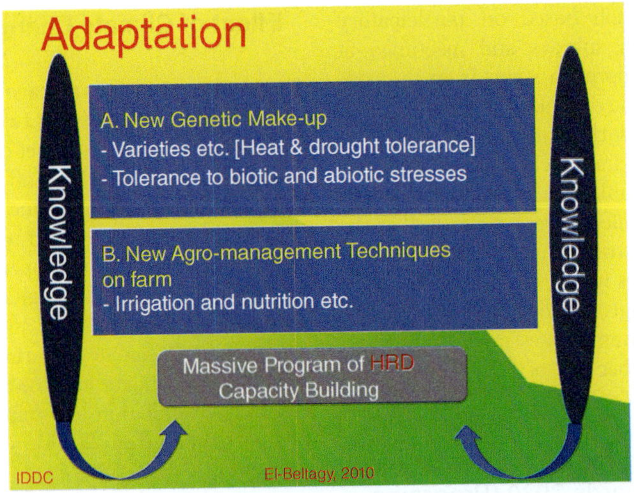


**Fig. 14.4.** Main elements of knowledge-based adaptation.

unpredictable, with serious impacts on agricultural productivity (Chapman, 2012; Oszako and Nowakowska, 2015; Omran and Negm, 2020).

While the overall impacts of CC on agricultural systems are expected to be negative, the effects may vary both in type and magnitude across geographical areas. Developing countries in dry areas are likely to be most affected, not only because of their predominantly low input, rainfed cropping systems, which rely on somewhat regular weather patterns, but also because of their more rapid population growth, which places greater pressures on agricultural production and creates more serious food insecurity risks (Battisti and Naylor, 2009; Lotze-Campen, 2011; Turner and Meyer, 2011; IPCC, 2014; Malhi *et al.*, 2021).

Adaptation to CC is the practice and process of adjusting to climate-induced adverse effects. In agriculture, adaptation to CC is about farmers' and other stakeholders' responses to environmental disturbances that affect their cropping systems. This may involve a broad array of alterations from livelihood and agronomic strategies to policy changes. Among those strategies that focus on the production system itself, CC adaptation strategies are commonly divided into two pathways and an overarching element:

**Agronomic management** (which tends to be short term) strategies encompass changing cultivation practices (timing or location of cropping activities, techniques of soil management, weed/pest/disease management) and the adoption of new varieties or shifting to alternative crops (figs, olives, dates and pomegranates) (Fig. 14.5) or crop combinations from plant species known to grow in the dryland to be utilized as food, feed, or fodder (Becerra and Garcia, 1994; Fucikovsky, 1994; Gardufio, 1994; Mazhar *et al.*, 2002).

**Genetic improvement** (a long-term approach) involves the development and adoption of new, better-adapted varieties of the crop of interest. While farmers have always conducted genetic improvement within their informal seed networks, a crucial role has been played over the last century by public (and, increasingly, private) plant breeding. Indeed, 'modern', scientific plant breeding has contributed to massive and rapid yield increases in many crops, as well as to increased tolerance to a variety of biotic and abiotic stresses; however, with more severe and frequent challenges from aggravated CC, plant breeders are being called to place extra efforts in improving and accelerating the tools and working strategies they use, to provide farmers with adapted varieties, particularly in more climate-vulnerable developing countries (Harlan *et al.*, 1973; Zohary, 2004; Mercer and Perales, 2010; Turner and Meyer, 2011; Vigouroux *et al.*, 2011; Ceccarelli, 2015).

**A massive program for human resource development** to cater for and provide knowledge generated through research to cope with the dynamic change of agro-ecological zones.

**Fig. 14.5.** Alternative crops suitable for dry areas. a) Figs (image: @Silverije, CC BY-SA 4.0); b) olives (image: @SABENCIA Guillermo César Ruiz, CC BY-SA 4.0); c) dates (image: @Mmelouk, CC BY-SA 4.0); d) pomegranates (image: @Kritzolina, CC BY-SA 4.0).

## Adaptation of Genetic Resources to CC

The West Asia North Africa (WANA) region, including Egypt, stands out as a cradle for globally significant crops. Three of Vavilov's eight identified centers of crop origin are in this region, with crops like barley, wheat, lentil, and vetch having been domesticated in the Near East. The International Centre for Agricultural Research in the Dry Areas (ICARDA) has highlighted the persistence of wild progenitors and relatives of these crops in the region, emphasizing their vulnerability to desertification (El-Beltagy *et al.*, 2005). While fencing can protect biodiversity, it is not a panacea, especially when soil seed banks are compromised. Grazing, if managed judiciously, is an integral part of rangeland management (El-Beltagy, 2008a, b).

In Egypt, the implications of CC for genetic resources in agriculture are multifaceted. Genetic resources for food and feed, both for livestock and aquaculture, are not only at risk but are also integral to mitigating and adapting to CC (Cheng *et al.*, 2022; Kovak *et al.*, 2022; Sandström *et al.*, 2022). The Commission on Genetic Resources for Food and Agriculture (FAO) serves as an intergovernmental forum for policy development, playing a pivotal role in addressing the challenges posed by CC to global food security (FAO, 2015).

Desertification and CC pose significant threats to plant biodiversity. Efforts to counteract this include storing germplasm in arctic permafrost facilities, an initiative driven by Scandinavian countries and the Global Crop Diversity Trust (El-Beltagy, 2008a, b).

The arid drylands face immense challenges due to CC, prompting the question of how science and technology can enhance the capacity of inhabitants to cope. Emerging tools such as biotechnology, genetic engineering, and gene editing offer substantial promise. The Agriculture Genetic Engineering Research Institute (AGERI), established in 1990 in Egypt, exemplifies the commitment to leveraging innovative technologies for CC adaptation (El-Beltagy, 2002; Madkour, 2003; El-Beltagy, 2008a, b; El-Beltagy and Madkour, 2012).

Egypt's commitment to the International Treaty on Plant Genetic Resources for Food and Agriculture and the establishment of the National Gene Bank (NGB) in 2004 underscore its dedication to sustainable genetic resources.

Water stress caused by drought and salinity is the most important abiotic factor limiting plant growth and crop productivity in arid lands. Therefore, enhancing osmotic stress tolerance in crops grown under such conditions could provide a solution to this problem (Madkour, 2003).

Active research at AGERI at the Agricultural Research Center in Egypt has led to the production of biotic and abiotic stress-tolerant crops, examples of which include: drought- and salt-tolerant local wheat varieties, tomato plants resistant to TYLCV virus, virus-resistant squash and cucumber, insect-resistant potato varieties, corn plants resistant to stem borers, and insect-resistant Egyptian barbadense cotton. Moreover, AGERI developed, through its own international patent, a biopesticide formulation to combat different pests attacking cotton, potatoes, and other crops.

High genetic diversity exists for the components of abiotic stress response in genetic resources collected from the dry areas. This resource has to be properly evaluated and documented for future use. In addition to field evaluation in stressed environments, where the individual stress effect cannot usually be separated from the multiple-stress response, germplasm characterization in controlled conditions using morpho-physiological criteria may reveal new evidence on different components of abiotic stress response. Moreover, advances in molecular biology open new opportunities for understanding the physiological and genetic mechanisms involved in plant response to different stresses and provide tools for molecular screening and genetic engineering applications in crop improvement for increased abiotic stress tolerance. A thorough evaluation and exploration of genetic diversity existing in the crop wild relatives, primitive forms, and landrace collections may be a strategic first step in germplasm development in the extensive arid drylands.

For the structural characterization of the germplasm collection, microsatellite-based markers today are most promising for genome analysis in cereal crops and leguminous crops. Microsatellites, also known as simple sequence repeats (SSRs), either genomic or EST-derived SSRs, are ideal DNA markers for population studies and genetic mapping because of their abundance, high level of polymorphism, dispersion throughout diverse genomes, ease of assay by polymerase chain reaction (PCR), and ease of dissemination of the technology among laboratories.

High-throughput genotyping has become an integral part of most advanced breeding programs. Single nucleotide polymorphisms (SNPs) serve as biallelic genetic markers, and they are used because of their high occurrence in genomes. Technological improvements make the use of SNP markers attractive for high-throughput EST mapping, the integration of genetic and physical maps, and use in marker-assisted breeding (Baum *et al.*, 2003; El-Beltagy and Madkour, 2012).

Development in large-scale, high-throughput technologies and robotics now allows researchers to simultaneously profile vast numbers of different genes or proteins. Microarrays, functional genomics, and gene editing will permit researchers to screen tens of thousands of genes in multiple tissues under myriad experimental conditions and would therefore improve the efficiency of breeding for abiotic stress (Madkour, 2003; Madkour, 2006; El-Beltagy and Madkour, 2012).

To safeguard plant genetic resources and use them optimally to help cope with CC, greater emphasis needs to be given to:

- enabling gene banks to respond to increased demands for germplasm that can be used in CC adaptation by including distinct characteristics in screening processes;
- reviewing breeding strategies and priorities crop-by-crop and region-by-region to ensure that crop improvement programs will be relevant to the challenges the world will be facing when the products are ready for release at the end of the crop improvement cycle;

- consolidating collections of wild species, including crop wild relatives, because of the increased likelihood that narrowly adapted and endemic species will become extinct;
- ensuring that collections give sufficient emphasis to stress-adapted genetic material that can contribute to adaptation to CC; and ensuring that policies facilitate international exchange of the genetic resources needed to adapt farming systems to shifts in the distribution of climatic agro-ecological zones;
- coping with uncertainty, recognizing that there will be shifting baselines as further assessment of CC is done;
- reviewing and strengthening policies to promote dynamic seed systems, including the establishment of community-based seed enterprises at strategic locations within the country and implementing measures to enable the exchange of seeds between farmers;
- increasing investment in research and information systems, which is absolutely essential (El-Beltagy and Madkour, 2012; FAO, 2015).

The Crop Trust was established in 2004 by FAO and Biodiversity International on behalf of CGIAR with the sole mission to safeguard the world's crop diversity for future food security and make it available for global use, forever and for the benefit of everyone through the Global Genebank Partnership (GGP), which is intended to provide a platform for partners to strengthen a cost-efficient global system that can provide scalable solutions and genetic resources data, which will empower plant breeders nationally and globally to continuously introduce new genetic resources for ever-changing agro-ecology. An early warning system for the impact of CC on the shift of agro-ecology will require continuous effort to match new genetic resources suitable for this changing agro-ecology.

## Agro-management Techniques

This section encompasses advanced tools aimed at optimizing water-use efficiency, such as the National Program for Modernizing Farm Irrigation. It also covers the intensification of production, protected agriculture, urban agriculture, and soilless agriculture. Additionally, various agro-management techniques, including raised beds and the utilization of smart agriculture and precision methods, are highlighted.

## Water Resources

The National Program of Optimization of Water Resources was initiated in 2008 (El-Beltagy and Abou-Hadid, 2008) and included as a major program of SADS 2030, concentrating on optimizing water productivity and production per unit area/m$^3$ of water. Components of this program include:

**a.** the National Program for On-Farm Water Management, which aimed for a 5.0-million-acre target area, saving an estimated 13 billion m$^3$ per year, which could reclaim 3.5 million acres of new land;
**b.** implementation of protected agriculture, roof gardening for urban agriculture, and soilless culture; and
**c.** utilization of Agriculture 4.0 tools, precision agriculture, e-agriculture, and various AI programs (Wang *et al.*, 2023).

### National program for on-farm water management

The predominant traditional surface irrigation system in Egyptian agriculture, with an efficiency of 45%, exerts considerable strain on the agriculture drainage system. Agriculture drains and irrigation canals occupy 10% of the agricultural area, exceeding 500,000 acres in the Nile delta and valley. Enhancing on-farm water systems promises water savings, additional land, and improved agricultural productivity.

Based on the prices of 2009, the costs of this project amount to 157 billion Egyptian pounds in total, of which 50 billion is to improve irrigation in the existing lands, 60 billion to reclaim 3 million acres of new land, in addition to 2 billion pounds for extension, training, and research programs, in partnership between all parties, including the government, the private sector, farmers, and grants and loans from international institutions.

Anticipated returns within ten years: approximately 200 billion Egyptian pounds covering costs and yielding high economic returns, coupled with social benefits like job creation and enhanced rural living standards.

### Project implementation

Grounded in experiences from the preceding two decades of the 20th century and international insights, the project anticipates various outcomes, including:

**1.** adding half a million acres for open irrigation and drainage canals;
**2.** creating enough jobs that will impact the livelihood to sustain 15–20 million citizens, to serve over 3 million additional acres by 2030;
**3.** enhancing the properties of old lands and elevating their productivity;
**4.** boosting agricultural intensification and modernization in the Nile valley;
**5.** improving health and minimizing endemic waterborne diseases among farmers;
**6.** reducing water losses in agricultural drainage systems;
**7.** reducing energy consumption for pumping irrigation and drainage; and
**8.** preserving the environment by curbing water pollution and encouraging civil and industrial facilities to upgrade sewage and industrial systems.

## Project Implementation at the Ministry of Agriculture

The project was funded in 2009 by the World Bank, IFAD, and Agence Française de Développement (AFD) and targeted Mesqa and Marawa (third- and fourth-level canals in the Nile river system) across five governorates (pilot 200,000 acres). Delays occurred due to socio-political challenges in Egypt from 2011 to 2014.

Expected results:

- 25% reduction in irrigation water usage in the project areas;
- development of a pilot demonstration model for horticultural lands in the valley and delta;
- a comprehensive human resource program to implement on-farm irrigation systems nationwide; and
- 15–20% increase in farm productivity and reduced operational costs.

### Protected agriculture (since 1986)

Egypt with limited water resources opted to increase the cropping areas through intensification and therefore, starting from the 1980s, initiated several activities related to protected agriculture research and development. Three main centers were involved in this effort: the Ministry of Agriculture, Ain Shams University, and Kafr El-Sheikh University. At Ain Shams Dryland Lab., an active research program was developed with Prof. A.R. Persson, at the Agriculture University of Norway AS. This joint program under the name of EGNO was developed from 1974 to support human resource development and physical infrastructure within the context of a major research and development program and extension (El-Beltagy and Persson, 1986; Maksoud and Persson, 1986), and these activities had further support from USAID with several programs: CALAR1, CALAR2, NARP, and ATUT.

The team of Ain Shams Arid Land Lab. had cooperation from the following institutions: Glasshouse Crop Research Institution (GCR), UK; Naaldwijk, The Netherlands (Kooistra *et al.*, 1975; Cooper, 1980); University of Arizona, Tucson Arid Land & Environmental Lab., Salinity Lab.; USAID; University of California; Wye College, UK; University of London; University of Wales, Aberystwyth, UK; and the International Centre for Advanced Mediterranean Agronomic Studies (CIHEAM), Paris.

A national protected agriculture committee was formed in 1987 with different national working teams in various specialized domains including agro-management techniques, irrigation mechanization, automation, basic infrastructure, pest control, breeding programs, marketing and economic studies, with support from the Deputy Prime Minister of Agriculture. Within this context, cooperation with UNDP and FAO was established.

Five Nile delta centers of excellence trained producers and university graduates in protected agriculture, supporting human capacity. Over 3000 specialists graduated from this program in all aspects related to protected agriculture.

The program introduced modern protected cultivation practices to Egypt and the Middle East, including greenhouse improvements to suit projected climate changes. Comprehensive training manuals and extension booklets were provided along with research publications (El-Beltagy and Persson, 1986) on root media (El-Beltagy *et al.*, 1993), efficient hydroponic techniques like nutrient film (Abou-Hadid *et al.*, 1988b), and sandponics (Ibrahim *et al.*, 1989) for high-value crops, reducing agricultural water consumption (Abou-Hadid *et al.*, 1987; Abou-Hadid *et al.*, 1988a; Moustafa *et al.*, 1998; Abou-Hadid, 2013), and impact on stress (Atta-Aly *et al.*, 1993).

## Soilless culture as a technique for intensive agriculture

Soilless culture grows plants in nutrient solutions with or without substrates like sand, gravel, vermiculite, rockwool, peat moss, or sawdust for root support. Open systems discard surplus solutions while closed systems recover, replenish, and recycle them (Zayed *et al.*, 1989; Abou-Hadid and El-Behairy, 1999).

Hydroponics are widely used for flowers, foliage, bedding, and high-value food crops, with vegetables being the most common soilless crops in Egypt (El-Behairy *et al.*, 2001a, b). Soilless culture enables greenhouses to overcome pending methyl bromide bans (Schnitzler *et al.*, 2006).

Soilless techniques have progressed as intensive horticulture production systems, now optimized by AI tools like neural networks, IoT, and Big Data linked to markets (El-Beltagy *et al.*, 2021).

## Urban agriculture (green roof)

Rapid Egyptian population growth pressures cities, with buildings and facilities replacing gardens and reducing green space. This increases air pollution from cars, factories, and paving while removing natural plant purification. Respiratory illnesses, immune suppression, and mental health problems affect public behaviors because of overcrowding, low oxygen, and lack of vegetation (El-Behairy and Abou-Hadid, 2008).

Green roofs can compensate for lost urban green space. Lightweight, non-leaking systems replace traditional trash-filled rooftops that compromise building aesthetics and health. Soilless culture is ideal. Benefits include:

- air purification – 1 $m^2$ of green roof removes 100 g of pollutants annually;
- increased oxygen and lower carbon dioxide – 1.5 $m^2$ provides one person's annual oxygen; photosynthesis consumes carbon dioxide and emits oxygen;
- cleaner roofs without pests like mosquitoes, flies, mice, cockroaches, and lizards that impact health; and
- reduced urban heat island effects – green roofs absorb less heat than concrete and asphalt, decreasing temperatures by 8–10°F, balancing ecosystems, and lowering smog and dust.

## Coping and adaptive capacity

The Ministry of Agriculture has been supporting the study of the effects of the phenomenon of CC for some 30 years. Egypt began to study the sensitivity of the agricultural sector to future CC and explore how to mitigate and adapt to the negative effects. This includes:

**1.** Restructuring and initiating new institutions. The SADS strategy and its business plan indicate the need to restructure the agriculture institutions, including the Ministry of Agriculture, cooperatives, and agri-business to optimize the linkages and harness the synergy of the system to improve the food chain and the bio-economy and resilience of Egypt.
**2.** Transforming and modernizing the agriculture curriculum at universities and agriculture schools.
**3.** Introducing new enabling legislation.

During these three decades, research committees and entities were established:

- The Central Laboratory for Agricultural Climate (CLAC): established to research climate–agriculture interactions, CLAC, founded by Abou-Hadid in 1999, focuses on current and future conditions' impact on agriculture and food security. It provides insights into accurate irrigation scheduling, early pest and disease forecasting, and optimal sowing dates. The real-time weather-data system spans 30+ agro-meteorological stations (Abou-Hadid, 1999), supporting various models and knowledge-based systems. Information dissemination occurs through mass media and a dedicated web page. Integration with a geographical information system (GIS) enhances land utilization planning (Medany *et al.*, 2001; Abou-Hadid, 2006).
- Climate Change Information Center (CCIC): founded in 2009, CCIC gathers and disseminates research findings on CC's impact on Egypt's agriculture. It offers technical recommendations for mitigating GHG emissions and explores economic implications, such as reduced yields and increased pest infestation, and reviews and broadcasts them through its website on the internet.

## Linking Farmers to Markets

A major program in the Sustainable Agriculture Development Strategy (SADS) towards 2030 involves restructuring the Egyptian food marketing chain to increase agricultural production and improve small farmers' livelihoods while minimizing food losses. The program encourages small farmers and producers to form regional cooperatives and companies to collectively market their production. Small landholders with less than one acre represent about 44% of Egyptian farmland owners. These cooperatives have shifted power back to farmers, who, by avoiding intermediaries, have increased revenues.

The program is expected to transform Egyptian agriculture with the following outcomes: (i) increased farm gate prices through enhanced organization, negotiation, and access to information by smallholder farmers; (ii) reduced production losses through financing for new technologies, marketing infrastructure (collection and cooling centers, packhouses, refrigerated trucks, etc.), and market access; (iii) increased assets and employment from enhanced on- and off-farm job opportunities. The program enhances farmers' capacities for organization and collective marketing through establishing farmer groups and associations. It builds skills for interpreting and responding to market signals, expands domestic and international market access through links with traders, processors, and exporters, and provides appropriate financial services for rural and agricultural enterprise growth across the value chain. It will also pilot innovative opportunities like inland fisheries financing, milk collection, and sales, etc.

Two successful examples are the El-Shams and SALASEL projects, which consolidated small landholdings and prevented fragmentation by coordinating cooperative crop planting over larger areas and marketing the product jointly.

El-Shams is a four-year agri-business project that started in September 2003, linking small landholding farmers and their associations to private sector agri-businesses, larger trade associations, traders, exporters, and investors. This helped increase high-value horticultural exports, domestic sales, income, and agricultural and non-farm jobs, targeting 24,000 small and medium farmers in upper Egypt. The project, financed by USAID and CARE Egypt, aimed to increase on- and off-farm rural jobs and income in upper Egypt by building small and medium farmer capacities for improved horticultural production, processing, and marketing (Abel Al *et al.*, 2007).

SALASEL is a pro-poor horticulture value chain in upper Egypt project, funded by MDG-F and implemented from 2010 to 2013 by four UN agencies: UNDP, UNIDO, ILO, and UN WOMEN (Fig. 14.6). It supported small upper Egyptian farmers to create jobs, increase incomes, and apply sustainable solutions using advanced technologies, intensive human resource development, and advanced AI-like computer expert systems. Two companies were established, and new crops were introduced in upper Egypt (Willot, 2012; Caldeyro and Khafagy, 2013; MDG Fund, n.d.; MDG Fund Egypt Joint Programme Fact Sheet, n.d.).

**Fig. 14.6.** SALASEL activities.

## Transformation and Modernization of Agriculture Curriculum at Universities and Agriculture Schools

In the 2019 amendment of the Egyptian Constitution of 2013, Article 21 states that the government is committed to spending not less than 2% of GDP on education and will encourage scientific research with not less than 1% of its GDP with a future proportional increase. In 2018, the Committee of Agriculture of the Supreme Universities' Council started revising, updating, and modernizing the curriculum of the agriculture college in Egypt with an emphasis on encompassing the new scientific innovations and the outcome of the latest research and technology. Several updated programs were introduced (i.e. related to agriculture intensification, protected agriculture, hydroponics, urban agriculture, biotechnology and bioengineering, food safety, food nutrition, pest control, seed production, inland and sea fisheries (fishery sciences), post-harvest technology, plant breeding for adverse environmental conditions, medical plants production and processing, organic agriculture/conservation agriculture, agriculture bio-system engineering program, etc.). These programs were linked with a massive plan for human resource development in collaboration and cooperation with universities in OECD countries, university staff member exchange, and joint research programs with international institutions. SADS has emphasized the importance of linking the modernization of technical agricultural schools within the context of the effort to modernize the whole sector of agriculture. The implementation of this program was going to be supported mainly by the government and private sector/cooperatives.

## Enabling Policy and Legislation

Enacting legislation is essential to strengthen existing institutions and establish new ones. Furthermore, international cooperation tied to a community of scientific practice across various domains will be instrumental in refining knowledge acquisition processes.

Several pieces of legislation have been suggested, endorsed, and enacted into law:

The following endorsed presidential decisions/laws:

1. **Agriculture Solidarity Law:**
    - created a mechanism for disbursing compensation for environmental disasters impacting the agricultural sector
    - provided crop insurance against biological risks that could sharply reduce the productivity of insured crops.
2. **Law amending provisions of the 1980 Agricultural Cooperation Law (2014 Presidential decree):**
    - allowed agriculture cooperatives to jointly finance projects with other legal entities for member benefit
    - gave cooperatives the right to establish joint-stock companies, enabling production and marketing enterprises needed for agricultural development
    - increased cooperative society capital to reach the economic sizes of business enterprises.
3. **2014 Health Insurance Law for Peasants:**
    - brought small farmers and landless under the national health insurance system
    - enabled adequate health services for small farmers and landless
    - provided treatment and rehabilitation services to farmers for accidents.
4. **2015 Law establishing the Contractual Farming Center:**
    - enabled farmers to contract future products at preset prices, avoiding unexpected price fluctuations
    - developed standard product specifications, increasing export capacity
    - created a mechanism to register contracts between producers, marketing, export, and manufacturing companies
    - created a dispute resolution mechanism between farmers and purchasing companies
    - used contracts as collateral with the Agricultural Credit Bank.
5. **2015 Law amending intellectual property rights protection:**
    - preserved plant breeder rights, encouraging new high-productivity plant variety research
    - produced varieties responding to CC and market needs, facilitating access to foreign plant strains and varieties
    - encouraged researchers, research centers, and universities to expand plant breeding and propagation biotechnology applications by preserving rights.

The following Laws are prepared and under review:

1. **Law on fishing and fish farms regulation:**
    - aims to preserve and develop Egyptian waterbody fish stocks
    - aims to expand marine fishing areas to 200 nautical miles exclusive economic zone from current territorial waters
    - aims to confront lake encroachment and pollutant dumping into waterways and lakes and ban lake drainage
    - aims to regulate fish-farm establishment, providing investor stability
    - aims to integrate plant culture and fish farming in desert lands by using saline groundwater for fish farming before plant culture.
2. **Farmers' pensions law (with the Ministry of Social Solidarity):**
    - aims to secure old-age income with an appropriate pension level at age 55
    - aims to determine minimum pension to ensure decent farmer living standards in retirement.
3. **General farmer union law (with State Council and Ministry of Justice):**
    - aims to cater to farmer interests and provide necessary assistance
    - aims to coordinate various agriculture-related insurance and social coverage funds
    - aims to adopt a national farmer illiteracy eradication project
    - aims to discuss laws related to agriculture before legislative approval

- aims to engage research centers to encourage the use of research results for agricultural sector development.

4. **Law on agricultural land encroachment:**
   - aims to preserve agricultural lands for agriculture, banning non-agricultural use
   - aims to protect the agricultural environment from pollution due to non-agricultural uses
   - aims to prevent building randomness, restricting construction to approved urban areas
   - aims to maintain agricultural production and self-sufficiency
   - aims to increase penalties via land withdrawal or imprisonment plus fines.
5. **Law on water use in field irrigation:**
   - aims to determine water quantities based on crop needs with ascending tax based on water use per approved tables for each crop.
6. **Dairy producers' union law:**
   - aims to support the dairy sector and preserve milk safety from producer to factory to consumer.

## Resilience and Adaptation

Climate change will impact Egyptian agriculture via increased temperature fluctuation of precipitation, reduction in yield of crops, animals, and fisheries, SLR, soil salinization, and hence induce desertification. This will disrupt food security and the livelihood of the rural inhabitants and induce outmigration, and hence it will cause threats to the economy. Major efforts should be taken to enhance both coping and adaptive capacity in Egypt through massive reform for policies, institutions, legislation, and clear outlining of the role of the different stakeholders in implementing agriculture-related CC adaptation programs. Identifying capacity-building needs is essential for facilitating education, training, and awareness raising. It also supports institutional strengthening and sustainability of skills at the national level.

Early-warning systems are a vital requirement for vulnerable agro-ecology and therefore the predicted dynamic impact of CC has to be assisted, this will require navigation through scales, global, regional to local (different agro-ecological zones in Egypt).

Enhancing the human and the physical infrastructure to fulfill, monitor, implement, and communicate with the farming community to plan for the adaptive and coping measurement that is required. Involving farmers and rural inhabitants, drawing on traditional knowledge, is crucial. Bridging the gap between science, local knowledge, and stakeholder awareness, including among farmers, is essential for more effective planning for adaptation (Omar *et al.*, 2015; Oranu *et al.*, 2018).

A recent study by Abou-Ali *et al.* (2023) highlighted the vital link between climate action and sustainable development. Egypt demonstrated commitment by hosting COP27, finalizing its National Climate Change Strategy, and updating its nationally determined contributions (NDCs) (Arab Republic of Egypt, 2022). However, Egypt should not only manage climate risks but also seize opportunities and encourage private sector engagement for a green transition, leveraging experiences from other countries.

Egypt's vulnerability to adverse climate impacts is evident across sectors like coastlines, agriculture, food security, water resources, poverty, inequality, health, and infrastructure. Aligning the NDC with the Sustainable Development Goals (SDGs) presents an opportunity to expedite progress on both agendas simultaneously (Mohieldin *et al.*, 2023). While Egypt's climate plans intersect with all 17 SDGs, practical interlinkages between the NDC and SDG1 need reinforcement through a holistic strategy.

Addressing these challenges necessitates urgent adaptation to climate impacts. Egypt has an opportunity to transition towards a sustainable, inclusive, and resilient economy by prioritizing investments that create jobs, protect biodiversity, boost resilience, and advance decarbonization. Establishing a comprehensive environmental monitoring system or database is crucial for effective SDG target implementation, requiring enhanced data acquisition, coordination, and legislative reforms for stronger governance (Hassan and Amin, 2022).

In Egypt, the agriculture sector is highly dependent on data quality and quantity to adapt and build resilience to CC. It is therefore of high priority to develop early-warning systems to better predict seasonal and daily changes in rainfall and flows in the Nile basin, both for improved water management practices and to improve preparedness for floods or extended dry seasons. Mapping Egypt's agricultural products is also important for more effective land use and future resource management.

## Implementation Capacity and Pathways to Egypt's Green Transition

### Implementation capacity

The implementation of Egypt's green transition requires a robust capacity that encompasses various tools, including adoption subsidies, carbon pricing, intellectual property legislation, direct financing, and regulatory instruments (IPCC, 2022). Building this capacity is fundamental for formulating policies, accessing finance, integrating adaptation and mitigation into sectoral planning processes, and fostering knowledge transparency. Identifying capacity building needs is essential for facilitating education, training, awareness raising, and peer-to-peer learning. It also supports institutional strengthening and sustainability of skills at the national level. However, the current fragmentation and coordination issues among ministries necessitate better coordination in the execution of CC actions and policies.

Moreover, while decentralization is crucial, empowering local-level institutions is equally important, offering a nuanced approach that combines the advantages of both centralization and decentralization.

### Pathways to Egypt's green transition

Egypt's green transition is pivotal in three key sectors: agriculture, the least contributor to CC; power; and transport, the latter two responsible for about 65% of the country's GHG emissions (Monforti *et al.*, 2021). The agriculture sector presents opportunities for resilience and adaptation, while the power and transport sectors offer avenues for mitigation.

Egypt's agriculture sector faces vulnerabilities such as heatwaves, SLR, soil salination, rainfall fluctuations, and desertification. These challenges impact food production, farmer incomes, and food security, posing threats to the economy, health, and well-being. To address these, a CC adaptation policy framework is essential, outlining principles, actions, roles, and financing recommendations for engaging stakeholders in implementing agriculture-related CC adaptation programs.

However, effective planning for climate hazards and assessing implementation progress is hindered by limited and inadequate data availability, low data accuracy, and limited resources. Public awareness remains a challenge, and weak implementation of laws and policies persists due to limited institutional capacities. Overlapping laws and environmental standards coexist with inadequate mechanisms to integrate environmental dimensions into sectorial plans. Limited funding and private sector participation remain major obstacles (Abou-Ali *et al.*, 2023).

## The Role of the Private Sector

The private sector in Egypt plays a dual role in climate action: as a provider investing in modern technologies, sustainable practices, and adaptation services, and as a purchaser requiring adaptation services for sectors like production systems and infrastructure. Adopting a holistic approach that addresses adaptation needs, accelerates mitigation investments, and supports Egyptians' transition is imperative. Seizing the opportunity for progress requires urgent action, leveraging available elements for greater advancement (Abou-Ali *et al.*, 2023).

## Major Elements of Adaptation

The core of the CC adaptation framework involves scaling up best practices in technologies, research, policies, capacity building, and financing to reduce vulnerability and enhance resilience. The agriculture sector should focus on easy, low-cost adaptation measures aligned with local conditions and sustainable development requirements.

To improve scientific capacity, Egypt should seek technical support from academics and international institutions, catalyze investments, and engage the private sector in developing and deploying earth observation sensors and IoT technologies. Data science, machine learning, and AI tools can enhance data transparency, standardization, and completeness. R&D investment in agriculture should be a top priority, aiming for a 10% share in national investments (Egypt Ministry of Agriculture, 2009).

## Conclusion

It is clear that the nation faces a critical juncture. Egyptian agriculture, pivotal to the country's economy and food security, is under significant strain due to CC, limited natural resources, and rapid population growth. The threats of SLR, diminishing water supply from the Nile, and soil salinity are compounded by a burgeoning population, making sustainable agricultural practices and innovative solutions more vital than ever.

To navigate these challenges, Egypt has embarked on a transformative journey, integrating scientific advancements and sustainable practices into its agricultural sector. The Sustainable Agriculture Development Strategy Towards 2030 is a testament to this commitment, focusing on optimizing water use, enhancing crop and livestock genetics, and strengthening market linkages. The strategy emphasizes the critical role of human resource development, institutional support, and robust legislation in building a resilient agricultural sector.

The integration of smart-agriculture technologies, precision farming, and advanced genomics, along with a focus on agro-industry and value-chain development, are pivotal steps towards sustainability. The strategy's alignment with the Sustainable Development Goals and the potential leveraged from the outcomes of COP27 demonstrate Egypt's proactive stance in addressing global environmental challenges.

In conclusion, Egypt's path forward in agriculture is one of resilience, innovation, and adaptability. By harnessing the power of science and technology, fostering institutional reforms with an emphasis on human resource development, and engaging in comprehensive policy development, Egypt aims not only to overcome its current challenges but also to secure a sustainable, prosperous future for its agricultural sector and its people. This journey, while fraught with challenges, holds the promise of transforming Egyptian agriculture into a model of sustainability and resilience in the face of global CC.

**Fig. 14.7.** Burial chamber of Sennedjem. Scene: plowing farmer. C.1200 BCE, from the Tomb of Sennedjem. (Wilkinson and Marsha, 1983)

## References

Abd-Elmabod, S.K., Fitch, A.C., Zhang, Z., Ali, R.R. and Jones, L. (2019) Rapid urbanisation threatens fertile agricultural land and soil carbon in the Nile delta. *Journal of Environmental Management* 252, 109668. DOI: 10.1016/j.jenvman.2019.109668

Abd-Elmabod, S.K., Jiménez-González, M.A., Jordán, A., Zhang, Z., Mohamed, E.S. *et al.* (2022) Past and future impacts of urbanisation on land surface temperature in Greater Cairo over a 45 year period. *The Egyptian Journal of Remote Sensing and Space Science* 25(4), 961–974. DOI: 10.1016/j.ejrs.2022.10.001

Abegaz, N.T., Tsidu, G.M. and Arsiso, B.K. (2023) Spatiotemporal variability of the Lake Tana water quality derived from the MODIS-based forel-ule index: the roles of hydrometeorological and surface processes. *Atmosphere* 14(2), 289. DOI: 10.3390/atmos14020289

Abel Al, A., Hill, B., Patterson, G., Schar, H. and Siddik, I. (2007) EL SHAMS Final Evaluation 2003–2007. Care Egypt. Cairo. Available at: https://www.careevaluations.org/wp-content/uploads/evaluations/elshams-final-evaluation.pdf (accessed 22 July 2024).

Abou-Ali, H., El-Ayouty, A. and Mohieldin, M. (2023) Climate action in Egypt: challenges and opportunities. In: Bhattacharya, A., Kharas, H. and Mcarthur, W.J. (eds) *Keys to Climate Action: How Developing Countries Could Drive Global Success and Local Prosperity*. Working paper #180-1. The Brookings Institution, Washington, DC, pp. 57–84.

Abou-Hadid, A.F. (1999) The use of climatic data in integrated crop management. *Proceedings of the 1st Congress on Recent Technologies in Agriculture.* Cairo University, Facuality of Agriculture, 27–29 November.

Abou-Hadid, A.F. (2006) Assessment of impacts, adaptation, and vulnerability to climate change in North Africa: food production and water resources. In: *A Final Report Submitted to Assessments of Impacts and Adaptations to Climate Change (AIACC).* Project No. AF 90. International Start Secretariat, Washington, DC. Available at: http://www.start.org/Projects/AIACC_Project/Final%20Reports/Final%20Reports/FinalRept_AIACC_AF90.pdf (accessed 22 July 2024).

Abou-Hadid, A.F. (2013) Protected cultivation for improving water-use efficiency of vegetable crops in the NENA region. In: *Good Agricultural Practices for Greenhouse Vegetable Production in the South and Eastern European Countries.* Plant Production and Protection Paper 230. FAO, Rome, pp. 137–149.

Abou-Hadid, A.F. and El-Behairy, U.A. (1999) Soilless culture, a tool for maximizing the utilization of arid land resources. In: Choukr-Allah, R. (ed.) *Protected Cultivation in the Mediterranean Region.* CIHEAM/IAV Hassan II, Paris.

Abou-Hadid, A.F., El Beltagy, A.S. and Burrage, S.W. (1987) Comparison of hydroponic systems for tomato production in Egypt. *Egyptian Journal of Horticulture* 14(2), 119–127.

Abou-Hadid, A.F., El Beltagy, A.S. and El Saied, H.M. (1988a) Potential evapotranspiration for different regions and water consumptive use patterns under plastic house conditions in Egypt, *Egyptian Journal of Horticulture* 15(1), 1–12.

Abou-Hadid, A.F., Zayed, A.M., El Behairy, U.A. and El Beltagy, A.S. (1988b) A comparison between nutrient film technique (NFT) and soil for tomato production under protected cultivation in Egypt, *Egyptian Journal of Horticulture* 16(2), 111–118.

Arab Republic of Egypt (2022) Egypt's first updated nationally determined contribution. In: United Nations Framework Convention on Climate Change (UNFCCC). Available at: https://unfccc.int/sites/default/files/NDC/2022-07/Egypt%20Updated%20NDC.pdf.pdf (accessed 22 July 2024).

Atta-Aly, M.A., El-Beltagy, A.S. and Saltveit, M.E. (1993) Effect of salt stress on the vegetative growth and development of tomato plants. *Acta Horticulturae* 323, 401–410. DOI: 10.17660/ActaHortic.1993.323.37

Barakat, H.N. and Hegazy, A.K. (1997) Reviews in Ecology: Desert Conservation and Development: A Festschrift for Prof. M. Kassas on the Occasion of His 75th Birthday. UNESCO, Cairo. Arid Zone Research Series, 1953–1969, No. 4). Available at: https://unesdoc.unesco.org/ark:/48223/pf0000110184 (accessed 23 July 2024).

Battisti, D.S. and Naylor, R.L. (2009) Historical warnings of future food insecurity with unprecedented seasonal heat. *Science* 323, 240–244. DOI: 10.1126/science.1164363

Baum, M., Grando, S., Backes, G., Jahoor, A., Sabbagh, A. *et al.* (2003) QTLs for agronomic traits in the Mediterranean environment identified in recombinant inbred lines of the cross 'Arta' × *H. spontaneum* 41–1. *Theoretical and Applied Genetics* 107, 1215–1225. DOI: 10.1007/s00122-003-1357-2

Becerra, S.C. and Garcia, V.D. (1994) Production of halophyte forage plants and their use in semi desert like conditions. In: Anaya-Garduño, M., Pascual-Moncayo, M.A. and Zárate-Zárate, R. (eds) *Proceedings of the IV International Conference on Desert Development: Sustainable Development for our Common Future*. International Desert Development Commission, México.

Beyene, T., Lettenmaier, D.P. and Kabat, P. (2010) Hydrolic impacts of climate change on the Nile River Basin: implications of the 2007 IPCC scenarios. *Climate Change* 100, 433–461. DOI: 10.1007/s10584-009-9693-0

Caldeyro, M. and Khafagy, M. (2013) Independent joint programme final evaluation: pro-poor horticulture value chains in Upper Egypt. MDG-F Secretariat. Available at: http://www.mdgfund.org/sites/default/files/Egypt%20-%20Private%20-%20Final%20Evaluation%20Report.pdf (accessed 22 July 2024).

CAPMAS (2019) *National Statistical Report for Monitoring SDGs in Egypt*. (in Arabic) Available at: https://capmas.gov.eg/pdf/SDG.pdf (accessed 23 July 2024).

Cathcart, R.B. (1985) What if we lowered the Mediterranean Sea? *Speculations in Science and Technology* 8, 7–15.

Ceccarelli, S. (2015) Efficiency of plant breeding. *Crop Science* 55(1), 87–97. DOI: 10.2135/cropsci2014.02.0158

Chapman, S.C., Chakraborty, S., Dreccer, M.F. and Howden, S.M. (2012) Plant adaptation to climate change: opportunities and priorities in breeding. *Crop Pasture Science* 63, 251–268. DOI: 10.1071/CP11303

Chattrairat, K., Jattomart, A. and Leclasantitham, A. (2022) Estimating water runoff from downscaling climate change scenarios using soil and water assessment tool and machine learning: a case study of Lake Tana basin in Ethiopia. *Science, Engineering and Health Studies* 16, 22020012. DOI: 10.14456/sehs.2022.62

Cheng, L., Zhang, X., Reis, S., Ren, C., Xu, J. *et al.* (2022) A 12% switch from monogastric to ruminant livestock production can reduce emissions and boost crop production for 525 million people. *Nature Food* 3(12), 1040–1051. DOI: 10.1038/s43016-022-00661-1

Cloern, J.E., Knowles, N., Brown, L.R., Cayan, D., Dettinger, M.D. *et al.* (2011) Projected evaluation of California's San Francisco Bay-Delta River system in a century of climate change. *PLOS ONE* 6, e24465. DOI: 10.1371/journal.pone.0024465

Conway, D. (2005) From headwater tributaries to international river: observing and adapting to climate variability and change in the Nile basin. *Global Environmental Change* 15, 99–114. DOI: 10.1016/j.gloenvcha.2005.01.003

Cooper, A.J. (1980) *The ABC of NFT: Nutrient Film Technique: The World's First Method of Crop Production Without a Solid Rooting Medium*. Grower Books, London.

Dettinger, M.D., Anderson, J., Anderson, M.L., Brown, L.R., Cayan, D.R. *et. al.* (2016) Climate change and the Delta. *San Francisco Estuary and Watershed Science* 14(3), Art. 5. DOI: 10.15447/sfews.2016v14iss3art5

Egypt Ministry of Agriculture (2009) Sustainable Agriculture Development Strategy (SADS): towards 2030. El-Beltagy, A. *et al.* (eds). Agricultural Research and Development Council, Cairo.

Egypt Ministry of Agriculture (2010) *Arabic Report on Monitoring the Land Use Changes in the Nile Delta and Valley Using Remote Sensing and GIS Techniques during the Period 1984–2007*. Technical Committee for Sustainable Soil Usage, Cairo.

Egypt Ministry of Agriculture and Ministry of Scientific Research (2015) Sustainable agricultural development business plan (2014/15–2017/18). In: El-Beltagy, A., Aboulnaga, A., Salem, H., Sheta, A., Sabri, S. *et al.* (eds) *The Frame of Sustainable Agricultural Development Strategy 2030*. (in Arabic)

El-Behairy, U.A. and Abou-Hadid, A.F. (2008) Impact of green roofs on global warming in urban areas. In: *Proceedings of the Ninth International Conference on Dry Land Development: Sustainable Development in the Dry Lands: Meeting the Challenge of Global Climate Change*. Alexandria, Egypt.

El-Behairy, U.A., Abou-Hadid, A.F., Medany, M.A. and Awad, M.M. (2001a) The effect of different cultivars, orientation and soilless culture systems on production and quality of strawberry. *Acta Horticulturae* 548, 59–64. DOI: 10.17660/ActaHortic.2001.548.4

El-Behairy, U.A., El-Shinawy, M.Z., Medany, M.A. and Abou-Hadid, A.F. (2001b) Utilization of 'A-shape' system of nutrient film technique (NFT) as a method for producing some vegetable crops intensively. *Acta Horticulturae* 559, 581–586. DOI: 10.17660/ActaHortic.2001.559.85

El-Beltagy, A. (2002) Reversing desertification: the potential for research. Desert and dryland development: challenges and potential in the new millennium. In: Ryan, J. (ed) *Proceedings of 6th International Conference on the Development of Dry Lands, 22–27 August*, Cairo, p. xi–xiv.

El-Beltagy, A. (2008a) Enhancing resilience in dry areas to cope with vagaries of climate change. In: *Proceedings of the 9th International Conference on the Development of Drylands*. IDCC, Alexandria, Egypt, pp. 69–76.

El-Beltagy, A. (2008b) Future challenges to the sustainable use of natural resources in the dry areas. In: El-Beltagy, A., Saxena, M. and Tao, W. (eds) *Human and Nature – Working Together for Sustainable Development of Drylands. Proceedings of the Eighth International Conference on Development of Drylands*, Bejing. ICARDA, Aleppo, Syria, pp. 1–8.

El-Beltagy, A. and Abou-Hadid, A.F. (eds) (2008) *The Basic Pillars of the National Program for Maximizing the Benefit from Water in Ancient Lands* (2nd edn). Ministry of Agriculture and Land Reclamation. (in Arabic)

El-Beltagy, A. and Madkour, M.A. (2012) Impact of climate change on arid lands agriculture. *Agriculture and Food Security Journal.* 1, 1–12. DOI: 10.1186/2048-7010-1-3

El-Beltagy, A.S. and Persson, A.R. (1986) Symposium on tomato production on arid land. *ISHS Acta Horticulture* 190.

El-Beltagy, A., Abou-Hadid, A.F., Gaafer, S.A. and Youssef, S.M. (1993) Interaction between root media and fertilizer for cucumber crops grown in greenhouses. *ISHS Acta Horticulturae* 323, 235–240. DOI: 10.17660/ActaHortic.1993.323.21

El-Beltagy, A., Valkoun, J., Amri, A., Thomas, R. and Shideed, K. (2005) Biodiversity, land degradation and poverty alleviation in dry and sub-humid lands. In: *Success Stories in Implementation of the Programme of Work in Dry Lands and the Global Taxonomy of 11th Meeting of the Subsidiary Body on Scientific, Technical and Technological Advice of the Convention on Biological Diversity.* Secretariat of the Convention on the Biological Diversity, Montreal, Canada, no. 21, pp. 13–20.

El-Beltagy *et al.* (2010) *Sustainable Agricultural Development Business Plan Overview (2010–2017).* ARC Press.

El-Beltagy, A., Sheta, A., El Fouli, M.M., Sabri, S., El-Gendy, A.M. *et al.* (2017) *Road Map: Technical, Social and Economic Studies for the National Egyptian Project for Reclaim and Development of 1.5 million Acre*. Academy of Scientific Research and Technology, Egypt. (in Arabic)

El-Beltagy, A., Madkour, M., Aboulnaga, A., El-Fouly, M., Sheta, A. *et al.* (eds) (2021) *Sustainable Agricultural Development 4.0: Smart Agriculture/Precision Agriculture/E-Agriculture*. Academy of Scientific Research and Technology, Egypt. (in Arabic)

FAO (2015) *Coping with Climate Change – The Roles of Genetic Resources for Food and Agriculture*. FAO, Rome. Available at: https://www.fao.org/3/i3866e/I3866E.pdf (accessed 23 July 2024).

Fucikovsky, L. (1994) Important diseases and pests of cactus and agave in Mexico. In: Anaya-Garduño, M., Pascual-Moncayo, M.A. and Zárate-Zárate, R. (eds) *Proceedings of the IV International Conference on Desert Development: Sustainable Development for Our Common Future*. International Desert Development Commission, México.

Gardufio, M.A. (1994) Socio-economic and ecological potential of Kochia as a fodder crop for dry regions. In: Anaya-Garduño, M., Pascual-Moncayo, M.A. and Zárate-Zárate, R. (eds) *Proceedings of the IV International Conference on Desert Development: Sustainable Development for Our Common Future*. International Desert Development Commission, México.

Global Carbon Project (2021) Supplemental data of Global Carbon Project (1.0) (dataset). Global Carbon Project. DOI: 10.18160/gcp-2021

Harlan, J.R., De Wet, J.M.J. and Price, E.G. (1973) Comparative evolution in cereals. *Evolution* 27(2), 311–325. DOI: 10.1111/j.1558-5646.1973.tb00676.x

Hassan, H. and Amin, E. (2022) Data and data systems. In: Mohieldin, M. (ed.) *Financing Sustainable Development in Egypt Report*. League of Arab States, pp. 22–42.

Ibrahim, A., El Abd, S.O., Mostafa, M.A. and El Beltagy, A.S. (1989) A possible role of cobalt in salt tolerance of plants. *Egyptian Journal of Soil Science* (Special Issue), 359–371.

IEA (2023) *Climate Resilience for Energy Transition in Egypt*. IEA, Paris. Available at: https://www.iea.org/reports/climate-resilience-for-energy-transition-in-egypt (accessed 23 July 2024).

IPCC (Intergovernmental Panel on Climate Change) (2014) Climate change 2014: impacts, adaptation, and vulnerability. Part A: Global and sectoral aspects. In: Field, C.B., Barros, V.R., Dokken, D.J., Mach, K.J., Mastrandrea, M.D. *et al.* (eds) *Contribution of Working Group II to the Fifth Assessment Report of the Intergovernmental Panel on Climate Change*. Cambridge University Press, Cambridge, UK and New York.

IPCC (Intergovernmental Panel on Climate Change) (2022) Climate change 2022: impacts, adaptation, and vulnerability. In: Pörtner, H.O., Roberts, D.C., Tignor, M., Poloczanska, E.S., Mintenbeck, K. *et al.* (eds) *Contribution of Working Group II to the Sixth Assessment Report of the Intergovernmental Panel on Climate Change*. Cambridge University Press. Cambridge University Press, Cambridge, UK and New York. DOI: 10.1017/9781009325844

Kebedew, M.G., Tilahun, S.A., Zimale, F.A., Belete, M.A., Wosenie, M.D. *et al.* (2023) Relating lake circulation patterns to sediment, nutrient, and water hyacinth distribution in a shallow tropical highland lake, *Hydrology* 10, 181. DOI: 10.3390/hydrology10090181

Kingston, D.G. and Taylor, R.G. (2010) Sources of uncertainty in climate change impacts on river discharge and groundwater in a headland catchment of the Upper Nile Basin, Uganda. *Hydrology and Earth System Sciences* 14, 1297–1308. DOI: 10.5194/hess-14-1297-2010

Kooistra, E., van den Ende, J. and Koornneef, P. (1975) 75 Years research at Naaldwijk: twelve papers on current research. *Netherlands Journal of Agricultural Science* 23(3). Available at: https://edepot.wur.nl/384650 (accessed 23 July 2024).

Kovak, E., Blaustein-Rejto, D. and Qaim, M. (2022) Genetically modified crops support climate change mitigation. *Trends in Plant Science* 27(7), 627–629. DOI: 10.1016/j.tplants.2022.01.004

Kummu, M., Heino, M., Taka, M., Varis, O. and Viviroli, D. (2021) Climate change risks pushing one-third of global food production outside the safe climate space. *One Earth* 4, 720–729. DOI: 10.1016/j.oneear.2021.04.017

Lotze-Campen, H. (2011) Climate change, population growth, and crop production: an overview. In: Yadav, S.S., Redden, R.J., Hatfield, J.L., Lotze-Campen, H. and Hall, A.E. (ed.) *Crop Adaptation to Climate Change*. Wiley-Blackwell, Chichester, UK, p. 1–11. DOI: 10.1002/9780470960929.ch1

Madkour, M. (2003) Biotechnology and its application in agriculture and food production: the Egyptian experience. In: Serageldin, I. and Persley, G. (eds) *Biotechnology and Sustainable Development: Voices from the South and North*. CAB International, Wallingford, UK, pp. 55–66.

Madkour, M. (2006) Harnessing new science to meet the challenges of drought. In: Serageldin, I. and Masood, E. (eds) *Changing Lives.* Biovision Alexandria. Bibliotheca Alexandria. Egypt, pp. 299–310. Available at: https://www.bibalex.org/Attachments/Publications/Files/Biovision2006_new.pdf (accessed 23 July 2024).

Maksoud, M. and Persson, A.R. (1986) EGNO Egyptian Norwegian tomato project. *Acta Horticulturae* 190, 59–70. DOI: 10.17660/ActaHortic.1986.190.4

Malhi, G.S., Kaur, M. and Kaushik, P. (2021) Impact of climate change on agriculture and its mitigation strategies: a review. *Sustainability* 13(3), 1318. DOI: 10.3390/su13031318

Mazhar, M., Arif, A., Chriyâa, A., El Mzouri, L. and Derkaoui, M. (2002) Cactus protects soil and livestock in rhamna region. *Acta Horticulturae* 581, 329–332. DOI: 10.17660/ActaHortic.2002.581.38

MDG Fund (n.d.) Egypt: pro-poor horticulture value chains in Upper Egypt. Available at: http://www.mdgfund.org/content/propoorhorticulturevaluechainsupperegypt (accessed 23 July 2024).

MDG Fund Egypt Joint Programme Fact Sheet (n.d.) Available at: http://www.mdgfund.org/sites/default/files/Egypt%20Joint%20Programmes%20Fact%20Sheet.pdf (accessed 23 July 2024).

Medany, M.A., Wadid, M.M. and Abou-Hadid, A.F. (2001) New extension system for modern irrigation management in Egypt. *Acta Horticulturae* 562, 407–411. DOI: 10.17660/ActaHortic.2001.562.49

Mercer, K.L. and Perales, H.R. (2010) Evolutionary response of landraces to climate change in centers of crop diversity. *Evolutionary Applications* 3(5–6), 480–493. DOI: 10.1111/j.1752-4571.2010.00137.x

Michaelson, R. (2018) Houses claimed by the canal life on Egypt's climate change frontline. *The Guardian*. Available at: https://www.theguardian.com/global-development/2018/aug/29/alexandria-little-venice-egypt-climate-change-frontline (accessed 23 July 2024).

Mohieldin, M., Wahba, S., Gonzalez-Perez, M.A. and Shehata, M. (2023) *Business Government, and SDG: The Role of Public-Private Engagement in Building a Sustainable Future*. Palgrave Macmillan, Cham, Switzerland. DOI: 10.1007/978-3-031-11196-9

Monforti, F.F., Crippa, M., Guizzardi, D., Muntean, M., Schaaf, E. *et al.* (2021) EDGAR v6.0 greenhouse gas emissions. European Commission, Joint Research Centre (JRC) [Dataset]. Available at: http://data.europa.eu/89h/97a67d67-c62e-4826-b873-9d972c4f670b (accessed 23 July 2024).

Moustafa, A., Al Mohammadi, A., Abou-Hadid, A. and Peacock, J.M. (eds) (1998) Protected agriculture in the Arabian Peninsula: summary proceedings of the international workshop, 15–18 February. Doha, Qatar. International Center for Agricultural Research in the Dry Areas (ICARDA), Aleppo, Syria. Available at: https://pdf.usaid.gov/pdf_docs/pnacf851.pdf (accessed 23 July 2024).

NASA Earth Observatory (2022) A grand new dam of the Nile: keeping a satellite eye on GERD. Available at: www.earthobservatory.nasa.gov./images/149699 (accessed 21 June 2023).

Nigatu, Z.M. (2013) Hydrological impacts of dynamic climate change on Lake Tan's water balance. Master's thesis, University of Twente. DOI: 10.13140/RG.2.2.13917.54240

Omar, A.R., Nawar, M.H., El Bendary, A.T. and Abou Bakr, H.E. (2015) Farmers' perception of climate change in rural Damietta Governorate. *Egyptian Journal of Agricultural Sciences* 66, 1–9. DOI: 10.21608/ejarc.2015.213106

Omran, E.S. and Negm, A.M. (eds) (2020) *Climate Change Impacts on Agriculture and Food Security in Egypt*. Springer Water, Switzerland AG. DOI: 10.1007/978-3-030-41629-4

Oranu, C.O., Achike, A.I., Zenebe, A. and Teklehaimanot, A. (2018) Comparative evaluation of farmers' perception and adaptation strategies to climate change and variability in Bako Tibe, Ethiopia and Abeokuta Nigeria. *American Journal of Climate Change* 7, 611–623. DOI: 10.4236/ajcc.2018.74038

Ortiz-Bobea, A., Ault, T.R., Carrillo, C.M., Chambers, R.G. and Lobell, D.B. (2021) Anthropogenic climate change has slowed global agricultural productivity growth. *Nature Climate Change* 11, 306–312. DOI: 10.1038/s41558-021-01000-1

Oszako, T. and Nowakowska, J.A. (2015) Climate change and food security: challenges for plant health, plant breeding and genetic resources. *Folia Forestalia Polonica. Series A. Forestry* 57(3). DOI: 10.1515/ffp-2015-0019

Pearce, F. (2010) Does Egypt own the Nile? A battle over precious water. Yale Environment 360. Available at: https://e360.yale.edu/features/does_egypt_own_the_nile_a_battle_over_precious_water (accessed 23 July 2024).

Sandström, V., Chrysafi, A., Lamminen, M., Troell, M., Jalava, M. *et al.* (2022) Food system by-products upcycled in livestock and aquaculture feeds can increase global food supply. *Nature Food* 3(9), 729–740. DOI: 10.1038/s43016-022-00589-6

Schnitzler, W., Leonardi, C., Tüzel, Y., El Behairy, U., Medany, M. *et al.* (2006) Efficient water use through environmentally sound hydroponic production of high-quality vegetables for domestic and export markets in Mediterranean countries: final report. Technical Report. Union Européenne.

Simonett, O. (2005) Potential impact of sea level rise: Nile Delta. UNEP/GRID-Arendal. COP5 Emission Graphics. Available at: https://www.grida.no/resources/7046 (accessed 23 July 2024).

Solyman, A. and Abdel Monem, T. (2020) Mapping Egypt vulnerability to sea level rise scenarios. In: Ewis Omran, E.S. and Negm, A. (eds) *Climate Change Impacts on Agriculture and Food Security in Egypt*. Springer Water, Cham, Switzerland. DOI: 10.1007/978-3-030-41629-4_9

Sörgel, H. (1932) *Atlantropa* (3rd edn). Fretz and Wasmuth Piloty and Loehle, Zürich, München.

Susnik, J., Vamvakeridou-Lyroudia, L.S., Savic, D.A. and Kapelan, Z. (2013) Integrated modeling of a coupled water-agricultural system using system dynamics. *Journal of Water and Climate Change* 4(3), 209–231. 10.2166/wcc.2013.069

Susnik, J., Vamvakeridou-Lyroudia, L.S., Baumert, N., Kloos, J., Renaud, F.G. *et al.* (2014) Interdisciplinary assessment of sea-level rise and climate change impacts on the lower Nile delta, Egypt. *Science of the Total Environment* 503–504, 279–288. DOI: 10.1016/j.scitotenv.2014.06.111

Turner, N.C. and Meyer, R. (2011) Synthesis of regional impacts and global agricultural adjustments. In: Yadav, S.S., Redden, R.J., Hatfield, J.L., Lotze-Campen, H. and Hall, A.E. (eds) *Crop Adaptation to Climate Change*. Wiley-Blackwell; Chichester, UK, pp. 156–165. DOI: 10.1002/9780470960929.ch12

Vigouroux, Y., Mariac, C., Mita, S.D., Pham, L., Gérard, B. *et al.* (2011) Selection for earlier flowering crop associated with climatic variations in the Sahel. *PLOS ONE* 6(5), e19563. DOI: 10.1371/journal.pone.0019563

Wang, T., Jin, H. and Sieverding, H.L. (2023) Factors affecting farmer perceived challenges towards precision agriculture. *Precision Agriculture* 24, 2456–2478. DOI: 10.1007/s11119-023-10048-2

Wilkinson, C.K. and Hill, M. (1983) Egyptian wall paintings: the Metropolitan Museum of Art's Collection of Facsimiles. Metropolitan Museum of Art, New York. Available at: https://www.metmuseum.org/art/collection/search/548354 (accessed 23 July 2024).

Willot, P. (2012) Independent joint programme mid-term evaluation: pro-poor horticulture value chains in Upper Egypt. MDG-F Secretariat. Available at: http://www.mdgfund.org/sites/default/files/Egypt%20-%20Private%20Sector%20-%20Mid-term%20Evaluation%20Report.pdf (accessed 23 July 2024).

Wondyfraw, M., Suneetha, P. and Krishna, T.V. (2019) The impact of infrastructure development on Lake Tana water quality. *International Journal of Emerging Technologies and Innovative Research* 6(2), 434–443. Available at: http://www.jetir.org/papers/JETIR1902563.pdf

Wong, C. (2023) Earth just had its hottest year on record-climate change is to blame. *Springer Nature*. DOI: 10.1038/d41586-023-03523-3

World Bank Group (1993) Egypt – an agricultural strategy for the 1990s. World Bank Country Study. Washington, DC. Available at: http://documents.worldbank.org/curated/en/845871468770946331/Egypt-An-agricultural-strategy-for-the-1990s (accessed 23 July 2024).

Worldometer (2023) Egypt Population, 2023. Available at: https://www.worldometers.info/world-population/egypt-population/ (accessed November 2023).

Yasuda, M., Fenta, A., Berihum, M.L., Inosako, K., Kawai, A. *et al.* (2022) Water level change of Lake Tana, source of the Blue Nile: prediction using teleconnections with sea surface temperature. *Journal of Great Lakes Research* 48(2), 468–477. DOI: 10.1016/j.jglr.2022.01.006

Zayed, A., Abou-Hadid, A.F., El-Behairy, U.A. and El-Beltagy, A.S. (1989) The use of Nutrient Film Technique for the commercial production of greenhouse tomato in Egypt. *Egyptian Journal of Horticulture* 16(2), 101–110.

Zohary, D. (2004) Unconscious selection and the evolution of domesticated plants. *Economic Botany* 58(1), 5–10. DOI: 10.1663/0013-0001(2004)058[0005:USATEO]2.0.CO;2

# 15 Humanity and Nature

**Aliaa R. Rafea[1,2*], Ismail Serageldin[3], Sharon Mijares[4,5], and Dena Merriam[6]**
*[1]Women's College, Ain Shams University, Egypt; [2]Chair, Human Foundation, Egypt; [3]Co-Chair, Board of Nizami Ganjavi International Center (NGIC); [4]California Institute for Human Science, Encinitas, USA; [5]National University, San Diego, USA; [6]Chair/Founder, Global Peace Initiative of Woman, New York, USA*

**Abstract**

This chapter delves into the intricate relationship between humanity and nature and its consequential effects on the Earth's ecology, culminating in the current climate change (CC) predicament. The Industrial Revolution marked a pivotal juncture; on the one hand, it engendered a cultural shift where humans stand as dominators over nature, severing the profound human–nature bond; on the other hand, it precipitated ecological imbalances through escalated fuel emissions.

Confronting the challenges of CC, Serageldin underscores that our special responsibility lies in acknowledging that we are an integral part of nature. Mijares expresses a way to combine science with spirituality, focusing on the rights of nature. Dena Merriam implores humanity to attune themselves to the myriad voices of nature, fostering a profound communion with all living beings. Rafea advocates a transformative scientific and cultural shift, drawing attention to inner well-being and nature to rekindle our bond with it.

## Introduction

### *Aliaa R. Rafea*

In this chapter, the contributors navigate the intricate dynamics between humanity and nature, illuminated by the rich tapestries of their various backgrounds and outlooks. Each narrative, though distinct, converges at intersections that form a cohesive mosaic of insights. These shared threads underscore the harmonious blend of these perspectives, enhancing the dialog without diminishing the authenticity of each voice.

Ismail Serageldin provides an encompassing perspective, illustrating the beautiful equilibrium within the web of life, linking all organisms on Earth, which have intricately evolved over millennia. The human footprint, dating back to the Holocene epoch, has imprinted transformative changes upon ecosystems. He emphasizes pivotal transformations in the last few centuries that have wrought destructive consequences, notably the combustion of fossil fuels for energy, spewing unprecedented volumes of greenhouse gases into the atmosphere, escalating global temperatures, destabilizing weather systems, and melting polar ice. He explains how the ecological damage induced by human actions is catastrophic.

Serageldin proposes mechanisms and strategies to potentially stabilize and enhance our current situation, intertwining scientific and

*Corresponding author: aliaa.readah.rafea@women.asu.edu.eg

 *Climate Change and Sustainable Agro-ecology in Global Drylands* (A.S. El-Beltagy, R. Lal and K. Malik eds.)
DOI: 10.1079/9781800624870.0015

ethical obligations. Fundamentally, he champions a shift in societal values towards a more integrated perception of our position within the life web. He contends that global acknowledgment of our collective responsibility to redress these disturbances and prioritize sustainable, inclusive strategies that protect humanity and our planetary home's stability is crucial. He concludes that humans, being the species that possess self-awareness of their actions, hold the power to alter their trajectory.

Mijares explores the nexus between spirituality and hierarchical stratifications, notably patriarchy, by asserting that an array of variables, from economics to politics and ideologies influence this relationship. Navigating through the historical evolution from gatherer-hunter communities, transitioning through agricultural to industrial societies, she sheds light on pivotal shifts that reoriented human worldview from a horizontal relationship with nature to a hierarchical one, marked by subjugating nature and devaluing women. She tackles spirituality in relation to religions, explaining that while numerous religions propagate the principle of caring for nature, their adherents often fail to perceive nature as a living being deserving of rights and consideration for its own intrinsic value. Mijares encapsulates her section with a notion that humanity's salvation hinges on adopting strategies that comprehend how to coexist and collaborate with nature while sustaining respect for all life on Mother Earth.

Merriam underscores the imperative of cherishing all life forms on Earth, emphasizing that humans are merely part of a vast family that includes myriad species across land, sea, and air. She observes that many CC proponents prioritize human well-being over that of other species. Such advocates often overlook the extinction of other species, failing to recognize them as sentient beings, focusing predominantly on human interests. Merriam's profound connection with whales is both awe-inspiring and commendable. Her spiritual interactions with these majestic creatures remind us of our limited understanding of ourselves and the vast tapestry of life around us.

Aliaa Rafea chose 'Healing nature by healing our nature' as the title for her section in this chapter. In this section, she grapples with the dichotomy of the advanced scientific and technological efforts to combat CC and the political will that drives them, juxtaposed against humanity's heedless actions that exacerbate the issue and impede strategic global initiatives. A core theme she explores is the cultural values propagated through modern times, where humans perceive themselves as the nucleus of the universe, with nature subserviently catering to their whims. This anthropocentric mindset is often bolstered by misconstrued religious beliefs purporting that humans are divinely ordained to dominate nature. She advocates for the urgent need to foster an alternative global ethos that rejuvenates our sense of interconnectedness with nature, re-envisioning our coexistence as a harmonious family. To this end, she argues that our detachment from our spiritual essence has estranged us from nature's manifold splendors. This estrangement has relegated us to states of isolation, imprisoned within the confines of our ego. In concluding this section, she underscores a fundamental truth: our journey to healing the environment is intrinsically tied to our journey of self-healing and realization.

## Nature, Humanity, and the Web of Life

***Ismail Serageldin***

*All living things on this Earth are interconnected in the web of life. From the oceans to the mountains, from plants animals, as well as bacteria and archaea, all fit together in a magnificent equilibrium developed over eons.*

The intricate and harmonious interconnectedness of all living things on Earth is a testament to the extraordinary tapestry of life that has been woven over countless millennia. From the deepest trenches of the oceans to the towering peaks of the mountains, from the tiniest microorganisms like bacteria and archaea to the grandest and most majestic creatures, every element of our planet's biosphere is interlinked. This delicate balance, finely tuned through the mechanisms of evolution and ecological interactions, forms what scientists often refer to as the 'web of life'.

Within this web, each species and ecosystem plays a crucial role. Predators help control prey populations, herbivores shape plant communities, and decomposers break down organic matter to recycle nutrients. Through photosynthesis, plants provide sustenance for herbivores and generate the oxygen essential for most life forms. The interconnectedness extends beyond species interactions; it encompasses the intricate relationships between organisms and their environment, including the Earth's climate, geology, and chemistry. This interconnected web of life reminds us of the importance of biodiversity conservation and the need to protect the fragile equilibrium that sustains all living things on our planet. As stewards of this precious world, we are responsible for understanding, appreciating, and safeguarding the intricate connections that make life on Earth possible.

*Humanity developed in the last 100,000 years, and civilizations grew in a climatically very stable period in the last 20,000 years, where the climate was relatively stable, resources were abundant, water was plentiful, and the yearly seasons gave a predictable rhythm to human activity.*

About 20,000 years ago, the Earth was already beginning its slow journey out of this glacial period. The ice sheets started to retreat, and the planet began to warm, leading to the Holocene, the interglacial period we are in today. This transition was punctuated by various events, including periods of rapid warming and episodes of cooler conditions.

The assertion that humanity's development coincided with a relatively stable climate over the last 20,000 years holds profound significance in understanding the trajectory of human civilization. Our species underwent remarkable transformations during this period, known as the Holocene epoch. This climatic stability, characterized by relatively mild and consistent conditions, allowed early human communities to transition from nomadic hunter-gatherer lifestyles to settled agricultural societies. With dependable access to resources, abundant water sources, and predictable seasonal cycles, humans could establish permanent settlements, cultivate crops, and raise livestock, marking the birth of organized civilizations.

The stability of the Holocene climate also facilitated the development of complex social structures, technological advancements, and cultural achievements. With surplus food production and the ability to store it for lean times, societies could support growing populations, leading to the rise of cities, specialized labor, and the spread of knowledge. This period witnessed the emergence of great civilizations like Mesopotamia, Egypt, and the Indus Valley, which laid the foundations for our modern world. However, it is crucial to recognize that our current era, characterized by climate change and resource depletion, presents new challenges that call for responsible stewardship of our planet to ensure a sustainable future for humanity.

*Human activity, especially in the past several hundred years when our technology has enabled us to magnify the impact of our actions many times over, is now causing disruptions in the delicate balance of the web of life and in supporting essential ecosystems from atmosphere to water. Examples abound...*

In the last few centuries, the exponential growth of human technology and industrialization has drastically altered the dynamics of the delicate web of life that sustains our planet. Human activity, driven by the desire for progress and economic growth, has led to significant disruptions in ecosystems and the Earth's climate system. Examples of these disruptions are both numerous and alarming. The burning of fossil fuels for energy production has released unprecedented amounts of greenhouse gases into the atmosphere, causing global temperatures to rise, destabilizing weather patterns, and melting polar ice caps.

Deforestation, driven by agriculture and urban expansion, has led to the loss of critical biodiversity and disrupted the balance of carbon dioxide and oxygen exchange that forests provide. Overfishing has depleted marine populations and disrupted oceanic ecosystems. Pollution from industrial processes, agricultural runoff, and plastic waste has contaminated water sources, affecting aquatic life and threatening human health. These disruptions jeopardize countless species' survival and endanger the systems that sustain human life, from clean air and water to stable climates and fertile soil.

Addressing these disruptions is an urgent global imperative. It requires a fundamental shift in human behavior, a transition to sustainable and environmentally friendly practices, and a commitment to preserving the delicate balance of the web of life. Recognizing our interconnectedness with all living things and our responsibility to safeguard the planet for future generations is crucial in mitigating the harm caused by our past actions and charting a more sustainable course for the future.

*The disruptions humans are causing in the ecological balance and the web of life are exacerbating the existential problems we confront.*

From CC to biodiversity loss to changing patterns of the ocean currents themselves, the damage that these disruptions are causing requires us to reflect on our actions, reform our behavior, and really move our activities toward the path of sustainable development.

The disruptions wrought by human activities on the delicate ecological balance and the interconnected web of life have far-reaching consequences that exacerbate our current existential problems. CC, driven by the excessive emission of greenhouse gases, is leading to rising temperatures, extreme weather events, and the displacement of vulnerable communities. Biodiversity loss, largely due to habitat destruction, overexploitation, and pollution, threatens the intricate relationships that sustain ecosystems and diminishes our planet's resilience to change. Even the altering patterns of ocean currents, driven by a warming climate, can disrupt weather patterns and marine ecosystems, affecting everything from global weather systems to fisheries.

Addressing these disruptions requires profound reflection on our actions and a dedicated commitment to reform our behavior. Transitioning towards sustainable development practices that prioritize the well-being of both the planet and its inhabitants is essential. This entails reducing our carbon footprint, conserving natural habitats, adopting cleaner energy sources, and implementing responsible resource management. It also necessitates a shift in societal values towards a more holistic view of our place within the web of life, recognizing that our survival and prosperity are intimately linked to the health of the planet. Only through concerted efforts to mitigate and reverse the disruptions we have caused can we hope to secure a sustainable and harmonious future for ourselves and the myriad species with which we share this remarkable planet.

*It is important to recognize that the human disruptions we are concerned about do not limit their negative consequences to the communities whose actions are causing these disruptions. The consequences suffered by others frequently far removed in space and time.*

Cumulative effects over time also lead to humanity crossing planetary boundaries of risk for different planetary systems (climate, water, etc.). That is a serious concern for all of humanity.

Indeed, it is crucial to acknowledge that the negative consequences of human disruptions extend far beyond the immediate communities responsible for these actions. The interconnected nature of our planet means that environmental and ecological changes triggered in one region can have ripple effects that reverberate across the globe and impact communities, ecosystems, and even future generations that are geographically distant. CC, for instance, does not discriminate; its consequences are felt worldwide through rising sea levels, extreme weather events, and disruptions to agriculture and food security.

Moreover, the cumulative effects of these disruptions pose a grave threat to humanity as a whole. When we push planetary systems, such as the climate or water resources, beyond their natural boundaries, we jeopardize the stability and resilience of the entire planet. Crossing these boundaries poses immense risks that can lead to irreversible changes in the Earth's systems, potentially triggering catastrophic consequences for all living beings. As a global community, it is imperative that we recognize the collective responsibility to address these disruptions and prioritize sustainable, inclusive solutions that benefit not only our immediate communities but also safeguard the well-being of all humanity and the integrity of the planet we call home.

*As we continue to stress the various Earth systems, we must be concerned about reaching potential tipping points where systems crash with potentially catastrophic effects on all of humanity.*

The concept of reaching potential tipping points in Earth's systems is a deeply concerning and widely recognized issue in the realm of environmental science and policy. These tipping points represent critical thresholds within ecological and climatic systems, beyond which abrupt and often catastrophic changes can occur. When we stress Earth's systems through activities like deforestation, excessive greenhouse gas emissions, and overexploitation of resources, we increase the risk of pushing these systems to the brink.

One of the most prominent examples is the potential tipping point related to climate change. If we continue to warm the planet and surpass certain temperature thresholds, we risk triggering feedback loops like the release of methane from thawing permafrost or the collapse of ice sheets, which can lead to accelerated sea-level rise and extreme weather events. These consequences would have catastrophic effects on human societies, especially in low-lying coastal areas.

Additionally, ecological systems can also experience tipping points. For instance, the loss of biodiversity in an ecosystem can reach a point where it disrupts critical ecological functions, making the system more vulnerable to disease outbreaks or invasive species. These shifts can have cascading effects on agriculture, water resources, and human health. Given the interconnectedness of Earth's systems and the potential for these systems to interact in complex ways, it is imperative that we recognize the urgency of addressing these stressors and take decisive action to prevent the crossing of critical tipping points that could have devastating consequences for all of humanity.

*Beyond seeking technological fixes to the problems we confront, many of which are due to our own misguided approach to development, we must seek to work in harmony with nature, for that is the true path to sustainable development.*

Indeed, seeking technological solutions alone is often insufficient to address the complex challenges we face as a result of our past and current development approaches. While technology can play a critical role in mitigating environmental and societal problems, it should be viewed as just one piece of a broader picture. Many of the issues we grapple with today, such as CC, biodiversity loss, and resource depletion, are fundamentally rooted in our unsustainable exploitation of natural resources and ecosystems. Thus, we must embrace a holistic approach that harmonizes human development with the natural world to achieve lasting and meaningful change.

Working in harmony with nature represents a paradigm shift in our thinking and behavior. It involves recognizing that the Earth's systems are intricately interconnected and interdependent. When we exploit natural resources without regard for their limits, we disrupt these delicate balances and risk long-term consequences that harm both the environment and human societies. Sustainable development, on the other hand, seeks to meet the needs of the present without compromising the ability of future generations to meet their own needs, recognizing that the health of ecosystems is inseparable from human well-being.

Embracing the true path to sustainable development means reevaluating our priorities, adopting regenerative practices that restore and protect the environment, and fostering a deep sense of respect and stewardship for the planet. It calls for responsible resource management, the preservation of biodiversity, and a commitment to reducing our ecological footprint. By aligning our actions with the principles of harmony with nature, we can strive towards a future where human development thrives while safeguarding the integrity and resilience of the natural world that sustains us.

*In discussing the global issues of climate change, biodiversity loss, or how to deal with water as a common public good, we must go beyond the hard calculus of economics, or the cynical calculus of political gains and losses.*

We must see the human face of the poor and the vulnerable. We must reflect on our common humanity, and thus our responsibility to engage, help, and empower the weak and the marginalized.

When addressing pressing global issues like CC, biodiversity loss, and equitable access to clean water, it is essential to move beyond the cold calculations of economics and the often-serving calculations of political gain. Instead, we should center our discussions on the human face of those most vulnerable to these challenges. These issues disproportionately impact marginalized communities, and it is our moral imperative to recognize and act upon the inherent value of every human life, irrespective of their socio-economic status or geographic location.

By recognizing our common humanity, we can cultivate a deeper sense of responsibility towards the less fortunate and vulnerable among us. In a world interconnected as never before, we must understand that the consequences of our actions ripple far and wide, affecting the lives of people who may have contributed the least to the problems at hand. Our collective responsibility to engage, help, and empower these communities goes beyond charity; it is about promoting justice and fairness in a world where resources and opportunities are often distributed unevenly.

In essence, addressing global challenges requires a shift in perspective, one that sees the interconnectedness of humanity and the embeddedness of humanity and its activities within nature's ecosystems, and hence the interdependence of our fates. By embracing our shared responsibility and working together to uplift the weak and marginalized, we can forge a path towards a more just and sustainable future for all, where the well-being of each individual is seen as integral to the well-being of the entire human family.

*Behind the statistics of suffering are real human beings. Communities try to cope with the vagaries of nature, and the institutional weakness of their societies. They are our brothers and sisters, and we must extend our hands to them.*

Behind the stark statistics of suffering are the faces of real human beings, individuals, and communities struggling to endure the harsh realities imposed by nature's unpredictability and the vulnerabilities within their societies. These are not just distant numbers on a spreadsheet; they represent our brothers and sisters in the global family of humanity. In moments of hardship and adversity, it becomes even more essential for us to extend our hands in solidarity and support.

Recognizing the shared human experience that unites us all, we must stand together to offer assistance and resources to those who find themselves on the front line of environmental and social challenges. Whether it is about communities facing the devastating impacts of natural disasters, the consequences of CC, or the burden of institutional weaknesses within their societies, our collective responsibility to help and uplift one another knows no borders. This ethos of compassion and empathy is the cornerstone of building a more compassionate and just world where the welfare and dignity of every individual are upheld as sacred.

In extending our hands to those in need, we not only provide immediate relief but also sow the seeds of hope and resilience. By working collaboratively to address the root causes of suffering and vulnerability, we can collectively pave the way for a more equitable and sustainable future where no one is left behind and the bonds of our shared humanity are strengthened.

We must learn to see the problem holistically in terms of the gossamer web of life and our special responsibility to protect it. For we are the first species that is conscious of its actions and the implications of these actions, as well as the first species that has the means to amend its behavior. In such an endeavor, science is the essential means to help chart our actions toward a path of sustainable development.

To address the pressing environmental challenges of our time, we must adopt a holistic perspective that recognizes the intricate interconnectedness of all life on Earth. Our planet is a delicate and complex web of life, where every species, ecosystem, and natural process plays a vital role in maintaining the balance of our biosphere. As the species at the apex of the Earth's ecosystems, humans hold a unique responsibility to protect and preserve this web of life. Our consciousness of the consequences of our actions and our ability to modify our behavior distinguish us from all other species on this planet.

This consciousness carries both a great burden and a profound opportunity. We are the first species to comprehend the far-reaching implications of our actions on the environment, from deforestation to pollution to CC. With this awareness, we also possess the capacity to make deliberate choices that benefit not only ourselves but the entire web of life. Our capacity to amend our behavior, adapt our practices, and pursue sustainable solutions empowers us to become responsible stewards of our planet. It is a defining moment in our history when we must rise to the occasion and use our unique consciousness and capabilities to protect the precious web of life that sustains us all.

These capabilities include an ever more insightful and vastly more powerful scientific enterprise. The new scientific insights are making enormously more powerful technologies possible, whose deployment requires thoughtful and ethical approaches. To the hard results of the empirical natural sciences, we need the social sciences' insights and the humanities' wisdom. This broadened perspective of going beyond information and knowledge towards wisdom to guide our actions will enable us to design the science-based policies and evidence-based regulations that should allow all of humanity to go back onto a path of sustainable development.

Our special responsibility lies in acknowledging that we are not separate from nature but an integral part of it. By embracing our role as custodians of the Earth, we can forge a path toward a more harmonious and sustainable coexistence with the gossamer web of life, ensuring a thriving planet for us, future generations, and all living beings that share this remarkable home.

## From Animism to the Anthropocene: What Happened?

***Sharon Mijares***

Many forces have contributed to our present environmental crises. For instance, our economic, political, and religious ideologies and related gender practices have all made contributions to the obvious environmental catastrophes. This problem is so complex and humongous that technological solutions are not adequate to effectively address it. Therefore, religious and spiritual approaches must also be used to promote stewardship of finite, precious, and fragile soil and other natural resources. Among human dimensions, gender equality and cultural issues are critical and need to be addressed. Furthermore, if this history is not considered, we are doomed to repeat the past mistakes and perpetuate the tragedy.

History illuminates how gender harmony is a major factor in healing the results of imbalanced choices. This section describes the human developments and processes occurring from the onset of the Holocene era as many have led to our current predicament. It asks us to live in the ways emphasized by our spiritual and religious traditions. This could open humanity to creative solutions and better ways of living.

### Pre-historical

Imagine being at one with the natural beauty and power of this world. This is the way that early nomadic humans lived. To be nomadic required specific sensitivities; it encompassed far more than the modern-day interpretation of simply moving from place to place. Instead, it speaks of cultures that were more in tune with the subtle forces of nature in land, water, and air. One could sense the changing of the seasons through these forces. Thus, the nomadic movement was in accordance with seasonal changes, changes the people experienced as an extension of themselves. They were not separate from nature but experienced themselves as one with it. Everything was imbued with life. Everything was alive with spirit. This is *animism*. Imagine living with a sense of being at one with the bushes, rocks, and trees, the rippling rivers and ponds, animals and insects, the air above you, and the earth beneath your feet. Their world was both dangerous and wondrous. Their sense of the spiritual was not dualistic, as everything was part of the Earth's breath.

Humanity foraged for foods, including seeds, nuts, fruits, roots, and plants. They lived in small tribal communities, with males and females working together. Men hunted the larger game that often required following an animal or a herd for longer periods of time, while women hunted closer smaller game to ensure caretaking of the young (Fuentes, 2012). Most of the protein sustenance came from hunting. Earlier hunter-gatherers had rituals to prepare for hunting as it was a sacred event. They thanked the animals for giving up their lives, recognizing them to be part of a great pattern of life. No doubt, the lives of these earlier human beings were fraught with environmental dangers, but there was far more respect for all that shared this cyclic way of life and death than seen in modern *Homo sapiens*. Unlike the future age of the Anthropocene, these threats within the cycles of life were *not* human created (Grinspoon, 2016).

Paleolithic and Neolithic human beings left images that have enabled modern researchers to understand more of our ancestors' lives. Artistic renderings were left for future generations to discover. For example, France's Chauvet Caves have abundant portrayals of ancient people's communal lives, supporting a spiritual relationship with hunting. Based on the analyses, it 'seems to be generally agreed that this picture of the "Dancing Sorcerer" was a cult object of great significance to the community which used the cave' (Brandon, 1959, p. 112). The images depicted their way of life (Clottes, 2016). Due to the differing bone structures of males and females, it has been proposed that most handprints on the walls were those of women (Hughes, 2013). This theory supports hypotheses of earlier gender harmony as well as the perspective that ordinary life endeavors, such as hunting, were spiritual events.

## The onset of economic processes

Life and humans' relationship to nature began changing following the Ice Age approximately 12,000 years ago as the Holocene epic began. During the Holocene era, humans discovered ways to practice agriculture, planting, and harvesting (History of Agriculture, 2023). Our living styles changed, and we became more sedentary. It was the onset of urban development. We developed 'selective hunting, herding, and settled agriculture' (Gupta, 2004, p. 58).

The human population went from around 6 million people at the start of the Holocene, some 10,000 years ago, to at least 100 million a mere 8000 years later. Agriculture, particularly annual grain crops and domesticated animals, accelerated human population growth and led to concentrated and sedentary living, eventually in state societies, in a relatively short period of time (Vitek, 2020, p. 29).

This also appears to be the onset of current economics, representing a 'distinct, momentous change that set the stage for the emergence of capitalism's world-ecology' (Vitek, 2020, p. 29). For example, one can well imagine that some farmers were successful, whereas others did not do as well, enabling a paradigm with social-economic differences. This also contributed to hierarchical ways of thinking. Did the loss of a felt connection with our environment as well as with all species sharing this planet contribute to the movement from the Holocene era to the Anthropocene age? (Certini and Scalenghe, 2015).

## The emergence of the world religions and patriarchal ideologies

Early humanity had a horizontal perspective of life; life itself was sacred. It differed from later beliefs. Religious historian Geoffrey Parrinder (1971) explained that during the transitional time to patriarchy, 'many cosmologies tell of the forcing apart of sky and earth, who were regarded as united in sexual union'. These changing mythologies reflected the beginning of the split between mind and body, male and female – and the loss of any remnants of paradise (authentic unity). Increasing emphasis would be placed on the disembodied mind and spirit. The emerging shift to a vertical perspective required that one look upwards towards an ideal, a hero, hierarchical figure of a god (Berman, 2000; Mijares *et al.*, 2007). One consequence was that 'we lost our foundation in the here and now' (Mijares *et al.*, 2007, p. 29). Nomadic tribes had shared creation stories (Eliade, 1978), based on their awareness of the movements in earth and sky. They passed on these stories via oral traditions, person to person. Support for the idea of the sacred is validated by the numerous flower petals found in early Neolithic gravesites, suggesting an honoring of death and perhaps belief in an afterlife. Examinations of burial practices suggest early spiritual practices, including the Paleolithic era, when floral grave linings were used to ordain the buried body (Nadel *et al.*, 2013).

Archeologists and anthropologists have found considerable evidence that the first representations of a creator deity were often represented as female; for example, it is estimated that the Venus of Willendorf figurines (approximately 4.4 inches, ~11 cm), 25,000–30,000 years ago were sacred icons. Examinations of ruins suggest they were placed in areas of relevance. Other representations of goddess figurines were plentiful (Goddess Worship: Theoretical Perspectives, 2023) in ancient Egyptian and Sumerian traditions. Often, there was a relationship between a feminine deity, agriculture, and nature itself. (Modern-day women have welcomed this knowledge after thousands of years of repressive treatment related to masculine ideologies.)

Evolution was occurring on every level in response to environmental changes and interactions. For example, *Homo sapiens*' brains had significant development (650 cc–1500 cc) during the *pre*-Holocene ice age (Ben-Dor and Barkai, 2021). This evolutionary development led to the ability to track animal herds for hunting. The human mind was also opening to new possibilities and developing cognitive perspectives of the surrounding world. At the beginning phases of this transition, spiritual practices were still associated with nature, but the observer of the new story or the myth was no longer at one with these spiritual forces. As the stories began to change, new ideas accompanied them. A separate, observing self was now present. (Yet, it is important to keep in mind that understanding the past and related hypotheses is still a process (Burja, 2021), and really nothing is written in stone.)

The world's religious traditions began developing approximately 5000 years ago with Oriental and Greek philosophies. Thoughts and beliefs about life were becoming increasingly abstract. Ultimately, the idea of a heavenly God above the natural world emerged. Thus, the divine was separated from actual life. In his book *Wandering God: A Study in Nomadic Spirituality*, Morris Berman points out (cited in Mijares *et al.*, 2007, p. 29): '...this new ideal of looking upward represented the onset of what he calls the *sacred authority complex*. The sacred could only be found outside of, and hierarchically above, life – thus, a split between heaven and earth was proclaimed.'

This was the beginning of significant changes in spiritual understanding. The emergence of observed forces in nature became externalized deities –interpreted as male and female gods. This is evidenced in Egypt (Parrinder, 1971) and Sumer (Kramer, 1972) as well as in numerous Earth-based religious groups, including the Vedic culture (Danielou, 1987; Mijares, 2009). These older stories were changed to support the newer beliefs (Kramer, 1963). The new versions would soon be recorded – eventually to impact people living in very different times and conditions. It appears that only the Earth-based indigenous and shamanic cultures throughout the ages have maintained a more holistic and unifying view of life and spirit.

The impetus to communicate is an ongoing human endeavor. It has enabled humans to record inspired and sacred texts and to express our thoughts and creativity. Writing began as the Mesopotamian cultures (in the area of present-day Iraq) developed cuneiform (wedge-shaped characters inscribed on clay). Soon after, the Egyptians developed hieroglyphics (a system of pictorial writing). Not long after these civilizations developed, the ancient Chinese *cradle of civilization* manifested in the area of what is known as the Yellow River valley. The Chinese writings were inscribed on turtle shells and bones. Creation myths began appearing in written form in all these mentioned lands. I would guess that the phrase 'written in stone' denotes that certain creativity and flow were lost once this process began as knowledge was considered finalized.

A new age was emerging. This led to the development of abstract thought and then to the inscription of the world's religious traditions. This was the beginning of an increasing intellectual development.

As the world's religious traditions emerged, beliefs in spiritual forces of nature were neglected as patriarchal ideals became dominant, encouraged by the newer creation stories, now considered religious fact. What is significant to the purpose of this section is that humanity was rapidly creating patriarchal ideologies, meaning that women and the feminine nature were quickly losing any power or influence in this newly developing paradigm. The following demonstrates how this imbalance began taking place – leading and relating to an ever-growing neglect and respect for the natural world.

Patriarchy refers to the bloodline of the father, thus claiming a male lineage rather than a female one. Men were supposedly ordained by God, Allah, Nirvana (or whatever language was used for naming the Absolute) with divine dominance in ordaining human behavior, including conception and birthright. This led to favoring the male in most, if not all, circumstances. Patriarchy is hierarchical; a supreme deity governs life, and the next lower rung of the ladder gives men the supposed God-given power to govern women.

The late French feminist and civil rights activist Francoise d'Eaubonne (1920–2005) recognized that both women and nature were simultaneously demeaned. She addressed this in her 1974 book *Le Feminism ou la Mort* (d'Eaubonne, 1974). She coined ecofeminism, encouraging women's efforts toward environmental protection and gender equality. It is imperative that we recognize the steps in which the current imbalances led to the disastrous results manifesting throughout the planet. Regardless of scriptural edicts directing humans to care for the Earth, in practice nature existed to supply human needs. It was no longer seen and experienced as a living being.

The online edition of the United Nations Environment Program notes that all developed religions have included an understanding of the human role in protecting nature in their sacred scriptures. Most religions teach that we are stewards of the natural world. This differs from recognizing it as a living presence with rights and needs of its own. Taoism sees nature as *the way*, emulating water by flowing in harmony with all of life's movements, yet this religious belief system has been unable to influence its homeland, China, one of the world's largest polluters. Humanity's next step is to earth these ideals and demonstrate how human beings live their daily lives.

Most interpretations of the Judeo-Christian scriptures strongly note that we are 'stewards' of the Earth, and all interpretations confirm our 'dominion over it' (Genesis 1: 26–31). Stewardship denotes caretaking, and we have failed to do this; the greater focus has been to have dominion (dominance) over it. Religious groups differ in their ideas for the caretaking of nature. These new religious revelations were interpreted by men inscribing these imbalances in their writings and sermons, convincing their followers that the following patterns accompanied religious obedience, leading to a favored outcome (a place in heaven or the freedom inherent within a divine meditative state).

The accepted Judeo-Christian version of differing creation stories was the one having Eve born out of one of Adam's ribs (Genesis 2: 21), immediately placing the woman as second best. But the third religion of the Abrahamic tradition, Islam, notes Adam (male archetype) and Eve (woman archetype) were created equally. All religions have extremist, conservative, or mystical versions of such stories, but regardless, women have not been treated equally in the Abrahamic traditions or other world traditions. Both women and nature have been demeaned, thus validating d'Eaubonne's proclamation. Gender and care of the Earth must accompany one another.

## Industrial and technological developments

In the human pursuit of growth, new ways of living with one another and with the Earth were enacted. The Industrial Age (1760–1840) initiated manufacturing practices, including the use of coal, metals, and oil. It also founded the 'nation state', which, according to Wimmer and Feinstein (2010), began changing political governments around the world. It began after the French and American Revolutions, embracing an idea of similar citizens encompassing sovereign nations with homogeneous people. The focus on production and profit was given far more relevance than nature. It has led to robbing nature of its resources (oil, minerals, gas etc.). This also led to steel and iron manufacturing processes.

Technologies require metals (e.g. cobalt, copper, lithium etc.). Because our technologies are increasingly expanding, all of the above and more are taken from the Earth to sustain them. At the same time, the movement favoring intellectual development has favored the human brain over its heart.

The 20th century brought two world wars. They required vast natural resources from nature and at the same time the weapons increasingly threatened the environment as well as all species on the Earth. The Marshall archipelago, namely Bikini Atoll islands in the Pacific Ocean, were used to test the effects of nuclear weapons, for instance.

After the displacement of the local inhabitants, 23 nuclear tests were carried out from 1946 to 1958. The cumulative force of the tests in all the Marshall Islands was equivalent to 7000 times that of the Hiroshima bomb. Following the use of nuclear bombs at Hiroshima and Nagasaki, the Bikini tests confirmed that humankind was entering a 'nuclear era'. The many militaries bear witness to the beginnings of the Cold War, the race to develop weapons of mass destruction, and a geopolitical balance based on terror (UNESCO World Heritage Center, n.d.)

The above is reflected in the history featured on the UNESCO site. It can evoke deep rage and grief that human beings would enact such atrocities against our world. But if our hearts are truly troubled, we can learn from this and create a healthier future for all.

A similar atrocity occurs with North Korea's bomb tests as they regularly fall into the ocean without regard for sea life or the impact on precious water. Military practices throughout the world are contributing to the devastation of land, water, and air. Such practices need to be seriously examined, considering the results.

## Conclusions

The modern agricultural, farming, economic, political, and religious institutions need to change. The transformation requires gender harmony and recognition of the equal rights of all peoples and the world we live in, our Mother Earth. Humanity's choices and evolutionary discoveries have led to nature's demise. Thus, the hope for humanity lies with approaches that understand how to live and work with nature while maintaining respect for all life on Earth. One example of this comes from Bolivia in 2012, when its government, led by indigenous Evo Morales, created the 'Mother Earth Law'. One clause within it notes:

> The expressed objective of the law is to establish the vision and fundamentals of integral development in harmony and balance with Mother Earth to Live Well, guaranteeing the continued capacity of Mother Earth to regenerate natural systems, recuperating, and strengthening local and ancestral practices, within the framework of rights, obligations, and responsibilities. (Framework Law of Mother Earth and Integral Development to Live Well, 2012)

It is time for humanity to wake up and discover the sacredness of what is possible in life. The Earth is reflecting this loss of spiritual connection and harmony. The destructive behaviors and confusion taking place in increasing numbers of human beings reflect this loss. The forces of Air are responding with hurricanes and tornadoes. The lack of caretaking of Water is manifesting in flooding or drought. Wildfires are burning forests and land. Earth's soil is polluted and degraded, destroying its capacity to create. These forces, the sacred elements, have been ignored. Destructive forces in nature reflect the destructive forces manifesting within humanity itself, for human choice has led to this predicament.

Yet, there are options that can put humanity on a path to live in symbiosis and harmony with nature. This path combines scientific innovations with spirituality, religion, and stewardship, focusing on the rights of nature. It is also impacted by caring for all of life: women, men, and people of color, of differing religions and creeds, must unite for the good of all. The goal is to protect, restore, and sustain life-support processes of nature through the translation of scientific and spiritual beliefs into action. With this approach, humanity has a bright future through respect for nature and living in symbiosis with it rather than trying to conquer it.

## Unheard Voices

### *Dena Merriam*

It was the fall of 2018, and I was on my way to Poland to attend COP24, stopping in Paris overnight, when I had one of those rare dream visions, just before awakening. In the dream I was standing by the window in my Manhattan apartment overlooking the East River in New York City when a large whale swam up the river and stopped right in front of my building. It looked directly up at me, and we made eye contact; an exchange took place. Looking directly at the whale, I found myself saying, 'I hear you'. After speaking those words internally, I opened my eyes, returning to the external world, feeling deeply touched by the experience. I knew I had just made a commitment, a commitment to listen and hear.

A few weeks earlier I had been in Japan at a CC meeting with a small group of religious leaders. An environmentalist from the Dominican Republic, who had extensive contact with whales, was also there. He spoke about the song of the whales, and their ability to communicate across thousands of miles; how they gather at a certain part of the ocean near the Dominican Republic at the same time each year, point their noses to the ground and sing to the Earth. Despite the assaults they are under, they still sing, while we humans have forgotten how to sing to the Earth and have forgotten how to hear the voices of other life forms.

My experience that night in Paris gave me a personal interaction with a whale, and a powerful transmission took place. The next day I found myself at the COP meeting, thinking, 'I am here to represent the whales for they have spoken to me'. I went to all the sessions I could find on the seas but not one mentioned the effects of CC on animal life. Much of the conversation centered on us humans; how we would be impacted in the future by the warming oceans and changes to ocean currents. The life forms of the seas were not represented at COP.

Last year, off the western coast of the USA, a usual heatwave occurred, causing some human deaths, but during that heatwave over a billion sea animals lost their lives. Earlier this year, 45 whales were found dead off the coast of New York, the cause unknown. This summer, around 100 whales beached themselves in Australia in what is called a 'mass stranding'. Many of them were young. Fifty-one of the whales died immediately, but officials were able to get 45 back into deep water. But those rescued whales beached themselves again, re-stranding. They all died. What is the message that we are not hearing? One must ask: Did they sacrifice themselves for a specific purpose?

I have not forgotten my dream experience of the whale. It still haunts me. But it is not just the whales who are trying to communicate with us. We have tuned out and lost the ability to hear the messages coming from all parts of the natural world.

In New York state, a third of the birds have disappeared over the last few decades. When I first moved to my home north of New York City in 1971, I would be awakened in the early morning to the humming chatter of a multitude of birds. Birdsong still can be heard in the early morning hours, but there are far fewer birds to sing to us. Recent studies have shown that North America has lost a billion birds over the last decades. Do we not care that our children and grandchildren may not have the joy and comfort of these fellow companions? It must be our priority to ensure that they do. New studies show a relationship between birdsong and mental health, and there is an increasing mental health crisis among the youth.

Each life in the wide and complex web of the natural world is an essential component of the whole. With each loss, our inner being knows and feels it. We are not immune to this disappearance of life. We will suffer along with all else. The science of CC is the foundation for us to transition to a more sustainable world, but it is not enough. Along with that science must come a change in consciousness, a reconnection, a re-bonding with the many life forms that share this Earth with us, enabling us to become more mindful, more attentive, more aware, more caring, more awake. We must become the transmitters of love that we are meant to be.

Whenever I hear of an incident with a whale, I feel it keenly and think of my whale friend who was inspired to approach me. That experience will never leave me. I told the whale, 'I hear you', but hearing must lead to action. I am not a scientist or government official, but I can be present and communicate internally with the whale community, just as I can show appreciation for the birds, take time to listen to their song, and thank them, express gratitude, and sorrow at what they are enduring. It is time to befriend again all that lives in the natural world.

We have reached a time when we must expand our consciousness to include more than our family, our friends, or the human community. Our survival depends on this expansion for we cannot proceed alone into the unknown era that we face. We need our animal companions, just as they need us. We need our forests and plant life, our rivers and soil, our air. We must see and live in the awareness of the whole because one part cannot survive without the health of the whole.

Each one of us has a part to play in shaping our collective future. We can advocate for sustainable policies, pressure governments and businesses to act, but at the end of the day, if we do not care to listen to the chirping of the birds, feel the grandeur of the whales and honor their love for the Earth, we will not get very far. Love is what will get us through this crisis we face, because from love, action follows.

I will close with another dream/vision of a few weeks ago. I was body surfing in the ocean, something I have never done, drifting further and further away from the shore, when I found myself surrounded by sharks, who were swirling around me. The thought entered my mind that this is surely the end. There is no way that I can escape them. Just as I was resigned to my death, an animal from below came and lifted my surfboard, carrying me safely to shore. It was a whale who had come to save me. What is the message? If we have the foresight, vision, and love to save our animal companions, they will surely come to save us.

## Healing Nature by Healing Our Nature

***Aliaa R. Rafea***

### Current situation

Though Earth's climate has experienced fluctuations throughout its history, the present warming trend is occurring at an unprecedented pace in the last 200 years since the onset of the Industrial Revolution, with a particularly sharp increase in the last decades. The Intergovernmental Panel on Climate Change (IPCC) states, 'Since the commencement of systematic scientific assessments in the 1970s, the impact of human activities on the climate system's warming has transitioned from being a theoretical notion to an established reality (Global Climate Change, n.d.).

For over five decades, humanity has focused on the environment and its profound influence on human existence. However, our progress toward achieving the envisioned goals for our planet and human well-being needs to catch up. Had the global community truly prioritized the environmental predicament, our situation would undoubtedly have been more favorable.

While equipped with the knowledge and tools to confront these challenges, the global landscape continues to be marred by the depletion of natural resources, rampant deforestation, the encroachment of deserts, and the escalating grip of poverty. These factors drive populations to uproot from their communities for basic survival.

Science indicates pathways to avert impending disasters for humanity. Yet, the intricate interplay between nature and human actions – such as consumption patterns, food production, and demography – and other intricate relationships necessitates pioneering scientific approaches and adjustments in human behavior (Falk *et al.*, 2022). Regrettably, prevailing valuation systems prioritize short-term gains, disregarding potential long-term catastrophes. This neglect places vulnerable communities and the entire world in continued jeopardy.

The pressing question arises: Are we up to the task of surmounting these challenges?

## Common scientific and cultural paradigms

The situation might appear disheartening, yet the world must cling to hope. As Einstein once remarked, 'You cannot solve a problem with the same mind that created it'. Viewing and analyzing nature solely for human exploitation needs to change. A shift in our consciousness towards nature is crucial to effectively address our present challenges (Rafea, 2019).

Thomas Kuhn, who introduced the concept of scientific paradigms, posits that scientific fields progress through phases where competing theories vie for dominance. Eventually, one theory proves more adept at solving the field's critical problems and becomes the prevailing paradigm. However, this dominant paradigm is not necessarily the ultimate truth but rather the most efficient approach at that particular time (Kuhn, 2012).

The prevailing scientific paradigms arose from the advent of modernity, heavily influenced by cultural values and priorities that assert that history progresses linearly from primitiveness to civilization. The overarching belief was that through science, humanity could comprehend and dominate nature. This scientific approach deeply influenced cultural values, inadvertently leading to the depletion of natural resources.

At the same time, it is evident that a significant driver behind the West's heedless environmental exploitation can be traced back to its foundational Judeo-Christian values. As per these doctrines, God fashioned man in his likeness and bestowed upon him dominion over the Earth. Under this perspective, nature's value is solely predicated on its usefulness to humans, fostering the notion that it can be indiscriminately exploited without consequences (Whites, 1967). Yet this paradigm is not exclusive to Judeo-Christian traditions; similar interpretations surface in common understandings of Islam. Such interpretations, however, deviate from the genuine teachings of the religion.[i]

The hierarchical perspective on the relationship between God and nature extends into dominant attitudes towards race, religion, nationalism, and gender, fostering a sense of superiority in these domains (Mijares *et al.*, 2007). Wars waged in the name of God arise from notions of hierarchical superiority. Similarly, colonialism, driven by European countries, derided and dismissed the traditions and cultures of other regions, reflecting this same sense of dominance and supremacy.

To restore our bond with nature, healing the human soul is essential. We must dispel the deep-seated notion of superiority, shatter illusory boundaries, and truly embrace our interconnectedness with all that exists. This is a lesson which we can learn from endogenous people, and Ancient Egyptian Civilization[ii] (Rafea, 2009).

## New paradigms

The perspective of eco-semiotics approach emphasizes that our perception of nature actively molds our interactions with it, resulting in a secondary version of nature with a human face.[iii] That is why anthropologists are fascinated by studying various worldviews, according to which humans innovate and create cultural systems that bind them and help them interpret who they are in relation to the cosmos.

Humans naturally question and form interpretations of their observations and express them through symbols, including language and art. Upon engaging with our surroundings, we naturally categorize, define, and interpret our observations. This act of perception leads to the reconstruction of nature within our minds; the abstract and the tangible converge in the process of cognition. What is considered 'truly real' for one community might differ from another community's perspective (Geertz, 1973). In ancient civilizations, people were able to interact intimately with nature; they created their language, which included rites, rituals, actions, and more.

We live in a different epoch with various cultures where most of them suffer from the hegemony of consumerism and individualistic approaches. The global culture persists in neglecting nature as a vibrant entity, of which humans constitute an integral part. Altering our perspective of nature holds the key to addressing CC and cultivating a fresh understanding of our interconnectedness as human beings, intertwined with one another and the entire cosmos. The historian Thomas Berry calls for nature rights, to place them equally as important as other rights (Berry, 2015). Humanizing Nature, and acknowledging her rights is part of a transformative worldview that should prevail as a global culture.

Grasping and deeply feeling our interconnectedness on both psychological and spiritual levels is crucial for forging a bond with nature, to allow it to guide and inspire us in addressing and healing its wounds. Humans' awareness of how it took the Earth countless millennia to shape an environment conducive to human existence should reveal this interconnectedness, not only physically, but also spiritually. Our Mother Earth was able to beget us, because she is alive. Our sustenance springs from her generosity, with the sun and air playing vital roles in our survival. By undergoing psychological and spiritual detachment from nature, humanity places itself in a state of self-alienation. Acknowledging the extraordinary intricacies that paved the path for life's emergence on our planet, as comprehended through contemporary knowledge, carries importance in enhancing our understanding of our existence as offspring of the cosmos. This recognition fosters an invitation for gratitude to flow through our hearts.

The realization that everything is interconnected stems from our inherent reliance on both our surroundings and one another. Scientifically, this understanding has been formalized through the general theory of systems. The theory suggests that many systems, whether they are biological, social, or mechanical, share certain common principles and characteristics. This approach strives to comprehend the world as a collection of intricately interwoven systems that cannot be broken down further. It directs attention not only to the entirety but also to the intricate interconnections among the individual components that comprise it (Von Bertalanffy, 1950; Laszlo and Krippner, 1998).

This approach perceives nature not as a linear sequence of cause-and-effect relationships but rather as a nonlinear, unpredictable arrangement of interconnected elements that reference themselves and lead to emergent coordination. Within this framework, humans are inherently embedded within the ecosystem, unable to elude the profound interconnectedness that ties them to the cosmos, even though they may not be aware of it.

Insights from physics reveal that living organisms operate as open systems, reliant on interaction and exchange with their external environment for survival. The reciprocal interactions among these component systems bind them together and sustain the larger system.

The organism is the archetypal metaphor of all deeply interconnected and interrelated systems. It is firstly a holistic entity, a system within which all constitutive entities exist in meaningful relationship to all other entities. The entities comprising an organism carry nested self-similarity, and fractal properties, which is to say that observation at multiple scales reveals information about the larger self-organizing system. 'In the organismic frame our identity moves from that of a single finite human, to an expression of a developing principle of consciousness being variously expressed in the world' (Nelson, 2015, pp. 26, 32).

We are on the edge of new science with other postulates. Bringing the organismic paradigm to our consciousness means that we perceive the universe as a living open system, with self-generating processes. The contemporary mind is in the midst of a transition; toward an organismic lens on reality – apprehended as an evolving, self-generating, and ultimately living process (Stapp, 2001). As a result, in the context of our existence, as humans, our lives are intricately interwoven with the entirety of the system – physically, psychologically, and spiritually.

Recognizing our interconnectedness fosters a worldview that shapes our perspective towards both nature and our own selves. Beyond the realms of spiritual metaphysics, this heightened consciousness allows us to perceive our existence beyond the confines of our individual identities and connect to life as manifested in everything. Unlike the old scientific long-standing worldview of linear notions of scientific progress, there is a new view observing how transformative ideas do not arise from the day-to-day, gradual process of experimentation and data accumulation but that the revolutions in science, those breakthrough moments that disrupt accepted thinking and offer unanticipated ideas, occur outside of 'normal science' (Schwartz, 2018).

The newly born paradigm's shift goes in the direction of understanding that humans are more than just bodies and intellect; they are emotional beings, inspired by passion and love, appreciate beauty, and detest ugliness. Beyond what we know about ourselves, there is much more hidden within our psyche. Often, we neglect this inner domain, which Jung considered a key to the process of individuation or psychological integration. He proposed that the energy that drives conscious psychic life pre-exists consciousness and is initially unconscious. As it nears consciousness, it manifests as projections, taking forms like manna, gods, daimons, and so on, serving as a primary source of vitality (Jung and Jaffe, 1963, p. 346).

Being conscious of the unconscious correlates with experiencing interconnectedness with the whole at a profound level of the psyche, particularly during the process of individuation, which takes place in an inner journey. C.G. Jung saw psychology to heal both the soul and nature. By introducing the concept of the collective unconsciousness, he unveiled the underlying connections that bind humans to a greater whole, transcending individual experiences and bridging them to dimensions beyond specific time and space. Using his own life as a canvas for exploration and through his therapeutic work with patients, Jung deepened his grasp of the human psyche and its fundamental archetypes (Jung and Jaffe, 1963)

Through the collective unconscious, humans transcend the constraints of time and space, linking deeply not just with other humans but also with nature. This bond is not just a philosophical notion, or a purely intellectual concept built on theoretical foundations. Instead, it is a living state of being, a readiness to draw inspiration from the unconscious and bring it to the conscious realm. Sacred scriptures frequently use symbolism to depict the connection between humans and nature, such as the tale of Adam being formed from the earth. This symbolically anchors humans to Mother Earth at an archetypal level. Yet, taking these narratives at face value can give rise to misunderstandings and may challenge one's faith, especially when these literal interpretations clash with scientific perspectives.

Amazingly, scientific breakthroughs and discoveries sprang through unconscious images first that inspired scientists to find solutions to their questions. Kekule had discovered the shape of the benzene molecule through a dream, similarly to Watson and his discovery of RNA and DNA. Both sensed images and translated them into scientific inputs (Zwart, 2019).

## Conclusion

Despite their efforts, conventional conferences and UN assemblies cannot achieve their objectives without a fundamental shift in our consciousness toward a deeper connection with nature – evident in every aspect of our surroundings and within ourselves. This change in perspective could reshape our scientific methodologies and daily life practices. Science might move away from solely linear methods, increasingly embracing interdisciplinary, multidisciplinary, and transdisciplinary approaches. This evolving vision might see scientists pioneering a renewed understanding of the human–nature relationship.

Individuals must reconcile with their inner selves, cherishing the profound beauty and love inherent in interconnectedness. Such love drives us to maximize our potential, filling our hearts with empathy and compassion towards everything, viewing it all as extensions of ourselves.

Our inner journeys are not detached from the world outside or the knowledge we accrue. On the contrary, the deeper we venture into our souls, the more driven we become to bring about good for ourselves and the broader humanity. True spirituality reveals itself in acts of service, and each person can make a global impact by committing to excellence in their daily tasks.

It is unnecessary to affirm our spiritual essence encased in physical form. This truth is self-evident in the intricate web of connections we share with the grand expanse of the universe and even the tiniest known entities. Divinity is visible in the life we comprehend and remains an enigma beyond our grasp. Facing CC, we encounter both a daunting challenge and a unique opportunity. It becomes an opportunity if humanity unites against this looming threat, working together to ensure a peaceful, sustainable future.

I would like to end this section with a quote from Albert Einstein:

> A human being is a part of the whole, called by us 'Universe,' a part limited in time and space. He experiences himself, his thoughts, and feelings as something separated from the rest, a kind of optical delusion of his consciousness. This delusion is a kind of prison for us, restricting us to our personal desires and to affection for a few persons nearest to us. Our task must be to free ourselves from this prison by widening our circle of compassion to embrace all living creatures and the whole of nature in its beauty. Nobody can achieve this completely, but the striving for such achievement is in itself a part of the liberation and a foundation for inner security. (in: Eves, 2003)[iv]

## General Conclusion

***Aliaa R. Rafea***

In exploring the intricate relationship between humanity and nature, the diverse approaches of the authors converge on certain pivotal themes. One unmistakable consensus among the four presentations is the profound interconnection of humans with nature, with the Industrial Revolution earmarked as a significant inflection point influencing climate change. Collectively, they underscore the imperative for shared responsibility, emphasizing that those who inflict a greater harm upon nature must shoulder the burden of safeguarding vulnerable communities.

Yet, while they find common ground in numerous areas, there are distinct nuances that each author brings, underscoring their unique perspectives. Dr. Serageldin and Dr. Mijares highlight the Holocene epoch as the dawn of human-induced alterations to nature, which set in motion an irreversible trajectory. My perspective delineates between naturally occurring climate variations and those exacerbated by human actions.

Dr. Serageldin and Dr. Mijares, along with myself, stress the paramount importance of international collaboration in grappling with the complexities of CC. A shared sentiment between Dr. Serageldin and myself is the indispensable role of science in addressing and navigating the challenges of our changing climate.

Spirituality emerges as a recurrent theme, with Dr. Mijares, Hon. Dena Merriam, and I weaving it into our narratives, albeit through varied lenses. Dr. Mijares perceives spirituality as intrinsically linked to the universe's feminine essence. Dena Merriam shares her profound spiritual communion with a whale, urging humanity to listen and resonate with nature's myriad voices truly. My stance, Aliaa Rafea's, pivots towards introspection, positing that the path to self-realization is rooted in a spiritual quest that invariably deepens our bond with nature.

In summation, while each author offers a unique tapestry of thoughts, our collective aspiration remains unambiguous: the salvation of our planet. Nature, in all its splendor, beckons us to protect and heal its scars. A radical shift in our collective consciousness is imperative. We must unite, not just in thought, but in action, ensuring that future generations are not left to bear the brunt of nature's wrath.

## Notes

[i] For example, many verses in the Qur'an address Nature as a living entity: 'We did indeed offer the Trust to the Heavens and the Earth and the Mountains; but they refused to undertake it, being afraid thereof: but man undertook it; - He was indeed unjust and foolish' (Qur'an 33:72). In another verse, we read that other creatures in the universe are communities, just like human beings: 'There is not an animal (that lives) on the earth, nor a being that flies on its wings, but (forms part of) communities like you' (Qur'an 6:38) The Qur'an adeptly highlights the harmony between everything: 'Neither the sun can overtake the moon, nor can the night outpace the day. Each of them is floating in orbit' (Qur'an: 36:40).
[ii] The Ancient Egyptians had an intricate understanding of nature, considering it a cohesive whole in which they were but a part. Their perspective was not anthropocentric; they felt an authentic bond with the world around them. This connection was manifested in various aspects of their culture, from art and everyday living to philosophy and the divine qualities they assigned to elements of nature, such as animals and birds. Their artistic representations, notably their sculptures, serve as a striking testament to this harmony, as they skillfully combine human and animal forms into a unified whole.
[iii] Eco-semiotics is an interdisciplinary field that combines ecology and semiotics to explore the systems of signs and meanings in the natural environment. Semiotics studies signs and symbols and how they convey meaning, while ecology focuses on the interrelationships between organisms and their environments. Combining these two domains, eco-semiotics delves into how nature communicates through signs, symbols, and codes and how humans interpret and assign meaning to natural signs. This interpretation impacts our relationship with the environment. By understanding the semiotic systems in nature, eco-semiotics can contribute insights for conservation efforts and foster a deeper ethical consideration for the environment. Charles Sanders Peirce, Thomas A. Sebeok, and Jakob von Uexküll are some major scientists who presented the approach of eco-semiotics.
[iv] The *New York Times* published this quote in an article by Walter Sullivan (March 29, 1972, titled 'The Einstein Papers: A Man of Many Parts'. Available at: https://www.nytimes.com/1972/03/29/archives/the-einstein-papers-a-man-of-many-parts-the-einstein-papers-man-of.html (accessed 16 September 2023).

## References

Ben-Dor, M. and Barkai, R. (2021) Prey size decline as a unifying ecological selecting agent in pleistocene human evolution. *Quaternary* 4(1), 7. DOI: 10.3390/quat4010007

Berman, M. (2000) *Wandering God: A Study in Nomadic Spirituality*. State University of New York Press, New York.

Berry, T. (2015) Ethics and ecology. In: Seymour, M. (ed.) *Educating for Humanity*. Routledge, pp. 155–164.

Brandon, S.G.F. (1959) The ritual perpetuation of the past. *Numen* 6(2), 112–129. DOI: 10.2307/3269309

Burja, S. (2021) Why civilization is older than we thought. Palladium. Available at: https://www.palladiummag.com/2021/05/17/why-civilization-is-older-than-we-thought/ (accessed 25 July 2024).

Certini, G. and Scalenghe, R. (2015) Holocene as anthropocene. *Science* 349, 246–246. DOI: 10.1126/science.349.6245.246a

Clottes, J. (2016) *What Is Paleolithic Art? Cave Paintings and the Dawn of Human Creativity*. University of Chicago Press, Chicago, Illinois. DOI: 10.7208/9780226188065

d'Eaubonne, F. (1974) *Le Féminisme ou la Mort*. P. Horay, Verso Books, Paris. (in French)

Danielou, A. (1987) *While the Gods play: Shaiva oracles and predictions on the cycles of history and the destiny of mankind*. Inner Traditions International, New York.

Eliade, M. (1978) *A History of Religious Ideas, Vol. 1: From the Stone Age to the Eleusinian Mysteries*. Trans. Willard R. Trask. University of Chicago Press. Chicago, Illinois, pp. 148–149. DOI: 10.1017/S0360966900018041

Eves, H.W. (2003) *Mathematical Circles: Mathematical Circles Adieu and Return to Mathematical Circles*. Mathematical Association of America. California.

Falk, J., Attig-Bahar, F., Colwell, R.R., Behera, S.K., El-Beltagy, A. *et al.* (2022) Addressing our planetary crisis: consensus statement from the presenters and International Advisory Committee of the Regional Action on Climate Change (RACC) Symposium held in conjunction with the Kyoto-based Science and Technology in Society (STS) Forum. *Sustainability Science* 1–3. DOI: 10.1007/s11625-021-01059-x

Framework Law of Mother Earth and Integral Development to Live Well (2012) (Bolivia). Available at: https://climate-laws.org/document/the-mother-earth-law-and-integral-development-to-live-well-law-no-300_42c0 (accessed 25 July 2024). (in Spanish)

Fuentes, A. (2012) *Race, Monogamy, and Other Lies They Told You: Busting Myths about Human Nature*. University of California Press. Berkeley, California. Available at: http://www.jstor.org/stable/10.1525/j.ctt1pn993 (accessed 25 July 2024).

Geertz, C. (1973) *The Interpretation of Cultures: Selected Essays*. Basic Books, New York.

Global Climate Change (n.d.) Do we know climate change is real? There is unequivocal evidence that Earth is warming at an unprecedented rate. Human activity is the principal cause. Available at: https://climate.nasa.gov/evidence (accessed 31 August 2023).

Goddess Worship: Theoretical Perspectives (2023) *Encyclopedia of Religion.* Encyclopedia.com. Available at: https://www.encyclopedia.com/environment/encyclopedias-almanacs-transcripts-and-maps/goddess-worship-theoretical-perspectives (accessed 27 August 2023).

Grinspoon, D. (2016) *Earth in Human Hands: Shaping Our Planet's Future*. Grand Central Publishing, New York.

Gupta, A.K. (2004) Origin of agriculture and domestication of plants and animals linked to early Holocene climate amelioration. *Current Science* 87(1), 54–59. Available at: http://www.jstor.org/stable/24107979 (accessed 25 July 2024).

History of Agriculture (2023) Wikipedia. Available at: https://en.wikipedia.org/wiki/History_of_agriculture (accessed 25 July 2024).

Hughes, V. (2013) Were the first artists mostly women? National Geographic. Available at: https://www.nationalgeographic.com/adventure/article/131008-women-handprints-oldest-neolithic-cave-art (accessed 25 July 2024).

Jung, C.G. and Jaffe, A. (1963) *Memories, Dreams, Reflections.* Trans. Richard Winston and Clara Winston. Fontana Press. London.

Kramer, S.N. (1963) *The Sumerians: Their History, Culture, and Character*. University of Chicago Press, Chicago, Illinois.

Kramer, S.N. (1972) *Sumerian Mythology: A Study of Spiritual and Literary Achievement in the Third Millennium B.C.* University of Pennsylvania Press, Philadelphia, Pennsylvania.

Kuhn, T.S. (2012) *The Structure of Scientific Revolutions*, 4th edn. University of Chicago Press, Chicago, Illinois.

Laszlo, A. and Krippner, S. (1998) Systems theories: their origins, foundations, and development. In Jordan, J.S. (ed.) *Systems Theories and a Priori Aspects of Perception*. Elsevier Science, Amsterdam, pp. 47–74. DOI: 10.1016/S0166-4115(98)80017-4

Mijares, S. (2009) Ancient and modern Yoga: a science of breath, healing, and enlightenment. In: Mijares, S. (ed.) *The Revelation of the Breath: A Tribute to its Wisdom, Power, and Beauty*. State University of New York Press, New York, pp. 11–24.

Mijares, S., Rafea, A., Falik, R. and Schipper, J.E. (2007) *The Root of All Evil: An Exposition of Prejudice, Fundamentalism, and Gender Imbalance*. Imprint Academic, Exeter, UK.

Nadel, D., Danin, A., Power, R.C., Rosen, A.M., Bocquentin, F. *et al.* (2013) Earliest floral grave lining from 13,700–11,700-y-old Natufian burials at Raqefet Cave, Mt. Carmel, Israel. *Proceedings of the National Academy of Sciences USA* 110(29), 11774–11778. DOI: 10.1073/pnas.1302277110

Nelson, A.D. (2015) Origins of consciousness: how the search to understand the nature of consciousness is leading to a new view of reality. Lulu.com.

Parrinder, G. (1971) *World Religions: From Ancient History to the Present*. Facts on File, New York.

Rafea, A.R. (2009) Nature and man: past and present in Egypt. In: *Humankind/Nature Interaction: Past, Present and Future*. Atti Dell'IUAES Congress, Firenze, Luglio 2003. Alinea Editrice, Vol. 12.

Rafea, A.R. (2019) An integrative approach for facing climate change challenges. In: *Proceedings of the 13th International Conference on Development of Drylands, Converting Dryland Areas from Grey into Green,* 11–14 February, Jodhpur, India. International Dryland Development Commission (IDDC), pp. 425–435.

Schwartz, S.A. (2018) Kuhn, consciousness, and paradigms. *Explore: The Journal of Science and Healing* 14(4), 254–261. DOI: 10.1016/j.explore.2018.04.004

Stapp, H.P. (2001) Quantum theory and the role of mind in nature. *Foundations of Physics* 31(10), 1465–1499. DOI: 10.1023/A:1012682413597

UNESCO World Heritage Center (n.d.) Bikini Atoll Nuclear Test Site. Available at: https://whc.unesco.org/en/list/1339 (accessed 25 July 2024).

Vitek, B. (ed.) (2020) *The Perennial Turn*. New Perennials Publishing. Middlebury, Vermont. Available at: https://landinstitute.org/wp-content/uploads/2020/12/NP_PerennialTurn_ePDF.pdf (accessed 25 July 2024).

Von Bertalanffy, L. (1950) The theory of open systems in physics and biology. *Science* 111(2872), 23–29. DOI: 10.1126/science.111.2872.23

Whites, L. (1967) The historical roots of our ecologic crisis. *Science* 155(3767), 1203–1207. DOI: 10.1126/science.155.3767.1203

Wimmer, A. and Feinstein, Y. (2010) The rise of the nation-state across the world, 1816 to 2001. *American Sociological Review* 75(5), 764–790. DOI: 10.1177/0003122410382639

Zwart, H. (2019) Archetypes of knowledge: the relevance of Jung's psychology of scientific discovery for understanding contemporary technoscience. *International Journal of Jungian Studies* 12(2), 159–179. DOI: 10.1163/19409060-01102005

# 16 Protecting Vulnerable Lives and Livelihoods in Times of Climate Change

**Ismail Serageldin***
*Co-Chair, Board of Nizami Ganjavi International Center (NGIC)*

**Abstract**
The drylands are home to over 2 billion of the poorest and most vulnerable people on Earth, and climate change is further impacting their precarious existence. Measures for adaptation and resilience could reduce their vulnerability, especially when they address the gender dimension, and focus on developing human resources and institutional capabilities. Failures in promoting such measures to protect the lives and livelihoods of the vulnerable populations, could result in major migrations of environmental refugees that could reach some 200 million persons. Harnessing the scientific and technological revolution in both ICT and the new biology, such policies and programs could fit in a broader transformation of the food and agriculture sector from being a global emitter of greenhouse gases (GHG) to being a major force for carbon capture and sequestration (CCS), while saving lives and livelihoods everywhere. All this will require that we adopt science-based policies and evidence-based regulations to truly promote sustainable development.

## Introduction: The General Problem of Climate Change

Today, we can no longer deny the reality of climate change (CC) and its predicted manifestations, from extreme weather events such as hurricanes, tornadoes, and severe storms, to extended periods of record heat and cold, wildfires, and terrible cycles of droughts and floods. Less visible, but equally important, is the devastation of the Earth's biodiversity, and in the distance, we can also see the risks of sea-level rise.

### Beyond mitigation: building capacity for adaptation and resilience

More importantly, however, is the reality that the terrible consequences of CC are already here. The poor, who contributed least to creating the problem, are the ones suffering most from its consequences, and they are suffering now. So, as we continue to pursue the reduction of emissions and then move on to negative reduction of the legacy of emissions in the atmosphere, we must also think of promoting adaptation and resilience in the poorest areas of the planet.

In various discussions of CC, it has been recorded that, as a moral responsibility, the rich countries who built their wealth by emitting GHGs over centuries, owe the poor such assistance. A fund for adaptation projects in the global south was promised at $100 billion annually (a fraction of what is really needed), but even that promise was never wholly fulfilled.

But the problems of the poor go further. These countries lack the human technical expertise and

*Address for correspondence: Ismail@serageldin.com


DOI: 10.1079/9781800624870.0016

the institutional mechanisms that would enable them to effectively involve local communities with national plans and projects. Thus, capacity building in developing countries is going to be very important.

### The special vulnerabilities of the populations in the drylands

Some of the smallest islands among the developing states (the SIDS as they are called – small island developing states), are likely to completely disappear – for example, the highest point in the Maldives is only about eight feet above sea level, making it susceptible to increased flooding and coastal erosion due to CC; thus, the Maldives and Tuvalu – another small island in the Pacific Ocean – face the risk of being completely submerged in the future, which would displace their populations. The drylands (which have much larger populations than the SIDS) are the most vulnerable to the ongoing evolution of CC and its impacts.

Drylands are areas of the Earth's surface that receive low and erratic precipitation, resulting in limited water availability for plants, animals, and human populations. Drylands are characterized by arid, semi-arid, and dry regions such as deserts, steppes, and savannas. These regions are often harsh and challenging environments, with extreme temperatures, low humidity, and high winds, and are home to specialized plants and animals adapted to survive in these conditions (D'Odorico and Porporato, 2006). Human populations living in drylands also face unique challenges, including water scarcity, soil degradation, and limited economic opportunities.

According to the United Nations, although the drylands cover only about 40% of the world's surface, they are home to 80% of the world's poorest people. Furthermore, some of the highest levels of malnutrition and child mortality in the world are also found in the drylands.

These areas are also very vulnerable to the erratic rainfall patterns that alternate cycles of droughts and floods. And even though we tend to think of wildfires as occurring in forested areas, dryland ecosystems, including savannas and grasslands, are also subject to wildfires, as higher temperatures and drier conditions create more favorable conditions for fires to start and spread.

## Investing in Human and Institutional Capacity

Institutional development and capacity building are crucial for achieving sustainable development in poor countries (Sachs, 2005). Effective institutions can promote good governance, transparency, accountability, and the rule of law, which are essential for creating an enabling environment for development programs to succeed. Corruption, on the other hand, can undermine development efforts by reducing public trust, distorting markets, and diverting resources from their intended purposes. In summary, institution building and fighting corruption are essential components of any successful development program in poor countries.

But so is the investment in health and education that will empower youth to rise to the enormous challenges that their countries face.

Furthermore, much needs to be done in both health and education.

### Health status

Children in drylands may face health challenges related to drought, malnutrition, and lack of access to clean water and sanitation facilities. According to UNICEF, these issues may affect an estimated 160 million children in dryland areas worldwide (UNICEF, 2019; UN News, 2023). In particular, malnutrition can have long-term impacts on children's health and development, including stunted growth and impaired cognitive function (FAO *et al.*, 2021).

### Education status

In dryland areas, access to education can be limited due to factors such as poverty, distance to schools, and cultural attitudes towards education. According to UNESCO, around 32 million children of primary-school age and 34 million adolescents of lower-secondary school age in sub-Saharan Africa are out of school (UNESCO, 2021).

### Tertiary education and migration

Clearly, there are fewer opportunities for tertiary education in these very poor regions compared to industrialized countries. As a result, some graduates may choose to migrate to richer countries in search of better opportunities. This creates what has been referred to as 'brain drain', further undermining the institution building effort by depriving the poorest countries of their most promising talent.

## Migration: The Double-edged Sword

For many poor countries where there are not enough local opportunities for a youthful and growing labor force, migration becomes an unavoidable corollary to improved education and mobility. Yet the phenomenon of 'brain drain' is very real.

But across the board, the poor are migrating in search of the possibility for a decent life somewhere beyond the misery of their present conditions, whether that is a move within the same country (internal migration) or across political frontiers (international migration). We need to recognize the complexities of each of these three types of migration, and also to be prepared for a massive increase in the numbers of migrants as CC continues to impact the poorest countries most. So, let's look at each of the three facets of migration; brain drain, internal migration, and international migration.

### Migration and brain drain

Brain drain is a controversial issue because it can lead to a loss of human capital in developing countries, which can hinder economic development and exacerbate social inequalities. On the other hand, some argue that brain drain can also have positive effects, such as remittances and the transfer of knowledge and skills.

What is the best way for the developing countries experiencing brain drain to respond to the challenges that the phenomenon poses? For a long time now, eminent practitioners have been arguing that dealing with the emigration of highly talented individuals is best handled by rejecting the strategy of trying to forcibly retain highly skilled individuals domestically, despite the limitations that the local environment imposes on their ability to blossom. A more viable alternative is the 'diaspora' model, which incorporates both current and former citizens into a network of rights and responsibilities within the broader community and build links to their home country. This approach is preferable from a human rights perspective since it respects the right to emigrate instead of trying to limit it. Additionally, having dual loyalty is now more widely viewed as acceptable rather than otherwise. This option is also becoming more feasible, as more and more countries now permit dual citizenship (Bhagwati, 2003).

Yet beyond brain drain, the forced migration of millions of very poor people trying to flee their terrible conditions in their home countries is equally true.

That migration is going to be enormously increased by the negative consequences of CC, and it occurs through both internal displacement inside their country of origin, and internationally by crossing international frontiers in search of a better life.

Our world is well on its way to a harsh climate where devastating droughts, floods, extreme weather, destruction of natural resources (particularly lands, soils, and water), and severe living conditions, including famine and starvation, can affect up to 1 billion people (UNEP, 2012).

Although scientific projections are not yet certain, the rapid pace at which CC is occurring could result in this scenario by 2050. If this were to happen, approximately one in every nine individuals would be compelled to migrate (internally or internationally).

Current predictions range from 25 million to 1 billion environmental migrants by 2050. These individuals may relocate permanently or temporarily, either within their own countries or across borders. According to a study conducted by the Institute for Environment and Human Security of the United Nations University in 2015, the most commonly cited estimate is 200 million (Brown, 2008).

It is in the interest of the rich countries of the north to help and to fund the global south generally and the drylands more specifically in designing better approaches and projects for adaptation and the promotion of resilience.

And then, together, they may be able to reduce the predicted tidal wave of international migration, or 'environmental refugees' as they have been named.

But what kind of approaches and projects could promote adaptation in the drylands?

## Designing Approaches and Projects for Adaptation

Some look at the pervasive poverty in the drylands, the very limited infrastructure, the almost total dependence of large parts of these poor populations on an increasingly erratic rainfall that alternates from severe drought to massive floods, and express doubts that much can be done. However, there are many promising approaches and encouraging projects for adaptation that are taking shape in the drylands. The following are some examples of successful approaches to designing policies and projects for adaptation in the poorest areas, especially in the drylands.

### Farmer-managed natural regeneration (FMNR)

FMNR is a technique that involves the systematic regeneration of trees and shrubs in degraded land areas. FMNR has been successfully used in many countries, including Niger, Mali, Burkina Faso, and Ethiopia, to increase agricultural productivity and improve the livelihoods of rural communities.

### The Great Green Wall

The Great Green Wall is an African-led initiative that aims to plant a wall of trees and vegetation across the Sahel region of Africa to combat desertification and improve the resilience of dryland communities to CC. The initiative involves 11 countries and has received support from various international organizations.

### Community-based adaptation (CBA)

CBA is an approach that involves engaging local communities in the design and implementation of CC adaptation measures. CBA has been successfully implemented in many dryland regions, including the semi-arid regions of Kenya, where it has been used to improve water management, increase agricultural productivity, and reduce poverty.

### Sustainable land management (SLM)

SLM involves the adoption of land management practices that improve soil health, increase water availability, and promote biodiversity. SLM has been successfully implemented in many dryland regions, including the Loess Plateau in China, where it has been used to improve soil quality, increase crop yields, and reduce erosion.

### Livestock insurance

Livestock insurance has been successfully used in many dryland regions to help farmers cope with the impacts of CC, such as drought and floods. For example, the Index-Based Livestock Insurance (IBLI) program in Kenya and Ethiopia uses satellite imagery and rainfall data to provide insurance to pastoralist communities.

It is also notable that some of these same measures suggested as adaptation measures can also be helpful in improving the resilience of dryland communities and promoting sustainable development in these regions. But resilience requires more.

## Promoting Resilience

Resilience is the ability of an existing system to withstand an external shock and 'bounce back' to its original functionality. Nations are likely to receive ever more such shocks as CC brings ever more extreme weather events and, *inter alia*, transforms the expected patterns of rain and drought.

In 2015, at the same time that humanity adopted the Sustainable Development Goals (SDGs) and the Paris Agreement, the world adopted an agreement to promote 'resilience' by reducing the risk of the disasters that are likely to strike various nations. It is known as the

Sendai Framework for Disaster Risk Reduction 2015–2030 or simply the Sendai Framework (UNDRR, 2021).

But the efforts of building capacity for disaster risk reduction (DRR) are also directed at building capacity to confront slow-onset disasters (Cancun Agreement – COP16) such as desertification and increasing salinity.

### A broader view of resilience

Resilience is a difficult concept to pin down. It does involve the national technical, economic, and institutional capabilities. Clearly, the rich industrialized countries with their sophisticated institutions and advanced technologies, communications, infrastructure, and health facilities have much more capacity to respond to any disasters that hit their populations, whereas the poorest countries, especially in the drylands, suffer profoundly. Compare floods in Germany and drought in Somalia.

But resilience also involves the social and human links – what has been referred to as 'social capital' (Nan Lin, 2001) – and the presence of strong social solidarity at the local community level is a major determinant of coping with consequences of disaster. Internal conflicts based on ethnic, religious, or other lines of cleavage in a country undermine the capability of society to respond. Women, victimized by society and disproportionately affected by climate and other disasters, play an important role in promoting the social capital in a community and more broadly in a society, thereby increasing resilience.

## Transforming the Food and Agriculture Sector

### A broad perspective

As we engage with increasing the adaptation and resilience of agriculture in the arid zones, we can see that work as part of a bigger transformation of the entire food and agriculture sector (FAS) globally, transforming the FAS so that it goes from being a major emitter of GHGs to being a major contributor to the solutions that humanity brings to meeting the challenge of CC (Smith *et al.*, 2008). Furthermore, that transformation would also enable the sector to be much more efficient and effective in its mission to reduce poverty and abolish hunger, even as it protects our ecosystems.

This broad perspective requires that we address many different aspects of the complex multifaceted reality of the FAS (Smith *et al.*, 2006). These aspects include: changing landscapes; deforestation, afforestation and reforestation; agroforestry; better water management; working with nature not against it; moving to precision farming; scientific management; land-use changes; sustainable land management; carbon sequestration in soils; urban green spaces; urban agriculture; the right kind of seeds; and reducing waste.

The revolution in ICT (including AI technologies) and in the new biology gives us a toolkit that enables us to do much about each of these topics.

### Land-use changes

Half of the world's habitable land (Ellis *et al.*, 2010) is used for agriculture (Ritchie, 2019). More than three-quarters of this is used for livestock production, despite meat and dairy making up a much smaller share of the world's protein and calorie supply (Poore and Nemecek, 2018); and this very frequently is a primary reason for deforestation.

### Deforestation

Globally, we deforest around 10 million ha of forest every year. That is an area the size of Portugal. Around half of this deforestation is offset by re-growing forests, so overall we lose around 5 million ha each year. Nearly all – 95% – of this deforestation occurs in the tropics.

The result of this rate of deforestation is that over the past 60 years, the global forest area has declined by 81.7 million ha, a loss that contributed to more than a 60% decline in global forest area per capita.

But we are fighting deforestation everywhere. The following are a few examples of land-use changes that can help respond to CC.

### Afforestation and reforestation

Planting trees and increasing forest cover can help sequester carbon dioxide from the atmosphere, reduce the risk of wildfires, and improve soil quality.

### Conservation agriculture

Conservation agriculture involves the adoption of practices such as reduced tillage, crop rotation, and the use of cover crops to improve soil health, increase carbon sequestration, and reduce GHG emissions.

### Agroforestry

Agroforestry involves the integration of trees into agricultural landscapes to provide multiple benefits, such as improving soil health, increasing biodiversity, and reducing GHG emissions.

### Sustainable land management

Sustainable land management involves the adoption of practices such as improved grazing management, water harvesting, and soil conservation to improve soil health, increase productivity, and reduce GHG emissions.

### Urban green spaces

Increasing the number of green spaces in urban areas can help reduce the urban heat island effect, improve air quality, and sequester carbon dioxide from the atmosphere.

These are just a few examples of land-use changes that can help respond to CC. The specific land-use changes needed will depend on local conditions and the particular challenges faced in different regions.

### Water management

Agriculture is responsible for about two-thirds of global water withdrawals, and in many of the poorest countries, that figure can exceed 80%. In the drylands generally, including the Middle East, the world's most water-stressed region, CC and global warming have severely impacted water resources and their quality, resulting in frequent droughts and depleted water reserves.

Insufficient rainfall and desertification have also led to political instability in this volatile area, exacerbating the risk of local conflict – not to mention the possibility of failed states. Beyond agriculture, the combination of extreme climate conditions, mismanagement of water resources, and inadequate infrastructure has also deprived the growing population of safe drinking water, food, shelter, and stability. This is contributing to expanded desertification and could also cause mass migration and displacement.

In light of these challenges, the drylands must prioritize sustainable water resource management by adopting advanced and innovative technologies, implementing good governance practices, promoting social empowerment, and safeguarding aquatic ecosystems. Although advanced water conservation, consumption monitoring, data management, and public participation measures are already available in large parts of the Middle East, and some other parts of the drylands, their implementation on a large scale throughout these lands, along with initiatives such as water reuse, resource recovery, and desalination, must be pursued.

Globally, we must recognize that some 260 international rivers are shared by two or more countries. These riparian countries need to establish general frameworks of cooperation with each other as rainfall becomes more erratic and droughts become more common, to avoid the rise of potential conflicts between them (Pearce, 2012).

### The right plants

It is important that the traits that we breed in plants reflect the current challenges we have to face. Thus, we want plants that are resistant to drought and salinity, have deep roots, and a short growing season. They must of course be high-yielding and have a high nutritional content. Ideally, we should also be concerned about senescence to reduce post-harvest losses.

Today, the new biological techniques, such as CRISPR/Cas9 and other genome-editing techniques, give us the tools we need. Not only should we be able to produce the archetype described above but, also, they should be suitably refined to respond to the needs of various regional and local agro-ecologies. Tying that with a good agricultural extension service and global scientific collaboration should enable the deployment of the appropriate seeds to the right places in time. Accordingly, the poor smallholder farmers in many of the regions at risk, starting with the dry areas, should be able to benefit from the fruits of the new science, whose results are imbedded in the seed that is made available in various locations.

### Carbon storage in the soil

While everyone is familiar with the fact that plants and trees consume $CO_2$ and exhale oxygen, less understood is that the soils in which the plants, shrubs, grasses, and trees are planted in are themselves major stores of carbon, and that these soils play an important role in carbon capture and sequestration. Professor Rattan Lal has invested years in this extremely valuable research (Lal and Bruce, 1999; Lal, 2004; Lal, 2010; Lal, 2016), and in the last decade these ideas have come to be accepted in mainstream thinking about agriculture and in environmental circles as well (Schwartz, 2014).

### Reducing methane emissions from livestock and better farming

GHGs are emitted by livestock and dairy operations through the digestive processes of ruminants, primarily cattle and sheep, as well as the decomposition of manure (Gerber *et al.*, 2013). By modifying livestock feeds, methane emissions can be reduced by livestock managers. To capture methane produced during manure storage, dairy and hog producers can install digesters, which can then be used to generate electricity. Digesters not only convert methane emissions from manure disposal into less-potent $CO_2$ but they also produce energy, thereby replacing $CO_2$ emissions that would have resulted from fossil fuel-based electricity.

### Enhancing carbon sequestration on agricultural land

To boost carbon sequestration, farmers can implement practices that increase the retention of crop residues in the fields and enhance the storage of organic matter in the soil (Eve *et al.*, 2002). These practices encompass transitioning from conventional tillage to conservation tillage or no-till approaches, minimizing or eliminating fallow periods in planned crop rotations, shifting from annual to perennial crops, augmenting field residues through irrigation, fertilization, planting cover crops or hay, and introducing additional organic materials like manure.

### Fish and aquatic resources

The discussions in the preceding paragraphs have not addressed the importance of fish and aquatic resources that also contribute significantly to the protein diet of populations in many countries (De Silva and Funge-Smith, 2005).

The significance of fish and aquatic resources in ensuring global food and nutritional security cannot be overstated. Though fish comprise a small amount of global protein intake (6.7%), they are an important source of animal protein, providing 17% of the world's meat consumption. Fish play an outsized nutritional role for many people; 3.1 billion people rely on fish for 20% of their daily protein intake, with some coastal communities reliant on fish for upwards of 70%.

### Increasing efficiency and eliminating waste

Eminent sources state that, as of 2021, one third of the world's food is wasted, which is equivalent to about 1.3 billion tons, which translates to a cost of approximately 1 trillion dollars (Mahroof and Sivarajah, 2021).

According to FAO, the food wastage includes up to 45% of all fruits and vegetables, 35% of all fish and seafood, 30% of all cereals, 20% of all dairy products, and 20% of all meat and poultry.

In the poor countries, in many parts of Africa, there are some 35% post-harvest losses. Inadequate infrastructure from storage to transport systems is at least partly responsible for a large part of these post-harvest losses. In the rich countries, wastage tends to be on the consumption side. As much as 40% of food produced, processed, and transported in the USA is wasted and ends up in landfill.

The UNEP Food Waste Index Report 2021 points out that the substantial amounts of food produced but not eaten by humans has substantial negative impacts, environmentally, socially and economically. Estimates suggest that 8–10% of global GHG emissions are associated with food produced but not consumed (UNEP, 2021).

### Overall reductions of GHG emissions in all parts of the food and agriculture sector

To transform the FAS from being a major emitter of GHGs into being a major force for reducing such emissions is feasible and should be the goal of much of what we do in this area. After all, the plants are nature's system of $CO_2$ capture and sequestration, and we must learn to work with nature, not against it. Some may be concerned that conservation and production of food may be in conflict, but the fear of such a trade-off is well known (Smith *et al.*, 2013), and we now believe that both objectives can be pursued together. This will be achieved by many actions on many fronts, and all this will enable the sector to better respond to the challenge of global poverty and hunger (including in the drylands).

## Nurturing Technological Innovation

We need innovations that can lead to reduced waste, increased water-use efficiency at the level of plants as well as farms, not to mention at the scale of significant irrigation systems, while limiting the continued depletion of underground reservoirs. Crops that not only have higher yields but are also more nutritious and more resistant to drought and salinity, with shorter growing seasons and deeper roots and reduced senescence are within reach. They will not only help meet humanity's needs, they would also improve the economic conditions of farmers, especially the smallholder farmers in the low-income countries (LICs) and the lower-middle-income countries (LMICs).

We are living in a period of scientific and technological revolutions covering everything from AI to supercomputing, to genome editing to nanotechnology to new materials, and many other areas. Each of these opens avenues that are filled with promise and fraught with risks, but also where sagacious management can lead to more innovative approaches to handling existing problems. Innovation in dealing with agriculture, water, and capacity building issues is also badly needed. We must learn to deploy such innovations in ways that work with nature not against it.

### Contributions from the new biology

#### *Gene editing*

This has emerged in the past decade as a platform technology with enormously broad potential. It is a powerful new toolkit for developing clean energy and climate (Giddings *et al.*, 2020). It has already developed solutions that policymakers have so far underemphasized. Among these is the ability of using gene editing to enhance the efficiency of photosynthesis, reduce methane emissions from cows and rice paddies, optimize biofuel crops, and solve many other climate challenges.

The revolution in biology is helping agriculture cope with many problems, and it is also helping with emissions more generally (ISAAA, 2020). Indeed, the new gene-editing tools are now unlocking new strategies to improve natural and agricultural carbon sinks, limit emissions from agriculture and other major sectors contributing to GHG emissions, and revolutionize biofuels (Hart *et al.*, 2021). These techniques when applied to soils are also capable of helping in combating desertification, while they can also be deployed to water bodies to enhance $CO_2$ capture by microorganisms such as cyanobacteria, green algae and some autotrophic bacteria which could potentially fix $CO_2$ more efficiently than higher plants, due to their faster growth (Rossi *et al.*, 2015).

Furthermore, although traditionally legumes are the plants that could promote nitrogen fixation at their roots, new research shows that seeds coated with fertilizer–generating microbes could emulate that process, and even grow in harsh and salty soils (Chandler, 2023).

Even without getting into the new techniques of genome editing, some have already recognized that the best way to promote carbon capture and sequestration is to maximize the use of nature's own system for doing so: plants, including trees. And since we need to combat deforestation, planting trees was seen as a great way to respond to the challenge.

One of the solutions presented is the Trillion Trees Initiative (Pomeroy, 2020), which recognized the plant's unique ability to use photosynthesis to collect carbon, though the process is not very efficient. Thus, experts are using gene-editing tools to improve it (Osborne, 2023). When these new techniques turn out successful technologies, it is projected that this will lead to 50% productivity gains in major crops, while slashing emissions significantly and boosting farmers' incomes (Hamilton, 2022). Generally speaking, the genetic tools that have helped create resilient crops could boost carbon removal, too (Crownhart, 2022).

Most of the agricultural GHG emissions are due to decomposition and transport of wasted food. Gene-edited soybean oil with longer shelflife and potatoes that resist bruising are now commercially available, contributing to waste reduction. Gene editing has the potential to enable low-carbon transport by optimizing microbes that can efficiently process cellulose.

About 6% of global agricultural GHG emissions are contributed by cattle, particularly due to the methane in their burps. Some cattle have been found to emit lower amounts of methane due to specific microbial populations found in their digestive tracts. With gene editing, this trait can be transferred to herds and thus contribute to reducing emissions.

### *Deploying the best management and using ICT and remote sensing to help the vulnerable populations*

Improving management of a farm even without the benefit of the new ICT or biological tools brings about notable results in improved output. Enhancing people's ability to cope with various risks and limitations faced by vulnerable populations can be achieved through biologically astute farm management. Although macro-socio-economic interventions can provide assistance, they are not the sole dependency for implementing these measures. One example of such management is the system of rice intensification (SRI), which demonstrates that cropping systems can be made somewhat resistant to climate-related challenges. This is accomplished by promoting the growth and health of plant root systems and nurturing the strength and fertility of the soil systems in which plants thrive. SRI unveils significant productive potentials present in existing rice plant genomes and soil systems, which have been overlooked and underutilized within the prevailing 'modern agriculture' paradigm that relies heavily on chemical fertilizers and pesticides (Uphoff, 2007).

But beyond these general principles of good management, we now have a new toolkit: remote sensing, small drones, and handheld mobiles or tablets all help open dramatically improved possibilities.

Remote-sensing technologies provide an important tool to aid site-specific management of crops. Remote sensing has the potential to provide real-time analysis of the attributes of a field in terms of nutrients and water availability. Easily obtainable commercial satellite imagery can be supplemented by small inexpensive drones that can give a granularity of squares up to 20 cm across. With a little bit of assistance from the local national Agricultural Research and Extension Service, and distribution of handheld phones or tablets, even smallholder farmers will be able to usefully improve their administration of water and nutrients and other requirements for an effective precision farming approach. It has been shown through schemes such as M-Pesa in Kenya that even the poorest people can manage using handheld mobile phones effectively for complex banking operations and transfers. M-Pesa now serves more than 51 million customers across seven countries in Africa.

### *Mobilizing AI to help the poorest*

It is clear, then, that the promising new biological technologies, and the new ICT-related technologies are indeed bringing us closer to our dual objective of:

(i) transforming the food and agricultural sector; and
(ii) bringing precision agriculture and its benefits to the smallholder farmers in the impoverished communities of the difficult areas of the drylands.

And now, in addition to the revolutions in biology and ICT, AI has been deployed in new and revolutionary ways: the Farmer.CHAT.

AI GPT4 technology has now been mobilized to help the smallholder farmer. Speaking in his or her own language, the farmer can ask and get answers through the Farmer.CHAT app and get trusted answers spoken in the same language and/or videos with voice-overs, again in the same language. AI in this fashion has dramatically extended the reach of the agricultural extension worker, while also allowing governments and agricultural extension services to have a much broader monitoring of conditions in the field in real time by these enhanced interactions with the farmers themselves.

On 12 April 2023, Farmer.CHAT was featured at the United Nations 77th General Assembly in New York, as part of the 2023 UN General Assembly briefing on 'Science-based evidence in support of sustainable solutions'.

All these promising innovations need to be deployed throughout the world, starting with the most promising programs for increasing adaptation and resilience for the poorest and most vulnerable populations, especially in the drylands. But can we mobilize the necessary funding to do that?

## Finance

The conventional mechanisms of aid and international finance need a major overhaul if they are going to help shift the world from its unsustainable path and seriously cope with the existential risks of CC. This is equally true of the need for major investments to achieve the proper management of water in our lives. We need to rethink funding structures, deployment vehicles, and finance mechanisms to maximize the effectiveness of public funds used to ensure that they support and accelerate private sector engagement.

We have to remember that in 2020, assets of financial institutions world-wide already amounted to US$ 468.7 trillion (Statista Research Department, 2022). The use of new instruments such as guarantees, insurance, and other tools could very significantly tap into that enormous amount of finance in order to increase the amount of financing available for projects that accelerate the energy transition as well as help fund projects for adaptation and resilience.

The conventional mechanisms of aid and international finance need a major overhaul if they are going to help shift the world from its unsustainable path and seriously cope with the existential risks of CC. This is equally true of the need for major investments to achieve the proper management of water in our lives. The current estimates of the investments required to achieve the energy transition globally, thereby reducing GHG emissions to net zero by 2050, and the needed investments for adaptation and increased resilience to confront the likely consequences of CC in the developing countries and to address the emerging water crisis globally, will amount to somewhere in the range of \$2–5 trillion, annually, for several decades.

Against these huge needs, we note that total Official Development Assistance (ODA) from the OECD countries amounts to about \$200 billion, annually, and that is for all sectors, not just climate-related ODA. The various multilateral Green Funds deploy another \$30 billion or so. It is clear that simply demanding that this funding come from these sources is unreasonable. There is no possibility for a tenfold expansion of ODA, and there is no possibility for the countries of the global south to generate such resources out of their overstretched government budgets.

We need to define a portfolio of policies and projects that would benefit the global south in responding to the challenges of CC with proper mitigation, adaptation, and resilience interventions.

We need to rethink funding structures, deployment vehicles, and finance mechanisms to maximize the effectiveness of public funds used to ensure that they support and accelerate private sector engagement. The public funds should be used to prepare the projects, and then de-risk such projects by financing guarantees, insurance, and hedge facilities for currency variations, so that they become attractive to private sector financing. If such de-risked projects are then assembled into portfolios with regional or sectoral characteristics and are actively marketed in a sufficiently liquid fashion to allow investors to buy fractions of what would be portfolios of collateralized debt obligations appropriately mediated by the international banking system, the trillions of dollars needed become more achievable.

The question remains as to what institutions are capable of developing these portfolios and marketing them to investors worldwide. Some would argue that the IFIs of the Bretton Woods System could be given an expanded mandate to do so. But other knowledgable experts think that to undertake such an enterprise effectively requires a new institution, a public-private global Green Bank (Ghanem, 2023). These and other proposals are all going beyond the simple repetition of demands for more funds, demands that are not likely to be met, towards promising avenues of the kind of transformative finance the world needs to confront the challenges of CC.

Finally, there are new ideas emerging in how to handle the direct contributions of the rich countries towards the existing facilities (including the Loss and Damage Fund approved in principle at COP27 in Sharm el Shaikh, Egypt, in 2022). These include assessments that would take into account the various contributions of the rich countries to the creation of the problem through GHG emissions over many decades. Fairness and justice also require that these approaches be given serious consideration as we continue to support the existing public-financed vehicles for support to the vulnerable among the human family.

## Addressing the Gender Dimensions

The IPCC and other research bodies have acknowledged that CC has differential impacts on women compared to men, particularly in the context of droughts and floods. This is especially true in the drylands where women carry a disproportionate share of dealing with such things as fetching water, cooking, coping with the diseases of the children, and so much more. But more importantly, in many societies, it is women who produce much of the food and who participate in its marketing.

But women are also the primary builders of community solidarity and the social capital that these poor societies need to nurture in order to increase their resilience and promote their sustainable development.

Thus, if we look at the gendered impacts of CC, we can identify a number of serious issues.

### Disproportionate vulnerability

Women often face higher risks and vulnerabilities due to social, economic, and cultural factors. They may have limited access to resources, information, decision-making power, and financial assets, which can hinder their capacity to cope with and adapt to CC impacts (IPCC, 2014). Furthermore, when the political situation deteriorates, their burden in moving (internally or internationally) is huge, as they invariably also have to take care of the children.

When agricultural failures, droughts, floods, or fires devastate people's means of sustenance and residence, individuals move with the aspiration of bettering their circumstances. However, this movement often exposes them – especially women and children – to risks of being victimized by predatory criminals engaging in human trafficking, forced labor, sex slavery, and various human rights abuses.

### Increased workload and time burden

In many societies, women are responsible for securing water, food, and fuel for their households. Climate change can exacerbate these responsibilities as droughts and floods affect water availability, crop productivity, and fuelwood accessibility. This leads to increased workloads and time burdens on women, limiting their opportunities for education, income generation, and participation in decision-making processes.

### Health impacts

Women's health can be disproportionately affected by CC-related droughts and floods. Limited access to clean water and sanitation facilities during droughts can increase the risk of waterborne diseases, affecting women who are responsible for water collection and sanitation. Flooding can also lead to increased health risks, including reproductive health issues and the spread of waterborne diseases (Watts *et al.*, 2020).

### Displacement and migration

CC-induced droughts and floods can result in population displacement and migration. Women, particularly those who are marginalized, may face specific challenges as they often bear the burden of caring for family members, including children and the elderly, during displacement. They may also be exposed to gender-based violence and discrimination in new environments (Brody *et al.*, 2008).

### Adaptation and resilience

Despite facing numerous challenges, women play a crucial role in CC adaptation and building resilience within their communities. Their knowledge of local ecosystems, traditional practices, and resource management can contribute to effective adaptation strategies.

To address these gendered impacts, it is important to promote gender equality, women's empowerment, and women's participation in decision-making processes related to CC mitigation, adaptation, and disaster risk reduction. Additionally, incorporating gender analysis into climate policies and programs can help ensure that the specific needs and perspectives of women are taken into account (UN Women, 2018).

But whatever the analyses, it is clear that the increased education of girls and empowerment of women are prime instruments to promote these desirable outcomes of amplified social solidarity and improved community resilience in the face of sudden shocks, as well as slow-onset disasters.

## Conclusions: An Existential Challenge

### The Reality of Climate Change

We are facing an existential challenge. The IPCC report (AR6) of 2021 shows that '...we've currently reached 1.1°C above the average temperatures from 1850–1900. Not in 125,000 years has Earth seen a time period with such a high average temperature for such a prolonged period of time, indicating that this is more than just a natural heat spike...This is coupled with reporting that atmospheric $CO_2$ is at a 2,000,000-year peak and that other greenhouse gases like methane and nitrous oxide are at 800,000-year peaks' (IPCC, 2021).

Scientists at the US National Oceanic and Atmospheric Administration (NOAA) published the following 'highlights' to underline the seriousness of the current situation:

- Each year, human activities release more carbon dioxide into the atmosphere than natural processes can remove, causing the amount of carbon dioxide in the atmosphere to increase.
- The global average carbon dioxide set a new record high in 2021: 414.72 parts per million.
- Despite the continued economic drag of the COVID-19 pandemic, the increase over 2020 was tied for fifth-largest one-year increase in the 63-year record.
- The annual rate of increase in atmospheric carbon dioxide over the past 60 years is about 100 times faster than previous natural increases, such as those that occurred at the end of the last ice age 11,000–17,000 years ago.
- The ocean has absorbed enough carbon dioxide to lower its pH by 0.1 units, a 30% increase in acidity.

Beyond the extreme weather events we are witnessing, beyond the cycles of droughts and floods, the wildfires and the melting ice, CC will have an enormous cumulative economic effect on the global economy. CC could reduce global economic output by up to $23 trillion by 2050 in comparison to growth levels that would occur in the absence of CC. This represents a potential decrease of 11–14% in annual global economic output worldwide due to the impact of CC. This sobering analysis is reported by Swiss Re, one of the world's largest providers of insurance to other insurance companies.

## The need for promoting adaptation and resilience to protect the lives and livelihoods of the vulnerable poor

We also see that the impacts of CC are already with us now. One shudders to think what they would be like if the trend of continued growth in emissions and increase in global warming continue. But even now, in many of the areas of the global south, the poor are suffering from the impacts of CC today. Efforts to curb emissions must be accompanied by measures to promote adaptation and resilience in the global south. The vulnerable communities, especially in the drylands, need assistance and funding to promote adaptation and resilience in their local environments. But this will not be just for projects; there must be substantial levels of investment in human and institutional capacity. Failure to do so will probably translate into a massive increase of 'environmental refugees' that could result in massive migrations that could reach 200 million persons or more.

## The role of women

The promotion of adaptation and resilience must be accompanied by a major emphasis on the gender dimension of both the challenges confronted and the solutions proposed. Women are central to building social capital and are frequently the major producers of food in many of the poorest settings. Therefore, educating girls and empowering women remain the central tenet that must be pursued. Thus empowered, women will go from being victims today to becoming the architects of a better future.

## The transformation of the food and agriculture sector

Agriculture fulfills the fundamental needs of humanity by providing food and fiber. Over the past century, the implementation of novel farming methods, exemplified by the Green Revolution, has enabled agriculture to meet the escalating demand for food and agricultural products and address the needs of a growing population.

Nevertheless, the future holds challenges such as increased food requirements, population growth, and higher income levels (accompanied by increased consumption of animal proteins), all of which are likely to place additional strain on natural resources. Considering that the global population is projected to exceed 9 billion by 2050, a 60–70% increase in food production will be necessary to sustain us if current global trends in diet and population continue (Alexandratos and Bruinsma, 2012).

But with an increasing recognition of the detrimental environmental effects of current modes of agricultural production, it is necessary to adopt new techniques and approaches that can fulfill future food demands while mitigating or minimizing the environmental impact of farming. But we can go further. We can envisage the transformation of the food and agriculture sector from being a net contributor to GHG emissions to becoming part of the solution by reducing such emissions as the plants, trees, and natural landscapes absorb and sequester carbon. And by working with nature, not against it, we can achieve more, much more.

## Envoi

In the preceding discussions, we have seen that we have the tools to do that. What is more, innovations that expand that toolkit from the new biology to digital transformations, to applications of AI are all happening at a bewildering pace. We can and must adopt and deploy these innovations. We can and must mobilize the requisite financing. We can meet the existential challenge of CC. We can redeem the promise of sustainable development and get humanity back to the implementation of the SDGs.

We can do all that by remembering that development, real development, is when the weak, the poor, and the destitute become the producers of their own welfare and bounty, not the recipients of charity or the beneficiaries of aid. We can achieve that by adopting policies and projects that reflect real priorities, nurture innovation, and empower people.

The world needs science-based policies and evidence-based regulations, not political posturing.

## References

Alexandratos, N. and Bruinsma, J. (2012) World agriculture towards 2030/2050: the 2012 revision. ESA Working Paper No. 12–03. FAO, Rome. Available at: https://www.fao.org/3/ap106e/ap106e.pdf (accessed 25 July 2024).

Bhagwati, J. (2003) Borders beyond control. *Foreign Affairs* 82(1), 98–104. DOI: 10.2307/20033431

Brody, A., Demetriades, J. and Esplen, E. (2008) *Gender and Climate Change: Mapping the Linkages – A Scoping Study on Knowledge and Gaps*. BRIDGE, Institute of Development Studies, Brighton, UK. Available at: http://www.adequations.org/IMG/pdf/GenderAndClimateChange.pdf (accessed 25 July 2024).

Brown, O. (2008) Migration and climate change. In: IOM Migration Research Series, No. 31. Executive Summary. United Nations, p. 9. Available at: https://publications.iom.int/system/files/pdf/mrs-31_en.pdf (accessed 25 July 2024).

Chandler (2023) *Seeds Coated with Fertilizer-Generating Microbes May Enable Agriculture on Marginal Land*. Massachusetts Institute of Technology, SciTechDaily. Available at: https://scitechdaily.com/seeds-coated-with-fertilizer-generating-microbes-may-enable-agriculture-on-marginal-land/ (accessed 4 June 2023).

Crownhart, C. (2022) *Why Scientists Want to Help Plants Capture more Carbon Dioxide: Genetic Tools that Have Helped Create Resilient Crops Could Boost Carbon Removal, Too*. MIT Technology Review. Available at: https://www.technologyreview.com/2022/10/19/1061931/plants-crispr-carbon-dioxide-removal/ (accessed 3 June 2023).

De Silva, S.S. and Funge-Smith, S. (2005) A review of stock enhancement practices in the inland water fisheries of Asia. FAO publication. Asia-Pacific Fishery Commission, Bangkok. Available at: https://www.fao.org/3/ae932e/ae932e02.htm (accessed 5 June 2023).

D'Odorico, P. and Porporato, A. (2006) *Dryland Ecohydrology*. Springer, Dordrecht, The Netherlands. DOI:10.1007/1-4020-4260-4

Ellis, E.C., Goldewijk, K.K., Siebert, S., Lightman, D. and Ramankutty, N. (2010) Anthropogenic transformation of the biomes, 1700 to 2000. *Global Ecology and Biogeography* 19, 589–606. DOI: 10.1111/j.1466-8238.2010.00540.x

Eve, M.D., Sperow, M., Howerton, K., Paustian, K. and Follett, R.F. (2002) Predicted impact of management changes on soil carbon storage for each cropland region of the coterminous United States. *Journal of Soil and Water Conservation* 57(4), pp. 196–204. Available at: https://www.jswconline.org/content/57/4/196 (accessed 25 July 2024).

FAO, IFAD, UNICEF, WFP and WHO (2021) The state of food security and nutrition in the world 2021. In: *Transforming Food Systems for Food Security, Improved Nutrition and Affordable Healthy Diets for All*. FAO, Rome. DOI: 10.4060/cb4474en

Gerber, P.J., Steinfeld, H., Henderson, B., Mottet, A., Opio, C. *et al.* (2013) *Tackling Climate Change through Livestock – A Global Assessment of Emissions and Mitigation Opportunities*. Food and Agriculture Organization (FAO), Rome. Available at: http://www.fao.org/docrep/018/i3437e/i3437e.pdf.

Ghanem, H. (2023) The world needs a Green Bank. Policy Brief. Policy Center for the New South, Rabat. Available at: https://www.policycenter.ma/publications/world-needs-green-bank (accessed 25 July 2024).

Giddings, V., Rozansky, R. and Hart, D. (2020) Gene editing for the climate: biological solutions for curbing greenhouse emissions. *Information Technology and Innovation Foundation (ITIF)*. Available at: https://itif.org/publications/2020/09/14/gene-editing-climate-biological-solutions-curbing-greenhouse-emissions/ (accessed 20 May 2023).

Hamilton, E. (2022) Altered gene helps plants absorb more carbon dioxide, produce more useful compounds. *News Media*. Available at: https://news.wisc.edu/altered-gene-helps-plants-absorb-more-carbon-dioxide-produce-more-useful-compounds/ (accessed 1 June 2023).

Hart, D., Rozansky, R. and Giddings, V. (2021) Gene editing and agriculture: challenges and opportunities to significantly address climate change. The Gene Literacy Project. Available at: https://geneticliteracyproject.org/2021/12/09/gene-editing-and-agriculture-challenges-and-opportunities-to-significantly-address-climate-change/ (accessed 4 June 2023).

IOM (2019) *World Migration Report 2020*. McAuliffe, M. and Khadria, B. (eds). Geneva. Available at: https://publications.iom.int/system/files/pdf/wmr_2020.pdf (accessed 2 June 2023).

IPCC (2014) Climate change 2014: Impacts, adaptation, and vulnerability. Part B: Regional aspects. In: Barros, V.R., Field, C.B., Dokken, D.J., Mastrandrea, M.D., Mach, K.J. (eds) Contribution of Working

Group II to the Fifth Assessment Report of the Intergovernmental Panel on Climate Change. Cambridge University Press, Cambridge, UK and New York. DOI: 10.1017/CBO9781107415386. Available at: https://www.ipcc.ch/site/assets/uploads/2018/02/WGIIAR5-PartB_FINAL.pdf (accessed 3 June 2023).

IPCC (2021) Climate change 2021: the physical science basis. In: Masson-Delmotte, V., Zhai, P., Pirani, A., Connors, S.L., Péan, C. (eds). *Contribution of Working Group I to the Sixth Assessment Report of the Intergovernmental Panel on Climate Change*. Cambridge University Press, Cambridge, UK and New York. DOI: 10.1017/9781009157896

ISAAA (2020) Genome editing offers solutions to GHG emission problems. Available at: https://www.isaaa.org/kc/cropbiotechupdate/article/default.asp?ID=18419 (accessed 2 June 2023).

Lal, R. (2004) Soil carbon sequestration impacts on global climate change and food security. *Science* 304(5677), 1623–1627. DOI: 10.1126/science.1097396

Lal, R. (2010) Enhancing eco-efficiency in agro-ecosystems through soil carbon sequestration. *Crop Science* 50(Supplement_1), S-120–S-131.xx. DOI: 10.2135/cropsci2010.01.0012

Lal, R. (2016) Beyond COP21: potential and challenges of the '4 per 1000' initiative. *Journal of Soil and Water Conservation* 71(1), 20A–25A. DOI: 10.2489/jswc.71.1.20A

Lal, R. and Bruce, J.P. (1999) The potential of world cropland soils to sequester C and mitigate the greenhouse effect. *Environmental Science and Policy* 2(2–3), 177–185. DOI: 10.1016/S1462-9011(99)00012-X

Mahroof, K. and Sivarajah, S. (2021) One third of food produced globally does not get consumed – so how can we reduce food waste? The World Economic Forum (WEF). Available at: https://www.weforum.org/agenda/2021/08/one-third-of-food-produced-globally-does-not-get-consumed-how-can-we-reduce-food-waste/ (accessed 5 June 2023).

Nan Lin (2001) Social capital: a theory of social structure and action. In: Granovetter, M. (ed.) *Structural Analysis in the Social Sciences*. Cambridge University Press, Cambridge, UK and New York.

Osborne, M. (2023) Genetically modified trees are taking root to capture carbon. *Smithsonian Magazine*. Available at: https://www.smithsonianmag.com/smart-news/genetically-modified-trees-are-taking-root-to-capture-carbon-180981675/ (accessed 25 July 2024).

Pearce, F. (2012) A global treaty on rivers: key to true water security. Yale Environment 360. Available at: https://e360.yale.edu/features/a_global_treaty_on_rivers_key_to_true_water_security (accessed 2 June 2023).

Pomeroy, R. (2020) One trillion trees. The World Economic Forum. Available at: https://www.weforum.org/agenda/2020/01/one-trillion-trees-world-economic-forum-launches-plan-to-help-nature-and-the-climate/ (accessed 1 June 2023).

Poore, J. and Nemecek, T. (2018) Reducing food's environmental impacts through producers and consumers. *Science* 360, 987–992. DOI: 10.1126/science.aaq0216

Ritchie, H. (2019) Half of the world's habitable land is used for agriculture. Our World in Data. Available at: https://ourworldindata.org/global-land-for-agriculture (accessed 4 June 2023).

Rossi, F., Olguín, E.J., Diels, L. and De Philippis, R. (2015) Microbial fixation of CO2 in water bodies and in drylands to combat climate change, soil loss and desertification, *New Biotechnology* 32(1), 109–120. DOI: 10.1016/j.nbt.2013.12.002

Sachs, J.D. (2005) *The End of Poverty: Economic Possibilities for Our Time*. Penguin, London.

Schwartz, J.D. (2014) Soil as carbon storehouse: new weapon in climate fight? Yale Environment 360. Available at: https://e360.yale.edu/features/soil_as_carbon_storehouse_new_weapon_in_climate_fight (accessed 25 July 2024).

Smith, P., Martino, D., Cai, Z., Gwary, D., Janzen, H. *et al.* (2006) Policy and technological constraints to implementation of greenhouse gas mitigation options in agriculture. *Agriculture, Ecosystems and Environment* 118(1–4), 6–28. DOI: 10.1016/j.agee.2006.06.006

Smith, P., Daniel, M., Zucong, C., Daniel, G., Henry, J. *et al.* (2008) Greenhouse gas mitigation in agriculture. *Philosophical Transactions of the Royal Society, Biological Sciences* 363(1492), 789–813. DOI: 10.1098/rstb.2007.2184

Smith, P., Haberl, H., Popp, A., Erb, K.H., Lauk, C. *et al.* (2013) How much land-based greenhouse gas mitigation can be achieved without compromising food security and environmental goals? *Global Change Biology* 19(8), 2285–2302. DOI: 10.1111/gcb.12160

Statista Research Department (2022) Assets of financial institutions worldwide 2002–2020. Available at: https://www.statista.com/statistics/421060/global-financial-institutions-assets/ (accessed 2 June 2023).

UN News (2023) World 'at a crossroads' as droughts increase nearly a third in a generation. Available at: https://news.un.org/en/story/2022/05/1118142 (accessed 25 July 2024).

UN Women (2018) Gender and climate change: turning promises into action: gender equality in the 2030 Agenda for Sustainable Development. Available at: https://www.unwomen.org/en/digital-library/publications/2018/2/gender-equality-in-the-2030-agenda-for-sustainable-development-2018#view (accessed 25 July 2024).

UNDRR (2021) What is the Sendai Framework for Disaster Risk Reduction? Available at: https://www.undrr.org/implementing-sendai-framework/what-sendai-framework#:~:text=The%20Sendai%20Framework%20focuses%20on,existing%20risk%20and%20increase%20resilience (accessed 25 July 2024).

UNESCO Institute for Statistics (2021) Out-of-school children and adolescents of lower secondary school age, by region. Available at: http://data.uis.unesco.org (accessed 5 June 2023).

UNICEF (2019) Fact Sheet: *The Climate Crisis is a Child Rights Crisis*. Available at: https://www.unicef.org/press-releases/fact-sheet-climate-crisis-child-rights-crisis (accessed 25 July 2024).

UNEP (United Nations Environment Programme) (2012) Global environment outlook 5: environment for the future we want. Available at: https://www.unep.org/resources/report/global-environment-outlook-5-environment-future-we-want (accessed 20 May 2023).

UNEP (United Nations Environment Programme) (2021) *Food Waste Index Report 2021*. Nairobi. Available at: https://www.unep.org/resources/report/unep-food-waste-index-report-2021 (accessed 5 June 2023).

Uphoff, N. (2007) Reducing the vulnerability of rural households through agroecological practice: considering the system of rice intensification (SRI). *Mondes en Développement* 2007/4 (n. 140), pp. 85–100. DOI: 10.3917/med.140.0085

Watts, N., Amann, M., Ayeb-Karlsson, S., Belesova, K. *et al.* (2020) The *Lancet* countdown on health and climate change: from 25 years of inaction to a global transformation for public health. *The Lancet* 397(10269), 129–170. DOI: 10.1016/S0140-6736(17)32464-9

# 17 The Day after COP28

**Adel S. El-Beltagy[1,2], Rattan Lal[3], and Kauser Malik[4]**

[1]*Chair, International Dryland Development Commission (IDDC);* [2]*Arid Land Agricultural Graduate Studies & Research Institute (ALARI), Ain Shams University, Egypt;* [3]*Rattan Lal Center for Carbon Sequestration and Management, The Ohio State University, USA;* [4]*Forman Christian College (A Chartered University), Lahore, Pakistan*

This chapter addresses and discusses what is required to employ the best of science and technology, policy, enhancing coping and adaptive capacity, and ethical consideration to save mankind and humanity and cater for the most vulnerable populations.

The world population presently is 8.1 billion and projected to reach 10 billion by 2050 (Worldometer, 2024). Seventy percent of the increase between now and 2050 will be in 24 of the poorest countries of the world. Global action is required to cater for drylands where 80% of the Earth's poorest people live with a total population of 2.5 billion.

The concentration of global population growth in the poorest countries presents a considerable challenge to governments in implementing the 2030 Agenda of the United Nations for Sustainable Development Goals (SDGs), which seeks to end poverty (SDG 1) and zero hunger (SDG 2), expand and update health and education systems (SDG 4), achieve gender equality and women's empowerment (SDG 5), reduce inequality and ensure that no one is left behind (SDG 10) (United Nations Department of Economic and Social Affairs, 2017).

The Earth just had its hottest year on record. Around 7.3 billion people face temperatures strongly influenced by global warming. The European Union's Copernicus climate change service predicted that 2023 was the hottest calendar year on record, with average temperatures up to October being 1.43°C above the pre-industrial average (Wong, 2023) and this will continue to 2024 (Witze, 2024).

This increase of temperature will impact the flora and fauna, including aquatic marine life forms, trigger zoonotic disease, and impact soil microbes and viruses which could cause a hazard for human health.

Meanwhile, the projected increase of temperature beyond 1.5°C, which may reach 3°C (UNEP, 2023), will reduce food production by 20–50% by year 2050. An additional 2 billion increase in population will call for the need to produce more food, by as much as 80%, while reducing the environmental footprint of food systems in general but that of agriculture in particular.

This book, with its chapters contributed by eminent scholars from around the world, addresses the need for a major transformation of agriculture systems. These include food production, food chain, minimizing food waste, and enhancing the capacity for adaptation and mitigation (carbon sequestration) with specific focus on the global drylands (GDLs).

As mentioned in Chapter 2, the outcome of international efforts towards UN Framework on Climate Change, starting from COP1 in Berlin 1995 and until COP28 in Dubai in 2023, was not fully translated into action. This situation

 *Climate Change and Sustainable Agro-ecology in Global Drylands* (A.S. El-Beltagy, R. Lal and K. Malik eds.)
DOI: 10.1079/9781800624870.0017

has increased the lack of trust from the most vulnerable countries as they strive to minimize and avoid the damage caused by climate change (CC) induced by activities in the industrial countries. A selected major outcome of COP27 (Sharm El-Sheikh) and COP28 (Dubai) is presented in Box 17.1, below.

Box 17.2 contains extracts from the remarks of the World Bank Group President at its annual meeting in Marrakesh, October 2023.

**Box 17.1.** UN Secretary-General, Antonio Guterres at the conclusion of COP27 in Sharm El-Sheikh, November 2022: 'The Conference has been driven by two overriding themes: Justice and Ambition.'

'Justice for those on the frontlines who did so little to cause the crisis – including the victims of the recent floods in Pakistan that inundated one third of the country. Ambition to keep the 1.5-degree limit alive and pull humanity back from the climate cliff. This COP has taken an important step towards justice. I welcome the decision to establish a loss and damage fund and to operationalize it in the coming period. Clearly this will not be enough, but it is a much-needed political signal to rebuild broken trust.'

'The voices of those on the frontlines of the climate crisis must be heard.'

'Finally making good on the long-delayed promise of $100 billion a year in climate finance for developing countries.'

'Changing the business models of multilateral development banks and international financial institutions.'

'The world still needs a giant leap on climate ambition. The red line we must not cross is the line that takes our planet over the 1.5-degree temperature limit.'

'Finally, justice and ambition require the essential voice of civil society. The most vital energy source in the world is people power. That is why it is so important to understand the human rights dimension of climate action. Climate advocates – led by the moral voice of young people – have kept the agenda moving through the darkest of days.'

**'We can and must win this battle for our lives.'**

(United Nations, 2022)

**Box 17.2.** The World Bank President Ajay Banga (Banga, 2023).

- 'The world is fraught with profound forces and rapid changes. We face declining progress in our fight against poverty, an existential climate crisis, food insecurity, fragility, a fledgling pandemic recovery, and are feeling the effects of conflicts beyond the front. A perfect storm of intertwined challenges and geopolitical complexity that taken together exacerbate inequality. Economic growth in much of the developing world is retreating. Falling from 6 percent to 5 percent in two decades, and on track for just 4 percent over the next 7 years. With each lost percent, 100 million people are pulled into poverty and another 50 million people are pushed into extreme poverty. Dig deeper and you'll find people struggling to provide for themselves and their families as incomes have stagnated. In Sub-Saharan Africa per capita income is the same as it was 14 years ago. Meanwhile, debt has increased throughout emerging markets – doubling in Africa – shackling countries to the ground just as they're trying to rise.'
- 'Beneath the surface, a growing mistrust is pulling the Global North and South apart, complicating the prospect of progress. **The Global South's frustration is understandable. In many ways they are paying the price for the prosperity of others.**'
- 'We must find a way to finance a different world where our climate is protected, pandemics are manageable – if not preventable – food is abundant, and fragility and poverty are defeated.'
- 'We can't help farmers expand crop yields and feed growing populations using techniques that were not designed for years long droughts.'
- 'With a job comes dignity, pride, the ability to provide for yourself and your family. Without a job – or the hope of one – human despair turns to anger, in those moments people grasp for any hand that offers a way out. Those costs can deliver irreparable harm on a society – or – on entire generations. The urgency and importance cannot be overstated. Our own estimates project that in the next

*Continued*

**Box 17.2.** Continued.

10 years, 1.1 billion young people across the Global South will become working age adults. Yet, in the same period and same countries, we are only expected to create 325 million jobs. The cost of inaction is unimaginable – this demographic dividend must not be allowed to become a demographic challenge.'

- 'We are developing a new approach to track climate outcomes based on impact. We plan to share more about our progress at COP28 in Dubai.'
- 'Knowing that when development is delayed – development is denied. By the numbers, one year without education reduces a person's future income by 10 percent annually. Depriving a child of proper nutrition in their first 1,000 days, decreases their income by around 17 percent every year – for the rest of their life.'
- 'We do not believe quality conflicts with speed; we believe we can deliver both quality and speed.'
- 'Our hope is to refocus the World Bank to confront challenges not just as a funding mechanism, but as a knowledge mechanism.'
- 'First, bringing our knowledge to the forefront of our country driven model – and Country Partnership Frameworks. Sitting as partners with governments, working to craft a focused development plan that is a marriage of their ambition and our expertise. Second, helping to create and cultivate bankable projects and implement them. Our knowledge teams, working with our country teams to bring all the World Bank offers to bear – being a force multiplier for governments when additional capacity is needed.'
- 'Embedded within the vault of the knowledge bank are our eight global challenges – Adaptation and Mitigation; Fragility and Conflict; Pandemic Prevention and Preparedness; Energy Access; Food and Nutrition Security; Water Security and Access; Enabling Digitalization; and Protecting Biodiversity and Nature.'
- 'We have inherited decades of knowledge and benefitted from the generosity of every nation, now we are being called upon to lead, and we have never been better positioned to deliver the progress that is demanded.'

(World Bank News, 13 October 2023)

## One Earth, One Family, One Future

### G20 New Delhi Leaders' Declaration, 9–10 September 2023

The G20 leaders (G20, 2023) commit to concrete actions to address global challenges, acknowledging persistent headwinds to economic growth and stability. Challenges include increasing greenhouse gas emissions, climate change, biodiversity loss, pollution, and rising commodity prices. They emphasize the interconnectedness of poverty, inequality, CC, pandemics, and conflicts, especially affecting women and children. The declaration aims to accelerate actions for sustainable development, climate challenges, and debt vulnerabilities in developing countries. Central banks prioritize price stability, and there is a focus on digital transformation, AI, and data advances to achieve the Sustainable Development Goals (SDGs).

*Key areas of focus:*

1. Eliminating hunger and malnutrition:
   - commitment to enhance global food security and nutrition;
   - emphasis on research cooperation for climate-resilient grains and strengthening local fertilizer production;
   - acceleration of innovations to reduce food loss, improve marketing, and build sustainable agriculture.
2. Delivering quality education:
   - commitment to inclusive, high-quality education and skills training;
   - emphasis on expanding access to Technical and Vocational Education and Training (TVET) and promoting scientific collaboration.
3. Circular economy and resource efficiency:
   - recognition of the critical role of circular economy and extended producer responsibility;
   - commitment to environmentally sound waste management and reducing waste generation by 2030.
4. Climate action and finance:
   - call for increased global investments to meet Paris Agreement goals;
   - recall of the commitment to mobilize US$ 100 billion climate finance per year by 2025.

5. Conserving ecosystems:
   - commitment to restore at least 30% of degraded ecosystems by 2030;
   - emphasis on implementing the Kunming-Montreal Global Biodiversity Framework (GBF).

### African Leaders' Nairobi Declaration on Climate Change, 4–6 September 2023 (The African Union, 2023)

*Key points:*

1. Climate impact on Africa:
   - Africa emphasizes a unified approach and political leadership for a shared African vision, particularly in addressing loss and damage.
   - The Intergovernmental Panel on Climate Change (IPCC) confirms Africa is warming faster than the rest of the world, leading to adverse impacts on economies, societies, and human well-being.
   - Africa faces prolonged droughts, floods, storms, wildfires, and a humanitarian crisis, affecting various sectors such as economies, health, education, peace, and security.
2. Oceans and climate action:
   - recognition of the critical importance of oceans in climate action, acknowledging commitments made in various fora.

*Call to action:*

- urgent call to the global community to reduce emissions, fulfill climate finance commitments, and support Africa in addressing climate change;
- specific calls to accelerate emission reduction efforts, honor the $100 billion annual climate finance commitment, and operationalize the Loss and Damage facility agreed at COP27.

*Commitments:*

- commitment to climate-aware and restorative agricultural practices, biodiversity protection, and land degradation neutrality;
- leadership in developing global standards, metrics, and market mechanisms to accurately value and compensate for nature protection and climate services;
- finalization and implementation of the African Union Biodiversity Strategy and Action Plan toward the 2050 vision of living in harmony with nature;
- integration of climate, biodiversity, and ocean agendas into national plans for sustainable development;
- building effective partnerships for financial, technical, and technological support and knowledge sharing for climate change adaptation;
- strengthening early-warning systems and climate information services, embracing indigenous knowledge, and citizen science.

*Additional calls to action:*

- emphasizing the opportunity to contribute to equality and shared prosperity through decarbonizing the global economy;
- urging a comprehensive response to the incipient debt crisis and proposing a global carbon taxation regime, including a carbon tax on fossil fuel trade, maritime transport, and aviation;
- proposing the establishment of a new financing architecture responsive to Africa's needs, including debt restructuring and relief;
- establishing the Africa Climate Summit as a biennial event to set the continent's new vision on global climate and development issues;
- using the Declaration as the basis for Africa's common position in the global CC process, especially leading up to COP28.

During and pre-COP27 and COP28, several webinars, workshops and meetings related to the impact of CC on dry areas organized by the IDDC were presented globally and regionally. The four pre-COP meetings were organized by the International Dryland Development Commission (IDDC); the Science and Technology in Society forum (STS); Regional Action on Climate Change (RACC), Japan; Nizami Ganjavi International Center (NGIC); African League of

Young Masters (ALYM); and hosted by Ain Shams University (2022) and Alexandria University (2023), Egypt (IDDC, 2022, 2023a); STS/RACC webinar held in Kyoto, Japan (2022/2023) (Falk *et al.*, 2023; STS, 2023; Falk *et al.*, 2024); 2022 United Nations Climate Change Conference (COP27) held in Sharm El-Sheikh, Egypt, 6–18 November 2022 (UNFCC, 2022); 2023 United Nations Climate Change Conference (COP28), Dubai, 30 November–13 December 2023 (SDG Knowledge Hub, 2023).

These meetings are projecting efforts needed to bridge the gap, especially after the COP27 breakthrough agreement on the new 'loss and damage' fund for vulnerable countries and the establishment of the fund in COP28.

The following are highlights and major points raised during these meetings

## Webinar at Ain Shams University, 2022

Special guest statement by Dr. Mahmoud Mohieldin, UN Climate Change High-Level Champion for COP27 and former Executive Director of the International Monetary Fund):

- emphasized the urgency of addressing CC and its impact on food production, particularly in dry areas;
- called for immediate implementation of existing adaptation measures;
- highlighted the need for developed countries to fulfill their financial commitments to support developing countries in implementing adaptation measures;
- stressed that reduction of agriculture production and disruption of the food chain will occur worldwide because of CC;
- advocated a comprehensive approach to face the challenge, with emphasis on climate adaptation;
- emphasized that the implementation of pledges already made by the industrialized countries in the previous global forums is the immediate need and not more negotiations.

*Introductory address: 'Navigating through uncertainties: agro-ecosystems affected by dynamic impact of climate change' by Prof. Dr. Adel El-Beltagy, Chair of the International Dryland Development Commission; Professor, Arid Land Agricultural Graduate Studies & Research Institute, Ain Shams University, Cairo*

- The current world population is 7.94 billion in 2022, expected to reach over 10 billion by 2050, requiring doubling of food production amid changing climate conditions.
- IPCC's Sixth Assessment Report warns of a potential 4°C temperature increase, which could reduce food production by 50% by the century's end.
- Global climate agreements like the Paris Agreement, Sendai Framework, and SDGs face challenges in political implementation.
- CC impacts agro-ecosystems, affecting the distribution of terrestrial and marine species, with significant consequences for cropping systems.
- Overcoming barriers to land-based mitigation options requires addressing socio-economic, institutional, technological, financing, and environmental challenges, with a focus on local-level assessments.
- Adaptation options aligned with sustainable development can enhance resilience, food and water security, disaster risk reduction, health, ecosystem services, poverty reduction, and equality.
- Urgent need for an action-oriented knowledge network to assess CC's impact, predict agro-ecosystem sustainability, and develop locally applicable solutions.
- Regional Action in Climate Change (RACC) initiated in 2009 aims to connect global science, technology, and policy communities for locally applicable solutions.
- Collaboration is essential to address CC threats and secure a better future for the 2.5 billion people in dry areas, especially when exceeding a 2°C temperature rise poses a high risk to their livelihoods.

### *Keynotes*

'FAO EFFORTS IN BUILDING RESILIENCE AND IMPROVING LIVELIHOODS OF LOCAL COMMUNITIES IN DRYLANDS', BY DR. ISMAHANE ELOUAFI, CHIEF SCIENTIST OF THE FOOD AND AGRICULTURE ORGANIZATION (FAO) OF THE UNITED NATIONS:

- focused on FAO initiatives in drylands, home to 2 billion people, facing challenges like land degradation, water scarcity, and CC impacts;

- discussed the Global Programme on Sustainable Dryland Agriculture using science-based approaches of protected agriculture, precision farming, and technology to enhance productivity and resilience;
- aims to strengthen decision-making, support innovation adoption, and reinforce FAO capabilities to empower vulnerable communities.

'FARMING CARBON IN GLOBAL DRYLANDS', BY PROF. DR. RATTAN LAL, DISTINGUISHED UNIVERSITY PROFESSOR OF SOIL SCIENCE AND DIRECTOR OF THE CFAES, RATTAN LAL CENTER FOR CARBON MANAGEMENT AND SEQUESTRATION AT THE OHIO STATE UNIVERSITY:

- highlighted the significance of drylands, covering 41–47% of global land and hosting 38–39% of population, projected to increase;
- proposed sequestering atmospheric $CO_2$ in drylands through conservation agriculture and complex rotations;
- introduced 'carbon farming' focused on managing carbon pools at farm level for mitigation and fair compensation.

'GENETIC RESOURCES FOR ADAPTATION TO CLIMATE CHANGE IN DRYLANDS', BY DR. RAJ PARODA, CHAIRMAN, TRUST FOR ADVANCEMENT OF AGRICULTURAL SCIENCES (TAAS); FORMER SECRETARY, DEPARTMENT OF AGRICULTURAL RESEARCH AND EDUCATION (DARE); AND DIRECTOR GENERAL, INDIAN COUNCIL OF AGRICULTURAL RESEARCH (ICAR):

- discussed CC's impact on agriculture, emphasizing adaptation and climate-smart agriculture and the crucial role of genetic resources in addressing climate-related stresses;
- discussed the successful adaptation strategies including shifting to stress-tolerant varieties and adopting resilient production systems;
- discussed lessons from COVID-19, stressing the need for resilience in agriculture;
- emphasized genetic resources for carbon sequestration, especially in dryland soils;
- outlined the way forward, focusing on exploring and conserving diverse genetic resources;
- advocated for national policies, research networks, collaboration, and capacity development;
- raised concerns about technology access in the discussion;
- underscored the importance of public-private partnerships, citing the success of Bt cotton in India.

'CGIAR'S AGRI-FOOD SYSTEMS CLIMATE ADAPTATION – RESPONDING TO THE GLOBAL FOOD CRISIS', BY ALY ABOUSABAA, REGIONAL DIRECTOR, CWANA AND DIRECTOR GENERAL, INTERNATIONAL CENTER FOR AGRICULTURAL RESEARCH IN THE DRY AREAS (ICARDA); FORMER VICE PRESIDENT OF THE AFRICAN DEVELOPMENT:

- the global food crisis has been exacerbated by a 65% surge in prices since the start of the COVID-19 pandemic and disruptions in Ukraine;
- CGIAR advocates for a transformative shift in agri-food systems, including climate-resilient crops;
- CGIAR's innovations, such as climate-resilient crops, offer solutions to the global food crisis;
- the Regional Integrated Initiative for CWANA focuses on innovation, biodiversity, and sustainable farming;
- CGIAR addresses weaknesses like price volatility and export restrictions, emphasizing capacity building, policy research, and collaboration;
- the organization also highlights the importance of the global north in agricultural collaboration and efforts in genetic enhancement and small ruminant research.

'SIMULATING THE IMPACTS OF CLIMATE CHANGE ON THE CROP AND SOIL PROCESSES', BY DR. HEMANSHU PATHAK, DIRECTOR GENERAL, INDIAN COUNCIL OF AGRICULTURE RESEARCH (ICAR) AND SECRETARY, DEPARTMENT OF AGRICULTURE RESEARCH AND EDUCATION (DARE), GOVERNMENT OF INDIA:

- discussed impacts of CC on crops and soil processes, including increased weather variability and extremes;
- advocated using simulation modeling as a tool to assess CC's impacts on agriculture;
- provided examples from India applying models to predict crop yield changes under climate scenarios.

*Keynote: Dr. Jeffrey Sachs, Former Director, the Earth Institute, Columbia University; UN Senior Advisor for Sustainable Development*

- underscored the urgent need to increase financing and investments in developing countries to address CC;

- called for substantial funding increases in key sectors like education, health, energy, and infrastructure;
- proposed utilizing multilateral development banks to provide long-term, low-interest loans totaling around 2–3 trillion dollars.

### *Presentations*

'CLIMATE CHANGE EFFECTS ON WEST ASIA AND NORTH AFRICA DRYLANDS', BY PROF. DR. MAHMUD AYED DUWAYRI, FORMER MINISTER OF AGRICULTURE, JORDAN; PRESIDENT, AJLOUN NATIONAL UNIVERSITY, AJLOUN, JORDAN; FORMER DIRECTOR, PLANT PRODUCTION AND PROTECTION DIVISION, FAO, ROME:

- CC disproportionately affects agriculture and poor farmers globally;
- industrialized countries, major contributors to emissions, cause widespread impact;
- rising temperatures, reduced precipitation, and extreme events threaten ecosystems and biodiversity;
- Jordan faces climate challenges; a World Bank-supported plan aims for sustainable agricultural productivity;
- adapting to CC requires gene identification, biodiversity conservation, and innovative water use;
- Jordan's water scarcity is worsened by refugee influx; sustainable water management and smart agriculture are crucial;
- COP27 should prioritize implementing decisions to assist developing countries and prevent a rise in climate refugees.

'RISING TEMPERATURES AND INCREASING DEMAND CHALLENGE WHEAT SUPPLY IN SUDAN', BY DR. TOSHICHIKA IIZUMI, NATIONAL AGRICULTURE AND FOOD RESEARCH ORGANIZATION, TSUKUBA, JAPAN, AND VISITING PROFESSOR, ALRC, TOTTORI, JAPAN:

- Sudan's wheat demand is increasing, with 83% imported in 2018; population growth to 80 million by 2050 requires strengthening domestic wheat production;
- wheat production in Sudan faces challenges like cyclic fluctuations, economic factors, and CC;
- a study using the CYGMA crop model projects a decline in domestic wheat production share to 8.0–12.2% of anticipated demand by 2050, depending on warming levels (+1.5°C and +4.2°C);
- strategies include adopting more heat-tolerant varieties and shifting production to cooler regions;
- producers in cooler regions can focus on high-yielding varieties; heat-tolerance improvement is a priority in hot regions to offset warming effects;
- current wheat varieties are insufficient to counter the adverse effects of projected warming; breeding and disseminating heat-tolerant varieties are crucial;
- required improvement rate in yield for new varieties is 2.7% per year with a 4.2°C temperature rise, but decreases to 0.3% per year if limited to 1.5°C;
- climate mitigation is essential alongside adaptation to ensure capacity for adaptation measures, like wheat breeding, is not exceeded.

'IDENTIFICATION OF APPROPRIATE AGRO-MANAGEMENT TECHNIQUES TOWARDS CLIMATE CHANGE ADAPTATION IN TROPICAL ECOSYSTEMS', BY DR. CH. SRINIVASA RAO, DIRECTOR OF ICAR-NAARM, HYDERABAD, INDIA:

- global climate change has severe impacts on agriculture, leading to reduced production, food shortages, and increased poverty, especially in vulnerable regions like India's rainfed drylands;
- India aims to double food production by 2050 to meet the growing population, making adaptation to CC crucial for SDGs;
- over 54% of India faces high to extremely high water stress, emphasizing the importance of rainwater management and water-smart technologies;
- crop management techniques, including pulse-based cropping, intercropping, and cultivating stress-tolerant varieties, are essential for climate adaptation;
- soil management practices like conservation agriculture, crop residue incorporation, and biodiversity conservation contribute to soil health and climate resilience;
- multiple stress-tolerant genotypes for crops and livestock have been developed and provided to farmers in affected areas;
- climate-resilient agriculture requires a comprehensive approach, including R&D,

education, capacity building, community sensitization, policy changes, and village-level implementation;
- in discussions, addressing land erosion in arid/semi-arid areas involves promoting practices such as planting legume cover crops and leaving crop residues on the surface.

### Panels

PANEL DISCUSSION ON 'RETHINKING NATIONAL STRATEGIES FOR CONFRONTING CLIMATE CHANGE IN THE LIGHT OF THE CHANGING POLITICAL REALITIES'

*MODERATOR*

Dr. Ismail Serageldin, Founding Director of the Bibliotheca Alexandrina, former Vice-President of the World Bank, and Co-Chair of NGIC Board of Trustees

*PANELISTS*

H.E. Mr. Amre Moussa, Former Secretary General of the Arab League, Former Minister of Foreign Affairs of Egypt
H.E. Ms. Eka Tkeshelashvili, Former Deputy Prime Minister (2010–2012), and Former Minister of Foreign Affairs of Georgia (2008)
H.E. Ms. Ameenah-Gurib Fakim, Former President of Mauritius (2015–2018)

*DISCUSSION*

- Mr. Amre Moussa emphasized the role of civil society cooperation globally to pressure member states and UN leadership to confront CC above political divisions. Proposed establishing a tribunal to address climate violations and compensate victims. He stated the UN Secretary General should rally countries.
- Ms. Eka Tkeshelashvili highlighted the urgent need for climate action, linking it to geopolitical considerations and the requirement for fair evaluation of coping needs across all countries. Urged Europe to recognize energy independence and proposed solutions to mobilize global public opinion for adaptation and resilience.
- Ms. Ameenah-Gurib Fakim lamented the failure to meet commitments from COP21 and stressed immediate action by empowering institutions like the IMF to provide substantial assistance to developing countries facing climate challenges. Suggested naming and shaming non-compliant countries and increasing awareness for sizable green funds.

PANEL DISCUSSION ON 'BEYOND MITIGATION: MOBILIZING THE REQUISITE RESOURCES FOR EFFECTIVE PROGRAMS OF ADAPTATION AND RESILIENCE'

*MODERATOR*

H.E. Zlatko Lagumdzija, Former Prime Minister of Bosnia and Herzegovina (2001–2002), Former Deputy Prime Minister (1993–1996, 2012–2015)

*PANELISTS*

H.E. Maria Fernanda Espinosa, 73rd President of the United Nations General Assembly (UNGA), Former Minister of Foreign Affairs (2017–2018), Minister of National Defense of Ecuador (2012–2014)
H.E. Chiril Gaburici, Former Prime Minister of Moldova (2015)
H.E. Mr. Petre Romano, Former Prime Minister of Romania (1989–1991)

*DISCUSSION*

- Ms. Maria Fernanda Espinosa called for balancing investments between mitigation and adaptation efforts globally. Advocated for multilateral, systemic solutions to urgently address climate emergencies and support developing countries.
- Mr. Chiril Gaburici discussed exploring legal frameworks at national and international levels to manage the complex issue of climate refugees and develop proactive approaches.
- Dr. Ismail Serageldin emphasized reforming and expanding the role of multinational development banks and financial institutions to mobilize sufficient resources for climate action in developing countries. He fielded audience questions about potential Security Council reforms.

In the discussion and Q&A portions, panelists urged unity, consciousness shifts, and mobilization of diverse groups to address systemic climate challenges.

## Webinar at Alexandria University, 2023

### *Background statement – pre-COP28 webinar*

Extreme weather events, such as hurricanes and droughts, are intensifying, affecting communities globally. Rising sea levels threaten coastal areas, disrupting ecosystems and displacing populations. Immediate global cooperation is vital to implement sustainable policies and practices effectively.

Poor and vulnerable populations in dryland regions, relying on agriculture and fishing, face worsening conditions due to CC. Erratic rainfall patterns, droughts, and floods disrupt their livelihoods, exacerbating poverty. A multifaceted approach involving sustainable practices and international aid is essential to address their specific needs.

Remote sensing and biotechnology revolutions offer solutions for soil and water management and developing climate-resilient crops. Equitable access to these technologies for small-scale farmers is crucial. Governments, international organizations, and the private sector must collaborate to ensure affordability and responsible use.

Water scarcity and erratic rainfall patterns continue to be major global challenges. Sustainable water management strategies, including efficient infrastructure and conservation practices, are crucial. Holistic approaches are needed to address these challenges and ensure equitable access to water resources.

Addressing climate challenges in poor dryland nations is crucial for sustainable development. Investing in climate adaptation measures, sustainable water management, and renewable energy will enhance resilience and break the cycle of poverty.

A comprehensive adaptation strategy for poor and vulnerable populations in dry areas is needed, requiring significant international financing. International collaboration is crucial for mobilizing resources and implementing effective strategies, recognizing CC as a shared global responsibility.

**Fig. 17.1a.** Pre-COP28 webinar: 'Impacting climate change – navigating through scales'. (On screen) Prof. Jeffrey D. Sachs, University of Colombia, UN advisor of sustainability; (on podium, from left to right) Dr. Ismail Serageldin, Co-Chair of the Board of Nizami Ganjavi International Center, Founding Director of the Bibliotheca Alexandrina, and Former VP of the World Bank; Dr Ayman Abou Hadid, Former Minister of Agriculture, Emeritus Professor, Arid Land Agricultural Graduate Studies and Research Institute (ALARI); Ain Shams Former President, Dr. Adel El-Beltagy, Professor at Ain Shams University and Chair of the International Dryland Development Commission (IDDC). (From IDCC, 2023b)

**Fig. 17.1b.** Pre-COP28 webinar: 'Impacting climate change – navigating through scales'. Young Master's colleagues presenting. African League of Young Masters (ALYM). (From IDDC, 2023b)

**Fig. 17.2a.** Pre-COP28 webinar: 'Impacting climate change – navigating through scales'. (On screen) Prof. Wang Tao, Chinese Academy of Science; (standing on podium) Dr. Ismail Serageldin, Co-Chair of the Board of Nizami Ganjavi International Center, Emeritus Librarian of Alexandria, and Former VP of the World Bank. (From IDDC, 2023b)

The pre-COP28 event in Alexandria, focusing on CC, energy transition, food security, water, pollution, biodiversity, vulnerable drylands, youthful perspectives from Africa, international collaboration politics, meeting financing requirements, and concluding remarks, aims to promote global awareness and action.

**Fig. 17.2b.** Pre-COP28 webinar: 'Impacting climate change – navigating through scales. (On screen) Dr. Ameenah Gurib-Fakim, sixth (and first female) President of the Republic of Mauritius (2015–2018) and a renowned researcher and scientist in the fields of biodiversity; (on podium, from left to right) Dr. Ismail Serageldin, Co-Chair of the Board of Nizami Ganjavi International Center, Founding Director of the Bibliotheca Alexandrina, and Former VP of the World Bank; Dr. Adel El-Beltagy, Professor at Ain Shams University and Chair of the International Dryland Development Commission (IDDC); Dr. Omar El Arini, Former First Chief Officer of the Multilateral Fund for the Implementation of Montreal Protocol, assisted the Global Environment Facility (GEF) in setting up the Adaptation Fund Board, and member of the Green Climate Fund. (From IDDC, 2023b)

## RACC14 Symposium: 'An Urgent Need for COP27: Confronting Converging Crises', Kyoto, Japan

A consensus statement from the symposium (Falk *et al.*, 2023) warned: 'the Earth's fires have been compounded by interaction with impacts from multiple pandemics (including COVID-19 and cholera) and the Russia–Ukraine war. As a result, new connections are becoming visible between CC and human health, large vulnerable populations are experiencing food crises, climate refugees are on the move, and the risks of water, food, and climate disruption have been visibly converging and compounding. As the planet warms, dryer areas are becoming dryer. Wetter areas are becoming wetter. This will impact crop production, especially in some of the most populated and poorest countries. Countries in the dry area (such as Egypt, Jordan, and Morocco) will suffer from acute water poverty. In the highly populated countries, the vulnerable populations of the planet, food is running short.

Over its 14 years, RACC has advocated the need to build 'knowledge-action networks' that can tailor knowledge from the natural and social sciences to specific adaptation needs as identified by local communities.

A wide range of technical assistance is needed, from biotechnology (for example, development of crop varieties that are resilient to regional climate changes) to applications of IT, AI, and renewable energy and storage (El-Beltagy, 2019).

Adaptation requires a change to agricultural and food systems. As proposed at the RACC co-sponsored Egypt-based drylands webinar (IDDC *et al.*, 2022), economic and other measures to assist the most vulnerable are required. Food needs to be grown with greater water efficiency. Food waste – some 14% before retail (FAO, 2019) and 17% of what is then sold (UNEP, 2021) – needs to be significantly reduced. Climate-smart agriculture should be a national priority with increased R&D funding for scaling innovations around adaptation and mitigation.

In summary, at all scales, adaptation requires globally available local data, reliable analytic techniques, community capacity to plan adaptation strategies, and the resources (scientific, technical, cultural, and economic) to implement them. To date, the rate of growth of the support for CC resilience lags behind the rapid growth of cascading and converging risks. Greater emphasis on adaptation is now needed.

A pre-COP28 symposium of RACC15 held in Kyoto, Japan, has issued a statement entitled: 'Critical hydrologic impacts from climate change: addressing an urgent global need' (Falk *et al.*, 2024). The statement stressed efforts at mitigating GHG must accelerate in all sectors including agriculture, land use, and land-use change and forestry, with restoration and conservation of ecosystems key to restoring the hydrologic cycle and water resources. But adaptation is no longer a secondary option. Governments at all levels need to anticipate converging threats and invest in economic, scientific, and policy responses. In particular, much greater and more urgent attention needs to be applied to the severe and escalating hydrologic impacts being driven by CC, especially for the most vulnerable communities. Needed actions include expanding access to local information about climate impacts, new financial investments and tools for local adaptation, education around both risks and adaptation strategies, global support to de-risk initiatives that invest in resilience where it is most needed, and provision of the resources to implement on-the-ground actions consistent with justice. Current efforts are inadequate at all scales given the rapid rate of changes now occurring and the extensive and widespread risks. In particular, beginning at COP28, it is vital that national leaders pledge to protect, resource, and support actions to help the most vulnerable populations, notably in the drylands, to build resilience to the mounting challenges of CC.

**Box 17.3.** UN Secretary-General's statement at the close of the UN Climate Change Conference COP28, December 2023.

'COP28 occurred at a decisive moment in the fight against the climate crisis – a moment that demands maximum ambition both in the reduction of greenhouse gas emissions and climate justice.

Science tells us that limiting global heating to 1.5 degrees will be impossible without the phase-out of all fossil fuels on a timeframe consistent with this limit. And developing countries must be supported every step of the way.

The era of fossil fuels must end – and it must end with justice and equity. COP28 offered some other building blocks for progress – including the operationalization of the Loss and Damage Fund, even though financial commitments are very limited. Pledges to the Green Climate Fund now total a record $12.8 billion for the most climate-vulnerable communities.

The new framework on adaptation action provides a measurable set of new targets to drive adaptation action and implementation. They must also be underpinned with credible climate policies and regulations, including a price on carbon and ending finance for fossil fuel. And both the preparation and implementation of these plans must be fully funded and supported. But much more is needed to keep the hope of the 1.5-degree limit alive and deliver climate justice to those on the frontlines of the crisis.

Many vulnerable countries are drowning in debt and at risk of drowning in rising seas.

It is time for a surge in finance, including for adaptation, loss and damage and reform of the international financial architecture.

We need increased capital and reform of the business model of multilateral development banks to massively increase direct support – and to leverage far more private finance at reasonable costs for climate action in developing countries.'

(United Nations, 2023)

**Box 17.4.** Statement of EU President Von der Leyen on the outcome of COP28, Brussels, 13 December 2023.

'2023 has been the hottest year on record, and so it is fitting that we also make it the most ambitious year on record for climate action. Parts of Southern Europe have reached almost 30 degrees Celsius again this week, in the middle of December, and we are not alone in facing such extreme weather. Climate adaptation is becoming increasingly important. That is why the European Union is also standing in solidarity and working in partnership with countries all around the globe. We are dedicating record amounts of money to international climate finance, including adaptation finance. The European Union has also helped to operationalize a new loss and damage fund at COP28, and with our Member States we have contributed over two-thirds of the initial funding pledges.

'We stand ready to do more, and we know that more must be done. For example, COP28 has also been the opportunity to discuss carbon pricing with other parties, so that more countries start to put a price on pollution. And we have also been able to lay the ground for broader financial reforms, new innovative sources of funding, and aligning all financial flows with the Paris Agreement.'

(European Commission, 2023)

At COP28, 159 world leaders endorsed the COP28 UAE Declaration on Sustainable Agriculture, Resilient Food Systems and Climate Action (Davey, 2023). Together, their nations represent 68% (530 million) of the world's farmers, 75% of the global population, 77% of global food production, 81% of the world's arable land, 83% of global emissions from the food system, and 83% of the world's agricultural GDP.

The Declaration commits these nations to scale up adaptation and resilience for farmers, fishers, and food producers; to promote food security and nutrition through social protection systems, school feeding programs and more; and to support workers in agriculture and food systems to maintain inclusive, decent work.

Critically, it includes an explicit call for nations to strengthen the integrated management of water in agriculture and food systems at all levels to ensure sustainability. And it emphasizes the need to maximize the climate and environmental benefits associated with agriculture and food systems while containing and reducing harmful impacts such as by sustainably boosting productivity, protecting and restoring land and natural ecosystems, enhancing soil health and biodiversity, and shifting from higher GHG-emitting practices to more sustainable production and consumption. This includes reducing food loss and waste and promoting sustainable aquatic 'blue foods':

- Integrate agriculture and food systems into their national climate plans, known as 'nationally determined contributions' (NDCs), as well as national adaptation and biodiversity plans and long-term climate strategies.
- Revisit and reorient policies and government support – including the $700 billion of agricultural subsidies the world spends every year – towards practices which reduce GHG emissions, bolster resilience and human, animal and ecosystem health, and reduce ecosystem loss and degradation.
- Scale up and enhance access to all forms of finance in food systems, including finance for smallholder farmers.
- Increase investment in science and innovation, including innovations arising from local and indigenous communities.
- Strengthen the multilateral trade system in support of the goals of the Declaration.

The Declaration concludes with a commitment by nations to bring their relevant ministries – including ministries of agriculture, health, climate, environment, energy, and finance – together to discuss these goals and to report back on their progress at COP29.

The Paris Agreement, in its preamble, recognizes 'the fundamental priority of safeguarding food security and ending hunger, and the particular vulnerabilities of food production systems to the adverse impacts of climate change'.

The FAO roadmap sets out a broad range of options, including the need to sustainably boost crop and livestock yields; reduce agricultural GHG emissions and build resilience; reduce food loss and waste; and shift diets in rich countries away from meat and towards plant-based foods – all amidst a changing climate and a continuing biodiversity crisis.

This roadmap is a significant development. It complements WRI's flagship World Resources Report, *Creating a Sustainable Food Future*, and the Food and Land Use Coalition's *Growing Better* report, which lay out a menu of solutions from farm to fork for achieving global food security and climate goals. With the promise of further instalments in 2024, the FAO roadmap has the potential to become a guide for climate-friendly food systems, similar to the IEA *Net Zero Roadmap to 2050* in the energy sector.

Finally, governments, foundations, and others committed financial support for sustainable food systems. This included $519 million from the UAE, the Bill and Melinda Gates Foundation, the Bezos Earth Fund, and others. These funds are destined for the next phase of the Consultative Group for International Agricultural Research (CGIAR) as well as for investments in reducing emissions from livestock and food loss and waste.

- Will wealthier nations provide funding to poorer nations at scale to support them in achieving a just rural transition and greater resilience for smallholder farmers?
- Will the Loss and Damage fund established at COP be channeled in part to losses and damages experienced in the food system?
- Will societies as a whole – including farmers, consumers, the private sector, government officials, and political leaders – rally to deliver lasting movements for change in food systems, consistent with the spirit of the declarations signed at COP28?

Negotiations will resume in Azerbaijan (Baku) – COP29.

We need to revisit the following international principles in the face of the drastic impact of CC. Selected major elements which require swift attention:

- The anticipation that CC will have major impacts on agro-ecosystems requires continuous dynamic assessments, globally, regionally, and at the local level where the major action of adaptation would have to occur. The assessment will require international cooperation, national capacity building, and a solid early-warning system.

We need to link all the parts of puzzle together in an integrative way seeking to mobilize the existing knowledge science and indigenous knowledge and utilize it to establish an early-warning system based on a real assessment for the impact of CC on a local level where the livelihoods of people are at risk.

- Shifting agro-ecological zones (the habitats) and crops, livestock, fisheries. There is a massive need for cooperation between all gene banks, worldwide, national and international (CGIAR, Svalbard), which require funding to support the enhancement of:
  a. gene bank passports, in addition to georeferencing and morphological descriptions we need to add molecular characteristics; and
  b. to continue collection missions.

This step is important for a program which will utilize the potential of genetic resources to be introduced to agro-ecological zones with new features imposed by CC (temperature, canopies, soil, moisture, wind, precipitation, etc.).

A global breeding program is needed to cater for the introduction of new varieties with superior traits for biotic and abiotic stresses.

Parallel to this, there will be a need for an expert system and AI tools to handle the agronomy of the new genetic make-ups which will be introduced to the new agro-ecological zones within the new context of changing agro-ecological zones.

This will require the establishment of an international platform from the CGIAR/FAO/UNEP/UNESCO/WMO/GFAR to form a vibrant network for knowledge and material exchange to cater for the most vulnerable agro-ecologies (as suggested in Chapter 2).

- Inject knowledge into the different agro-management techniques including precision agriculture, climate-smart agriculture, agriculture 4.0, regenerative agriculture, intensive agro-ecology etc., using the best of science innovation including AI, Big Data, neural networks, expert systems, drones, sensors, robots etc., optimization of water-use efficiency, biodiversity, soil.
- All the global programs will require an intensive human resource development program in line with the ideas discussed in Copenhagen and Cancun until COP28,

which related to the technology fund and knowledge sharing.

- Restructuring the present institutions and linking them nationally, regionally, and internationally to increase the coping adaptive capacity.
- Political instability and conflict could be induced by disruption of water supply, and/or increased water demand, which in turn can undermine national, regional, and, hence, global security (El-Beltagy, 2000; Gleick and Iceland, 2018).
- The use of transboundary rivers requires an international legal framework which allows preventing conflicts and protecting the ecosystems.
- Transboundary harm (international environmental law): 'States should prevent activities on their territory causing significant harm or damage to the environment of another State or areas beyond national control'.
- Within the context of the impact of CC, to develop new international laws and different codes of conduct to face and contain anticipated disruption.
- International law of outmigration (environmental refugees).
- A law to maintain and protect the oceans beyond national control as a common resource.
- Attention should be given to circular and bioeconomy (knowledge-based production and use of biological resources). Bioeconomy, overall in Europe, reached €2.43 trillion, and presently in India $70 billion, aiming at $300 billion by 2030; and there is a call to increase bioeconomy activities and to link them to urban and pre-urban food production/processing (Muscat *et al.*, 2021; von Braun, 2023). Circular bioeconomy merits more attention in COP29 in Baku.
- A number of countries stepped forward with roughly $700 million for the Loss and Damage Fund. But this is a drop in the ocean compared to estimated requests for $1–5 trillion in years to come. New modalities for finance including regional, international banks and the private sector should be developed and introduced to COP29 in Baku for endorsement.
- A result-based program to assess the loss and damage with clear indicators related to vulnerability in different agro-ecological zones.
- Adaptation merits more attention as the vulnerable people cannot wait for the long process of mitigation. Adaptation finance is widening. We need significant scaling up to enhance adaptation finance.
- We need a combination of nature-based solutions blended with new technology, i.e. regenerative agriculture, organic agriculture, climate-smart agriculture, agriculture 4.0, precision agriculture, intensive agro-ecology.
- We need to stress that people-centered climate effects across agri-food systems/justice with sustainable development/poverty eradication and inclusivity objectives will lead to a more just resilience future for all.
- UN rights-based approach to global challenge/human rights/food for all. The right to food (the theme of the meeting in Doha, 6–7 February 2024).
- Translating the concept of leaving no one behind necessitates a massive cooperation and support program between nations to minimize the unequal burdens of CC impact and risk of livelihood of the most vulnerable population mainly in the developing countries (dry areas).
- A new vibrant international system for law and order with implementing power is urgently needed (drastic UN reform).

## Concluding Statement

From the inaugural COP in Berlin in 1995 to COP28 in Dubai in 2023, significant resolutions reflecting the hopes of the most vulnerable communities affected by CC have languished without implementation. The core issue at the heart of this predicament lies in the stark absence of political will, which, in turn, stems from the neglect of those most affected by the ravages of CC. This code of conduct exemplifies a refusal to acknowledge the rights of others.

In light of these challenges, a pressing need arises for establishing a novel multilateral international platform empowered to enact measures beneficial to humanity and the environment, safeguarding our precious natural resources

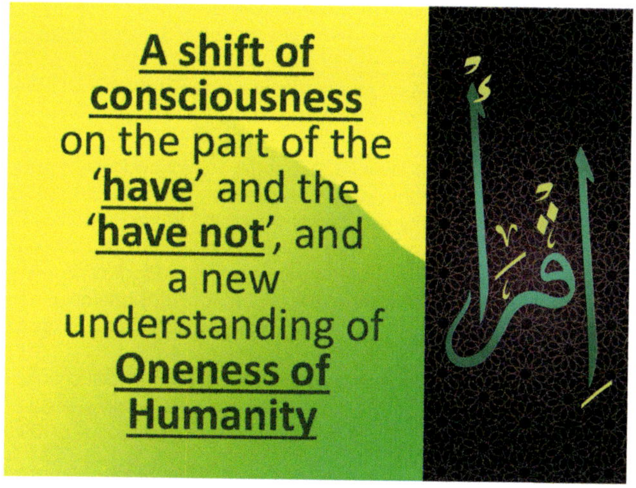


**Fig. 17.3.** The shift of consciousness.

and the sustenance of humankind. Such an endeavor calls for a transcendence of gender and racial boundaries. We require a worldview that recognizes our intrinsic interconnectedness as members of the human family, forging connections that bridge the divides constructed by race, nationality, and civilization.

This transformation does not entail obliterating our unique cultural identities or imposing a single set of values upon others. On the contrary, it celebrates the respect and appreciation of our rich diversity, dismantling the dominance of one culture over the rest. In turn, this fosters a spirit of cooperation and the willing sharing of scientific knowledge, all in the pursuit of a common goal – effectively addressing the challenges posed by CC. This book, with its multifaceted contributions spanning various realms of knowledge, serves as a vital step toward realizing this vision.

## References

African Union (2023) The African leaders' Nairobi Declaration on Climate Change and call to action. Available at: https://au.int/sites/default/files/decisions/43124-Nairobi_Declaration_06092023.pdf (accessed 26 July 2024).

Banga, A. (2023) Remarks by World Bank Group President Ajay Banga at the 2023 Annual Meetings Plenary. World Bank. Available at: https://www.worldbank.org/en/news/speech/2023/10/13/remarks-by-world-bank-group-president-ajay-banga-at-the-2023-annual-meetings-plenary (accessed 26 July 2024).

Davey, E. (2023) 6 major food breakthroughs at COP28 and what comes next. World Resources Institute. Available at: https://www.wri.org/insights/food-system-breakthroughs-cop28-whats-next (accessed 26 July 2024).

El-Beltagy, A. (2000) Strategic options for alleviating conflicts over water in dry areas. In: *The Crawford Fund Sixth Conference 'Food, Water and Wars: Security in a World of Conflict'*. ACIAR monograph series. Parliament House, Canberra, 73, 67–82.

El-Beltagy, A. (2019) Navigation through uncertainties: agro-ecosystems affected by dynamic impact of climate change. In: *Proceedings, 13th International Conference on Development of Dry Land*. International Dryland Development Commission and Arid Zone Research Association of India, Jodhpur, India, 11–14, pp. 3–12.

European Commission (2023) Statement of President von der Leyen on the outcome of COP28, Brussels. Available at: https://ec.europa.eu/commission/presscorner/detail/en/statement_23_6582 (accessed 26 July 2024).

Falk, J., Colwell, R.R., Behera, S.K., El-Beltagy, A.S., Gleick, P.H. *et al.* (2023) An urgent need for COP27: confronting converging crises. *Sustainability Science* 18, 1059–1063. DOI: 10.1007/s11625-022-01253-5

Falk, J., Gleick, P.H., Asayama, S., Attig-Bahar, A., Behera, S. *et al.* (2024) Critical hydrologic impacts from climate change: addressing an urgent global need. *Sustainability Science* 19, 241–244. DOI: 10.1007/s11625-023-01428-8

FAO (2019) *The State of Food and Agriculture: Moving Forward on Food Loss and Waste Reduction*. UN Food and Agriculture Organization, Rome. Available at: https://www.fao.org/3/ca6030en/ca6030en.pdf (accessed 26 July 2024).

G20 (2023) G20 New Delhi Leaders' Declaration, G20 Summit 2023 New Delhi. Available at: https://www.mea.gov.in/Images/CPV/G20-New-Delhi-Leaders-Declaration.pdf (accessed 26 July 2024).

Gleick, P. and Iceland, C. (2018) Water, security, and conflict. Issue brief. World Resources Institute, Washington, DC. Available at: https://www.wri.org/research/water-security-and-conflict (accessed 26 July 2024).

IDDC, RACC, and ICARDA (2022) *Impact of Climate Change on Food Production in the Dry Areas*. Hybrid webinar of climate change on food production in the dry areas, 3–5 September, Ain-Shams University, Cairo. Available at: https://www.icarda.org/media/events/impact-climate-change-food-production-dry-areas (accessed 26 July 2024).

IDDC (International Dryland Development Commission) (2022) Impact of climate change on food production in the dry areas. Background note. RACC/IDDC/ICARDA/ALARI hybrid webinar. Available at: http://www.drylanddevelop.org/uploads/6/1/7/8/61785389/background_note__as_of_30_august_2022_.pdf (accessed 26 July 2024).

IDDC (International Dryland Development Commission) (2023a) The Pre-COP28 conference, Alexandria. Statement. Available at: http://www.drylanddevelop.org/uploads/6/1/7/8/61785389/pre-cop28_statement.pdf (accessed 26 July 2024).

IDDC (International Dryland Development Commission) (2023b) The Pre-COP28 conference, Alexandria. Photo Gallery. Available at: http://www.drylanddevelop.org/pre-cop-28-photo-gallery.html (accessed 26 July 2024).

Muscat, A., de Olde, E.M., Ripoll-Bosch, R., Van Zanten, H.H.E., Metze, T.A.P. *et al.* (2021) Principles, drivers and opportunities of a circular bioeconomy. *Nature Food* 2, 561–566. DOI: 10.1038/s43016-021-00340-7

Science and Technology in Society (STS) Forum (2023) RACC 2023. Available at: https://www.stsforum.org/racc2023/ (accessed 26 July 2024).

SDG Knowledge Hub (2023) 2023 UN Climate Change Conference (UNFCCC COP 28). 30 November–12 December 2023, Dubai, UAE. International Institute for Sustainable Development (IISD).

United Nations (2022) Statement by the Secretary-General at the conclusion of COP27 in Sharm el-Sheikh. Available at: https://www.un.org/sg/en/content/sg/statement/2022-11-19/statement-the-secretary-general-the-conclusion-of-cop27%C2%A0-sharm-el-sheikh%C2%A0%C2%A0 (accessed 26 July 2024).

United Nations (2023) Secretary-General's statement at the closing of the UN Climate Change Conference COP28. United Nations. Available at: https://www.un.org/sg/en/content/sg/statement/2023-12-13/secretary-generals-statement-the-closing-of-the-un-climate-change-conference-cop28 (accessed 26 July 2024).

United Nations Department of Economic and Social Affairs (2017) World population projected to reach 9.8 billion in 2050, and 11.2 billion in 2100. Available at: https://www.un.org/en/desa/world-population-projected-reach-98-billion-2050-and-112-billion-2100 (accessed 10 January 2024).

UNEP (United Nations Environment Programme) (2021) United Nations Environment Program, Food Waste Index Report 2021, p. 70. Available at: https://www.unep.org/resources/report/unep-food-waste-index-report-2021 (accessed 20 September 2024).

UNEP (United Nations Environment Programme) (2023) *Emissions Gap Report 2023*. In: *Broken Record – Temperatures Hit New Highs, Yet World Fails to Cut Emissions (Again)*. Nairobi. DOI: 10.59117/20.500.11822/43922

UNFCC (United Nations Framework Convention on Climate Change) (2022) Sharm el-Sheikh Climate Change Conference, 6–20 November 2022. Sharm El-Sheikh, Egypt. Available at: https://unfccc.int/cop27 (accessed 26 July 2024).

Von Braun, J. (2023) 1 bioeconomy: toward sustainable living and bio-sensitive cities. In: Schellnhuber, H. and Armillas Tiseyra, R. (eds) *Reconstructing the Future: Cities as Carbon Sinks*, pp. 18–25. DOI: 10.1515/9783035627015-003

Witze, A. (2024) Earth boiled in 2023 – Will it happen again in 2024? *Spring Nature* 625, 637–639. DOI: 10.1038/d41586-024-00074-z

Wong, C. (2023) Earth just had its hottest year on record – climate change is to blame. *Springer Nature* 623. DOI: 10.1038/d41586-023-03523-3

Worldometer (2024) World Population, 2024. Available at: https://www.worldometers.info/world-population/ (accessed 10 January 2024).

# Index

# CABI – who we are and what we do

This book is published by **CABI**, an international not-for-profit organisation that improves people's lives worldwide by providing information and applying scientific expertise to solve problems in agriculture and the environment.

CABI is also a global publisher producing key scientific publications, including world renowned databases, as well as compendia, books, ebooks and full text electronic resources. We publish content in a wide range of subject areas including: agriculture and crop science / animal and veterinary sciences / ecology and conservation / environmental science / horticulture and plant sciences / human health, food science and nutrition / international development / leisure and tourism.

The profits from CABI's publishing activities enable us to work with farming communities around the world, supporting them as they battle with poor soil, invasive species and pests and diseases, to improve their livelihoods and help provide food for an ever growing population.

CABI is an international intergovernmental organisation, and we gratefully acknowledge the core financial support from our member countries (and lead agencies) including:

UKaid from the British people
Ministry of Agriculture People's Republic of China
Australian Government Australian Centre for International Agricultural Research
Agriculture and Agri-Food Canada
Ministry of Foreign Affairs of the Netherlands
Schweizerische Eidgenossenschaft Confédération suisse Confederazione Svizzera Confederaziun svizra Swiss Agency for Development and Cooperation SDC

## Discover more

To read more about CABI's work, please visit: **www.cabi.org**

Browse our books or explore our online products at:
**https://www.cabidigitallibrary.org**

Interested in writing for CABI? Find our author guidelines here:
**www.cabi.org/publishing-products/information-for-authors/**

Printed and bound by CPI Group (UK) Ltd, Croydon, CR0 4YY

13/09/2026

14961117-0005